Food Science Text Series

The Food Science Text Series provides faculty with the leading teaching tools. The Editorial Board has outlined the most appropriate and complete content for each food science course in a typical food science program and has identified textbooks of the highest quality, written by the leading food science educators.

More information about this series at http://www.springernature.com/series/5999

Arun K. Bhunia

Foodborne Microbial Pathogens

Mechanisms and Pathogenesis

Second Edition

 Springer

Arun K. Bhunia
Molecular Food Microbiology Laboratory
Department of Food Science
Department of Comparative Pathobiology
Purdue University
West Lafayette, IN, USA

ISSN 1572-0330   ISSN 2214-7799   (electronic)
Food Science Text Series
ISBN 978-1-4939-9246-1   ISBN 978-1-4939-7349-1   (eBook)
https://doi.org/10.1007/978-1-4939-7349-1

Printed on acid-free paper

This Springer imprint is published by the registered company Springer Science+Business Media, LLC part of Springer Nature.
The registered company address is: 233 Spring Street, New York, NY 10013, U.S.A.

Dedicated to my wife Banashri and our son, Arni, and daughter, Irene

Preface, Second Edition

The first edition of this graduate-level *Foodborne Microbial Pathogens: Mechanisms and Pathogenesis* textbook was published in 2008, and it was also translated into the Russian language in 2014. Ever since, a significant advancement in the pathogenic mechanism of microbes, in particular those associated with foodborne diseases, necessitated the second edition. The first edition had 15 chapters and the new edition is expanded to 20. The five new chapters cover the description of diseases that are caused by viruses, parasites, molds and mycotoxins, fish and shellfish toxins, and opportunistic and emerging foodborne pathogens. Additionally, several pathogens such as Nipah virus, Ebola virus, *Clostridium difficile*, *Enterobacter sakazakii*, *Brucella abortus*, and *Aeromonas hydrophila*, among others, are added to this edition. This edition is rich in many more graphic illustrations for easy comprehension of the complex mechanistic process.

I am ever so grateful to my doctoral and postdoctoral mentors, Prof. Bibek Ray and Prof. Michael G. Johnson, for their constant encouragement and for instilling a deep-rooted positive influence in exploring, learning, and sharing the knowledge of microbiology. The students whom I interacted throughout the world and their love for microbiology were a constant inspiration for this book. Once again, my graduate students and professional colleagues came forward to review some of the book chapters. My sincerest appreciation goes to Prof. M.G. Johnson, Dr. Gary Richards, Prof. J.S Virdi, Dr. Peter Feng, Dr. David Kingsley, Dr. Kristin Burkholder, Rishi Drolia, and Taylor Bailey for their excellent input and insights of many chapters they reviewed. This textbook should be an excellent learning tool for graduate and undergraduate students, academicians, and food microbiology professionals to gain advanced knowledge and insights of the microbial pathogenesis of foodborne or zoonotic pathogens that are transmitted to humans through food. This book is dedicated to all the microbiologists who are passionate about exploring the intricacies and complexity of the life of a microorganism.

West Lafayette, IN, USA   Arun K. Bhunia

Preface, First Edition, 2008

Ever since my days in veterinary school, I was fascinated with the field of microbiology. I always wondered how such small microscopic organisms are capable of causing infections in other living organisms, big or small, young or old, and healthy or immunocompromised. The subject captured my imagination. Many of the same microorganisms that cause diseases in animals also infect humans. In recent days, pathogens of animal origin impose even greater concern with increasing threat of avian influenza to cause pandemic and spread of deadly bovine spongiform encephalopathy (mad cow disease) and many bacterial pathogens such as *Listeria*, *E. coli* O157:H7, *Salmonella*, *Yersinia*, and *Campylobacter*. I am especially intrigued by the cunning strategy pathogens employ for their survival in a host and their exploitation of host cellular machinery to promote their own invasion into the host. Pathogenic mechanism is complex, and unraveling that process requires great minds. Today, microbiologists, cell biologists, and immunologists employing many sophisticated molecular tools are unraveling that secret at a very fast pace. Thus, it requires a great deal of efforts to compile and update information in a textbook, and it was rather a monumental task. My goal with this book was to paint a bigger picture of pathogenic mechanism of foodborne pathogens, which are responsible for many of modern-day outbreaks and diseases worldwide, and narrate the subject with easy-to-comprehend illustrations. When I began teaching an advanced graduate-level food microbiology course that dealt with pathogenic mechanism of foodborne pathogens in the mid-1990s, there was hardly any textbook that covered different foodborne microorganisms and the depth of materials needed for the course, especially the mechanism of infection for foodborne pathogens. That necessitated the collection and review of great deal of literature to provide updated materials to my students. That was the beginning and was also the inspiration and motivation to write a textbook on the subject. In the last two decades, there had been a tremendous progress in the area of food microbiology especially the study of molecular mechanism of pathogenesis, and a great deal of efforts was placed to compile those information in the first edition of the current textbook. In this book, an introductory chapter highlights the significance of foodborne pathogens, epidemiology, and the reason for increasing cases of foodborne illnesses. In Chap. 2, a brief review of biology of microorganisms and the importance of structural components as those related to pathogenesis is provided. In addition, diseases caused by viruses, parasites, mycotoxins, and seafood toxins have been included. In Chap. 3, a comprehensive review of the digestive system, mucosal immunity, and the host immune system has been described. This chapter provides the basic foundation for the understanding of the complexity of disease production by foodborne pathogens. First of all, foodborne pathogens' primary site of action is the digestive tract; therefore, one must have adequate knowledge to understand the interaction of pathogens with host gastrointestinal tract. Second, host innate and adaptive immune responses dictate the progression of a disease. Moreover, some pathogens exploit the host immune system as part of their disease-producing mechanism. Therefore, it is essential to have some basic understanding of immune system in order to understand the disease process. As it is often said, "It takes two to tango," or "One needs two hands to clap"; thus, I believe that the knowledge of biology of a pathogen and the corresponding host immune response go hand in hand to comprehend the full picture of pathogenesis process. In Chap. 4, general mechanisms of foodborne pathogens have been included to provide the overall big picture of mechanism of infection and intoxication. In Chap. 5, a brief review of the animal and cell culture models as necessary tools to study pathogenesis is discussed. In Chaps. 6, 7, 8, 9, 10, 11, 12, 13, 14, and 15, sources, biology, pathogenic mechanism, prevention and control, and detection or diagnostic strategies for individual foodborne bacterial pathogens are described. In addition to traditional foodborne pathogens, descriptions of some of the key pathogens with bioterrorism implications such as *Bacillus anthracis* and *Yersinia pestis* have been included to provide unique perspective. In this book, I am pleased to generate both digital and hand-drawn artworks to illustrate the pathogenic process, and I hope these illustrations will aid in better understanding of the mechanism of pathogenesis with greater enthusiasm. I also hope this textbook would be a valuable resource not only for food microbiology graduate or undergraduate students but also for the medical microbiologists, microbiology professionals, and academicians involved in food microbiology and food safety-related research or teaching. I would like to convey my gratitude to my current and former postdoctoral research associates and graduate students particularly Kristin Burkholder, Jennifer Wampler, Pratik Banerjee, and Ok Kyung Koo for their assistance in collecting literatures and reading the draft chapters. The comments and inputs provided by the students of FS565 over the years were extremely helpful in developing the course contents for this book. Finally, my sincerest and humble gratitude goes to my professional colleagues for their generous time and efforts in reviewing the chapters and providing expert comments and critiques: Chap. 1, Prof. M. Cousin, Purdue University; Chap. 2, Prof. M.G. Johnson, University of Arkansas; Chap. 3, Prof. R. Vemulapalli, Purdue University; Chap. 4, K. Burkholder, Purdue University; Chap. 6, Dr. P. Banada, Purdue University; Chap. 7, Prof. A. Wong, University of Wisconsin, and Prof. J. Mckillip, Ball State University; Chap. 8, Dr. G.R. Siragusa, USDA-ARS, Athens, GA, and Dr. V. Juneja, USDA-ARS, Wyndmoor, PA; Chap. 10, Prof. B. Reuhs, Purdue University; Chap. 11, Prof. S. Rickie, University of Arkansas, and

Dr. M. Rostagna, Purdue University; Chap. 12, Dr. R. Nannapaneni, Mississippi State University; Chap. 13, Prof. J.S. Virdi, University of Delhi South Campus, India; Chap. 14, Dr. G.B. Nair, Center for Health and Population Research, Dhaka, Bangladesh; and Chap. 15, Prof. C. Sasakawa, University of Tokyo, Japan.

West Lafayette, IN, USA   Arun K. Bhunia

Contents

Introduction to Foodborne Pathogens

Introduction to Food Microbiology

Food microbiology is a branch of microbiology that focuses on the study of microorganisms associated with food intended for human or animal consumption. Microorganisms use food as a source of nutrient for survival and growth or a vehicle for transmission to the human or animal host. Food microbiology is subdivided into three focus areas that include the study of beneficial, spoilage, and disease-causing microorganisms (Fig. 1.1). Beneficial microorganisms are those used in food fermentation to produce shelf-stable products with unique flavors, tastes, texture, and appearance. These include cheese, fermented meat (pepperoni), vegetables (pickled cucumber, olives), dairy products (yogurt), and ethnic fermented products such as kefir, sauerkraut, idli, kimchi, and so forth. In fermented products produced by natural or controlled fermentation, microorganisms metabolize complex substrates to produce enzymes, flavor compounds (diacetyl), acids (lactic acid, acetic acid, propionic acid, etc.), alcohols, and antimicrobial agents (bacteriocins, hydrogen peroxide) to improve product shelf life and to provide characteristic product attributes. Microorganisms also produce enzymes, which also help the breakdown of indigestible compounds to make the product more palatable and easy to digest. In addition, the beneficial microorganisms also serve as probiotics to impart direct health benefit by modulating the

immune system to provide protection against chronic metabolic disease (e.g., diabetes), bacterial infection, atherosclerosis, osteoporosis, and allergic reactions. Examples of beneficial microorganisms are *Lactobacillus acidophilus, L. rhamnosus, L. casei, L. plantarum, Lactococcus lactis, Streptococcus thermophilus, Pediococcus acidilactici,* and *Saccharomyces cerevisiae.* These microorganisms generally are safe and do not cause disease in humans, unless the opportunistic environment such as immunocompromised condition or other health complications facilitate systemic dissemination of the organisms.

Food spoilage microorganisms are those, which upon growth in food, that produce undesirable texture, odor, and appearance and make food unsuitable for human consumption. These microorganisms produce protease, lipase, and glycolytic enzymes (amylase, pectinase, etc.) to break down proteins, fats, and carbohydrates, respectively, and alter physical (such as slimy texture), chemical (color change, oxidation) and sensory properties. Sometimes uncontrolled growth of many of the natural fermentative microorganisms can also cause spoilage, such as the growth of lactic acid bacteria in ready-to-eat meats or pasteurized milk. It is estimated that 25% of all food produced (postharvest) are lost due to the microbial spoilage. Food spoilage is a serious issue in developing countries because of inadequate processing and refrigeration facilities. Controlling microbial food spoilage will continue to receive

© Springer Science+Business Media, LLC, part of Springer Nature 2018
A. K. Bhunia, *Foodborne Microbial Pathogens*, Food Science Text Series,
https://doi.org/10.1007/978-1-4939-7349-1_1

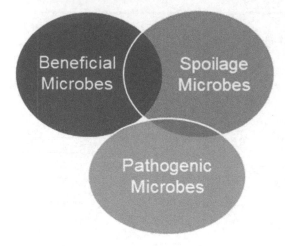

Fig. 1.1 Three branches of food microbiology. Beneficial and spoilage microorganisms, and spoilage and pathogens have some overlapping activity (*shaded area*)

increasing attention since food demand is expected to double in 2050 to feed estimated 9.3 billion people in the world. The current (2018) world population is about 7.6 billion. Examples of food spoilage microorganisms are *Alicyclobacillus*, *Brochothrix*, *Bacillus*, *Clostridium*, *Erwinia*, *Lactobacillus*, *Leuconostoc*, *Pseudomonas*, *Shewanella*, and some yeasts and molds (*Botrytis*, *Alternaria*, *Penicillium*, *Fusarium*, *Mucor*, *Rhizopus*). The spoiled food is generally of less concern from a public health standpoint, since many disease-causing microorganisms may not be able to survive in the microenvironment created in a spoiled food. Moreover, the pathogenic microorganisms are poor competitors and may not survive in food in the presence of a natural microbial community whose metabolic by-products such as acids, bacteriocin, hydrogen peroxide, and CO_2 are inhibitory to pathogens.

Foodborne pathogenic microorganisms (Table 1.1), on the other hand, when present or grown in a food may not alter the aesthetic or sensory quality of products, and thus the microbial safety may not be easy to assess based on product appearance alone. Thus, microbiological or analytical testing is required to assess pathogen contamination. Some foodborne pathogens such as proteolytic *Clostridium botulinum*,

Bacillus species, and some mold species are known food spoilage organisms, yet they can also produce toxins and cause disease. Based on the nature of the disease elicited by pathogens, foodborne diseases are again classified into three types: "intoxication," "toxicoinfection," and "infection." The knowledge pertaining to the mechanism of pathogenesis and the diseases caused by foodborne pathogens continues to augment our understanding of epidemiology and disease transmission process and to develop novel detection and diagnostic tool, to design novel vaccine and drugs, and to help formulate effective prevention and control strategies (Fig. 1.2).

Intoxication results when the preformed microbial toxin is ingested with food and water resulting in the onset of symptoms very quickly. Food intoxications are generally caused by toxins produced by *Staphylococcus aureus*, *Clostridium botulinum*, and *Bacillus cereus*. Intoxication is also associated with algal toxins ingested by fish or shellfish and mycotoxins produced by fungi. "Toxicoinfection" refers to the disease caused by the toxins produced inside the host by the live pathogens after ingestion of food. Toxins interact with the host epithelium and result in the gastrointestinal symptoms including diarrhea and occasional vomiting. Examples are *Clostridium perfringens*, enterotoxigenic *Escherichia coli* (ETEC), and *Vibrio cholerae*. "Foodborne infection" occurs when the live pathogenic microorganisms, which are generally invasive, are ingested and cause severe tissue damage. "Foodborne infection" is caused by bacterial pathogens such as *Salmonella enterica*, *Campylobacter jejuni*, enterohemorrhagic *Escherichia coli* (*E. coli* O157:H7), *Shigella* spp., *Yersinia enterocolitica*, and *Listeria monocytogenes*; viral pathogens including *Norovirus*, hepatitis A virus, and *Rotavirus*; and parasites including *Toxoplasma gondii*, *Cryptosporidium parvum*, *Cyclospora cayetanensis*, *Giardia intestinalis*, *Taenia solium*, and *Trichinella spiralis*.

Among the foodborne pathogens, *Norovirus* tops all infections with over a million people infected each year in the USA. Protozoan infections are mostly associated with water and fresh

Table 1.1 List of foodborne pathogens involved in outbreaks from contaminated food and water

Bacteria	*Virus*	*Parasite*
Aeromonas hydrophila	Astrovirus	Cryptosporidium parvum
Bacillus anthracis	Hepatitis A virus	Cyclospora cayetanensis
Bacillus	Hepatitis E virus	Entamoeba histolytica
cereus/subtilis/licheniformis	Norovirus	Giardia intestinalis
Brucella	Rotavirus	Cystoisospora belli
abortus/melitensis/suis		Taenia solium/saginata
Campylobacter jejuni/coli		Toxoplasma gondii
Clostridium botulinum		Trichinella spiralis
Clostridium perfringens	**Infective proteins**	**Molds and mycotoxins**
Cronobacter sakazakii	BSE (bovine spongiform	Aflatoxin, ochratoxin, fumonisin, patulin,
Escherichia coli	encephalopathy)	trichothecenes, DON (deoxynivalenol)
Listeria monocytogenes	TSE (transmissible	
Mycobacterium	spongiform	
paratuberculosis	encephalopathy)	
Salmonella enterica	**Emerging pathogens**	**Seafood toxins**
Shigella spp.	Clostridium difficile	Scombroid toxin, ciguatera fish poisoning, diarrhetic
Staphylococcus aureus	Brucella abortus	shellfish poisoning, paralytic shellfish poisoning
Vibrio spp.	Trypanosoma cruzi	(saxitoxin), pufferfish poisoning
V. cholerae non-01	Avian flu virus	
V. parahaemolyticus	Nipah virus	
V. vulnificus	Ebola virus	
V. fluvialis		
Yersinia enterocolitica		

Fig. 1.2 The interrelationship of foodborne pathogens and pathogenesis with other branches of microbiology

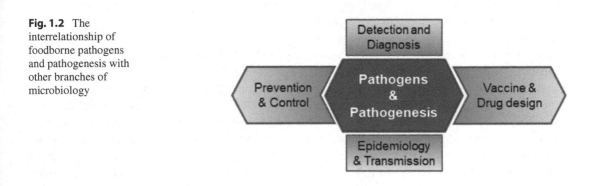

produce such as the fruits and vegetables. A large volume of these commodities is imported from countries where food production and processing are performed under inadequate hygienic practices, which may contribute to the increased incidences of contamination. Molds produce mycotoxins such as aflatoxin, ochratoxin, fumonisin, trichothecene, and DON (deoxynivalenol) that are mutagenic, carcinogenic, or hepatotoxic. Prolonged exposure to mycotoxins may result in serious and sometimes fatal diseases. The onset of symptoms due to mycotoxin intoxication is delayed and is not as dramatic as bacterial- or virus-induced diarrhea, vomiting, or neurological disorders. Thus, the significance of mycotoxin ingestion is often overlooked. In contrast, bacterial or viral pathogens affect a large population resulting in high morbidity and mortality, and their surveillance statistics are updated routinely. The importance of molds in food has rarely peaked consumers' or regulatory agencies' interest, especially in the USA; however, in recent years, increased emphasis has been placed on understanding the properties, synthesis, and pathogenesis of foodborne molds and their mycotoxins.

Microorganisms are ubiquitous and can survive and grow in extreme conditions in nature, in food, and in human or animal hosts. Some bacteria can grow in high temperatures and are called thermophiles (50–70 °C); some even grow in hot geysers such as extremophiles (>70 °C) or extreme cold (−20 °C). Most thermophilic and extremophilic microorganisms are nonpathogenic to humans. Some microorganisms grow at refrigeration to ambient room temperatures (1–25 °C), while mesophiles grow at 25–37 °C. Both psychrotrophs and mesophiles are capable of causing diseases in humans.

Based on their response to different pH, microbes are grouped as aciduric (<pH 7), alkaliphile (pH > 7.0), halophile (salt-loving), or barophiles (pressure-loving). Based on the oxygen requirements, bacteria are grouped as aerobic, obligate anaerobic, or facultatively anaerobic. Foods are prepared and stored under controlled environments where residual oxygen concentrations can dictate what type of microorganisms will survive and grow. Anaerobes can grow only in the vacuum-packaged or canned foods in which oxygen is removed mechanically or by heating. Aerobes grow on the surface of food where oxygen is abundant. A similar scenario applies for pathogens that cause disease in the gastrointestinal tract, where oxygen gradients vary from the upper part of the small intestine to the lower part of large intestine. The upper part is highly oxygenated, while the lower part is devoid of oxygen. Again, oxygen concentrations vary from the center of the lumen to the proximity of the epithelial lining where the oxygen concentration is higher because of cellular respiration. These environments select the types of pathogens that will colonize or be present in different parts of the intestine. Besides microorganisms being introduced into our body via food, water, or air, microorganisms also exist since birth as commensals in the digestive tract, skin, nasal passages, and reproductive and urogenital tract, and these microbes are generally beneficial to the host. However, under favorable conditions, commensals may behave as opportunistic pathogens.

Food is a complex milieu that contains salts, acids, ions, aldehydes, flavoring agents, and antimicrobial preservatives. Pathogenic microorganisms transmitted through foods, in most cases,

are able to adapt well to the harsh environments of food and maintain their pathogenic traits. During the transition to the host, pathogens may express a completely separate set of genes that ensure their survival and disease-producing capabilities in the host.

Food safety is essentially an ongoing problem in rapidly changing and growing food industry, due in part to consumer's increasing reliance on ready-to-eat convenience foods. Food is globally sourced and distributed; thus, contamination of a food with a pathogen will present a greater economic and social impact than ever before. Certain foods can also become vulnerable to malicious contamination with infective agents; thus, food defense is becoming an essential part of our education. Knowledge and understanding of pathogens in foods and their survival mechanisms in foods as well as in the human and food animal host should be important areas of focus to prevent food-related illnesses and increase the well-being of the population.

What Is a Pathogen?

In fact, only a small fraction of all microbes cause disease in humans or animals via the food- or feed-borne route. A pathogen is an organism that is able to cause cellular damage by establishing in tissue, which results in clinical signs with an outcome of either morbidity (defined by general suffering) or mortality (death). More specifically, a pathogen is characterized by its ability to replicate in a host, by its continued persistence of breaching (or destroying) cellular or humoral barriers that ordinarily restrict it, and by expressing specific virulence determinants to allow a microbe to establish within a host for transmission to a new susceptible host. Pathogens could be classified as zoonotic, geonotic, or human origin based on their transmission patterns and movement among different hosts and vectors. Zoonotic diseases are characterized by transmission of infective agents such as bacteria, viruses, parasites, and fungi from animals to humans. Examples of zoonotic pathogens are *Escherichia coli* O157:H7, *Staphylococcus aureus*,

Salmonella enterica serovar Typhimurium, *S. enterica* serovar Enteritidis, *Campylobacter jejuni*, *Yersinia enterocolitica*, *Mycobacterium tuberculosis*, Ebola virus, Nipah virus, *Toxoplasma gondii*, and *Trichinella spiralis*. Geonotic diseases are acquired from soil, water, or decaying plant materials. An example of geonotic pathogen is *Listeria monocytogenes*. Human origins are exclusively transmitted from person to person, and pathogens of human origin are *Salmonella enterica* serovar Typhi, *Vibrio cholerae*, *Shigella* spp., and hepatitis A. Different diseases caused by foodborne pathogens are summarized in Table 1.2.

Poverty, competition for food, crowding, war, famine, and natural disaster help pathogens to survive and spread in the environment. Domestication of animals also allowed pathogens to come in close contact with humans and, thus, helped foster human acquisition of these pathogens. In recent years, there is a great concern of possible pandemic outbreak of bird flu (avian influenza virus) in humans due to the transmission of the virus to the poultry handlers or people being exposed to the infected flocks. Bird flu virus (strain H5N1) was responsible for several fatal infections in Asia and other parts of the world, often via live poultry markets (see Chap. 6).

In the past few decades, industrialized countries have witnessed increased number of outbreaks of the foodborne pathogen, in part, due to massive expansion of food industry, inadequate hygienic and sanitary practices, and distribution of a food item to consumers globally (see section on "Why High Incidence of Foodborne Outbreaks"). Foodborne outbreaks may be characterized as an *epidemic* when a large population of consumers from multiple states is affected due to ingestion of pathogen originating from a single food item. In contrast, *sporadic* outbreak involves incidence of one or two cases in a community. The *pandemic* outbreak is defined as the incidence of disease that is associated with a highly infectious agent affecting people from multiple countries on a global scale with a large number of populations succumb to high mortality or

Table 1.2 Diseases and symptoms caused by foodborne pathogens

Disease or clinical symptoms	Pathogens/toxins involved
Vomiting, diarrhea, dysentery	*Staphylococcus*, *Bacillus*, *Cronobacter*, *Salmonella*, *Shigella*, *Yersinia*, *Vibrio*, *Norovirus*, *Rotavirus*, *Entamoeba*, *Cryptosporidium*, *Cyclospora*, *Giardia*, *Cystoisospora*, *Taenia*, *Trichinella*
Arthritis (reactive arthritis, Reiter's syndrome, rheumatoid arthritis)	*Campylobacter*, *Salmonella*, *Shigella*, *Yersinia*
Hemorrhagic uremic syndrome (HUS), kidney disease	Shiga toxin-producing *E. coli* (STEC); *Shigella* spp.
Hepatitis and jaundice	Hepatitis A virus (HAV), hepatitis E virus (HEV)
Guillain–Barré syndrome (GBS), Miller Fisher syndrome (MFS)	*Campylobacter*
CNS disorder: meningitis, encephalitis	*Listeria*, bovine spongiform encephalopathy (BSE), *Taenia* spp. (cysticercosis)
Miscarriage, stillbirth, neonatal infection	*Listeria monocytogenes*, *Toxoplasma gondii*
Paralysis	*Clostridium botulinum*, shellfish toxin (algal toxin), *Campylobacter*
Malignancies and autoimmune diseases	Mycotoxin (aflatoxin, deoxynivalenol)
Allergic response	Seafood toxin (scombroid toxin)

morbidity. Foodborne pathogens may not be a pandemic threat, but global distribution and trading of food may be a cause for concern in the future. *Epizootic* outbreak refers primarily to animal diseases and is linked to an infectious disease that appears as new cases in a given population of animals. If such infectious agents are found in food animals and if there is a potential for transfer of those infective agents to animal handlers and food producers/processors, it would be a major cause for concern to humans.

What Are the Attributes of Pathogenicity?

Some pathogens are designated as "primary pathogens," which regularly cause disease. Some are classified as "opportunistic pathogens" that infect primarily immunocompromised or at-risk populations. Nevertheless, both primary and opportunistic pathogens share similar attributes: entry and survival inside a host, finding a niche for persistence, able to avoid the host's defense (stealth phase), able to replicate to a significant number, transmitted to other host with high frequency, and able to express specialized traits within the host. For example, *Salmonella enterica* invades host cells when proper levels of O_2, pH, and osmolarity are maintained. This sends appropriate signals to the PhoP/Q regulon for expression of specific invasion-associated genes.

Several factors affect pathogens growth and survival inside a host such as O_2, CO_2, iron, nutrients, pH, bile salts, mucus composition, the balance of natural microflora, quorum sensing, and physiological status such as stress hormone like epinephrine or norepinephrine.

Pathogens are clonal and they are generally derived from a single progenitor. They are often selected by the environmental selective pressures such as heat, extremes of pH, and antibiotic treatments. However, the emergence of a new or a highly virulent form of pathogens suggests possible transfers of virulent genes are occurring among microorganisms. These novel genetic variants have arisen due to a point mutation, genetic rearrangement in the chromosome, and gene transfer between organisms through horizontal (*Shigella* to *E. coli*) or vertical (*E. coli* to *E. coli*) modes (Fig. 1.3). Plasmids, bacteriophages, transposons, and pathogenicity islands (i.e., a large piece of DNA carrying virulence gene cluster (see Chap. 4) can all serve as a vehicle for gene transfer.

Involvement of a specific gene or genes in pathogenesis can be studied in the laboratory by using various molecular tools such as mutation, genome sequencing, and cloning. In mutation, a specific gene is disrupted by chemical mutagenesis, transposon mutagenesis, site-specific muta-

tions (homologous recombination), or in-frame deletion. Genetic complementation is often done to restore the gene function and to confirm the involvement of that gene in pathogenesis. Transposon (Tn) elements are used routinely to induce mutation. The major benefit of using transposon is that they carry antibiotic markers that help in locating a specific gene on the chromosome. However, transposon could act as a transcriptional terminator. If Tn lands on the promoter or the first gene of an operon, it affects the transcription of the genes located downstream. In addition, if the transposon is inserted in a housekeeping gene, essential for bacterial growth or survival, that bacterium cannot be selected from that experiment. Transposons carry antibiotic resistance gene(s) and insertion sequences for integration into the host chromosome. They may also carry virulence genes. Transposons are generally specific for either Gram-positive (Tn*916*, Tn*917*, Tn*1545*) or Gram-negative (Tn*5*, Tn*7*, Tn*10*) bacteria. In-frame deletion is highly efficient and is used widely to create a mutant strain by employing a method called SOE (splicing by overlap extension). Genome sequence of many microorganisms is now available; thus, a series of polymerase chain reaction (PCR) methods is used to selectively remove a target gene by SOE to create an in-frame deletion mutant. Whole genome sequencing is also a powerful tool for studying pathogenesis. Sequencing of a gene or its product (amino acid sequence) and subsequent matching with the database will reveal the identity and function of the gene. Genetic cloning is also an important strategy for studying pathogenesis. In cloning experiments, a suitable vector is constructed with a gene of interest and, subsequently, transferred to an avirulent strain by electroporation or conjugation. The function of the gene in the new strain is analyzed by in vitro cell culture or in vivo animal bioassays.

Sources of Foodborne Pathogens

Many foodborne pathogens are ubiquitous in nature and generally found in soil, water, animals, and plants. Pathogens are introduced into a

Fig. 1.3 Gene transfer and acquisition in bacteria

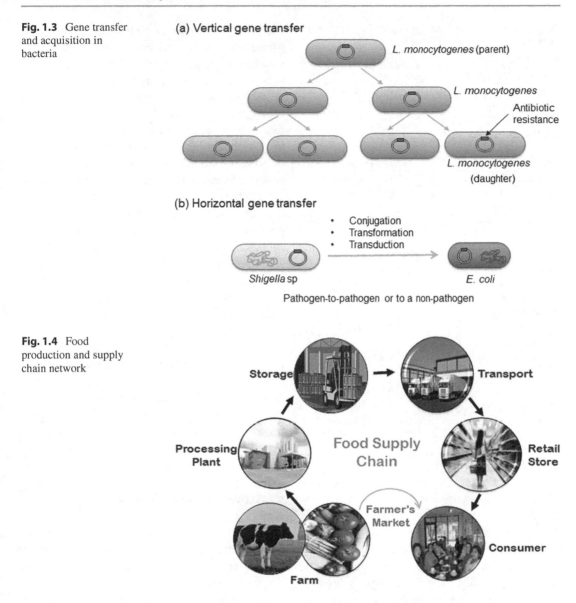

Fig. 1.4 Food production and supply chain network

food processing plant through raw materials (Fig. 1.4). Humans, air, water, and equipment may also bring the organisms to foods in the processing plant. Recontamination of processed food also frequently occurs and contributes to foodborne outbreaks and illnesses. Some pathogens can survive for a prolonged period on the inanimate objects and serve as a source. A list of foodborne microorganisms and their survival on inanimate surface is summarized in Table 1.3. Pathogenic microbes use various strategies to persist on food and in food production facilities. These strategies may include biofilm formation, quorum sensing, the viable but nonculturable state (VBNC), spore formation, stress response/ adaptation to intrinsic conditions in food such as pH, osmotic stress (salt), acid, temperature, and resistance to sanitizers (see Chap. 4).

When considering a product safety, it is important to know the type of microorganisms or toxins likely to be present, their numbers, and concentrations. In addition, their response to the heat, pH, salts, and other processing conditions is important to consider. The numbers and types of microorganisms present in a finished food product are influenced by the original source of the

Table 1.3 Persistence of pathogens on inanimate surfaces

Organisms	Duration of survival
Bacterial pathogens	
Campylobacter jejuni	Up to 6 days
Escherichia coli	1.5 h–16 months
Clostridium difficile (spores)	5 months
Listeria species	1 day–several months
Mycobacterium bovis	>2 months
Mycobacterium tuberculosis	1 day–4 months
Salmonella enterica serovar Typhi	6 h–4 weeks
Salmonella enterica serovar Typhimurium	10 days–4.2 years
Shigella species	2 days–5 months
Staphylococcus aureus	7 days–7 months
Streptococcus pyogenes	3 days–6.5 months
Vibrio cholerae	1–7 days
Viral pathogens	
Astrovirus	7–90 days
Adenovirus	7 days–3 months
Norovirus	8 h–7 days
Influenza virus	1 day–2 days
Rotavirus	6 days–60 days
Hepatitis A virus	2 h–60 days

Adapted from Kramer et al. (2006). BMC Infect. Dis. 6:130

food, its microbiological quality in the raw or unprocessed state, the sanitary conditions under which the product was handled or processed, and the conditions for subsequent packaging, handling, storage, and distribution (Fig. 1.4). In a raw agricultural product, generally, the maximum microbial load is on the food surfaces, whereas generally there is negligible to none inside. A list of foods involved in major confirmed outbreaks since the year 2000 is summarized in Table 1.4.

Meats, Ground Meat, and Organ Meats

Raw beef carries a large number of *E. coli* since they are natural inhabitants of intestines of mammals; therefore, during slaughter the carcass may be contaminated with fecal bacteria. Fresh meats also can be contaminated with *Salmonella* and *Staphylococcus aureus* originating from the skin,

hide, or feathers. Ground beef is made from meat, trimmings, and fat and has a large surface area that favors the growth of aerobic bacteria. The knives of a meat grinder, if not properly sanitized or washed, may be the source of contamination. Furthermore, one heavily contaminated piece of meat may be sufficient to contaminate an entire lot of ground meat. Pathogens such as *Clostridium perfringens*, *Bacillus cereus*, *Listeria monocytogenes*, and enterohemorrhagic *E. coli* (EHEC) are associated with these types of products. The liver, kidney, heart, and tongue may carry Gram-positive cocci, coryneforms, *Moraxella*, and *Pseudomonas*. Lymph nodes in food animals are secondary lymphoid tissues in which pathogens are supposed to be destroyed biologically but if there is survival can serve as a major source of pathogens. Mechanically deboned meat/poultry/fish generally carry lower microbial loads because of less handling and minimal human interventions. In addition, electrical stimulation converts glycogen to lactic acid, thus resulting in lowered pH, which suppresses bacterial loads. During rigor mortis, the release of cathepsin facilitates muscle tenderization, and lysozyme reduces the bacterial counts.

Vacuum-Packaged Meats

In products vacuum packaged in oxygen impermeable films, air is removed to increase the shelf life. Generally, the shelf life of vacuum-packaged meats is about 15 weeks. In another strategy, initially, an oxygen permeable packaging film allows the growth of indigenous *Pseudomonas* spp., but lactic acid bacteria (*Lactobacillus*, *Leuconostoc*, and *Carnobacterium*) become predominant after the *Pseudomonas* spp. remove oxygen and increase the CO_2 level inside the package, which in turn favors the growth of facultative anaerobes (lactic acid bacteria) and anaerobic microbes such as *Clostridium* species (*Clostridium botulinum*, *C. perfringens*). Later, lactic acid bacteria such as *Lactobacillus* and *Brochothrix* grow as well as pathogenic bacteria such as *Yersinia enterocolitica* and *Staphylococcus aureus*.

Table 1.4 Foods and pathogens involved in select outbreaks from 2000–2016

Year	Foods involved	Place	Pathogen	No. of illnesses	No. of deaths
2016	Live poultry and backyard flock	USA	*Salmonella* (multiple serovars)	895	3
2016	Raw scallops (sushi)	USA	Hepatitis A virus	292	2
2016	Frozen vegetables	USA	*L. monocytogenes*	9	3
2015	Ice cream	USA	*L. monocytogenes*	10	3
2014	Chicken	USA	*Salmonella* Heidelberg	634	–
2014–2015	Apple	USA	*L. monocytogenes*	35	7 deaths, 1 miscarriage
2012	Tuna fish from India	USA	*Salmonella* Bareilly	258	–
2011	Lettuce	USA	*E. coli* O157:H7	60	–
2011	Strawberry	USA	*E. coli* O157:H7	15	1
2011	Cantaloupe	USA	*L. monocytogenes*	148	33
2011	Sprouts	Germany	*E. coli* O104:H4	3911	48
2011	Papaya	USA	*Salmonella*	106	–
2011	Alfalfa sprouts	USA	*Salmonella*	140	–
2010	Celery	USA	*L. monocytogenes*	10	5
2010	Lettuce	USA	*E. coli* O145	26	
2010	Eggs	USA	*S.* Enteritidis	1939	–
2009–2010	Black and red pepper	USA	*S.* Montevideo	272	
2009	Alfalfa sprouts	USA	*Salmonella*	235	
2009	Frozen Raspberries	Finland	Norovirus	200	
2009	Peanut butter	USA	*S.* Typhimurium	714	9
2008	Mexican-grown peppers	USA	*S.* Saintpaul	1017	2
2007	Chicken and turkey pot pies	USA	*Salmonella*	401	–
2007	Peanut butter	USA	*Salmonella*	425	–
2007	Frozen ground beef patties	USA	*E. coli* O157:H7	40	–
2008	Ground beef	USA	*E. coli* O157:H7	45	–
2007	Milk or milk-related products	USA	*L. monocytogenes*	5	3
2006	Green onions	USA	*E. coli* O157:H7	67	–
2006	Spinach	USA	*E. coli* O157:H7	199	3
2002	Hamburger	USA	*E. coli* O157:H7	18	–
2002	Processed chicken	USA	*L. monocytogenes*	46	7
2000	Bean sprouts	USA	*Salmonella*	23	–
2000	Dairy farms	USA	*E. coli* O157:H7	56	–
2000	Delicatessen turkey	USA	*L. monocytogenes*	30	4 deaths, 3 miscarriages
2000	Homemade Mexican-style cheese	USA	*L. monocytogenes*	12	5 miscarriages

Poultry

Salmonella enterica serovars Enteritidis, Typhimurium, Infantis, Reading, Blockley, and Kentucky, *Clostridium perfringens*, *Campylobacter jejuni*, and *E. coli* are associated with intact and ground raw poultry. Workers may also be the source of other *Salmonella* serovars like – Sandiego and Anatum. Fresh poultry may be the source of *Pseudomonas*, coryneforms, and yeasts. Ground turkey also may carry fecal streptococci.

Fish and Shellfish

Microbial loads in shrimps, oysters, and clams depend on the quality of the water from which they are harvested and the months of the year. If untreated raw sewage is present, the microbial quality deteriorates. During handling, fecal coliforms, fecal streptococci, and *Staphylococcus aureus* may be incorporated into the product. *Salmonella* also is found in oysters possibly due to contaminated water. Fish and shellfish also are the source for *Pseudomonas* spp., *Clostridium perfringens*, *Listeria monocytogenes*, *Vibrio parahaemolyticus*, *Vibrio vulnificus*, *Salmonella enterica* serovar Enteritidis and Typhimurium, *Campylobacter jejuni*, *Yersinia enterocolitica*, and enteroviruses (hepatitis A). Smoked salmon and shrimp may carry pathogenic *L. monocytogenes*. Seafood may also be a major source of algal toxins or neurotoxins originating from seasonal feeding on dinoflagellate algal blooms. Certain allergic diseases such as scombroid poisoning are also associated with fish and shellfish (see Chap. 9).

Fruits and Vegetables

Sources of human pathogen contamination in fruits and vegetables at preharvest stage include soil, irrigation water, inadequately composted animal manure, dust, wild and domestic animals, human handling, and water used for pesticide spray, foliar treatments, and growth hormones (Fig. 1.5). The sources of contamination at postharvest stage include human handling (workers, consumers), harvesting equipment, transport containers, wild and domestic animals, pests, dust, wash and rinse water, sorting, packing, cutting and further processing equipment, crosscontamination, transportation, and sales.

Human pathogens interact with fresh produce not merely in a physical manner but attach to the surface by employing fimbriae, pili, and flagella involving biochemical signals and multiple compounds for such adherence. Furthermore, pathogens on the produce surface can be attached as aggregates, partially buried in the surface wax or in cracks in the cuticle, and are protected against environmental stress and surface disinfectants. Furthermore, human pathogens can also enter into tissues (xylem) of fresh produce, through

cuts, wounds, roots, insect, and animal bites. These pathogens residing inside the plant tissue will not stimulate plant defense system because they are not harming living tissues. In the absence of plant defense, these pathogens will not be exposed to plant antimicrobials since these are compartmentalized in specialized cells. Additionally, vacuum and hydrocooling such as showering or washing warm produce with chilled water may help create a partial internal vacuum, facilitating internalization of pathogens into produce.

Bacterial species commonly found on plant surfaces are *Enterobacter agglomerans*, *Pseudomonas syringae*, and *Pseudomonas fluorescens*. These microorganisms can coexist with the normal microflora on the fresh produce that is contaminated from the field. Plant pathogens and spoilage microorganisms such as *Agrobacterium* spp., *Erwinia* spp., *Clostridium* spp., *Bacillus* spp., *Klebsiella* spp., *Pseudomonas* spp., *Ralstonia* spp., *Xanthomonas* spp., and *Xylella* spp. can initiate interactions with the plant to have successful colonization on the plant. Human pathogens such as *E. coli* O157:H7, *Salmonella enterica*, *Clostridium perfringens*, *C. botulinum*, and *Listeria monocytogenes* from soil, manure, irrigation water, wild animals, and insect vectors can attach to the plant surface through secondary colonization. Protozoan species such as *Giardia intestinalis*, *Entamoeba histolytica*, *Cystoisospora belli*, and *Cyclospora cayetanensis* are also associated with produce and herbs. Examples of outbreaks associated with fruits and vegetables are *E. coli* O157:H7 in apple cider, spinach, lettuce, and sprouts; *E. coli* O104:H4 in fenugreek sprouts; *Salmonella enterica* in cantaloupes, watermelons, cilantro, tomatoes, sprouts, and almonds; *Cyclospora cayetanensis* in raspberries; hepatitis A in strawberries and green onion (scallion); and *L. monocytogenes* in cabbage, celery, cantaloupe, and apples. These events emphasize the need for further sanitation and processing or improved quality assurance to assure the safety of these minimally processed products. Postharvest and external contamination has been considered the major concerns in produce safety; however, recent outbreaks with

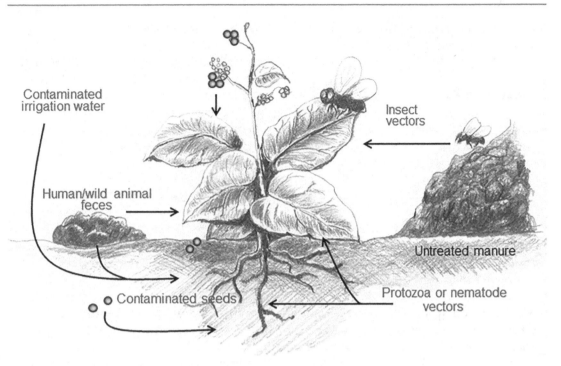

Contaminated
irrigation water

Insect
vectors

Human/wild animal
feces

Untreated manure

Contaminated seeds

Protozoa or nematode
vectors

Fig. 1.5 Mode of transmission of foodborne enteric pathogens in fruits and vegetables (Adapted and redrawn from Brandl 2006. Annu. Rev. Phytopathol. 44, 367–392)

lettuce and spinach suggest that these foods can take up pathogens internally on the field prior to harvest (Fig. 1.5). Indeed, in the 2006 spinach outbreak with *E. coli* O157:H7, feral pigs and cattle from the nearby ranch were responsible for the transmission of this pathogen. Experimental evidence showed that some pathogens like *E. coli* O157:H7 and *Salmonella* are able to survive and grow inside the veins of the above two plant tissues and tomatoes due to the abundant supplies of carbohydrates and moisture.

Dairy Products

Cow's udder, hide, and milking utensils may carry predominantly Gram-positive bacteria such as aerobic spore formers (*Bacillus* spp.), psychrotrophic *Pseudomonas* spp., and others including *Mycobacterium* and *Clostridium* species. Historically, cows with mastitis could produce raw milk infected with *Staphylococcus aureus* which if improperly pasteurized can cause foodborne illness. More recently, food-

borne outbreaks have occurred due to consumption of raw milk, homemade ice cream containing fresh eggs, or cheese made with unpasteurized cow's milk. In 1980–1981, 538 cases of *Salmonella* infection occurred with cheddar cheese and raw milk and certified raw milk. In 1980–1982, 172 cases of *Campylobacter* infection occurred with raw milk and certified raw milk. In 1995, a *Salmonella* outbreak occurred with ice cream when a tanker carrying raw liquid egg also transported pasteurized milk without proper cleaning and sanitization between products. In 1985, 2000, and 2006, *L. monocytogenes* infection occurred due to consumption of Mexican-style soft cheese (queso fresco). This ethnic product is made from unpasteurized milk and has a relatively non-acidic pH. In 2013, *Listeria monocytogenes* outbreak associated with cheese caused six illnesses and one death, and in 2015, outbreak associated with a nationally distributed brand of ice cream resulted in ten illnesses and three deaths. In 2016–2017, soft raw milk cheese was responsible for six illnesses and two deaths in the USA. *Yersinia enteroco-*

litica outbreaks also were associated with pasteurized and unpasteurized milk.

Delicatessen Foods

Microbial load and type of pathogens in delicatessen foods such as salads, sandwiches, and deli foods depends on the ingredients, meats, and vegetables and spices used to make them. Food handler's direct contact also can lead to increased incidences of *Staphylococcus aureus* via workers' nasal passages and hands. While considered to be relatively low-moisture ingredients, spices (see below) may carry spores of *Clostridium*, *Bacillus*, and molds which are able to survive in the presence of 8–12% moisture. Delicatessen foods also carry psychrotrophic *Listeria monocytogenes*. Dehydrated foods such as soups of chicken noodle, chicken rice, beef noodle, vegetable, mushroom, and pea may carry as high as 7.3 $\log_{10}$ g^{-1} ml^{-1} of *Clostridium perfringens* as well as coliforms.

Spices

Spices constitute seed, flower, leaf, bark, roots, or bulb and are used whole or ground or singly or mixed to enhance taste and flavor of food. Spices are an integral part of many ethnic cuisines and used in a variety of dishes. Spices include dehydrated onion, garlic, mustard seed, red pepper, sesame seed, black pepper, paprika, cinnamon, cumin seed, white pepper, oregano, poppy seed, ginger, allspice, anise, basil, bay leaves, cloves, fennel seed, sage, thyme, and turmeric. Spices are considered low-moisture ingredients, and they contain about 8–12% moisture. Some spices may contain microorganisms as high as 10^{6-7} cfu g^{-1} which include the spores of molds, *Aspergillus*, *Fusarium*, *Alternaria*, *Penicillium*, *Rhizopus*, *Cladosporium*, and *Trichoderma*, and bacteria, *Bacillus* and *Clostridium* spp. In addition, *Salmonella enterica* and *Staphylococcus aureus* have been found. In 2010, *Salmonella enterica* serovar Montevideo outbreak was associated with salami and contaminated black pepper as an ingredient was responsible for transmission.

Low-Moisture Products

These include chocolate, powdered infant formula, raw almonds, toasted oats breakfast cereal, dry seasonings, paprika-seasoned potato chips, dried coconut, infant cereals, peanut butter, and children's snacks made of puffed rice and corn. *Salmonella enterica* has been the primary pathogen associated with many of these products. Additionally, *Cronobacter sakazakii* has been associated with powdered infant formula. Even though these pathogens do not grow in these dried products, but a surviving low infectious dose allows their transmission to host causing disease when these foods are rehydrated.

Foodborne Pathogen Statistics and Socioeconomic Impact

Worldwide, foodborne pathogens cause numerous illnesses and deaths. The World Health Organization (WHO) estimated that worldwide, in 2005, 1.4 million people died from diarrheal diseases due to contaminated water and food. In 2012, the estimated global mortality rates due to foodborne disease ranged from 0.26 to 15.65 deaths per 100,000 populations. Globally, among the foodborne pathogens, *Norovirus* caused 125 million cases, while *Campylobacter* species caused 96 million cases. In Africa, Asia, and Latin America, there are about 1 billion cases of gastroenteritis per year in children under the age of 5, leading to 5 million deaths. In Mexico and Thailand, half of the children aged 0–4 years suffer from the *Campylobacter*-induced enteritis. In Europe, 50,000 cases per million population suffer from acute gastroenteritis. In the Netherlands, about 300,000 cases per million population occur yearly. In Northern Ireland and the Republic Ireland, about 3.2 million episodes of gastroenteritis are reported each year. In Australia, 5.4 million cases of foodborne gastroenteritis occur each year. In England, 20% of the population, i.e., 9.4 million people, suffer from acute gastroenteritis each year, and the primary contributing microorganisms are identified as *Norovirus*, *Campylobacter* species, rotavirus, and nontyphoidal *Salmonella* species.

In the USA, there are an estimated 48 million cases with 128,000 hospitalizations and 3000 deaths associated with foodborne infections each year. Of the 48 million cases, 9.4 million cases are caused by 31 known pathogens of bacterial (3.6 million cases, 39%), viral (5.5 million cases, 59%), and parasitic (0.2 million or 2%) origin. The majority of illnesses were caused by norovirus (58% illnesses), followed by nontyphoidal *Salmonella* (11%), *Clostridium perfringens* (10%), and *Campylobacter* spp. (9%). The leading causes for hospitalizations were *Salmonella* spp. (35%), *Norovirus* (26%), *Campylobacter* (15%), and *Toxoplasma gondii* (8%), while the most deaths were associated with nontyphoidal *Salmonella* (28%), *Toxoplasma gondii* (24%), *Listeria monocytogenes* (19%), and *Norovirus* (11%). Foodborne diseases of microbial origin have become the number one food safety concern among the US consumers and regulatory agencies, and this trend is probably true in most countries (Table 1.5).

Food animals and poultry are the most important reservoirs for many of the foodborne pathogens. Therefore, meat, milk, or egg products may carry *Salmonella enterica*, *Campylobacter jejuni*, *Listeria monocytogenes*, *Yersinia enterocolitica*, or *E. coli* O157:H7. Control of pathogens in raw unprocessed products is now receiving major emphasis to reduce pathogen loads before arrival at a processing plant. On-farm, pathogen-controlling strategies will help achieve that goal. However, the presence of pathogens in ready-to-eat (RTE) raw produce products is a serious concern since those products generally do not receive any treatments lethal to microbes before consumption. Also, many recent foodborne outbreaks resulted from consumption of undercooked or processed RTE meats (hotdogs, sliced luncheon meats, and salami), dairy products (soft cheeses made with unpasteurized milk, ice cream, butter, etc.), in addition to minimally processed fruits (apple cider, strawberries, cantaloupe, etc.) and vegetables (sprouts, lettuce, spinach, etc.).

Annual economic losses in the USA result from deaths, illnesses, loss of work, loss of manpower, and product loss account for about $78 billion. Besides acute gastroenteritis, the sequelae of the foodborne infections result in chronic rheumatoid conditions; ankylosing spondylitis–autoimmune disease (HLA); hemolytic uremic syndrome (HUS) due to Shiga toxin (Stx) from EHEC; atherosclerosis due to lipid deposition in arteries; Guillain–Barré syndrome and Miller Fisher syndrome from *Campylobacter* infections; reactive arthritis from *Salmonella*, *Shigella*, and *Campylobacter* infections; autoimmune disease such as allergic encephalitis; and autoimmune polyneuritis. Foodborne infections also vary between countries due to eating habits of the population. In Japan, high *Vibrio parahaemolyticus* cases are seen due to consumption of raw fish. Scandinavians and people from middle/eastern countries sometimes suffer from botulism due to consumption of improperly processed fish, meat, and vegetables.

Why High Incidence of Foodborne Outbreaks?

It is believed that some new pathogens are emerging, which are responsible for increased incidences of foodborne diseases (Table 1.1). Some are recognized recently whose ancestors probably caused foodborne illnesses for many thousands of years. An example is *E. coli* O157:H7, a new strain first reported on 1982, which is responsible for numerous outbreaks in recent years. This bacterium has evolved relatively recently from an enteropathogenic *E. coli* (EPEC) progenitor. Likewise, *E. coli* O104:H4, an enteroaggregative bacterium associated with the sprout outbreak in 2011 in Germany, has acquired bacteriophage-encoded *stx2* gene. It is thought to be a recently evolved serovar. In addition, many of the older pathogens are reemerging and contributing to the overall foodborne outbreak statistics. Besides, human sufferings and fatalities, the high number of foodborne outbreaks in recent years have had devastating economic impacts on food producers and processors. It has been a challenging task for scientists to figure out the reasons for the greater numbers of outbreaks in recent years. Some factors (Table 1.6) are discussed below which may

Table 1.5 Estimated yearly cases of foodborne diseases, related deaths, and associated cost in the USA

Pathogens	Cases	Hospitalizations	Deaths (%)	Cost (million dollars)
Bacteria				
Bacillus cereus	63,400	20	0	15
Brucella spp.	839	55	1 (0.3)	18
Campylobacter spp.	845,024	8463	76 (5.5)	6879
Clostridium perfringens	965,958	438	26 (0.4)	93
Clostridium botulinum	55	42	17.3 (0.2)	466
Shigatoxigenic *E. coli* (STEC), O157	63,153	2138	20 (2.9)	635
STEC, non O157	112,752	271	20	154
Enterotoxigenic *E. coli* (ETEC)	17,894	12	0	24
Listeria monocytogenes	1591	1455	255 (15.9)	2040
Salmonella spp. (nontyphoidal)	1,027,561	19,336	378 (0.5)	11,391
Shigella spp.	131,254	1456	10 (0.1%)	1254
Staphylococcus aureus	241,148	1064	6 (0.1)	168
Streptococcus group A	11,217	1	0	24
Vibrio parahaemolyticus	34,664	100	4	88
Other *Vibrio*	17,564	83	8 (1.7)	88
Vibrio cholerae	84	2	0	0.2
Vibrio vulnificus	96	93	36 (1)	268
Yersinia enterocolitica	97,656	533	29 (0.1)	1107
Parasites				
Giardia intestinalis	78,840	225	2 (0.1)	282
Toxoplasma gondii	86,686	4428	327 (20.7)	3456
Cryptosporidium parvum	57,616	210	4 (0.4)	168
Cyclospora cayetanensis	11,407	11	0	17
Trichinella spiralis	156	6	0	2
Viruses				
Norovirus	5461	14,663	149 (6.9)	3.677
Rotavirus	15,443	348	0	18
Astrovirus	15,443	87	0	19
Hepatitis A	1566	99	7	58
Total (known = 32,462 + unknown = 45,208)				77,671

Adapted from Scallan et al. (2011). Emerg. Infect. Dis. 17, 7–15

explain the plausible reasons for increased incidence: (1) increased surveillance and reporting, (2) changes in the food manufacturing and agricultural practices, (3) changes in consumer's habits, (4) increased at-risk populations, (5) improved detection methods, and (6) emerging pathogens with survivability in stressed conditions.

Surveillance and Reporting

In the past, foodborne incidence reporting was poor or underreported. Sometimes, persons suffering from illness also did not always consult their doctors, and sporadic cases were not reported routinely. In addition, the causative agents were not always identified because of lack of better methodologies. Sensitive detection methods are now available, and the epidemiological survey has been improved. Computer-based databases such as FoodNet and PulseNet in the USA, EC Enter-Net for *Salmonella* species and *E. coli* O157:H7 in Europe, and WHO Global Salm-Surv are now available to assess the trends, changes in expected numbers, and types from historical data (Table 1.7). These following databases are used as an alert mechanism for future outbreaks.

Table 1.6 Factors affecting the emergence of increased foodborne illnesses from food

1. *Increased surveillance and reporting*
2. *The food manufacturing and agricultural practices*
Centralized production facility
Distribution to multiple states/other countries
Intense agricultural practices
Minimal processing (produce and fruits)
Increased importation of fresh produce
3. *Consumer habits*
Eating more meals outside the home
Increased popularity of fresh fruits and vegetables
4. Increased at-risk populations (immunocompromised, elderly)
5. Improved detection methods and tracking of pathogens
6. Emerging pathogens with improved survivability in stressed conditions

Table 1.7 List of surveillance programs currently used in the USA and other countries

Program	Purpose
US surveillance and monitoring programs	
FoodNet	Foodborne Diseases Active Surveillance Network (FoodNet): routine surveillance of select foodborne pathogens in ten states in the USA (*see* Table 1.5)
PulseNet	DNA fingerprints of pathogens based on pulsed-field gel electrophoresis
CalciNet	Fingerprints of calciviruses
EHS-Net	Environmental health and cause of foodborne diseases
CAERS	CFSAN Adverse Event Reporting System for foods, cosmetics, dietary supplements
eFORS	Electronic Foodborne Outbreak Reporting Systems of CDC (Center for Disease Control and Prevention)
eLEXNET	Electronic Laboratory Exchange Network (data from FDA, USDA, DOD) from all 50 states
Global food industries and air travel	
EC Enter-Net	European Commission Enter Networks for *Salmonella* and *E. coli* O157:H7
OzFoodNet	Australian foodborne disease information for risk assessment and policy, training for foodborne disease investigation
WHO Global Salm-Surv	Surveillance resources and training in foodborne disease for participating countries

Agricultural Practices and the Food Manufacturing

Agricultural practices affect the incidence of microorganisms in the intestinal tract and on the surfaces of the food animal. High ambient temperature and moisture can encourage salmonellae growth in animal feed. Use of antibiotics in feeds generally kills the certain population of microorganisms including pathogens but encourage other resistant ones to grow. Transportation to the slaughterhouse in crowded trucks, in extreme weather conditions, creates stress and weakened immune systems, thus favoring microbial growth. Feed withdrawal practice in poultry, if exceeds more than 12 h before slaughtering, may cause leaky or thinned gut, vulnerable to rupture during evisceration and possibly the transfer of pathogens to other tissues. High-speed slaughter and evisceration also can result in the product contamination due to damaged gut. Intensive farming allows faster growth of pathogens, and recycling of animal waste products and animal by-products results in increased opportunities for transmission of pathogens to humans. Feeding animal products to another species allows one pathogen (e.g., a prion) to adapt and transmit to another host (see Chap. 6). Since 1997, the US government and several European countries have banned the use of animal products (MBM, meat bone meal) as an animal feed ingredient to control the spread of the prion agent that causes bovine spongiform encephalopathy (BSE) (see Chap. 6).

Changes in food manufacture and consumption practices are also contributing factors in increased foodborne diseases. Consumption of preprepared foods at home and outside the home, consumption of chilled and frozen foods, and increased consumption of poultry and fish as part of a healthy diet may increase the incidence of salmonellosis, *Campylobacter*-induced enteritis, and *Vibrio parahaemolyticus*-induced gastroenteritis. Cross-contamination of raw foods with cooked/processed foods and undercooking of meat and holding at a higher temperature also are contributing factors. Reduced use of salt, less use of food preservatives (sorbate and benzoate) for health reasons, and demand for more natural, fresher, healthier, and convenience meals may also serve as contributing factors. These types of food require greater care during production, storage, and distribution to ensure pathogen growth does not occur in these products. Tightening or enforcing strict hygienic measures in the food manufacturing facilities is needed to reduce the incidence of pathogens. Improved sanitary practices during cleaning of processing plant equipment with the rotation of sanitizers can help avoid the emergence of resistant microbes.

Ever larger centralized food processing facilities are thought to be a major contributing factor in recent years. The products produced by such processors are distributed widely to multiple states or many countries worldwide. Thus, such products if contaminated can affect large populations with devastating consequences. For example, in 1985 in a pasteurization plant in Chicago, raw unpasteurized milk contaminated finished pasteurized milk at packaging resulting in an estimated 15,000 cases of illnesses from salmonellosis. In January 1993, in a major fast-food restaurant chain, undercooking of hamburger meat resulted in *E. coli* O157:H7 outbreaks in which more than 600 persons were infected including many children. Several were hospitalized, 35 showed hemolytic uremic syndrome, and 3 died. The FDA then declared raw ground beef contaminated with this pathogen to be "adulterated," necessitating that it be recalled and destroyed. In 1997, *E. coli* O157:H7-tainted ground beef resulted in 25,000 pounds of ground beef recall. In 1998–1999, an outbreak of *L. monocytogenes* occurred due to consumption of hotdogs/lunchmeats resulted in 79 cases with 16 deaths and 3 miscarriages. In 2000–2001, consumption of Mexican-style soft cheese resulted in 12 cases of listeriosis in the USA. In 2002, consumption of sliced turkey meat caused a multistate outbreak of *L. monocytogenes* with 50 cases, 7 deaths, and 3 abortions. In 2003, raw milk cheese was responsible for an outbreak in Texas, and in 2005, a multistate outbreak involving consumption of turkey deli meat affected nine states and caused 12 illnesses. In 2006, spinach outbreak with *E. coli* O157:H7 resulted in

199 illnesses with 3 fatalities in 26 states in the USA. In 2006–2007, *Salmonella enterica* serovar Tennessee contaminated several peanut butter containing products resulting in 628 cases in 47 states in the USA. In 2011, 3911 people were infected and 47 deaths were associated with *E. coli* O104:H4 contamination of fenugreek sprouts in Germany and France. In 2010, *S. enterica* serovar Enteritidis in eggs were responsible for 1939 illnesses; *S. enterica* serovar Montevideo outbreak with salami was responsible for 230 illnesses and 43 hospitalizations. *L. monocytogenes* was responsible for several outbreaks between 2010 and 2015: 57 cases with 22 fatalities in Canada due to consumption of processed RTE meat products, 27 cases with 8 fatalities from Quargel sour milk curd cheese in Europe, 146 cases with 32 deaths from cantaloupe, 32 cases with 7 deaths from caramel apple, and 10 cases with 3 deaths from a national distributed brand of ice cream in the USA. The cantaloupe outbreak led the FDA to press criminal charges against the grower/owners of the farms and packing sheds from which this fruit was shipped.

Consumer Habits

Consumption of food outside the home also increases the chance of foodborne illnesses. Meal prepared and eaten at home also could cause disease because of poor hygienic practices during storage and preparation of food. Persons with underlying conditions also serve as the contributing factors: liver disease patients are susceptible to *Vibrio vulnificus* infection; therefore, these patients are advised not to eat raw oysters; and immunocompromised and pregnant women should avoid RTE meats, pâté, deli, or prepared meals and cheeses made with unpasteurized milk because of possible *Listeria monocytogenes* contamination. If outbreaks occur in public institutions such as restaurants, hotels, hospitals, cruise lines, manufacturing facilities, institutions, and so forth, the consequences are devastating because large numbers of people are at risk. This happens because of ignorance, poor management, sloppy practices, and lack of education or understanding principles and practices needed for safe handling of foods.

Increased at-Risk Populations

Populations that are susceptible to infections are increasing and people are living longer. Young, old, pregnant, and immunocompromised (YOPI) people are susceptible to various foodborne diseases. In addition, diabetes, cancer patients receiving chemotherapy, patients with organ transplants, and AIDS (acquired immune deficiency syndrome) patients are vulnerable to foodborne illnesses. Pathogen-free food may lead to increased susceptibility to diseases because subclinical infection may strengthen immunity. "Delhi belly," "Montezuma's revenge," or "traveler's diarrhea" affects only travelers and not the indigenous populations. Nutritional factors, physiological status, and concurrent or recent infection of intestinal tract also can favor increased infection.

Improved Detection Methods and Tracking of Pathogens

Improved detection methods are now capable of detecting low numbers of pathogens in products. Biosensors, PCR, immunoassays, mass spectrometry, and whole genome sequencing (WGS) techniques are very sensitive, and now low numbers of pathogens can be detected from products, which would normally give negative results with earlier less sensitive detection technologies.

Emerging Pathogens with Improved Survivability in Stressed Conditions

Newly identified and emerging pathogens are also a major contributing factor. Clinical, epidemiological, and laboratory studies have identified a number of so-called emerging and new pathogens (Table 1.1). Emergence due to unsafe food handling practices, undercooking of hamburger,

consumption of raw milk, ice cream contaminated with raw liquid eggs, consumer demands for fresher foods, and increased consumption of chicken that may be undercooked results in high numbers of enteritis cases. The emergence of antibiotic or preservative-resistant organisms also is a contributing factor. Antibiotic and acid-resistant *Salmonella enterica* and *E. coli* and also heat-resistant *Salmonella* and *Listeria* can survive certain sublethal processing conditions and persist in the product. Microorganisms acquiring virulent genes through vertical or horizontal transfer may become a new pathogen with highly virulent gene sets. For example, acquisition of *stx* genes by *E. coli* O157:H7 and *E. coli* O104:H4 strains has been shown. The viable but nonculturable organisms are also problematic such as *Norovirus*, *Campylobacter jejuni*, and *Vibrio* species that are difficult to detect.

Food Safety Authorities and Pathogen Control Acts in the USA

Federal food safety authorities in the USA include the US Department of Agriculture (USDA), Food and Drug Administration (FDA), Environmental Protection Agency (EPA), Department of Justice (DOJ), and Department of Defense (DOD) (Table 1.8). President Lincoln founded the USDA in 1862. The Pure Food and Drug Act and the Federal Meat Inspection Act (FMIA) were passed in 1906. The USDA's Bureau of Chemistry and Bureau of Animal Industry (BAI) were responsible for enforcing FMIA. In 1927, the USDA's Bureau of Chemistry was recognized, and it was renamed in 1931 as the Food and Drug Administration (FDA). In 1940, the FDA was separated from the USDA. In 1953, the FDA became a part of the Department of Health, Education, and Welfare and now the Department of Health and Human Services (HHS).

In 1938, the Federal Food, Drug, and Cosmetic Act was passed, and the FDA holds the authority to establish food safety standards. In 1957, the Poultry Products Inspection Act; in 1967, the

Table 1.8 Food safety authorities in the USA

Agency	Responsibility
US Department of Agriculture-Food Safety Inspection Service (USDA-FSIS)	Meat, poultry, and liquid egg products
Food and Drug Administration (FDA)	Fruits and vegetables, milk and dairy products, shell eggs, nuts, flours, spices, canned low-acid foods
Environmental Protection Agency (EPA)	Waterborne diseases
Animal and Plant Health Inspection Service (APHIS)	Zoonotic diseases, animal health
Department of Defense (DOD) and Department of Justice (DOJ)	Intentional contamination of products with pathogens, biothreat agents

Wholesome Meat Act; in 1968, the Wholesome Poultry Act; and in 1970, the Egg Products Inspection Act (EPIA) were passed. The FDA was responsible for inspection of whole egg and egg products, and since 1995, the FSIS (Food Safety Inspection Service) is the responsible agency for inspection of pasteurized liquid, frozen, or dried egg products, and the FDA assumed responsibility for shell egg safety. In 1996, the FSIS included Pathogen Reduction/HACCP (Hazard Analysis Critical Control Points) as a landmark rule to control and prevent microbial pathogens.

In 2011, the Food Safety Modernization Act (FSMA) was passed to prevent foodborne illnesses and deaths. The FSMA enables the FDA to focus more on preventing foodborne illnesses rather than reactive response after an outbreak happens. The key elements of the FSMA are to supply safe food to consumers by implementing critical preventive measures through five broad focus areas: prevention, inspection and compliance, response, imports, and enhanced partnership. (1) Prevention strategy is intended for science-based decision-making process to control pathogens across the food supply chain including food facilities, production and packaging system, and products. Incorporation of HACCP and HARPC (Hazard Analysis and

Risk-Based Preventive Controls) principles that are scientifically and technically sound should be an integral part of prevention. Furthermore, the FDA assumes the authority to prevent intentional contamination of products. (2) Inspection and compliance involve mandatory inspection and laboratory testing, mandated inspection frequency, access to records, and testing by accredited laboratories. (3) In response category, if a production facility is implicated in an outbreak, the FDA has the authority for mandatory recall, suspension of registration of the food production facility, enhanced product tracing abilities, and additional record keeping for high-risk foods. (4) Imports: the imported food must meet the US standard; third-party certification and the FDA have the authority to deny entry of food from a foreign producer who fails to comply with the US standard. (5) FSMA also requires the enhanced partnership and collaboration among all food safety agencies domestic or foreign for public health safety.

Global Concerns with Foodborne Pathogens

In 2018, human population in the world is about 7.6 billion. Worldwide, about 852 million people, one-sixth of the world population, are chronically hungry due to an extreme poverty. According to the Food and Agriculture Organization (FAO) of WHO, up to 2 billion people lack food security intermittently due to varying degrees of poverty. Each year about 6 million children die of hunger, i.e., about 17,000 per day. By 2050, the world population is expected to be 9.3 billion, and food demand is anticipated to double as compared to today's demand. Besides population growth, climate change (such as drought, high temperature, flooding), conflict and war, and social unrest will also impact farming and food production. In many economically impoverished countries, lack of infrastructure for food transportation, processing, unsanitary and unhygienic food production and preparation practices, insects, uncontrolled/unregulated use of pesticides, chemicals, and antibiotics also affect food production and supply of nutritious safe food to masses.

The health of food animals is also critical in food security since the food animal account for large quantities of protein (meat, milk, egg) in the daily diet. The Animal Production and Health Division (AGA) of FAO reports that globally several diseases could be problematic including rinderpest, avian influenza (H5N1), African swine fever, contagious bovine pleuropneumonia, foot-and-mouth disease (FMD), Rift Valley fever, rabies, and African trypanosomiasis.

The field of food microbiology has a critical role to play in securing foods to the masses, especially by preventing food losses from contamination of food spoilage and pathogenic microorganisms. Beneficial microorganisms, on the other hand, will help protect the food supply by converting nutritionally poor raw materials to rich nutritious products through active fermentation. Microbial fermentation processes can also help preserve seasonal foods to yield self-stable products for year-round supply. Single-cell proteins from algae and yeasts will also be able to supplement protein in the diet to curtail food shortage.

Alternate Protein Source

Consumption of meat from wildlife, often referred to as "bush meat," has increased in Central and West Africa and is perceived as a factor that can alleviate poverty. Wild animals such as elephant, gorilla, chimpanzee and other primates, crocodile, forest antelope (duikers), porcupine, pig, cane rat, pangolin, monitor lizard, guinea fowl, and anteaters have been smoked, dried, and often exported to other countries. This rate of harvesting would endanger wildlife population to extinction. Moreover, there is a concern for infectious disease transmission such as monkey poxvirus, Ebola virus, HIV-like virus, simian foamy virus, and Nipah virus to humans (see Chap. 6).

Case Study
Seventeen days in June 2008 (29 Air France flights from Central and West Africa arrived at Paris), 134 people were searched, 9 carrying bush meat and 83 had livestock or fish. Total amounts of product seized were 414 pounds. One passenger had 112 pounds of bush meat and no other luggage. Most of the bush meats were smoked and dried carcasses. Some were identifiable; others were not, so those animals were boiled to expose their skeleton for identification. Eleven types of bush meats were confiscated: monkeys, large rats, crocodiles, small antelopes, and pangolins (anteaters). Forty percent of which are listed as endangered species. It is estimated that about 5 tons of bush meat arrive in Paris each week. Likewise, bush meats are also imported to the USA. Kristine Smith, a wildlife veterinarian in the US Customs office said, "We get these big boxes of meat, sometimes you see primate heads or hands in there." R. Ruggiero of the US Fish and Wildlife Service said, "In Africa today, many wildlife populations are being eaten to extinction. The greatest impact to wildlife populations in Africa is the bush meat trade."

Climate Change and Global Warming

Water is called the environmental common since it links environment, agriculture, animals, and humans. Without water, life cannot be sustained. Yet, it can serve as an efficient vehicle for rapid transportation of pathogens from a local source to a global scale through tributaries, rivers, and oceans. The scarcity of water leads to drought, and that can cause a rise in many pathogens such as molds and mycotoxins. Global warming and climate change leading to temperature rise can have an impact on the persistence and occurrence of bacteria, viruses, parasites, and fungi and the corresponding foodborne diseases. Global warming can affect microbial ecology and growth, plant and animal physiology, and host susceptibility resulting in the emergence of plant and animal diseases and insect infestations, all of which could influence foodborne diseases and zoonoses. The rise in ocean temperature can increase toxic algal blooms, yielding high levels of biotoxins in marine waters and then their bioaccumulation in animals and fish. Hurricane (cyclone/tornado) and flooding increase waterborne diseases such as cholera and dysentery and fish and shellfish poisoning. Flash floods can cause increased runoff of soil nutrients such as nitrogen and phosphates from agriculture field into rivers, lakes, and oceans promoting increased toxic algal bloom which can contaminate potable water and fish and shellfish growing in that water.

Foodborne Outbreaks Associated with Imported Foods

According to the Economic Research Service (ERS) of the USDA, overall, 16 percent of all foods eaten in the US is imported with 85% of the seafood and up to 60% of fresh produce being imported. Increased outbreaks are associated with imported foods in the USA. According to CDC, between 2005 and 2010, there were 39 outbreaks linked to imported foods, fish, and spices, resulting in 2348 illnesses involving foods from 15 countries (Table 1.9). Fish was the most common commodity (17 outbreaks) followed by spices (6 outbreaks). Import from Asia was responsible for about 45% of the outbreaks.

Travel-Associated Foodborne Infections

The WHO estimates that about 15–20 million travelers to developing countries experience diarrhea annually. The CDC reported that between 2004 and 2009, travel-associated enteric infection cases were 8270 excluding enterotoxigenic

Table 1.9 Global concern of food and water associated organisms

Source	Pathogens of concern
Water	*Shigella, Vibrio,* protozoa, nematodes
Fresh fruits and vegetables	Protozoan species (*Giardia, Cyclospora*), *Shigella,* nematodes
Dairy products (milk)	*Campylobacter, Staphylococcus, Listeria*
Rice and pasta	*Bacillus cereus* (*Bacillus* spp.)
Meats	*E. coli, Salmonella enterica*
Fresh water fish and sea foods	*Vibrio, Aeromonas*
Street foods	*Staphylococcus, Bacillus,* hepatitis A, enteric viruses
Powdered infant formula	*Cronobacter sakazakii*
Ingredients (spices and herbs):	Molds and mycotoxins, *Salmonella, Bacillus,* and *Clostridium* spores

Table 1.10 Travel-associated enteric pathogen infection in returning Americans from abroad between 2004 and 2009

Pathogen	Travel-associated cases in the USA (2004–2009)	%
Campylobacter	3445	41.7
Salmonella	3034	36.7
Shigella	1071	13.0
Cryptosporidium	317	3.8
STEC	257	3.1
Cyclospora	54	0.7
Vibrio	44	0.5
Yersinia	32	0.4
Listeria	16	0.2
Total	**8270**	**100**

Adapted from Kendall et al. (2012). Clin Infect Dis 54, S480–S487

E. coli (ETEC, responsible for traveler's diarrhea) cases after returning to the USA. Forty-two percent cases were due to *Campylobacter*, 32% due to *Salmonella*, and 13% due to *Shigella* (Table 1.10). The greatest risks were in Africa (~76 cases per 100,000), Asia (~23 per 100,000), and Latin America and Caribbean countries (20 per 100,000). During this reporting period, the top travel destinations were Mexico, India, Peru, and the Dominican Republic.

Fig. 1.6 Street food safety

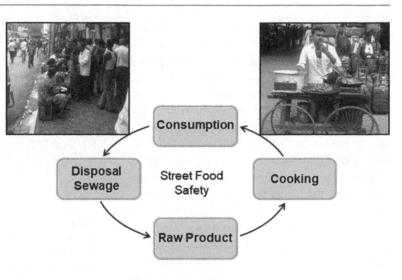

Fig. 1.6 Street food safety

Control of Foodborne Diseases in Economically Underdeveloped Countries

Several factors should be considered to improve food safety in economically poor countries: improvement in farming practices, infrastructure, and transportation of food products to market; availability of clean and potable water; improved food processing facility, hygienic practices during food preparation, refrigeration, and storage; personal hygiene; insect control; street food safety (Fig. 1.6); food safety training and education; and implementation of HACCP in food production.

Challenges and Concerns with Global Food Safety

Foodborne disease statistics and surveillance data are scanty. Underreporting of foodborne diseases, misdiagnoses, large susceptible populations, poverty, and lack of infrastructure hinder data collection by the public health sector. This lack of public health data can limit policymaking and the formation of more effective guidelines for proper farming, processing, cooking, distribution, and food safety practices. Malnourishment, undernourishment, immunological health, and poor education and communication can also hinder progress.

Summary

Food microbiology is a branch of microbiology that focuses on the study of microorganisms that are associated with food intended for human or animal consumption. Microorganisms use food as a source of nutrient for survival and growth or a vehicle of transmission to the human or animal host. Food microbiology is broadly classified into three focus areas: beneficial microorganisms, spoilage microorganisms, and pathogenic microorganisms. Beneficial microorganisms are those used for making traditional or ethnic fermented products, and as probiotics, which are gaining increased popularity because of their health-beneficial effects. Spoilage microorganisms, on the other hand, are responsible for product spoilage and place an economic burden on the producers, processors, and retail store owners for product losses. This is a serious issue in developing countries because of inadequate processing and refrigeration facilities. Foodborne pathogen contamination in foods presents a serious challenge which may result in severe diseases such as food intoxication, toxicoinfection, and infection. Mortality, morbidity, and product recalls are serious consequences of outbreaks caused by foodborne pathogens. Most foodborne pathogens grow in the mesophilic range and a few in the psychrophilic range and their growth does not usually alter the aesthetic quality of foods. Some pathogenic traits are sometimes acquired through plasmids, transposons, bacteriophages, or through pathogenicity islands. Foodborne pathogens can be zoonotic, geonotic, or human origin, and consumption of contaminated foods results in foodborne diseases. In order for a foodborne pathogen to cause disease, the microbe must be able to survive in food and, when transferred to human hosts, find niches, multiply, and express virulence factors to cause host cell damage. Worldwide, foodborne pathogens are responsible for large numbers of outbreaks, illnesses, and mortalities. Foodborne pathogens are a serious public health concern, and outbreaks are attributed to the emergence of new pathogens and reemergence of some old pathogens. The routine epidemiological and food product surveys are introduced by many countries in order to provide an accurate picture of global distribution and occurrence of foodborne diseases. The reasons for the emergence of increased foodborne diseases have been investigated. Several factors are thought to be responsible: improved survey system and the creation of the database for various pathogens; changes in agricultural and food manufacturing practices; consumer's habits of food consumption and preparation; the increased population of the susceptible group; improved survival and adaptation of pathogens in harsh food environments; and improved detection methods. To control foodborne pathogen-related illnesses and deaths, the US government has passed the Food Safety Modernization Act (FSMA) in 2011 as a science-based proactive preventive strategy rather than a reactive passive approach to the food safety. Globally, food safety is a major concern due to increasing numbers of food- and water-associated illnesses and mortality, the emergence of highly infectious diseases from bush meats derived from wild animals, travel-associated intercontinental transfer of pathogens, and globalization of food supplies.

Further Readings

1. Baird-Parker, A. (1994) Foods and microbiological risks. *Microbiology* **140**, 687–695.
2. Bhunia, A.K. (2014) One day to one hour: how quickly can foodborne pathogens be detected? *Future Microbiol* **9**, 935–946.
3. Brandl, M.T. (2006) Fitness of human enteric pathogens on plants and implications for food safety. *Annu Rev Phytopathol* **44**, 367–392.
4. Callejón, R.M., Rodríguez-Naranjo, M.I., Ubeda, C., Hornedo-Ortega, R., Garcia-Parrilla, M.C. and Troncoso, A.M. (2015) Reported foodborne outbreaks due to fresh produce in the United States and European Union: Trends and causes. *Foodborne Pathog Dis* **12**, 32–38.
5. Crim, S.M., Griffin, P.M., Tauxe, R., Marder, E.P., Gilliss, D., Cronquist, A.B., Cartter, M., Tobin-D'Angelo, M., Blythe, D., Smith, K., Lathrop, S., Zansky, S., Cieslak, P.R., Dunn, J., Holt, K.G., Wolpert, B. and Henao, O.L. (2015) Preliminary incidence and trends of infection with pathogens transmitted commonly through food - Foodborne Diseases Active Surveillance Network, 10 US Sites, 2006-2014. *MMWR Morb Mortal Wkly Rep* **64**, 495–499.
6. Dewaal, C.S. and Plunkett, D.W. (2013) The Food Safety Modernization Act - a series on what is essential for a food professional to know. *Food Protect Trends* **33**, 44–49.
7. Falkow, S. (1997) What is a pathogen? *ASM News* **63**, 359–365.
8. Flint, J.A., Duynhoven, Y.T.V., Angulo, F.J., DeLong, S.M., Braun, P., Kirk, M., Scallan, E., Fitzgerald, M., Adak, G.K., Sockett, P., Ellis, A., Hall, G., Gargouri, N., Walke, H. and Braam, P. (2005) Estimating the burden of acute gastroenteritis, foodborne disease, and pathogens commonly transmitted by food: An international review. *Clin Infect Dis* **41**, 698–704.
9. Kendall, M.E., Crim, S., Fullerton, K., Han, P.V., Cronquist, A.B., Shiferaw, B., Ingram, L.A., Rounds, J., Mintz, E.D. and Mahon, B.E. (2012) Travel-associated enteric infections diagnosed after return to the United States, foodborne diseases active surveillance network (FoodNet), 2004–2009. *Clin Infect Dis* **54**, S480–S487.
10. Kirk, M.D., Pires, S.M., Black, R.E., Caipo, M., Crump, J.A., Devleesschauwer, B., Doepfer, D., Fazil, A., Fischer-Walker, C.L., Hald, T., Hall, A.J., Keddy, K.H., Lake, R.J., Lanata, C.F., Torgerson, P.R., Havelaar, A.H. and Angulo, F.J. (2015) World Health Organization estimates of the global and regional disease burden of 22 foodborne bacterial, protozoal, and viral diseases, 2010: A data synthesis. *PLoS Med* **12**, e1001921.
11. Kramer, A., Schwebke, I. and Kampf, G. (2006) How long do nosocomial pathogens persist on inanimate surfaces? A systematic review. *BMC Infect Dis* **6**.
12. Methot, P.-O. and Alizon, S. (2014) What is a pathogen? Toward a process view of host-parasite interactions. *Virulence* **5**, 775–785.
13. Ray, B. and Bhunia, A. (2014) *Fundamental Food Microbiology. Fifth edition.* Boca Raton, FL: CRC Press, Taylor and Francis Group.

14. Scallan, E., Hoekstra, R.M., Angulo, F.J., Tauxe, R.V., Widdowson, M.A., Roy, S.L., Jones, J.L. and Griffin, P.M. (2011) Foodborne illness acquired in the United States—major pathogens. *Emerg Infect Dis* **17**, 7–15.
15. Scharff, R. (2012) Economic burden from health losses due to foodborne illness in the United States. *J Food Prot* **75**, 123–131.
16. Silk, B.J., Mahon, B.E., Griffin, P.M., Gould, L.H., Tauxe, R.V., Crim, S.M., Jackson, K.A., Gerner-Smidt, P., Herman, K.M. and Henao, O.L. (2013) Vital signs: *Listeria* illnesses, deaths, and outbreaks - United States, 2009-2011. *MMWR Morb Mortal Wkly Rep* **62**, 448–452.
17. Tauxe, R.V., Doyle, M.P., Kuchenmueller, T., Schlundt, J. and Stein, C.E. (2010) Evolving public health approaches to the global challenge of foodborne infections. *Int J Food Microbiol* **139**, S16–S28.
18. Tirado, M.C., Clarke, R., Jaykus, L.A., McQuatters-Gollop, A. and Frank, J.M. (2010) Climate change and food safety: A review. *Food Res Int* **43**, 1745–1765.
19. Trevejo, R.T., Barr, M.C. and Robinson, R.A. (2005) Important emerging bacterial zoonotic infections affecting the immunocompromised. *Vet Res* **36**, 493–506.
20. Walsh, K.A., Bennett, S.D., Mahovic, M. and Gould, L.H. (2014) Outbreaks associated with cantaloupe, watermelon, and honeydew in the United States, 1973–2011. *Foodborne Pathog Dis* **11**, 945–952.
21. Zweifel, C. and Stephan, R. (2012) Spices and herbs as source of *Salmonella*-related foodborne diseases. *Food Res Int* **45**, 765–769.

Biology of Microbial Pathogens

2

Introduction

All organisms belong to one of the three domains of life: Bacteria, Archaea and Eukarya (Fig. 2.1). The phylogenetic classification of the organisms is achieved based on the sequence comparison of small subunit rRNA gene. The classical taxonomic ranks are a domain, phylum, class, order, family, genus, species, serovar, and strain. An example for each rank is provided for a strain of *Escherichia coli* (Fig. 2.2). In this chapter, general properties of microorganisms including bacteria, viruses, molds, parasites, and algae (the source of most seafood-associated toxins) that are responsible for foodborne diseases are reviewed. In addition, the morphological and structural characteristics of microorganisms in relation to their pathogenesis are discussed so that this background knowledge will aid in understanding the mechanism of pathogenesis and the host response to pathogens in the subsequent chapters.

Bacteria

Bacterial genera associated with foodborne infections or food intoxications are listed in Table 2.1. Morphologically, bacterial cells are rods (1–10 μm in length and 0.5–1 μm in diameter), spherical, curved, or spiral (Fig. 2.3). Some cells may form long chains depending on the growth environments or physiological conditions or some may exist in singlet or doublet. Some cells are motile and may display unique motility such as tumbling, corkscrew rotation, swimming, and swarming.

Nutrients, temperature, and gaseous composition of the environment influence bacterial growth and metabolism. The aerobic bacteria require oxygen for respiration and energy, while the anaerobic bacteria grow in the absence of oxygen. Obligate or strict anaerobes cannot withstand traces of oxygen, while aerotolerant anaerobes have tolerance for some amounts of oxygen.

Microbes are also classified based on their temperature requirements. Some microorganisms grow at refrigeration to ambient room temperatures (1–25 °C), and they are called psychrotrophs (cold tolerant). Their optimal growth temperature is above 15 °C and maximum is above 20 °C. Psychrophiles (cold loving) can grow and reproduce at −20 °C to +20 °C with an optimal temperature of growth at 15 °C. Psychrophiles are also found in the polar ice, permafrost, and at the bottom of an ocean. Mesophiles grow at 25–37 °C and both psychrotrophs and mesophiles are capable of causing diseases in humans. Some bacteria grow at high temperatures and are called thermophiles (50–70 °C); some even grow in hot geysers such as extremophiles (>70 °C). Maintenance of membrane fluidity is critical for bacterial survival and

© Springer Science+Business Media, LLC, part of Springer Nature 2018
A. K. Bhunia, *Foodborne Microbial Pathogens*, Food Science Text Series,
https://doi.org/10.1007/978-1-4939-7349-1_2

Fig. 2.1 Universal phylogenetic tree based on a comparison of small-subunit rRNA gene sequences (Adapted and redrawn from Eckburg et al. 2003. Infect. Immun. 71:591–596)

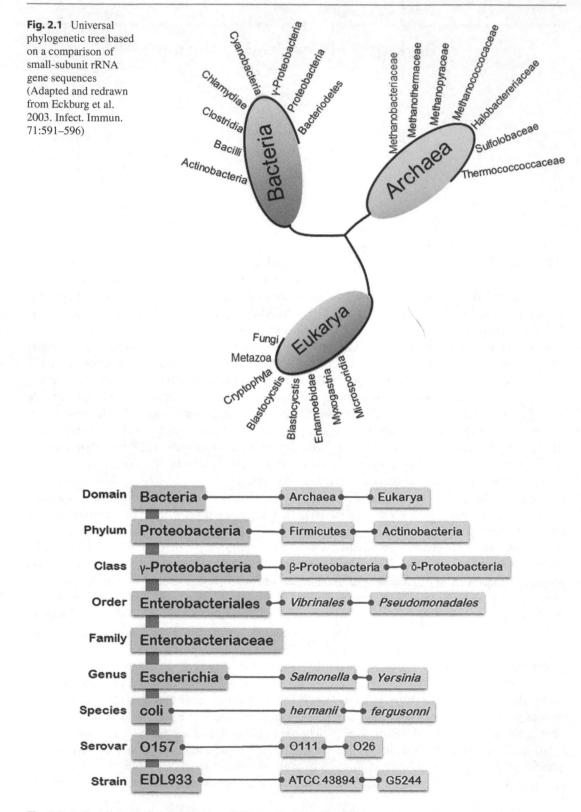

Fig. 2.2 A flow diagram showing the taxonomic classification of bacteria

Table 2.1 Bacterial genera associated with foodborne infections or food intoxication

Gram-positive	Gram-negative
Bacillus	Aeromonas
Clostridium	Arcobacter
Listeria	Brucella
Mycobacterium	Campylobacter
Staphylococcus	Cronobacter (Enterobacter)
	Escherichia
	Salmonella
	Shigella
	Vibrio
	Yersinia

growth under extreme conditions (both low and high temperatures) so that the enzyme activity, electron transport, ion pump, and nutrient uptake are possible. Altogether, only a very small fraction of all microbes is capable of causing diseases in humans or animals. Most of the pathogenic bacteria are mesophilic and a few are psychrophilic, while thermophiles are rarely pathogenic. Based on the microbial response to acidity (pH), bacteria are also grouped as aciduric (<pH 7) or alkaliphilic (pH 8.5–11 with an optimum of pH 10.0). *Thiobacillus ferrooxidans* is an example of

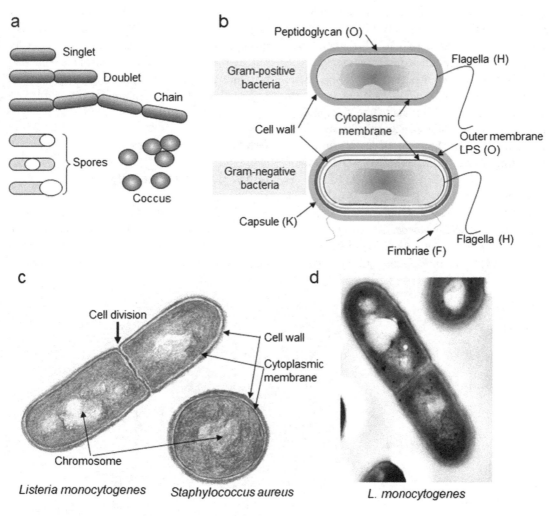

Fig. 2.3 Bacterial cells: (**a**) Morphology, (**b**) Structural differences in Gram-positive and Gram-negative bacteria, (**c**) Schematics and original (**d**) transmission electron microscopic cross section of Gram-positive *Listeria monocytogenes* and *Staphylococcus aureus* cells

extremophile aciduric bacterium that can grow at pH of ~1.5. Some bacteria also require salt for their growth, can tolerate moderate to high levels of salt (1.7–30%), and are called halophiles (examples are *Vibrio vulnificus*, *Staphylococcus aureus*). Some bacteria that can thrive at high pressure are called barophiles or piezophiles such as those found in deep sea or ocean floor and can withstand pressure exceeding 38 MPa.

Bacteria are divided into two groups based on their cell wall structures and staining characteristics: Gram-positive and Gram-negative, named after the Danish bacteriologist, Hans Christian Gram (1853–1938), who developed the Gram staining method in 1884. Gram-positive bacterial cell envelope consists of a thick rigid peptidoglycan polymer and retains crystal violet–iodine complex to appear purple to blue after Gram staining, while the Gram-negative bacterial cell wall is porous and does not retain this stain. Counter staining of the Gram-negative cells with safranin (carbol fuchsin) stains the cells pink to red. Acid-fast bacteria such as mycobacteria have mycolic acid and lipid on their cell envelope, rendering them unstainable with Gram stain. *Mycobacterium* retains carbol fuchsin after acid-fast staining also known as Ziehl–Neelsen staining. Examples of Gram-positive bacteria are *Listeria*, *Staphylococcus*, *Streptococcus*, and *Clostridium*; Gram-negative examples are *Salmonella*, *Escherichia*, *Campylobacter*,

Yersinia, and *Vibrio*; and the acid-fast bacteria are *Mycobacterium tuberculosis* and *M. bovis*.

The outermost layers of Gram-positive bacteria contain a thick rigid cell wall or peptidoglycan (PGN) structure. Cell wall also contains proteins and teichoic acid (TA) or wall teichoic acid (WTA), teichuronic acid, lipoteichoic acids (LTA), lipoglycan, and polysaccharides. The inner layer is a porous cytoplasmic membrane (CM) which consists of a lipid bilayer (Fig. 2.4).

Gram-negative bacteria, on the other hand, have an outer membrane (OM) layer, a thin peptidoglycan layer, and an inner cytoplasmic membrane (Fig. 2.5). The OM consists of a lipid bilayer, in which the lipopolysaccharide (LPS) is located on the outer leaflet of the bilayer. The major components of LPS are lipid A, which is a glycophospholipid consisting of β-1,6-D-glucosamine disaccharide. Phosphate and carboxylate groups of the lipid A provide a strong negative charge to the outer surface.

Gram-Positive Bacteria

Cell Wall and Peptidoglycan

The cell wall consists of a large number of molecules that have a multitude of functions. In addition, the cell wall protects cells from mechanical damage or osmotic lysis. The major component

Fig. 2.4 A schematic cross section of the Gram-positive bacterial cell wall

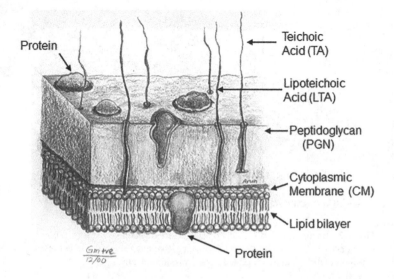

Protein

Teichoic Acid (TA)

Lipoteichoic Acid (LTA)

Peptidoglycan (PGN)

Cytoplasmic Membrane (CM)

Lipid bilayer

Protein

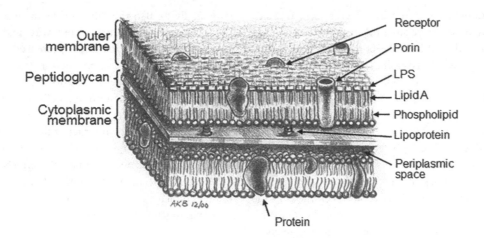

Fig. 2.5 A schematic cross section of Gram-negative bacterial cell wall and membranes

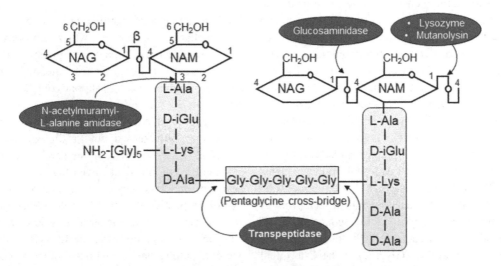

Fig. 2.6 The structure of cell wall (peptidoglycan) from a typical Gram-positive bacterial species (*Staphylococcus aureus*). Enzymes are marked by an oval circle showing their site of action. *NAG* N-acetyl-glucosamine, *NAM* N-acetylmuramic acid (Adapted and redrawn from Navarre and Schneewind. 1999. Microbiol. Mol. Biol. Rev. 63:174–229)

of the cell wall is peptidoglycan (about 20–80 nm thick) and is known as murein, which consists of peptides and sugar moieties (Figs. 2.4 and 2.6). PGN is highly complex and dynamic structure, which contains a disaccharide *N*-acetyl-D-glucosamine (NAG) and *N*-acetylmuramic acid (NAM) and linked by β-1,4-glycosidic linkage (GlcNAc-(β1–4)-MurNAc) and pentapeptide (Fig. 2.6). The enzyme transpeptidase (e.g., sortase) helps in the formation of peptide cross-link

with disaccharide GlcNAc-(β1–4)-MurNAc molecules and provides stability to the peptidoglycan backbone. In *Staphylococcus aureus*, the tetrapeptide, consisting of Ala, Glu, Lys, and Ala, forms a bridge with the pentaglycine (Gly$_5$) peptide. In *Listeria monocytogenes*, pentaglycine is absent, but the crossbridge is formed by an amide bond between the ε-amino group of a *meso*-diaminopimelic acid (*m*-Dpm) and the D-Ala of the adjacent cell wall (Fig. 2.6). Cell wall peptidoglycans with a low degree of cross-linking are much more susceptible to the degradation by the cell wall hydrolases than are those with a high degree of cross-linkers. Penicillin or other β-lactam antibiotics inhibit transpeptidases used for cell wall formation, hence affecting the bacterial cell wall growth in Gram-positive bacteria. In addition, β-lactam can activate autolysin, which degrades peptidoglycan. A carbohydrate-hydrolyzing enzyme, lysozyme (M_r. 14.4 kDa), breaks down peptidoglycan. Lysozyme present in body fluids, such as saliva and tears, and in the avian egg white is also known as β-1,4-N-acetylmuramidase and cleaves the glycosidic bond between the C-1 of NAM and the C-4 of NAG in the peptidoglycan (GlcNAc-(β1–4)-MurNAc). Lipoteichoic acid, teichoic acid, and surface proteins can prevent lysozyme action and protect the bacterium from lysis. Other enzymes that hydrolyze PGN are glucosaminidase, endopeptidase (e.g., lysostaphin from *S. aureus*), muramidase, amidase, and carboxypeptidase (Fig. 2.6). Cell wall peptidoglycan helps maintain the structural integrity and the shape of cells; therefore, when the PGN is totally removed, the bacterium forms a structure called *protoplast* while the bacterial cell wall with some remnants of PGN called a *spheroplast*.

Cell wall also carries several surface proteins containing an amino acid sequence motif consisting of L (Leucine), P (Proline), X (any), T (Threonine), and G (Glycine) in the C-terminal end, where X could be any amino acid. This LPXTG motif helps bacterial surface proteins to anchor to the peptidoglycan backbone. Examples of proteins that contain LPXTG motif are internalin in *Listeria monocytogenes*, the M protein in *Streptococcus pyogenes*, protein A in *Staphylococcus aureus*, and fibrinogen-binding protein in *S. aureus* and *S. epidermidis*. These proteins serve as adhesion factors or binding molecules for the host cell receptors. An immune response against Gram-positive bacterial pathogens often targets PGN as an antigen.

PGN performs multiple functions in a host. PGN is a strong vaccine adjuvant and activates dendritic or macrophages and monocytes for improved antigen presentation, cytokine production, and strong immune response against an antigen. PGN also initiates complement activation through the alternative pathway and suppresses appetite by inducing increased tumor necrosis factor (TNF-α) production (see Chap. 3). During innate immunity against bacteria, toll-like receptor (TLR) of immune cells such as the macrophage binds to PGN for recognition. TLR-2 was thought to interact with PGN; however, it was later determined that the nucleotide-binding oligomerization domain protein (Nod)-1 and Nod-2 serve as mammalian pattern recognition receptor for PGN (see Chap. 3). PGN can be cytotoxic to some host cells.

Teichoic Acid and Lipoteichoic Acid

Teichoic acid is a polyanionic polymer and has a polysaccharide backbone, which consists of glycerol or ribitol linked by phosphodiester bonds, i.e., sugar–alcohol–phosphate, and is buried in the peptidoglycan backbone (Fig. 2.4). Cell wall TA or WTA is uniformly distributed over the entire peptidoglycan exoskeleton. The function of TA is not fully known, but it is thought that the negatively charged TA captures divalent cations or provides a biophysical barrier that prevents the diffusion of substances and binds enzymes that hydrolyze peptidoglycan. TA or WTA also plays an important role in bacterial physiology, cell division, ion homeostasis, biofilm formation, bacterial pathogenesis, and host immune response including complement activation and phagocytosis (opsonization) (see Chap. 3).

Lipoteichoic acid being a polyanionic polymer is inserted into the lipid portion of the outer leaflet of the cytoplasmic membrane (CM), travels through the peptidoglycan, and is exported outside the cell wall (Fig. 2.4). Both WTA and LTA are unique for Gram-positive bacteria and are absent in Gram-negative bacteria. The function of LTA is unknown, but it serves as species-specific decorations of the peptidoglycan exoskeleton. LTA and the surface proteins provide unique serotype characteristics of a bacterium, for serological classification. LTA with different sugar molecules determines the serotype of the bacteria. This antigenic classification is called somatic antigen or O antigen.

Cytoplasmic Membrane

The cytoplasmic membrane consists of a lipid bilayer and carries transport proteins, which bind specific substrate and transport unidirectionally toward the cytoplasm. The proteins are responsible for secretion of periplasmic extracellular proteins, energy metabolism (electron transport system, ATPase), and cell wall synthesis

Gram-Negative Bacteria

Outer Membrane

The cell envelope of Gram-negative bacteria is surrounded by an outer membrane (OM), which is about 7.5–10 nm thick and consists of phospholipids, proteins, and lipopolysaccharide (LPS) (Fig. 2.5). The OM has a lipid bilayer arrangement; the outer leaflet consists of LPS and the inner leaflet has phospholipid. The LPS monomer has a strong lateral interaction with other LPS in the OM thus providing rigidity and gel-like appearance. LPS consists of three distinct regions: O side chain, core oligosaccharide, and lipid A (Fig. 2.7). The O side chain consists of repeating polysaccharide subunits of mannose, rhamnose, and galactose and is the most variable part of the LPS and can even show variability in strains within a species. The core oligosaccharide of LPS is covalently linked to the lipid A through the acidic sugar, 2-keto-3-deoxyoctanate (KDO). The lipid A is the most conserved region in the LPS and it secures LPS in the OM architecture. Lipid A is very toxic, and it is one of the most important molecules involved in bacterial pathogenesis and immune modulation. In Gram-negative bacterial pathogens when LPS is sloughed off, it is referred to as endotoxin or pyrogen. LPS induces host cytokine (TNF-α, IL-1) production by macrophages and increased body temperature (fever).

The OM is also embedded with many proteins and is collectively called OMPs (outer membrane proteins). The OMPs serve as a transporter of nutrients and ions across the OM to the periplasm and are residence sites for membrane-bound

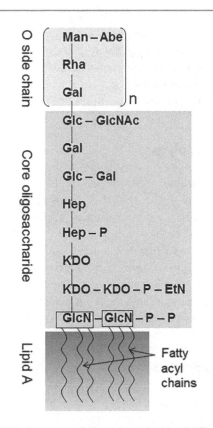

Fig. 2.7 Structure of lipopolysaccharide (LPS) from *Salmonella enterica* serovar Typhimurium. It has three regions: O side chain, core oligosaccharide, and the lipid A. *Abe* abequose, *Man* mannose, *Rha* rhamnose, *Gal* galactose, *Glc* glucose, *GlcNAc* N-acetylglucosamine, *Hep* heptulose, *KDO* 2-keto-3-deoxyoctonate, *GlcN* glucosamine, *P* phosphate, *EtN* ethanolamine

enzymes, such as protease and phospholipase. The OMPs serve as the adhesin for interaction with mammalian cells. The most abundant proteins in OM are porins, which are trimeric β-barrels forming channels. The pore diameter in porins varies from 6Å to 15Å, and porins serve as a selective transport system for small molecules (<600 Da). Porins are filled with water, and they help in the passive diffusion of small solutes without binding them. Nutrients can diffuse through porins, but if the concentration is very low, transport can occur via a substrate-specific active transporter system. Some porins may be substrate-specific such as the sucrose-specific porin, ScrY in *Salmonella*, and maltooligosaccharide-specific maltoporin, LamB in *E. coli*. Porins also play an important role in bacterial pathogenesis by activating an immune response, signaling pathway, and antibiotic resistance. Porins can activate both classical and alternative complement pathways and induce an inflammatory response.

Peptidoglycan

The peptidoglycan structure of Gram-negative bacteria is similar to that seen in Gram-positive bacteria, but it is much thinner (about 5–10 nm) and is less defined. It is attached to the OM by a lipoprotein, wherein the lipid moiety remains embedded in the inner leaflet of the OM. PGN is sensitive to lysozyme, but this enzyme is too large to pass through the porin in Gram-negative bacteria, which are thus resistant to lysozyme.

Periplasmic Space

Periplasmic space (PS) is the area or the compartment located between peptidoglycan and the cytoplasmic membrane. PS contains a concentrated gel-like matrix called periplasm and serves as the storage space for numerous enzymes (e.g., phosphatases), proteins that bind nutrients (sugars, amino acids, vitamins, ions) and direct them to the cytoplasmic membrane for transport, oligosaccharides (prevent changes in osmolarity), toxins, and some peptidoglycans which crosslinked into gel. Gram-negative bacteria also constantly release membrane vesicles (MVs) from OM. These MVs contain an OMP, LPS, phospholipids, and a periplasm containing protease, phospholipase C, alkaline phosphatase, and autolysin. MVs are released from both planktonic and biofilm forming cells and attack other vulnerable bacterial clones under nutrient starvation conditions providing selective growth or colonization advantages. MVs are thought to play a role in pathogenesis by delivering virulence factors such as phospholipase C, protease, and hemolysins directly to the host cells without being degraded by inactivating environmental enzymes. In many bacteria such as *Pseudomonas*, *Proteus*, *Serratia*, *Vibrio*, *Haemophilus*, *Neisseria*, *Escherichia* and others, these MVs participate in pathogenesis.

Protein Secretion Systems

Several nanomachine channels also known as secretion systems exist in both Gram-positive and Gram-negative bacteria to transport proteins, DNA, and certain virulence factors across the bacterial cell envelope (Table 2.2). In Gram-negative bacteria, six secretion systems facilitate protein export from the cytosol across the outer membrane. Those systems are type I secretory system (T1SS), type II (T2SS), type III (T3SS), type IV (T4SS), type V (T5SS), and type VI (T6SS). T1SS, T2SS, T3SS, T4SS, and T6SS span from cytoplasmic membrane (CM) to the outer membrane (OM), while T5SS spans only the OM. *T1SS* exports a large variety of protein substrates from cytosol to the extracellular milieu in both pathogenic and nonpathogenic Gram-negative bacteria. Examples of type I exporter are those proteins belonging to ATP-binding cassette (ABC), iron-binding proteins, and hemolysins. *T2SS* is also found in a wide range of pathogenic and nonpathogenic Gram-negative bacteria and helps in secretion of the protein from periplasm into the extracellular milieu. Examples are pseudolysin of *Pseudomonas aeruginosa* and cholera toxin of *Vibrio cholera*. *T3SS* is primarily found in pathogenic bacteria and exports effector proteins (virulence proteins) directly into the cytoplasm or the plasma membrane of target eukaryotic cells. T3SS is present in Gram-negative bacterial genera of *Salmonella*, *Shigella*, *Pseudomonas*, and *Yersinia* and enteropathogenic (EPEC) and enterohemorrhagic (EHEC) *Escherichia coli*. It is also known as a "molecular syringe," which directly delivers virulence proteins across the cytoplasmic membrane of the host without exposing the proteins to extracellular milieu. *T4SS* is present in both Gram-positive and Gram-negative bacteria and exports proteins and DNA–protein complexes from the bacterial cell into the cytoplasm of the recipient cell (both prokaryote and eukaryote). This system is ubiquitous and promiscuous and is capable of transporting DNA–protein complex to other bacteria, archaea, yeasts, molds, and plants. For example, VirB in *Agrobacterium tumefaciens* allows the transfer of DNA–protein complex in plant tissues causing cancerous growth. Other examples of

Table 2.2 Protein secretion systems in bacteria

Name	Source	Function
Type I secretion system (T1SS)	Gram-negative	Exports a large variety of protein substrates from cytosol to extracellular milieu in both pathogenic and nonpathogenic bacteria. Examples: proteins with ATP-binding cassette (ABC), iron-binding proteins, and hemolysins
Type II secretion system (T2SS)	Gram-negative	Secretes protein from periplasm into extracellular milieu such as pseudolysin of *Pseudomonas aeruginosa* and cholera toxin of *Vibrio cholera*
Type III secretion system (T3SS)	Gram-negative	Exports effector proteins (virulence proteins) directly into the cytoplasm or the plasma membrane of target eukaryotic cells
Type IV secretion system (T4SS)	Gram-negative and Gram-positive	Exports proteins and DNA–protein complexes from the bacterial cell into the cytoplasm of the recipient prokaryote and eukaryote cell
Type V secretion system (T5SS)	Gram-negative	Secretes virulence factors and involves in cell-to-cell adhesion and biofilm formation
Type VI secretion system (T6SS)	Gram-negative (*Proteobacteria*)	Translocates toxic effector proteins into both prokaryotes and eukaryotes
Type VII secretion system (T7SS)	Mycobacterium (Gram-positive	A specialized secretory system for export of virulence proteins in *Mycobacterium tuberculosis* and *M. bovis*
SecA, SecA2	Gram-negative Gram-positive	Export virulence proteins and enzymes

type IV system include CAG and ComB system in *Helicobacter pylori*, Ptl system in *Bordetella pertussis*, and Dot in *Legionella pneumophila*. T4SS is also responsible for the transfer of plasmid-borne antibiotic resistance gene. *T5SS* is a single membrane-spanning secretion system and is an autotransporter. It secretes virulence factors and is involved in cell-to-cell adhesion and biofilm formation. *T6SS* is widely distributed in *Proteobacteria*, and it translocates toxic effector proteins in both prokaryotes and eukaryotes. It plays an important role in bacterial pathogenesis and competition. An additional secretory system, *T7SS*, was originally isolated from mycobacteria and is a specialized secretory system for export of virulence proteins in *Mycobacterium tuberculosis* and *M. bovis*. Mycobacteria have mycomembrane consisting of free lipids and mycolic acid linked to the peptidoglycan via arabinogalactan, and T7SS translocates protein from CM to the extracellular milieu. Interestingly, T7SS gene clusters are also found in some non-acid-fast bacteria such as *Listeria monocytogenes*, *Bacillus subtilis*, and *Staphylococcus aureus*.

In Gram-positive bacteria, most secreted proteins are translocated across the cell envelope by Sec pathway consisting of membrane protein complex comprised of SecY, SecE, and SecG and are mediated by ATPase SecA. Precursor proteins acquire their folded structure after being translocated out of the cytosol by either SecA1 or SecA2. SecA2 is also present in Gram-negative bacteria. Most proteins are exported by SecA1, while only limited proteins including some virulence proteins are translocated by SecA2. Sec-dependent proteins contain a classical Sec signal sequence at the amino terminus; however, SecA2 can translocate proteins bearing no classical Sec signal sequence. The virulence factors that are translocated by SecA2 are superoxide dismutase (SodA) in *L. monocytogenes* and *Mycobacterium* and LAP (*Listeria* adhesion protein) in *L. monocytogenes*.

Accessory Structures in Gram-Positive and Gram-Negative Bacteria

The accessory extracellular structures including fimbriae, pili, flagella, and capsules provide structural integrity and facilitate bacterial colonization, motility, exchange of genetic materials, and survival in in vitro and in vivo environments. These structures also serve as important virulence factors for pathogenic bacteria by promoting bacterial adhesion, colonization, motility, and evasion of host immune system (see Chap. 4). Mutations in the genes encoding these accessory structures provide antigenic shift thus allowing pathogens to overcome host immune defense system. Furthermore, genetic and phenotypic properties of the accessory structures allow bacterial classification. The following key antigens are used as a target for serotyping or serogrouping.

Fimbriae (Pili) and Curli

Gram-negative bacteria carry cilia-like long appendages composed of glycoproteins, which can be used by bacteria to exchange genetic materials between organisms or used for attachment to a different substrate including the host cell. Fimbrial antigens are also called colonization factor antigens (CFA) and are important in bacterial adhesion, motility, and inter-bacterial communication (see Chap. 4 for details). Curli are extracellular amyloid-like protein fibers and are involved in bacterial adhesion, surface colonization, and biofilm formation.

Flagella

The flagellar antigen is designated "H," which stands for "Hauch" in German. Both Gram-positive and Gram-negative bacteria may carry one or more filamentous whiplike structures extending outward from their cell wall, called flagella. Flagella aid in bacterial motility or locomotion and reaching the nutrient-rich environment. Flagella may exist as either single polar flagellum or multiple flagella surrounding the bacterium. They are designated as monotrichous, a single polar flagellum such as in *Vibrio cholera*; amphitrichous, one on each pole, for example, are *Alcaligenes faecalis* and *Campylobacter jejuni*; lophotrichous, multiple flagella in each pole, for example, *Vibrio fischeri*; and peritrichous, multiple flagella surrounding the bacterial cell, for example, *Salmonella enterica* and *E. coli* (Fig. 2.8).

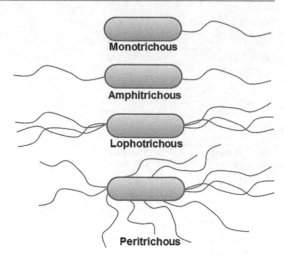

Fig. 2.8 Flagellar arrangements in bacteria

The flagellum, a complex self-assembling nanomachine, consists of a woven structure of flagellin proteins. There are three main structures: an engine, a propeller, and a universal joint that connects them to form the bacterial flagellum. The engine includes a rotor and a stator, which are embedded in the cytoplasmic membrane, and a rod that acts as a drive shaft, which extends from the rotor through the peptidoglycan layer to the outer membrane. The helical filament acts as a propeller for locomotion and movement. The proteins involved in the flagellar assembly are similar to the proteins seen in T3SS in Gram-negative bacteria. In some bacteria, flagella are considered important virulence factors aiding in adhesion and invasion by bacterial pathogens such as *Campylobacter jejuni*, *Legionella pneumophila*, *Clostridium difficile*, *Salmonella enterica* serovar Typhi, and *Vibrio cholera*. In some other pathogens, flagella are also involved in secretion of virulence factors.

O (Somatic) Antigen

In Gram-negative bacteria, O antigen represents O side chain, core oligosaccharide of the LPS located in the outer membrane (Fig. 2.8). O-polysaccharide is conserved, but it may be shared among different genera of Gram-negative bacteria. For example, *E. coli* O8 antigen is shared by *Klebsiella pneumoniae* and *Serratia marcescens*; *E. coli* O157 antigen is shared by *Citrobacter freundii* and *Salmonella enterica*. In addition, proteins in OM may contribute to O serotyping profile. In Gram-positive bacteria, O-antigenic profile is shared by peptidoglycan, TA or WTA, LTA, and surface proteins.

Capsule

The capsule is a thick layer of a gel-like structure composed of complex polysaccharides (i.e., sialic acid, N-acetylglucosamine, glucuronic acid, glucose, galactose, fucose) and located outside the cell envelope. In some bacteria, the capsule may be thin and is called glycocalyx. The capsule is generally visualized by negative staining (generally stained with India ink). It is also known as the K antigen and helps protect the cells against desiccation. The capsule acts as a virulence factor and it helps bacterial adhesion to the host cells. It protects cells from engulfment by macrophages and other phagocytic cells. The Gram-positive bacteria that produce capsules include *Bacillus cereus*, *Bacillus anthracis*, *Streptococcus mutans*, *Streptococcus pyogenes*, and *S. pneumoniae*, while the Gram-negative bacteria are *Haemophilus influenza*, *Pseudomonas aeruginosa*, *Escherichia coli*, *Agrobacterium tumefaciens*, and *Acetobacter xylinum*. The latter two are not normally pathogenic to man.

Endospore Formation

Some bacterial species forms spores as a survival mechanism. Only a few bacterial genera, namely, the Gram-positive *Bacillus*, *Alicyclobacillus*, *Clostridium*, *Sporolactobacillus*, and *Sporosarcina* and the Gram-negative *Desulfotomaculum*, are capable of forming spores. Among these bacteria, *Bacillus* (*B. cereus*, *B. anthracis*) and *Clostridium* (*C. perfringens*, *C. botulinum*) cause foodborne diseases in humans. Bacterial spores are called endospores since they are produced inside a cell and there is only one spore per cell. Endospores may be located terminal, central, or off-center, causing bulging of the cell. Under a phase-contrast microscope at 1000 × magnification, spores appear refractile spheroid or oval. The surface of a spore is negatively charged and hydrophobic; hence, it is sticky. Spores are much more resistant to physical and chemical antimicrobial treatments as compared to vegetative cells. This is because the specific structure of bacterial spores is quite different from that of vegetative cells from which they are formed (Fig. 2.9). In the core of the spore, protoplasm contains DNA, RNA, enzymes, dipicolinic acid (DPN), divalent cations, and very little water. The core is surrounded by an inner membrane, which is the predecessor of the cell cytoplasmic membrane. The germ cell wall surrounds the inner membrane, which is the predecessor of the cell wall in the emerging vegetative

Fig. 2.9 Drawing of an electron microscopic image of a clostridial bacterial spore, exosporium, coat, outer membrane, cortex, germ cell wall and inner membrane, and core

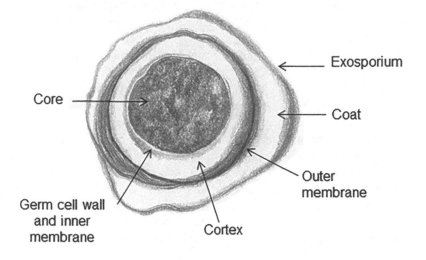

Exosporium

Core

Coat

Germ cell wall and inner membrane

Cortex

Outer membrane

cell. The cortex located surrounding the cell wall is composed of peptides, glycan, and an outer forespore membrane. The spore coats located outside the cortex and the membrane are composed of layers of proteins such as small acid-soluble proteins (SASP), which protect spore DNA from thermal denaturation and other harsh treatments. Spores of some species may have a structure called exosporium located outside the spore coat. During germination and outgrowth of spore, the cortex is hydrolyzed, and outer forespore membrane and spore coats are removed by the emerging vegetative cell.

The spores are metabolically inactive or dormant, and can survive for years, but are capable of emerging as vegetative cells (one cell per spore) under a suitable environment. The life cycle of spore-forming bacteria has a vegetative cycle (by binary fission) and a spore cycle, which goes through several stages in sequence, during which a cell sporulates and a vegetative cell emerges from a spore (Fig. 2.10). These stages are genetically controlled and influenced by different environmental parameters and biochemical events, which are briefly discussed here. Sporulation in bacteria is triggered by the changes in the environmental factors such as reduction in nutrient availability (particularly carbon, nitrogen, and phosphorous sources) and changes in the optimum growth temperature and pH. The transition from the cell division cycle to sporulation is genetically controlled, involving many genes. A cell initiates sporulation only at the end of completion of DNA replication and synthesis of adenosine bis-triphosphate (Abt) by spore-formers under carbon or phosphorous depletion. Sporulation events have several stages: (1) termination of DNA replication, alignment of chromosome in axial filament, and formation of mesosome; (2) invagination of cell membrane near one end and completion of the septum; (3) engulfment of prespore or forespore; (4) formation of germ cell wall and cortex, accumulation of Ca^{2+}, and synthesis of DPN; and (5) deposition of spore coats. In general, Gram-positive spore-formers in the *Bacillus* group can form mature spores in the absence of carbon and nitrogen sources on a substrate such as soil extract agar.

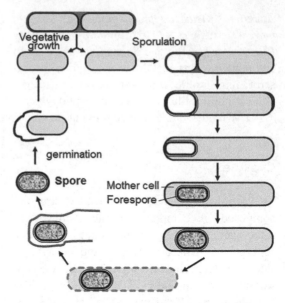

Fig. 2.10 Schematic presentation of the cycles of cell division and endospore formation, germination, and outgrowth of spore forming bacteria (Adapted and redrawn from Foster 1994. J. Appl. Bacteriol. 76:S25–S39)

Conversely, *Clostridia* usually require the presence of carbon and nitrogen substrates to produce mature heat-resistant spores.

Viruses

Virus means "poison" in Latin. Food virology includes the study of viruses that are transmitted via food and are commonly known as enteric. Enteric viruses are transferred via fecal–oral or fecal–water–oral routes and are highly infectious. Viruses are shed in large numbers (10^9 particles per gram) from infected patients through feces and a slightly lower numbers by vomitus. Person-to-person or fecal–oral transmission is a common mechanism for viral infection. The infectious dose is about 10–100 particles. Infection is also referred as "stomach flu" characterized by watery diarrhea, nausea, and vomiting and resembles bacterial food poisoning. Viruses are obligate intracellular parasites, i.e., they require a living host for replication and cannot grow outside their specific host nor in the food. Viruses are generally "host-specific," i.e., a virus can be a plant, animal, or human, or bacteria-specific and generally do not show cross-species infection. However, some zoonotic viruses are able to infect both humans and animals, and some viruses undergo genetic modifications to adapt themselves to different hosts. Viruses are metabolically dependent on the host. Animal cell culture or chick embryos are used for viral growth, replication, and isolation. Viruses replicate rapidly to yield prodigious titers in the host cells. Though there are a few antiviral drugs, viruses generally do not respond well to the conventional antibiotics. Hence, prevention strategies involve appropriate vaccination and proper hygienic practices.

Virus Classification/Taxonomy

Enteric viruses are generally nonenveloped RNA viruses and belong to the family of *Adenoviridae*, *Caliciviridae*, *Hepeviridae*, *Picornaviridae*, and *Reoviridae*. Viruses are classified based on the size, shape, structure, and nucleic acid content. Viruses contain either DNA or RNA. The nucleic acid may be single- or double-stranded nucleic acid molecules. DNA viruses generally have double-stranded DNA, and RNA viruses have single-stranded RNA molecule. Enteric viruses are found in the gastrointestinal tract, and small round structured viruses are called "SRSVs," which are found in feces. Based on the genetic elements, Dr. David Baltimore has classified viruses into seven types (1) dsDNA, (2) ssDNA, (3) dsRNA, (4) (+) sense ssRNA, (5) (−) sense ssRNA, (6) RNA reverse transcribing viruses, and (7) DNA reverse transcribing viruses. Later, Baltimore received the Nobel Prize for discovering that the RNA viruses reproduce by using a novel enzyme, RNA reverse transcriptase.

Virus Structure

Common particle shape of a virus may be cubical or icosahedral, i.e., a polyhedron with 20 triangular faces, 12 corners, and spherical or helical shape. The size of the virus is determined by an electron microscopy or virus's ability to pass through a defined membrane filter. Viruses are small particles generally ranging from 20 to 300 nm. The smallest virus is picornavirus (20–30 nm), a causative agent of foot-and-mouth disease (FMD). Some of the large viruses are the poxvirus or vaccinia virus (~300 nm), Nipah virus (500 nm), and Ebola virus (1200 nm). Virus structure varies – picornavirus has an icosahedral symmetry, tobacco mosaic virus (TMV) is helical, and herpesvirus, vaccinia virus, and poliovirus are spherical. The nucleic acid core is surrounded by a "protein coat" called nucleocapsid, which consists of capsomere. In some cases, protein coat is surrounded by an envelope made up of a lipid bilayer and accessory protein molecules. The envelope is sensitive to solvents. This envelope carries specific surface molecules that aid viral interaction with the host cell receptors. For example, hemagglutinin (H) and neuraminidase (N) in human influenza virus or bird flu virus bind to glycoside receptors on the host epithelial cells (Fig. 2.11).

Viral Replication

Environmental survival is critical for viral persistence and transmission and enteric viruses can survive in the stomach acid. Viruses are obligate intracellular parasites and require a living host for their replication. The virus life cycle has seven steps: (1) attachment to the host receptor, (2) penetration into the cell, (3) uncoating of DNA/RNA, (4) transcription and/or translation, (5) nucleic acid replication, (6) assembly of viral nucleic acid and coat proteins, and (7) release of matured virus particles (Fig. 2.12). RNA viruses encode genes for RNA-dependent RNA polymerase, a nonstructural enzyme (replicase), which is needed for RNA replication inside the host. On the other hand, DNA-dependent RNA polymerase is also called RNA transcriptase responsible for transcription of RNA from DNA leading to structural protein synthesis. The structural proteins such as capsid proteins are needed for viral packaging.

For gastroenteritis caused by Norovirus, the virus enters, replicates, and destroys mature enterocytes, triggering decreased absorption/

Fig. 2.11 Schematic viral structure (influenza virus)

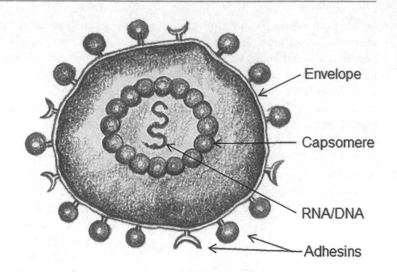

reabsorption resulting in epithelial inflammation and diarrheal symptoms. Even though the damaged enterocytes are replaced by the immature enterocytes, gastroenteritis continues since these immature cells are unable to perform a normal physiological function. Viruses are shed in high numbers and are stable outside the host environment. Norovirus and Rotavirus are true gut inhabitants, while poliovirus, hepatitis A virus (HAV), and hepatitis E virus (HEV) are found in nonenteric locations. An elaborate description of several key enteric viral diseases is described in Chap. 6.

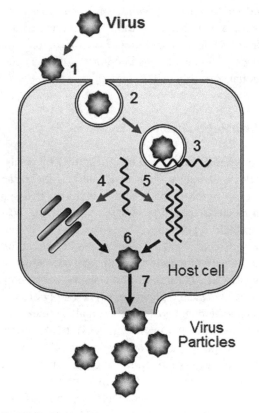

Fig. 2.12 Viral replication in a host cell: (*1*) attachment to the host cell receptor; (*2*) penetration, (*3*) uncoating, (*4*) transcription and/or translation, (*5*) nucleic acid replication, (*6*) assembly of viral proteins and nucleic acids, and (*7*) release of matured virus particles

Parasites

Parasites are another type of organisms that require living host cells in which to grow and reproduce by deriving nourishment and protection from the host. Parasites are unicellular or multicellular and can be transmitted by vectors such as flies, mosquitos, ticks, or food. Parasites may be ectoparasite such as ticks and mites, endoparasite such as *Trichinella*, and epiparasite that feeds on another parasite. Foodborne parasites (Table 2.3) are considered endoparasites including protozoa, cestodes (tapeworms), nematodes (roundworm), and trematodes (flatworms or flukes). Some endoparasites are unicellular (protozoa) and some are multicellular (metazoa). A majority of parasitic diseases are considered zoonotic. Immunocompromised people are highly susceptible to the intracellular protozoan parasites. Contamination of food products (generally fresh produce) with parasites often indicates inadequate sanitary hygienic practices employed during production and processing. Water, fresh fruits, and vegetables are known to be the major source of parasites. Globalization of food supply presents a favorable condition for spread and distribution of parasites thus presenting a major challenge in prevention and control of foodborne parasites.

Life Cycle and Growth Characteristics

Parasites are significantly larger than the bacteria and vary in their sizes. Many are intracellular. The general life cycle consists of a stage in which parasites are maintained as a cyst or oocyst (egg) in the environment (feces, soil, water, produce) and can contaminate food products. Parasites can be also transmitted through meat if the meat animal harbors the parasite. Oocysts are then transformed into the larvae stage called trophozoites, bradyzoites, or merozoites, which then mature into adults. The larval stage is the most infective stage. Parasites often require more than one animal host, i.e., an intermediate host to complete their life cycle. The parasites that complete their life cycle in a single host are known as *homoxenous* (Example, *Giardia lamblia*, *Trichinella spiralis*). Parasites that require more than one host to complete their life cycle are called *heteroxenous* (*Toxoplasma gondii, Taenia solium, Taenia saginata*). Parasites do not prolif-

erate in food; therefore, they are only detected by direct means using microscopy or molecular and antibody-based assays. Parasites cannot be cultured even in a rich growth medium but require a living host. Details of individual foodborne parasitic diseases are presented in Chap. 7.

Molds and Mycotoxins

Molds or fungi are ubiquitous in nature and cause diseases in humans, animals, and plants. Molds also cause significant food spoilage affecting grains, nuts, fruits, and vegetables. Some molds produce mycotoxins, which may be carcinogenic, teratogenic, and/or retard growth of the infected host (Table 2.4).

Mold Structure

Molds are multicellular microorganisms, and a typical mold possesses hyphae, conidiophore – consisting of stalk, vesicle, sterigmata, and conidia (spores) (Fig. 2.13). Often the identification of a mold species is done by examining the morphological shape and size of the spores under a light microscope. The cell wall of a mold (hypha) consists of a polymer of hexose chitin and *N*-acetylglucosamine (NAG). Mold spores or hyphae may be allergenic to some humans, and mold hyphae may proliferate in damp environments such as in buildings and walk-in coolers with high humidity and poor air circulation.

Table 2.3 Foodborne parasites

Group	Genus/species
Protozoa	*Giardia lamblia*
	Entamoeba histolytica
	Cyclospora cayetanensis
	Cryptosporidium parvum
	Cystoisospora belli
	Toxoplasma gondii
	Trypanosoma cruzi
Cestodes (tapeworms)	*Taenia saginata*, *T. solium*
	Echinococcus spp.
Nematode (roundworms)	*Anisakis* simplex
	Ascaris lumbricoides
	Trichinella spiralis
Trematodes (flatworms or flukes)	*Schistosoma mansoni*
	Fasciola hepatica

Table 2.4 Foodborne molds and mycotoxins

Mold	Mycotoxin
Aspergillus flavus *Aspergillus parasiticus*	Aflatoxin
Aspergillus ochraceus	Ochratoxin
Penicillium verrucosum	Ochratoxin
Aspergillus carbonarius	Ochratoxin
Fusarium verticillioides	Fumonisin
Fusarium verticillioides	Deoxynivalenol-DON
Fusarium graminearum	Zearalenone
Penicillium expansum	Patulin
Claviceps purpurea	Ergot (alkaloid)

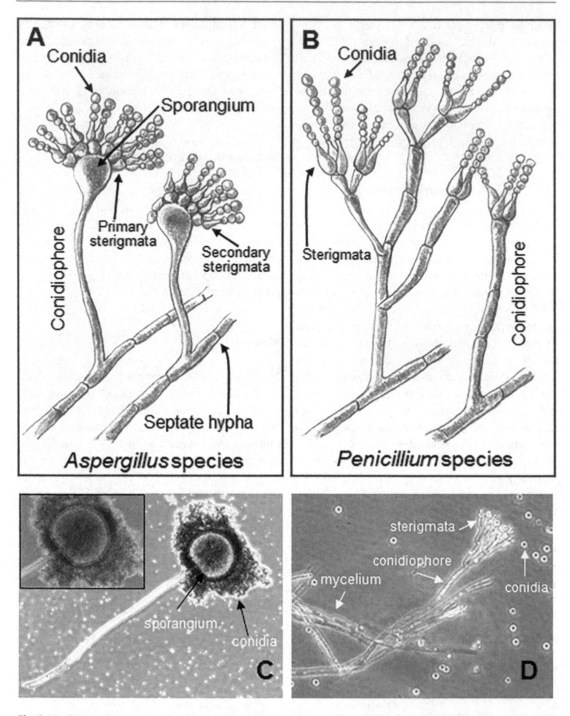

Fig. 2.13 Schematic structure of molds: (**a**) *Aspergillus* species and (**b**) *Penicillium* species. Panels (**c**) and (**d**) are the light microscopic photograph of *Aspergillus niger* (**c**) and *Penicillium citrinum* (**d**) (magnification 400×)

Mycotoxins

Toxigenic molds produce a variety of mycotoxins as secondary metabolites (Table 2.4), and the optimal condition for mycotoxin production is at a water activity (A_w) of 0.85–0.99 and a temperature range of 10–30 °C. Some of the same mycotoxins can be produced by different mold species.

The toxigenic molds are a major problem in agriculture commodities such as the grains, cereals, nuts, and fruits. Mycotoxins produced in these products can cause severe health problems in both humans and animals. Mycotoxins are produced as secondary metabolites that are low molecular weight molecules. Mycotoxins are highly stable and are difficult to destroy by traditional food processing conditions. Some mycotoxins emit fluorescence when exposed to the ultraviolet (UV) light; thus, UV can be used for screening of contaminated food/feedstuff. Mycotoxins can cause acute disease affecting the kidney and liver leading to chronic diseases such as liver cancer, birth defects, skin irritation, neurotoxicity, and death. Three general mechanisms of mycotoxin action are described: mutagenic, teratogenic, and carcinogenic. During the mutagenic action, the mycotoxin (e.g., aflatoxin) binds to DNA, especially the liver mitochondrial DNA resulting in point mutation or frameshift mutation due to deletion, addition, or substitution in DNA, and affects liver function (hence hepatotoxic). The teratogenic action causes DNA breakage and leads to birth defects, while the carcinogenic effect causes irreversible defects in cell physiology resulting in abnormal cell growth and metastasis. In recent years, the importance of mycotoxins has been highlighted for their potential use as a weapon for bioterrorism. Three major genera of molds, *Aspergillus*, *Fusarium*, and *Penicillium*, are of significant interest in food safety for their production of mycotoxins. Details of several key mycotoxins are presented in Chap. 8.

Fish and Shellfish Toxins

Fish and shellfish toxins are derived from single-cell marine plant, algae (marine dinoflagellates), bacteria (*Cyanobacteria*), or microbial actions originated from marine, brackish (a mixture of freshwater and seawater), and freshwater environments. In general, the seafood toxins are small molecules and are heat-stable. Six classes of toxins are significant: paralytic shellfish poisoning (PSP), diarrhetic shellfish poisoning (DSP),

neurotoxic shellfish poisoning (NSP), ciguatera fish poisoning (CFP), azaspiracid shellfish poisoning (AZP), and amnesic shellfish poisoning (ASP). These toxins are neurotoxins and are known as Na channel blockers. The symptoms vary from a tingling sensation in the lips, prostration, and a reduced heart rate, and depending on the severity of the disease, patients may suffer from paralysis and succumb to death. In addition, the scombroid toxin or histamine food poisoning is associated with scombroid fish such as tuna and mackerel due to the microbial metabolic action. First, microbial proteinases attack fish proteins to release free histidine, which is then decarboxylated to form biogenic amines: histamine, cadaverine, and saurine. These amines elicit a typical allergic response in humans. Details of the fish and shellfish toxins are discussed in Chap. 9.

Summary

All organisms belong to one of the three domains of life: Bacteria, Archaea, and Eukarya. As foodborne pathogens, general properties such as the morphological and structural characteristics of bacteria, viruses, molds, parasites, and algae are reviewed in this chapter. The cell walls of Gram-positive bacteria contains a thick peptidoglycan (PGN), which is highly complex and dynamic containing a disaccharide N-acetyl-D-glucosamine (NAG) and N-acetylmuramic acid (NAM) and linked by β- 1, 4-glycosidic linkage (GlcNAc-(β1–4)-MurNAc) and pentapeptide. The PGN not only protects the bacterial cells against mechanical or physical damages but also hosts a numerous structural and functional proteins for rigid exoskeleton and for functional attributes such as bacterial pathogenesis and induction of host immune response. The Gram-negative cell envelope consists of the outer membrane (OM) which carries LPS (an endotoxin), a thin PGN layer, and a cytoplasmic membrane. Due to the presence of OM, the protein secretion system in Gram-negative bacteria is mediated by several secretory pathways designated as tType I–Type VI secretory pathways,. Oof which tType

III secretion system is known as the molecular syringe that delivers bacterial virulence proteins directly to the interior of the host cell. Some bacteria produce endospores, which are essentially a long-term survival strategy for the bacteria.

Most foodborne viruses cause severe gastroenteritis and affect a large number of people every year. Foodborne enteric viruses are shed in large numbers (10^9 particles per gram) from infected patients through feces or vomitus. Person-to-person or fecal–oral transmission is a common mechanism for the spread of such viral infection. Since viruses are highly infectious, only a small dose of 10–100 particles is required to infect a person. Protozoan parasites are increasingly becoming a major concern due to their spread through fresh vegetables and fruits. Immunocompromised people are highly susceptible to the intracellular protozoan parasites. Water and soil tainted with feces generally serve as the major contamination sources, and the presence of these pathogens indicatesing breaches in the hygienic or sanitary practices during food production and harvest. Mycotoxins are small molecules produced as secondary metabolites by the toxigenic molds. Mycotoxins may exert carcinogenic, mutagenic, or teratogenic activities. Seafood toxins are generally associated with fish and shellfish that acquire toxins from ingested algae or are due to the bacterial enzymatic activities on fish proteins.

Further Readings

1. Beveridge, T.J. (1999) Structures of Gram-negative cell walls and their derived membrane vesicles. *J Bacteriol* **181**, 4725-4733.
2. Boneca, I.G. (2005) The role of peptidoglycan in pathogenesis. *Curr Opin Microbiol* **8**, 46-53.
3. Conlan, J.V., Sripa, B., Attwood, S. and Newton, P.N. (2011) A review of parasitic zoonoses in a changing Southeast Asia. *Vet Parasitol* **182**, 22-40.
4. Costa, T.R.D., Felisberto-Rodrigues, C., Meir, A., Prevost, M.S., Redzej, A., Trokter, M. and Waksman, G. (2015) Secretion systems in Gram-negative bacteria: structural and mechanistic insights. *Nat Rev Microbiol* **13**, 343-359.
5. Dawson, D. (2005) Foodborne protozoan parasites. *Int J Food Microbiol* **103**, 207.
6. Dorny, P., Praet, N., Deckers, N. and Gabriel, S. (2009) Emerging food-borne parasites. *Vet Parasitol* **163**, 196-206.
7. Eckburg, P.B., Lepp, P.W. and Relman, D.A. (2003) Archaea and their potential role in human disease. *Infect Immun* **71**, 591-596.
8. Feltcher, M.E. and Braunstein, M. (2012) Emerging themes in SecA2-mediated protein export. *Nat Rev Microbiol* **10**, 779-789.
9. Foster, S.J. (1994) The role and regulation of cell wall structural dynamics during differentiation of endospore-forming bacteria. *J Appl Bacteriol* **76**, S25-S39.
10. Galdiero, S., Falanga, A., Cantisani, M., Tarallo, R., Pepa, M.E.D., D'Oriano, V. and Galdiero, M. (2012) Microbe-Host Interactions: Structure and Role of Gram-Negative Bacterial Porins. *Curr Protein & Peptide Sci* **13**, 843-854.
11. Kalaitzis, J.A., Chau, R., Kohli, G.S., Murray, S.A. and Neilan, B.A. (2010) Biosynthesis of toxic naturally-occurring seafood contaminants. *Toxicon* **56**, 244-258.
12. Keates, S., Hitti, Y.S., Upton, M. and Kelly, C.P. (1997) *Helicobacter pylori* infection activates NF-kappa B in gastric epithelial cells. *Gastroenterology* **113**, 1099-1109.
13. Murphy, P.A., Hendrich, S., Landgren, C. and Bryant, C.M. (2006) Food mycotoxins: An update. *J Food Sci* **71**, R51-R65.
14. Navarre, W.W. and Schneewind, O. (1999) Surface Proteins of Gram-Positive Bacteria and Mechanisms of Their Targeting to the Cell Wall Envelope. *Microbiol Mol Biol Rev* **63**, 174-229.
15. Rodriguez-Lazaro, D., Cook, N., Ruggeri, F.M., Sellwood, J., Nasser, A., Nascimento, M.S., D'Agostino, M., Santos, R., Saiz, J.C., Rzezutka, A., Bosch, A., Girones, R., Carducci, A., Muscillo, M., Kovac, K., Diez-Valcarce, M., Vantarakis, A., von Bonsdorff, C.H., Husman, A.M.D., Hernandez, M. and van der Poel, W.H.M. (2012) Virus hazards from food, water and other contaminated environments. *FEMS Microbiol Rev* **36**, 786-814.
16. Schneewind, O. and Missiakas, D. (2014) Sec-secretion and sortase-mediated anchoring of proteins in Gram-positive bacteria. *Biochem Biophys Acta - Mol Cell Res* **1843**, 1687-1697.
17. Setlow, P. (2014) Germination of spores of *Bacillus* species: What we know and do not know. *J Bacteriol* **196**, 1297-1305.
18. Weidenmaier, C. and Peschel, A. (2008) Teichoic acids and related cell-wall glycopolymers in Gram-positive physiology and host interactions. *Nat Rev Microbiol* **6** (4), 276-287.

Host Defense Against Foodborne Pathogens

3

Introduction

Interaction of pathogenic microbes with their hosts leads to two consequences: a full-blown disease or a subclinical or no infection at all. Pathogen dominance results in either morbidity or mortality, while a successful protective host response averts disease. The immune system plays a key role in both of these outcomes. In 1798, Edward Jenner (1749–1823), a British scientist, first noticed that the milkmaids recovering from cowpox (vaccinia virus infection) never contracted smallpox. Later, he injected the content from a cowpox pustule (pox lesion) into an 8-year-old boy, who subsequently never developed smallpox after a challenge with the pathogen. He termed this phenomenon as "vaccine," a Latin word derived from "vaccinus" meaning "from cow." This work is regarded as the foundation of immunology. Later, Louis Pasteur (1822–1895) developed vaccines against anthrax and rabies. In addition, the pioneering works of Paul Ehrlich (1854–1915) and Emil von Behring (1854–1917) demonstrated the importance of passive vaccination or acquired immunity in disease defense. At the same time, Elie Metchnikoff (1845–1916) discovered that the innate immunity plays an important role against an infection. Collectively these early pioneering works laid the foundation for our understanding of the modern-day immunology and vaccinology.

The ability of the human body to protect against microbial infections is continuously evolving because of continued exposure to different microbes and their components. Ironically, microbes are also evolving in the same manner so that they can adapt themselves to the host. The human body has a rich source of nutrients, minerals, vitamins, and trace elements that support the growth of both pathogens and commensal microbes. Though the pathogens are the primary disease-causing organisms, the commensals can also cause disease but only under a favorable condition; hence, these are referred to as opportunistic pathogens. In most situations, the immune system restricts the spread of an infective agent; however, a breach in the immune system or an overt response toward the pathogen results in the onset of clinical symptoms. Therefore, to understand the disease process, one has to have a deeper knowledge of the host immune system and the underlying defense mechanism. The host immune system can be compared to "an impenetrable fort" impervious to attack by enemies and that is guarded by the soldiers like immune cells, armed with deadly weapons such as cytokines, toxins, antimicrobial peptides, complement proteins, and antibodies to stay at an "alert" position to combat sudden or deliberate attack by unsuspected enemies as pathogens.

Foodborne pathogens enter the body primarily through oral route; therefore, the immune response in the gastrointestinal (GI) tract is

© Springer Science+Business Media, LLC, part of Springer Nature 2018
A. K. Bhunia, *Foodborne Microbial Pathogens*, Food Science Text Series,
https://doi.org/10.1007/978-1-4939-7349-1_3

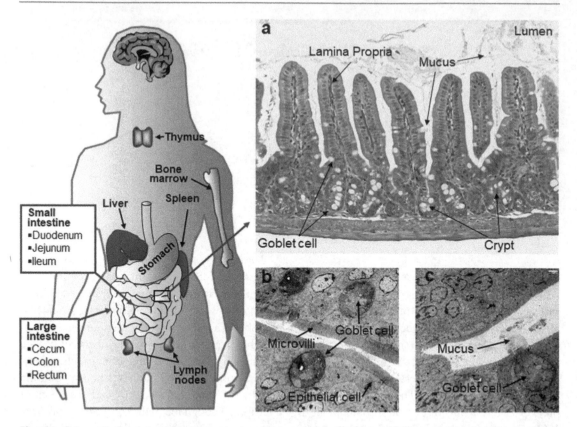

Fig. 3.1 Schematic drawing showing the gastrointestinal tract, various immunologic organs, and tissues in human. (**a**) Light microscopy image of mouse ileal section (H&E stained) showing villi, crypt, mucus, and mucus-secreting goblet cells (**b**) and (**c**). Transmission electron microscopy images of mouse ileum showing goblet cells and mucus secretion (Courtesy of Rishi Drolia and Arun Bhunia)

critical in preventing localized or systemic infection. Adult human GI tract is over 23 ft. long starting from the mouth to the anus, and the different parts, the mouth, esophagus, stomach, small intestine (duodenum, jejunum, and ileum), and large intestine (cecum, colon, and rectum), each possess unique physiological and immunological defense barriers against enteric pathogens (Fig. 3.1). Immunological defense is of two types, innate and adaptive, which are described below (Table 3.1).

Innate Immune Response

Innate immune response is naturally present in the body and is the first line of defense against a pathogen. It is constitutive but not antigen-specific. It can be activated but cannot be induced by subsequent exposure to invading agents. The major site of action of a foodborne pathogen or their toxins is the gastrointestinal tract; therefore, the onset of clinical symptoms such as nausea and vomiting could start as early as 30 min to 1 h from ingestion. The natural enteric defense is the most crucial against intoxication or infection. Enteric host defense is multifaceted, and several factors work in concert to achieve protection against a pathogen in the intestinal tract. These include intestinal epithelial barrier, mucus, complement proteins, antimicrobial peptides (AMPs), resident microbiota, and pattern recognition receptor of pathogens or pathogen-associated molecular pattern (PAMP) such as a toll-like receptor (TLR) and nucleotide-binding oligomerization domain (NOD) protein. PAMP, TLR, and NOD can initiate signaling events for enhanced phagocytosis to eliminate pathogens from the host.

Table 3.1 The major differences between the innate and adaptive immunity

Characteristics	Innate	Adaptive
Specificity	Lack specificity (fewer shared antigens)	High (microbial and nonmicrobial antigens)
Diversity	Limited (narrow)	Very large
Memory	None	Yes
Cellular/chemical barriers	Skin, mucosal epithelial, antimicrobials	Lymphocytes in epithelia, secreted antibodies on the surface
Proteins	Complement, PAMP, TLR, cytokines, antimicrobial peptides	Antibodies, cytokines
Cells	Phagocytes (macrophages, neutrophils, eosinophils, keratinocytes, Langerhans cells, dendritic cells), natural killer (NK) cells	Lymphocytes
Action	Immediate	Slow to develop, takes at least 4–7 days

In the intestinal architecture, immediately below the mucus layer, an epithelial cell lining provides protection against luminal microbes/pathogens from entering into the host. The epithelial lining consists of four cell lineages: absorptive epithelial cells, hormone-producing enteroendocrine cells, mucus-secreting goblet cells, and antimicrobial peptide-producing Paneth cells (Fig. 3.1). Lamina propria (LP) is located immediately beneath the epithelium, home to monocytes and macrophages, dendritic cells (DCs), and lymphocytes, which play an important role in innate and adaptive immunity. DCs extend their appendages in between the individual epithelial cells and samples the luminal content to initiate an innate response. DC also presents antigens to T and B lymphocytes for a subsequent adaptive immune response. Peyer's patch and lymphoid follicles located in the submucosal layer are part of the adaptive immune system and provide specific immune response against pathogens in the intestine.

Adaptive Immune Response

Adaptive immunity is also referred to as acquired immune response. It is induced or stimulated by the pathogen, and the response is generally pathogen-specific. The response increases in magnitude with successive exposure to an antigen, and the response is mediated by immune effector cells or molecules: cytotoxic T lymphocytes (CTL), antibodies, cytokines, and chemokines. The adaptive immune response is slow to develop and typically requires 4–7 days. During foodborne pathogen infection, the adaptive immune response may have a significant role only against a limited set of pathogens that require a long incubation period and cause systemic infection. These include *Listeria monocytogenes*, *Shigella* species, *Salmonella enterica*, *Yersinia enterocolitica*, hepatitis A virus, *Toxoplasma gondii*, *Trichinella* species, and some additional viral and parasitic pathogens. Adaptive immune response to some of the foodborne bacterial pathogens may also lead to adverse outcomes such as reactive arthritis and Guillain–Barré syndrome for life.

Innate Immunity of Intestinal Tract

The intestinal environment is a complex ecosystem and consists of gastrointestinal epithelium, immune cells, antimicrobial molecules, oxygen deficiency, and resident microbes. Pathogens overcome these barriers to cause infection. Pathogens first adhere and colonize the gut and then activate the signaling pathways to induce pathological effects. Pathogen colonization may be transient or long-term. Some may also use host cell cytoskeleton as a target to gain entry (see Chap. 4). The intestinal environment is also conducive for pathogen survival and growth. Roles of each innate or natural immune component in protection against pathogens are discussed below.

Skin

The skin is the first line of defense against many pathogens including the foodborne pathogens since food handlers may encounter pathogens during handling. Human skin is made up of epidermis and dermis layers. The epidermis consists of stratum corneum, stratum lucidum, stratum granulosum, stratum mucosum, and stratum germinativum. The dermis layer contains glands, adipose tissues, and lymphoid tissues. The skin has a slightly acidic pH (pH 5.0) and acts as a natural barrier for pathogen entry. The epidermis is dry; thus, it prevents bacterial growth since bacteria require moisture for active replication. In addition, the resident microflora on the skin prevents colonization of harmful bacteria. Lysozyme secreted from the sweat glands in the hair follicle has antibacterial effect by disrupting the peptidoglycan architecture of bacterial cell wall. Skin-associated lymphoid tissue (SALT) is located in the dermis and provides specific immune response against infective agents. Several phagocytic and antigen-presenting cells (APC), Langerhans cells (localized phagocytic cells), dendritic cells, and keratinocytes, are also located in the skin and may help eliminate the pathogen. The major APC of the epidermis are keratinocytes, which maintain the acidic environment, produce cytokines, and ingest and kill skin-associated bacteria such as *Pseudomonas aeruginosa* and *Staphylococcus aureus*.

Mucus Membrane

The mucus membrane (Fig. 3.1) forms the inner layer of the gastrointestinal tract, respiratory tract, and urogenital tract. Mucus membranes consist of epithelial cells, and depending on the location, the shape and organization of the cells may vary. Cells may appear as stratified, squamous, cuboidal, or columnar. There are four epithelial cell lineages in the gastrointestinal tract: epithelial cell, Paneth cells, goblet cell, and enteroendocrine cell.

The epithelial layer environment is moist, warm, and suitable for bacterial colonization and growth. Epithelial cells are tightly bonded to form a barrier against pathogens or microbes in the intestine. They are highly selective in translocating molecules across the membrane. This contrasts to the endothelial cells, located in the blood vessels, that are loosely bound and allow movement of blood cells and bacterial cells. The junction between two epithelial cells has four areas: tight junction (TJ), adherens junction (AJ), desmosome (Des), and gap junction (Gap) (Fig. 3.2). TJ consists of about 40 proteins including zona occludin (ZO-1, ZO-2, and ZO-3), membrane-associated guanylate kinase family, occludin, claudin, cingulin, and phosphoproteins. TJ proteins are regulated by classical signal transduction pathways such as heterotrimeric G proteins, Ca^{2+}, protein kinase C, and raft-like membrane microdomains. AJ consists predominantly of cadherin–catenin complex (E-cadherin, α-catenin, and β-catenin), vinculin, actin, and so forth. Desmosome architecture is shaped by desmoplakin, plakophilin and plakoglobin, desmoglein, and desmocollin. TJ and AJ maintain cell polarization; however, certain pathogens or toxins such as *Clostridium perfringens* enterotoxin (CPE) and *Vibrio cholerae* ZOT toxin target TJ domain to modulate intestinal epithelial permeability.

Epithelial cells also carry several receptor molecules for interaction with microorganisms. Those are called (1) pattern recognition receptor (PRR), (2) toll-like receptor (TLR), (3) nucleotide-binding oligomerization domain (NOD) protein, and (4) pathogen-associated molecular pattern (PAMP), which bind to bacterial lipopolysaccharide (LPS), peptidoglycan (PGN), and lipoteichoic acid (LTA).

Paneth cells located in the crypt, or folded area, of the mucosal layer are phagocytic cells and produce lysozyme and cryptdin, a 35-amino acid-long antimicrobial peptide. Cryptdin is inhibitory to *Listeria monocytogenes*, *Escherichia coli*, *Salmonella enterica*, and other pathogens. Goblet cells are located in between the epithelial cells. They secrete mucus and play a major role in the intestinal natural immunity as discussed below.

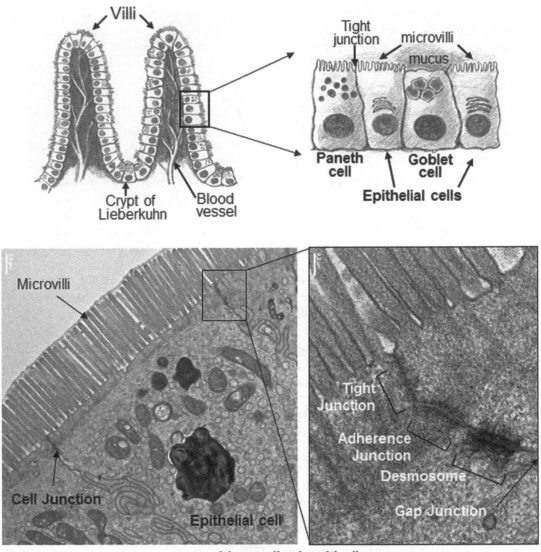

Fig. 3.2 Schematics of intestinal villi, epithelial cell lining, and cell junctions (TEM image; Courtesy of Rishi Drolia and Arun Bhunia)

Goblet Cells and Mucus

Mucus is an important natural protective barrier against microbial colonization and infection. The thickness of mucus layer varies throughout the gut. The stomach and colon have two layers of mucus while the remaining gut has one layer. Mucus thickness is greater in the large intestine than in the small intestine, and the thickness increases progressively distally along the intestine. Mucus is produced by the goblet cells, which are derived from the stem cells located in the crypt. As they mature, they ascend upward to the villus (Figs. 3.1 and 3.2). They are polarized cells containing apical and basolateral domains. Brush border (BB), also known as microvillus, is located on the apical side of the goblet cells. BB consists of actin filaments and actin-bundling proteins (villin and fimbrin), and it is responsible for vesicle trafficking. Mucin, after synthesis, is packaged in the secretory granule and is released into the lumen after 5–7 days during goblet cell exfoliation (death). Goblet cell proliferation and differentiation is regulated by Wnt and Hedgehog signaling pathways. Exfoliation occurs near the tip of the villi, which is initiated by a cell death program called "anoikis." Anoikis is mediated by focal adhesion kinase or β1-integrin-related events such as protein kinase signaling pathways: PI-3-kinase/Akt, mitogen-activated protein kinase, stress-activated protein kinase/Jun amino-terminal kinase, and Bcl-2. Goblet cell function and the composition of mucus are affected by numerous factors including changes in the resident microbial flora, the presence of harmful enteric pathogens, and nutritional deficiencies.

Mucus consists of mucin protein and polysaccharide. The polysaccharide makes mucus sticky. Mucin is a glycoprotein. The core protein moiety, apomucin, is cross-linked with carbohydrate chains or O-linked oligosaccharides (O-glycans) that are attached to the serine and threonine residues by a glycoside bond. Mucin forms a high-order structure through polymerization of several molecules resulting in a large molecule with molecular weights in millions of daltons. There are three subfamilies of mucin: gel forming, soluble, and membrane bound. Eighteen genes encode for mucin (MUC) synthesis, and the mucins are designated as MUC1, MUC2, MUC3A, MUC3B, MUC4, MUC5B, MUC5AC, MUC6–MUC13, and MUC15–MUC17. Among them, the MUC2, MUC5AC, MUC5B, and MUC6 exist in the secreted form and assemble via interchain disulfide bridges. Mucins on the membrane are found either as membrane-bound or the secreted form. Membrane-bound mucin is highly adherent and remains associated tightly with the membrane, while the secreted form is loosely attached. The secreted form remains associated with the membrane-bound mucin by forming covalent and noncovalent bonds.

Mucin secretion occurs by two processes: (1) immediate exocytosis, i.e., the mucin is produced and is released immediately from the cells by exocytosis, and (2) delayed release, i.e., mucin is first stored in a large vesicle after synthesis and then is released. The release is regulated by specific stimuli involving activation of signaling pathways. The activators may include acetylcholine, vasoactive peptides, neurotensin, IL-1, nitric oxide (NO), and cholera toxin. In contrast, *Clostridium difficile* toxin A downregulates exocytosis thus favoring bacterial colonization and survival in the intestine.

Mucin binds small molecules and proteins. The complex architecture of mucin performs several functions: (i) binds water (hygroscopic), (ii) selective transport of molecules toward the membrane due to gel-filtration effect, (iii) ion-exchange molecule, and (iv) sequestration function due to its ability to serve as a receptor for selectins, lectins, adhesion molecules, and microorganisms.

Mucus acts as a barrier network for pathogens. Mucus contains sIgA, lactoferrin, lysozyme, and lactoperoxidase. Lactoferrin binds iron intimately, thus sequestering it from bacterial utilization. Lysozyme breaks down peptidoglycan of bacterial cell wall, and lactoperoxidase produces superoxide "O" radical, which is toxic to bacteria. Ciliated cells in the mucus membrane also propel or sweep away invading pathogens. Mucus gel is a source of energy and nutrients for resident microflora and promotes microbial colonization. Mucus also protects the epithelial cells against inflammation and injury from pathogens or toxins. For example, mucin inhibits the adhesion of enteropathogenic *E. coli* (EPEC), *Yersinia enterocolitica*, *Shigella*, and rotavirus. However, some pathogens regulate mucin secretion and favor bacterial colonization. For example, *Helicobacter pylori* colonizes mucus and reduces mucin exocytosis. *L. monocytogenes* induces exocytosis of mucin through listeriolysin O (LLO) and inhibits bacterial entry. Probiotic bacteria such as *Lactobacillus* spp. induce mucin production and prevent pathogen attachment. Mucus also protects and lubricates epithelial surface. It helps in fetal development, epithelial renewal, carcinogenesis, and metastasis.

In the mucus membrane, M (microfold) cells located in the intestine as part of the gut-associated lymphoid tissue (GALT) or Peyer's patches deliver pathogens to the local lymphoid tissue by transporting them across the mucosa (Fig. 3.3). M cell is naturally phagocytic and serves as the gateway for translocation of intracellular bacteria such as *L. monocytogenes*, *Salmonella*, *Yersinia*, and *Shigella* across the epithelial barrier. M cells constitute only less than 0.1% of the epithelial cells in the lining of the intestine and are present in the follicle-associated solitary lymphoid structure throughout the intestine. Mucus-associated lymphoid tissue (MALT) is synonymous to GALT and is a secondary lymphoid tissue, which responds selectively to the pathogen.

Antimicrobial Peptides

Antimicrobial peptides (AMPs) are produced by many vertebrates including fish, frogs, and mammals. AMPs have broad-spectrum antimicrobial activity and are associated with the airways, gingival epithelia, and corneas. AMPs are also present in the reproductive, the urinary, and the GI tract. They play a major role in innate immunity and assist in pathogen elimination. AMPs are small peptides of 20–40 amino acids long and have α-helix and β-sheets. Intrachain disulfide bonds formed between cysteine residues stabilize the structure. AMPs are produced as inactive pre-propeptides and are activated after removal of the N-terminal signal sequence. Their antimicrobial activity is nonspecific. Peptides are inserted into

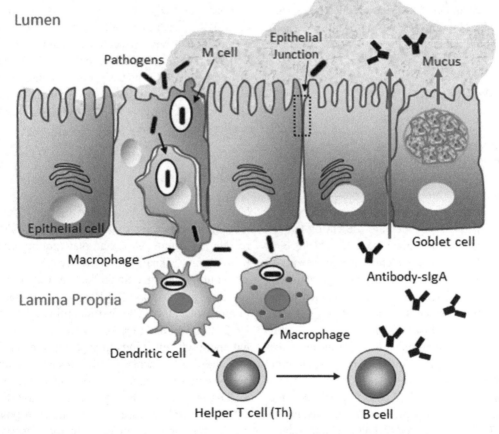

Fig. 3.3 Specific immunity in the mucosal surface. Bacteria are engulfed by M (microfold) cells and translocated across the mucosal barrier. Resident macrophages or dendritic cells can engulf the bacteria, process antigens, and present them to the T cells for the production of antibody (sIgA) by B cells. sIgA (secretory IgA) travels through the epithelial cells and is secreted into the lumen

the membrane to form pores by "barrel-stave," "carpet-like," or "toroidal pore" mechanism. They induce leakage of ions from cells and cause the collapse of the proton-motive force (PMF). In addition, the peptides may affect the cytoplasmic membrane, septum, cell wall, nucleic acid and protein, and enzyme activity in microbes. They are effective against bacteria, and enveloped viruses, but not against the nonenveloped viruses. Pathogens may develop resistance by expelling AMPs by energy-dependent pumps, altering membrane fluidity, and cleaving the AMPs with proteases.

AMPs are produced by epithelial cells, Paneth cells, neutrophils, and macrophages. Examples of AMPs are lysozyme, phospholipase A2, α1-antitrypsin, defensin, cryptdin, angiogenin 4, and cathelicidins. Human β-defensins 1, 2, and 3 are cationic and are cysteine-rich peptides of 3.5–4 kDa and are produced by intestinal epithelial cells. Human α-defensins 5 and 6 are produced by the Paneth cells.

Toll-like Receptor

Toll-like receptors (TLRs) are transmembrane proteins known as signaling molecules and were originally discovered in *Drosophila* (fruit fly). These evolutionarily conserved pattern recognition molecules are typically expressed in intestinal epithelial cells and immunocompetent cells in the lamina propria. They are present on the cell surface and in endosomal membranes (Fig. 3.4). Several TLRs are now identified, and they play a major role in innate immunity through activation of NF-κB, cytokine production, and recruitment of inflammatory cells. For example, TLRs are present in macrophages and are known to form a "microbial recognition" system. Toll-like molecules are analogous to the IL-1 receptor. Different TLRs are activated by different types of microbial factors (Table 3.2). For example, TLR-1 recognizes bacterial lipoprotein and unconventional LPS; TLR-4 is activated by bacterial lipoarabinomannan (LAM), bacterial lipoprotein (BLP), and yeast cell wall particles (zymosan). TLR-4/TLR-6 are activated by LPS and LTA, TLR-5 by bacterial flagella, and TLR-11 by profilin. TLR-2

was thought to recognize peptidoglycan, but it was shown that it interacts with lipoprotein or lipoteichoic acid often present in the peptidoglycan preparations as a contaminant. Peptidoglycan interacts with intracellular NOD-1 and NOD-2 receptors in mammalian cells. TLR-3 recognizes viral dsRNA; TLR-7 and TLR-8 recognize single-stranded viral RNA and small synthetic compounds and TLR-9 bacterial DNA. Following recognition of the microbes by TLR on macrophage or other immune cells, a cascade of signaling pathways is activated to allow transcription of proinflammatory immune response genes for IL-1, TNF, TGF-β, IL-6, IL-8, IL-12, IL-17, IL-18, and a variety of other chemokines. The transcription factor NF-κB regulates the expression of these cytokines. These cytokines eventually help eliminate the microbes through lysis of target cells or by inducing an adaptive immune response. For example, IL-1 and TNF activate T cells, IL-6 activates B cells, and IL-12 activates NK cells to provide protective immunity.

PML polymorphonuclear leukocytes, *DC* dendritic cells, *NK* natural killer cells, *NOD* nucleotide-binding oligomerization domain protein, *dsDNA* double-stranded DNA, *ssRNA* single-stranded RNA.

NOD Proteins

Nucleotide-binding oligomerization domain (NOD) proteins are also another class of PAMP-recognizing molecules that play an important role during innate immune response. The nucleotide-binding domain leucine-rich repeat-containing receptors (NLRs) serve as innate immune receptors. The first identified NLRs are NOD-1 and NOD-2. NOD-1 expressed in intestinal epithelial cells binds to peptidoglycan (DAP, diaminopimelic acid) of the invasive Gram-negative pathogen and other commensal bacteria. NOD-1 is thought to be highly active in intestinal cells in Crohn's disease patients and is responsible for an uncontrolled inflammatory response. NOD-2 is highly expressed in the Paneth cells and monocytes, and it recognizes muramyl dipeptide (MDP) of PGN of both Gram-positive and Gram-negative bacteria. NOD-2 promotes

Fig. 3.4 Schematic of toll-like receptor (TLR) location in a cell and their corresponding ligands. *DAP*, diaminopimelic acid; *MDP*, muramyl dipeptide; *PGN*, peptidoglycan; *LPS*, lipopolysaccharide; *NOD*, nuclear oligomerization domain; *LAM*, lipoarabinomannan

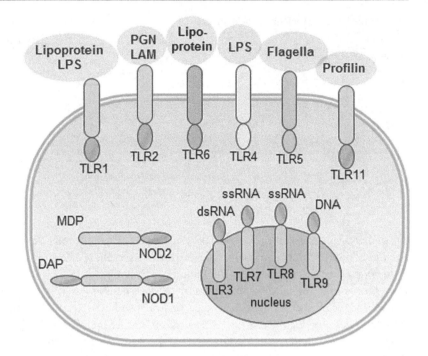

Table 3.2 Toll-like receptors (TLRs) and the corresponding ligands

Receptor	Cell type	Ligand
TLR1	Ubiquitous	Bacterial (triacyl) lipoproteins
TLR2	PML, DC, monocytes	Lipoteichoic acid from G(+) bacteria
TLR3	DC, NK	dsRNA
TLR4	Macrophages, DC, endothelial cell	LPS
TLR5	Monocytes, immature DC, epithelial cells, NK, and T cells	Flagellin
TLR6	High in B cells, lower in monocytes and NK cells	Bacterial (diacyl) lipoproteins
TLR7	B cells, DC	ssRNA recognition on endosomes (mouse)
TLR8	Monocytes, low in NK and T cells	ssRNA recognition on endosomes (human)
TLR9	DC, monocyte, macrophages, PML, NK, microglial cells	CpG (unmethylated DNA)
NOD-1	DC, macrophages, epithelial cells	Diaminopimelic acid (DAP) of peptidoglycan from Gram-negative bacteria
NOD-2	DC, monocytes, macrophages, paneth cells, epithelial cells	Muramyl dipeptide from peptidoglycan of both Gram + and Gram- bacteria

Paneth cell activation for increased cryptdin production, while DC and macrophage activation result in proinflammatory cytokine production and augmented bacterial killing.

Resident Microbiota

All mammals are born without any microorganism in their system; however, they begin to acquire microbes during and immediately after birth, and eventually, every surface exposed to the external environment becomes inhabited by commensal bacteria. Microbes are aerobic, facultative anaerobic, and anaerobic. In the GI tract, the facultative anaerobes are associated close to the epithelial layer for oxygen. The proportion of anaerobic bacteria increases from proximal to the distal part of the intestine. Contents of the stomach and duodenum contain approximately 10^1–10^3 cfu g^{-1}, of the proximal small bowel 10^4–10^8 cfu g^{-1}, of the terminal ileum to colon 10^{10}

Table 3.3 The major bacteria phyla in the gut

Firmicutes	Proteobacteria	Bacteroidetes
Approx. 255 genera; 2475 species	Approx. 366 genera; 1644 species	Approx. 20 genera; 130 species
Lactobacillus *Bacillus* *Bifidobacterium* *Clostridium* *Streptomyces* *Mycoplasma*	***Enterobacteriaceae* family**: *Escherichia* *Salmonella* *Shigella* *Proteus* *Enterobacter* *Klebsiella* *Serratia* **Non-*Enterobacteriaceae* family** *Vibrio* and nitrogen-fixing bacteria	*Bacteroides* genus (major)

cfu g^{-1}, and of the colon to rectum 10^{13} cfu g^{-1}. The major phyla present in the gut are *Firmicutes*, *Proteobacteria*, and *Bacteroidetes* (Table 3.3). The *Firmicutes* represent approximately 255 genera and 2475 species, and the major genera include *Lactobacillus*, *Bacillus*, *Clostridium*, *Streptomyces*, and *Mycoplasma*. The *Proteobacteria* represent approximately 366 genera and 1644 species, and the major genera belong to the *Enterobacteriaceae* (EB) family comprising *Escherichia*, *Salmonella*, *Shigella*, *Proteus*, *Enterobacter*, *Klebsiella*, and *Serratia* and non-EB such as *Vibrio* and nitrogen-fixing bacteria. Phylum *Bacteroidetes* comprises approximately 20 genera and 130 species, and *Bacteroides* is the major genus related to human gut microbiota.

Interestingly, the host requires the colonization by commensal microbiota for its development and maintenance of health. Indigenous (autochthonous) bacteria provide essential nutrients and help in the metabolism of indigestible compounds by synthesizing necessary enzymes. Microbiota also influence the release of biologically active gastrointestinal peptides, regulate intestinal endocrine cells, and contribute toward the development of intestinal architecture. They also influence nutrient absorption, mucosal barrier fortification, xenobiotic metabolism, angiogenesis, and postnatal intestinal maturation. They also prevent colonization by opportunistic pathogens and participate in innate and the adaptive immune response. The imbalance between microbiota and the pathogens may lead to dysbiosis or a disease state. During dysbiosis,

epithelial cells secrete IL-1 and IL-6, DCs secrete IL-12 and IL-23, Th1 and intraepithelial lymphocyte cells (IEC) release IFN-γ, Th17 secretes IL-17A, macrophages produce TNF-α and IL-12, and B cells secrete commensal-specific IgG. Epithelial barrier function is also compromised during dysbiosis. While the pathogens induce a proinflammatory response by activating NF-κB, the natural microbiota (example, *Lactobacillus* spp.) suppress unnecessary inflammatory response and help maintain immune homeostasis and prevent epithelial barrier dysfunction.

Microflora or their components such as LPS, PGN, and FMLP (formylated chemotactic oligopeptide) translocate actively through the mucosal barrier (mucus and epithelium) and are found in the lamina propria. There they activate macrophages, dendritic cells, neutrophils, NK cells, and T cells (Fig. 3.3). Microflora together with the immune system regulate intestinal motility, secretion, proliferation, villous length, and the crypt depth. Furthermore, the amount of IgA produced is directly dependent on the number of gut flora. For example, germ-free animals have low levels of plasma cells, IgA, and a decreased number of immune cells. Overall, the natural microflora help in the development of the immune system. Ironically, in some individuals, the immune response against natural microflora may lead to the onset of inflammatory bowel's disease (IBD): Crohn's disease (CD) and ulcerative colitis (UC).

Natural microflora in the gut act as probiotics and prevent invasion of enteric pathogens. For

Table 3.4 Role of intestinal flora in intestinal function and immune response

Influence the release of biologically active peptides, regulate intestinal endocrine cells and epithelial structure
Nutrient absorption, mucosal barrier fortification, xenobiotic metabolism, angiogenesis, postnatal intestinal maturation
Participate in innate and adaptive immune system
Pathogens induce proinflammatory response by activating NF-κB and by production of IL-1β, TGF-β, TNF-α, IL-6, IL-8, IL-12, IL-17, IL-18, IL-23, and other chemokines
Natural microbiota (e.g., *Lactobacillus* spp.) suppress unnecessary inflammatory response – help maintain immune homeostasis
Microflora or components, i.e., LPS, PGN, and FMLP, translocate actively through mucosal barrier (mucus and epithelium) and found in lamina propria
Activate macrophages, dendritic cells, neutrophils, NK, T cells
Microflora plus immune system regulate intestinal motility, secretion, proliferation, villous length, and crypt depth
Amount of IgA produced is directly dependent on the gut flora
Microbiota activate plasma cells to produce IgA
Decreased diversity of flora leads to decreased number and activation of immune cells
Bottom line: natural microflora help develop immune system

example, microcin produced by *E. coli* prevents invasion of *Salmonella*, *Shigella*, and *Listeria*. *Lactobacillus* and *Bifidobacterium*, which remain stable throughout life, can inhibit *Salmonella* invasion. *Lactobacillus* may reduce the pathogen-induced loss of epithelial integrity. Bacteriocins produced by some natural flora can kill certain pathogens in the gut. In summary, the AMPs and the microbiota together provide frontline defense by inducing mucus secretion and activating immune response thereby protecting the host against invading pathogens (Table 3.4).

Gut microbiota also play an important role outside the gut. A balanced gut microbiota is essential for brain development and psychological well-being and prevents the onset of many chronic diseases and disorders including obesity, type I diabetes, autoimmune disease, allergy and asthma, autism, atherosclerosis, and cancer.

Other Components of Innate Immunity

Transferrin is a glycoprotein abundant in the liver that can prevent bacterial growth by sequestering iron. Complement proteins, especially those activated in the alternative pathway by bacterial LPS or LTA or by viral proteins, are present in the blood circulation (see subsection on Complement). The complement activation product "C3b" acts as an opsonin and facilitates enhanced phagocytosis by macrophages, and the "membrane attack complex" (MAC) directly destroys microbes by exerting in the cytoplasmic membrane by forming pores. Macrophages, neutrophils, and NK cells also eliminate pathogens nonspecifically during the early phase of infection. Macrophage-derived cytokines, α and β interferons (IFN), and TNF are important during innate immune response. Interferons inhibit viral replication, and TNF initiates inflammation to protect against bacteria-induced damage.

Immune Response in the Gastrointestinal Tract

Mouth/Throat/Respiratory Tract

Saliva in the mouth contains lysozyme, lactoferrin, and sIgA and provides protection against pathogens. Resident bacteria compete for nutrients and the colonization sites with invading pathogens. Ciliated epithelial cells sweep away pathogens, and the mucus ball helps remove pathogens. Epithelial cells are sloughed off periodically to dispose of colonized bacterial cells on them as a strategy to eliminate pathogens. Large particles are sometimes expelled by coughing. Alveolar macrophages in the respiratory tract or nasopharynx also remove pathogens.

Stomach

Gastric juice has a low pH (2–2.5) due to the secretion of HCl from parietal cells, and most foodborne pathogens are inhibited by acid. Food can neutralize the pH of stomach juice to pH 4–pH 5; thus, acid-resistant or acid-tolerant pathogens are unaffected and can pass through the stomach to the small intestine where the pH is about 6.8. In addition, liquid food like milk, soup, and beverages can rapidly transfer bacteria through the stomach thus preventing longer exposure to acid. Psychosomatic disease or physiological abnormalities, such as achlorhydria, may also raise the stomach pH making the individuals vulnerable to a foodborne infection. The stomach also secretes some proteolytic enzymes that can destroy pathogens. Natural microflora such as *Lactobacillus*, *Streptococcus*, and some yeast are present at 10^1–10^3 cfu ml^{-1} in the stomach and may contribute toward the innate immunity.

Small Intestine

The small intestine has three regions: duodenum, jejunum, and ileum. In the small intestine, bile salts (sodium taurocholate and sodium glycocholate) are secreted from the gall bladder, and their detergent-like action helps destroy the cell walls of Gram-positive bacteria. There is a fast flow of mucus and cell sloughing, and peristaltic movement of the intestine may eliminate pathogens from the gut. As mentioned earlier, the Paneth cells located in the crypt of the mucosal layer (Fig. 3.2) in the small intestine also provide protection against enteric pathogens. These cells possess phagocytic activity and produce lysozyme and cryptdin that inhibit foodborne pathogens. GALT also removes bacteria by inducing a specific immune response. Natural microbiota present in the jejunum and ileum are in the range of 10^4–10^8 cfu ml^{-1}. The major genera present in this part of intestine are *Lactobacillus*, *Enterobacter*, *Streptococcus*, *Bacteroides*, *Bifidobacterium*, and *Fusobacterium*; they provide protection against pathogen colonization.

Large Intestine

The large intestine has three segments: cecum, colon, and rectum. The epithelial lining is covered with inner and outer mucus layers. The inner layer is highly adherent and rich in sIgA and AMPs, while the loosely attached outer layer carries resident microbes. The large intestine is home to an abundant resident microflora (10^{10-14} cfu g^{-1}), 97% of which are anaerobic. The microbial genera present in the large intestine include *Bifidobacterium*, *Lactobacillus*, *Escherichia*, *Clostridium*, *Bacteroides*, *Streptococcus*, *Peptostreptococcus*, *Ruminococcus*, *Fusobacterium*, *Pseudomonas*, *Veillonella*, *Proteus*, yeast, and protozoa. These microbes prevent colonization of pathogens in the gut by physical hindrance and by producing metabolic by-products such as short-chain fatty acids (SCFA: formic acid, acetic acid, butyric acid, propionic acids, and valeric acid) and bacteriocins that are inhibitory to the transient pathogens. Epithelial cell sloughing also occurs in the colon and, together with mucus, can prevent colonization and attachment of pathogens.

Adaptive Immunity

Adaptive immunity is characterized by specific response to a particular antigen resulting in the production of specific antibody or cytokines. The most important characteristics of adaptive immune response are that it remembers each encounter, i.e., a memory response, and that the responses are amplified with each successive encounter with the pathogen (Fig. 3.5). Specific immunity can be active or passive based on the form of immune response it generates. Active immunity is developed when the host immune system is stimulated by an antigen. For example, immunization with the tetanus toxoid, an inactive form of tetanus toxin, develops an antibody response that is capable of neutralizing tetanus toxin in the subsequent exposure. Passive immunity is achieved when the serum or cells from a previously immunized host are introduced to a recipient to confer immediate protection against the pathogen. Examples of passive immunity are anti-botulinum serum against botulism and anti-venom serum given after a snakebite.

Adaptive immunity is also classified based on the components of the immune system that provide the response, i.e., humoral or cell mediated. Humoral immunity involves the production of

Fig. 3.5 A graph showing the primary and secondary immune response against hypothetical antigens A and B

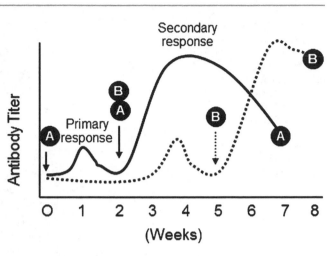

antigen-specific antibodies by B lymphocytes. Generally, extracellular antigens such as protein toxins or soluble protein antigens induce humoral immunity. Cell-mediated immunity (CMI) involves activation of T-lymphocyte subsets that are involved in the elimination of pathogens. Generally, intracellular pathogens such as bacteria, viruses, or parasites induce CMI.

Characteristics of Adaptive Immune Response

The adaptive (specific) immune response is highly specific, diverse, and self-regulated, remembers the first antigen encounter (memory response), and is able to discriminate self-antigens from nonself. Properties of specific immune response are:

1. The specific immune response is generally raised against structural components of complex proteins or polysaccharides. Antigenic determinants or epitope-specific lymphocyte clones are selected during an immune response, and these clones continue to produce specific antibodies or cytokines during their life spans.
2. The specific immune response can respond to a large number (~10^9) of antigenic determinants due to the broad "lymphocyte repertoire" in a host.

3. Specific immune response remembers the first encounter of an antigen, and the response is enhanced during the subsequent exposure due to the presence of memory cells, which survive for prolonged period even in the absence of an antigen. Memory cells respond to a very low concentration of antigen with increased immune response and with higher affinity. Memory response is larger, more rapid, and qualitatively different from the first.
4. The specific immune response is self-regulated, and it does not produce antibodies or cytokines in the absence of an antigen. Lymphocytes perform the initial response function for a brief period and then convert to the memory cell. Feedback mechanisms also control the immune response with the help of a blocking antibody or increased concentration of cytokines. Antigen–antibody complex also regulates T-cell response and alter cytokine cascades to control the immune response.
5. The specific immune response can discriminate self from nonself. It is an important feature of the specific immune response to be able to distinguish self-antigens from a foreign antigen. T or B cells that are stimulated by self-antigen are generally eliminated from the body. Self-antigens are also known as tolerogens since no immune response is developed against them. Immunologic unresponsiveness is called "anergy." Foreign antigens may act as tolerogens or immunogens depending on the

Fig. 3.6 Phases of immune response consist of activation, proliferation, and differentiation into antibody-secreting plasma cells or memory cells. Similar events also occur for T lymphocytes. Activation of a clone is antigen-specific

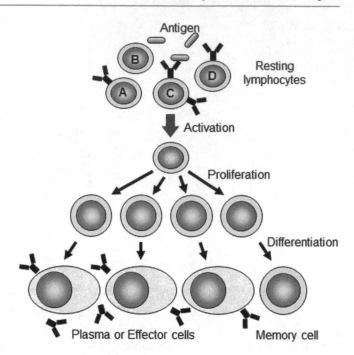

physicochemical state of the antigen, the dosage, and the route of administration. An antigen administered orally or intravenously may not elicit a response; however, if the same antigen is administered subcutaneously or intraperitoneally, it may induce an immune response. If an immune response is generated against a self-antigen, it leads to autoimmune disease. For example, systemic lupus erythematosus (SLE) is a condition where the antibody is produced against cellular nucleic acids resulting in glomerular nephritis and vasculitis. In myasthenia gravis, the antibody is developed against the acetylcholine receptor and thus interferes with the transmission of nerve impulses causing muscular weakness.

Phases of Immune Response

The introduction of an antigen initiates a cascade of events leading to the immunologic response and the elimination of the pathogen from a host (Fig. 3.6). Each antigen has the capacity to activate separate existing clones, and each clone can proliferate to produce a specific effector response. In the cognitive phase, initially an antigen binds to the membrane receptor of resting lymphocyte

where membrane-bound IgM and IgD serve as receptors. In the activation phase, the clones are activated and enlarged and begin to proliferate and differentiate into the effector and the memory cells. The effector cells produce antibodies or cytokines to initiate effector functions through complement activation or activation of phagocytes to neutralize antigens, while the memory cells remain dormant, circulate in the body, and are involved in the immune surveillance. When the memory cells encounter an antigen, they convert into effector cells and respond to the antigen with vigor. Memory cells can survive for a prolonged period depending on the antigen type and the frequency of exposure. The most effective vaccine against an infectious agent is the one that is capable of generating a robust long-lasting memory cell response.

Tissues and Cells of Immune System

Tissues

The primary lymphoid organs are bone marrow and the thymus in mammals. In avian species, the bursa of Fabricius, a specialized gland located in cloaca, is equivalent to the mammalian bone marrow. The secondary lymphoid organs are lymph nodes, spleen, and lymphoid tissues such as gut-associated lymphoid tissue (GALT, Peyer's patch) (Figs. 3.1 and 3.3), mucus-associated lymphoid tissue (MALT), and skin-associated lymphoid tissue (SALT).

Bone Marrow

Bone marrow contains stem cells which are responsible for hematopoiesis, i.e., all blood cells are originated from these stem cells. Bone marrow also serves as the site for B-cell growth and the early phase of maturation. Growth factors or colony-stimulating factors for B cells such as granulocyte–monocyte colony-stimulating factor (GM-CSF) and for macrophages such as granulocyte–colony-stimulating factor (G-CSF), monocyte colony-stimulating factor (M-CSF), IL-1, IL-2, and IL-7 are produced in the bone marrow. In birds, the bursa is the site for B-cell growth and maturation.

Thymus

The thymus is located adjacent to the larynx in the upper part of the thoracic cavity and provides the microenvironment for T-lymphocyte maturation and growth. It is a bilobed organ containing multiple lobules. The cortex of the lobule is the resident site for thymocytes (T lymphocytes) that are found at various stages of their maturation. Immature T cells do not carry any antigen receptors. Mature T cells migrate from the cortex to medulla and encounter "self-antigen" for the first time. Depending on the type of antigens they encounter, these T cells differentiate into either CD4$^+$ or CD8$^+$ T cells and express co-receptor molecules, CD4 or CD8, respectively. In the medulla, the epithelioid cells, also known as giant macrophages, and the resident macrophages present "self-antigens" to the T cells. During development in the thymus, only T cells that do not react to "self-antigens" are allowed to mature and enter the blood circulation.

Lymph Node

Lymph nodes are secondary lymphoid tissue and are part of the lymphatic system in mammals but are absent in birds. An adult human body may have 400–450 lymph nodes distributed throughout the body. Examples are cervical lymph nodes (neck area), axillary lymph nodes (armpit area), bronchial lymph nodes (bronchus), mesenteric lymph nodes (intestinal tract associated), inguinal lymph nodes (groin area), and so forth. For food-borne diseases, mesenteric lymph nodes (MLN) play an important role during infection, and they serve as a firewall to prevent live organisms from the systemic spread. In humans, there are about 100–150 MLN distributed in the mesentery. Structurally, a lymph node has three regions: an outer cortex, paracortex, and inner medulla. The cortex has lymphoid follicles containing a distinct germinal center in each follicle. Germinal centers, also called B-cell zones, are where B cells proliferate and differentiate into the antibody-secreting plasma cells. In the cortex, macrophages, dendritic cells, and interdigitating reticular cells are localized and can process and present antigens to B cells. The paracortex has the T-cell zone and is primarily occupied by the helper T cells (Th or CD4$^+$). Cytotoxic T cells (T$_C$ or CD8$^+$) are also found in this area. The medulla is home to the matured antibody-secreting B cells (plasma cells), Th cells, macrophages, and dendritic cells.

Spleen

The spleen is a secondary lymphoid organ and is the largest in the lymphatic system, present in virtually all vertebrates. It is located in the left upper quadrant just above the stomach. The spleen is brownish in color. It filters blood and helps remove dead blood cells from the circulatory system. It has two regions of interest: white pulp and red pulp. White pulp has periarteriolar lymphoid sheaths that contain lymphoid follicles. Lymphoid follicles contain germinal centers which are home to B cells. White pulp also contains CD4$^+$ and CD8$^+$ T cells. The red pulp

Fig. 3.7 Diagram showing the origin of various immune cells from the stem cells

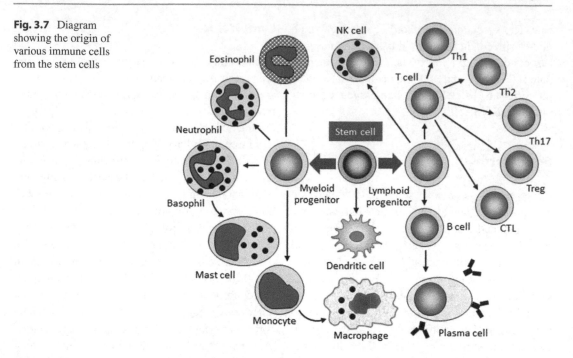

Lymphoid Tissue

The mucosal immune system is a regional lymphoid tissue such as gut-associated lymphoid tissue or Peyer's patch or solitary intestinal lymphoid tissue (SILT). It contains "M" cells that engulf bacteria and transport them to the subepithelial region in the lamina propria. Localized dendritic cells present antigens to T cells, which in turn activate plasma cells for production of immunoglobulins. Dendritic cells also transport the pathogens to the deeper tissues such as the liver, spleen, and lymph nodes. The primary immunoglobulin type found in the gut is secretory IgA (sIgA), which is transported to the lumen for neutralization of the antigen.

Cells of Immune System

Cells that are important in immune response are originated from pluripotent stem cells in the bone marrow and belong to two distinct lineages:

contains macrophages, dendritic cells, and the plasma cells, and the antigen-specific interaction takes place on this site.

myeloid lineage and the lymphoid lineage (Fig. 3.7). Cells originated from the myeloid lineage are granulocytes; neutrophil, basophil, eosinophil, and phagocytes; monocyte, and macrophage. The lymphoid lineage comprises T lymphocytes, B lymphocytes, and natural killer (NK) and natural killer T (NKT) cells.

Monocytes and Macrophages

Monocytes are originated from the bone marrow and then migrate to blood circulation. Matured monocytes differentiate into macrophages and reside in various tissues. The tissue-specific macrophages are known as histiocytes, epithelioid cells (skin), or multinucleated (polykaryon fused cells) giant cells. Macrophages are also present in various organs and differentiate into alveolar macrophages (lungs), Kupffer cells (liver), splenic macrophages (spleen), mesangial cells (kidney), synovial A cells (joints), microglial cells (brain), and Langerhans cell (skin).

The cytoplasm of macrophages appears granular due to phagocytic vacuoles (phagosomes) and lysosomes. Lysosomal contents have low pH and contain proteolytic enzymes, reactive "O" radicals, NO (nitric oxide), prostaglandin, defensin, and lysozyme. After phagocytic engulfment, the antigen is trapped inside the phagosome. The phagosome then fuses with the lysosome creating the phagolysosome. Lysosomal antimicrobial components inactivate and subsequently degrade the pathogen. Macrophages also possess surface receptors for antibody (FcR), complement, IL-2, and transferrin. During the oxidative burst, NO is produced by the macrophage and attacks metalloenzymes of bacteria to produce the superoxide peroxynitrite ($OONO^-$) which can also cause collateral damage to the host. Macrophages have three distinct functions: professional phagocytosis, antigen presentation, and production of cytokines.

Professional Phagocytosis

As the first line of defense, macrophages engulf pathogens and destroy them (Fig. 3.8). Macrophages can readily recognize antibody-coated pathogen using the Fc (fragment crystalline) part of an antibody binding receptor (FcR) for elimination. Antibody, complement, or lectin-like molecules that help macrophages to recognize antigens are called "opsonins," and the phagocytosis process is known as "opsonization." Macrophages remove dead or injured self-cells, tissues, tumor cells, and apoptotic cells. Thus, they prevent inflammation which may be triggered by the dead or injured cells and their contents. The phagocytosis process involves several steps. First, the macrophage recognizes an opsonin-coated target with or without using a specific surface receptor and then forms a pseudopod to trap it inside the phagosome. The phagosome fuses with the lysosome forming the phagolysosome. Lysosomal contents aid in the pathogen destruction and degradation. The vesicle-containing degraded products are eventually transported outside the cell by exocytosis.

Antigen Presentation

Macrophages are rich in the major histocompatibility complex (MHC) class II protein; hence, it is considered an efficient antigen-presenting cell (APC). Macrophages process the antigen (pathogen) by using the lysosomal proteolytic enzymes and then present the peptide fragments using MHC class II molecules to T or B cells for effector function. Macrophages found in the skin, lymph nodes, spleen, and thymus are involved in antigen presentation in these sites.

Production of Cytokines

After encountering an antigen, macrophage releases soluble effector molecules called cytokines that recruit inflammatory cells such as neutrophils, eosinophils, basophils, monocytes and other macrophages, and lymphocytes. Macrophages also produce growth factors for

Fig. 3.8 Steps showing phagocytosis and destruction of a pathogen by a macrophage

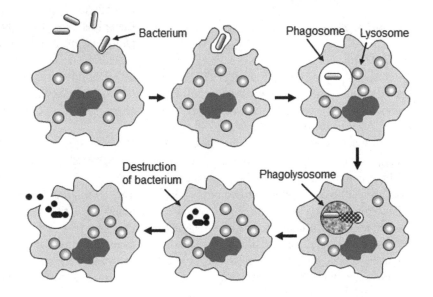

fibroblasts and vascular endothelium to repair injured tissues resulting from inflammation. The cytokines produced by macrophage are IL-1α, IL-1β, TNF-α, IL-6, IL-10, and IL-12.

Dendritic Cells

Dendritic cells (DCs) have a starlike appearance resembling the dendrites of a neuron. DCs are phagocytic and known antigen-presenting cells similar to macrophages. DCs originate from the bone marrow from the progenitor cells (stem cell). The immature DCs circulate throughout the body and convert into matured DCs in the presence of antigens and cytokines. They migrate to different organs and tissues including the skin, airways, intestine, and lymphoid organs (spleen and lymph nodes). They play important roles in both innate immunity and adaptive immunity. They are professional antigen-presenting cells; they phagocytose antigens, process them, and present them to T and B cells using MHC class II molecules on their surface. DCs express membrane markers CD40, CD70, and CD86, and they secrete IL-12 and interferons. During enteric pathogen infection, they are known to carry intracellular pathogens such as *Salmonella enterica* and *Listeria monocytogenes* from submucosal locations to extraintestinal sites.

Granulocytes

Neutrophil

Neutrophils are polymorphonuclear (PMN) leukocytes and contain granules and multilobed nuclei (Fig. 3.9). Granules known as lysosomes contain protease, myeloperoxidase, lysozyme, elastase, defensins (cysteine-rich cationic peptides), and acid hydrolases, such as β-glucuronidase and cathepsin B. Neutrophils also contain a receptor for antibody (FcR) and complement. They are attracted to the site of infection during inflammation where they act as professional phagocytes and aid in chemotaxis of other immune cells. Neutrophils are primarily effective against bacterial infection.

Eosinophil

Eosinophil, a polymorphonuclear leukocyte, responds during inflammation. The cytoplasmic granules (lysosome) stain intensely with an acidic red dye, eosin, hence the name eosinophil (Fig. 3.9). Granules contain acid phosphatase, peroxidase, cationic protein, and reactive oxygen species (ROS). They carry receptors for IgE (FcR$_ε$) and other immunoglobulins. Eosinophils are also professional phagocytes and assist in chemotaxis of other immune cells. They are effective against parasitic infection such as protozoa and helminths, which are often resistant to the degradation by lysosomal enzymes of neutrophils and macrophages. Eosinophil counts in the blood are high in patients showing immediate hypersensitivity (allergic) reactions or suffering from parasitic infections.

Basophil

The cytoplasmic granules of some leukocytes stain intensely with the basophilic dye, hematoxylin, and hence are given the name basophil (Fig. 3.9). Basophils are circulating counterpart of mast cells located in tissue, and the granules contain vasoactive amines such as histamine and serotonin. They also carry receptor (FcR) for IgE. Binding of IgE–antigen complex to the receptor of basophils or mast cells triggers a signaling cascade resulting in the release of vasoactive amines, which are key mediators of immediate (type I) hypersensitivity reactions. Basophils are primarily effective in removing allergens from the system by increasing the membrane permeability due to the action of vasoactive amines and allowing other cells such as NK and macrophages to reach to the sites. They are nonphagocytic in nature.

Lymphoid Tissue

The lymphoid lineage consists of two populations of lymphocytes: B lymphocytes and T lymphocytes (Fig. 3.10). B lymphocytes are derived from the bursa of Fabricius in birds and in an equivalent organ, bone marrow, in mammals. B lymphocytes differentiate into plasma cells which

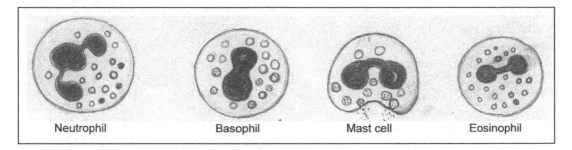

Neutrophil   Basophil   Mast cell   Eosinophil

Fig. 3.9 Granulocytes have a multilobulated nucleus and contain numerous granules. The granular contents are proteolytic enzymes, "O" radical, or amines, which aid in the destruction of microbes or induction of hypersensitivity reaction

Fig. 3.10 T- and B-lymphocyte lineages: origin, differentiation, and maturation

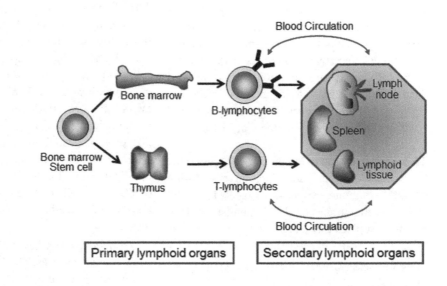

produce antibodies. T lymphocytes undergo their maturation phases in the thymus and produce interleukins, interferons, and cytokines. T lymphocytes differentiate into T-cell subsets: cytotoxic T cells (CTL or Tc), helper T cells (Th), and regulatory T cells (T_{reg}).

T Lymphocytes

T lymphocytes originate in the bone marrow. They migrate to the thymus where they differentiate into Th and CTL. They produce a plethora of cytokines. They also express MHC (classes I and II) molecules on their surface and recognize peptide antigens presented in association with MHC molecules with the T-cell receptor (TCR). They do not respond to soluble antigens. They have receptors for immunoglobulins (Ig). They carry specific surface markers which are designated as CD (cluster of differentiation) followed by a number. For example, the CD2 molecule is expressed on all T cells and binds to sheep red blood cell (RBC) forming a rosette-like appearance. Numerous CD molecules are identified so far, and T-cell subsets are classified based on the presence of different surface CD markers (Table 3.5).

Helper T Cells

Helper T cells are also designated as Th or CD4$^+$ T cells because of the presence of the unique CD4 surface marker. Th cells are classified into Th1, Th2, Th17, and regulatory T cells (T_{reg}).

Helper T cells recognize antigens when presented in association with an MHC class II molecule and subsequently secrete cytokines which activate B cells and macrophages.

Th1 In general, Th1 cells induce a cellular inflammatory type of response. Th1 cells respond to antigens when presented by macrophages, and in turn, they further activate macrophages (Fig. 3.11). They secrete IL-2, IL-3, interferon-γ (IFN-γ), lymphotoxin, GM-CSF, TNF-β, and lymphokines that promote B-cell proliferation. Th1 cells help B cells to synthesize IgG2a. Th1 cells are not stimulated by IL-1 (macrophage derived). Th1 cells help in cell-mediated immunity and delayed (type IV) hypersensitivity reactions. Th1 cells are preferentially developed during intracellular bacterial and viral infections. IL-12 and IFN-γ produced by macrophage and NK cell support the development of Th1 cell.

Th2 Th2 cells respond to the antigens presented by B cells and play an important role during humoral immune response (Fig. 3.11). Th2 expresses membrane CD30, a member of TNF receptor superfamily. The Th2 population is

developed during helminth infection and with exposure to common environmental allergens, and they are responsible for removal of extracellular infective agents. IL-4 production by other cell types favors the development of Th2 cell. Th2 cells secrete IL-3, IL-4, IL-5 (eosinophil activation factor), IL-6, IL-10, IL-13, and GM-CSF. They stimulate B-cell proliferation and isotype switching for IgE, IgM, IgG1, and IgA. IL-4 promotes IgE and IgG1 production by B cells (plasma cells).

Th17 The Th17 cell lineage is characterized by its capacity to produce IL-17, but not IFN-γ or IL-4. However, the development and the expansion of Th17 cells are dependent on IL-23. Th17 cells and the T_{reg} cells share common developmental pathways. Th17 cells carry a transcription factor, RORγt (RAR-related orphan receptor gamma), while T_{reg} carries FoxP3 (forkhead box P3) protein, also known as scurfin. TGF-β is required for the development of Th17 and T_{reg} cells from naïve T cells. In the presence of TGF-β or TGF-β plus IL-6, T cells can differentiate into FoxP3+ T_{reg} cells or RORγt+ Th17 cells, respectively.

Th17 cells are a key pathological marker for many human diseases. Th17 cells regulate inflammation by producing IL-17 and its family members: IL-17A, IL-17B, IL-17C, IL-17D, and IL-17E. IL-17E is also referred to as IL-25. In addition, Th17 cells produce TNF-α, IL-6, IL-21, IL-22, and IL-26. IL-17 has diverse biological functions: it recruits neutrophils, activates macrophages, and stimulates the production of

Table 3.5 Selected surface markers for T lymphocytes

Cluster designation	Distribution
CD1	Thymus
CD2	All T cells; rosette formation
CD3	Mature T cells
CD4	Helper T cells (Th1, Th2, Th17)
CD8	Cytotoxic T cells
CD4+/CD25+/Foxp3+	Regulatory T cells (Trcg)

Fig. 3.11 Differences in antigen presentation by helper Th1 and Th2 cells and their respective effector functions

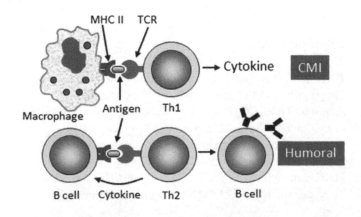

proinflammatory cytokines (IL-1α, IL-1β, IL-6, TNF-α, IFN-γ) and AMPs from a variety of immune and nonimmune cells. IL-17 can also enhance the capacity of CD4⁺ T cells to produce IL-2 and promotes the proliferation of both conventional T cells and T_reg cells. IL-17 is necessary for protective immunity against *Escherichia coli*, *Salmonella enterica*, *Klebsiella pneumoniae*, *Bordetella pertussis*, and other infective agents. Th17 cells are also involved in the pathology of many autoimmune diseases such as allergy, rheumatoid arthritis, multiple sclerosis, and inflammatory bowel disease.

T_reg A subpopulation of T lymphocytes carrying the transcription factor FoxP3 is now recognized as the regulatory T lymphocyte (T_reg), which controls the immune response in a specific manner. T_reg cells maintain tolerance to self-antigens and immune homeostasis. T_reg cells originate from thymus, and they are mostly CD4⁺ T cells and express CD25; hence, they are designated CD4⁺/CD25⁺/Foxp3⁺ T lymphocytes. They can suppress the activation, proliferation, and effector functions of CD4⁺ and CD8⁺ T lymphocytes, NK cells, B cells, and APC and can control autoimmune disease, immunopathology, allergic response, tolerance to tissue grafts, and fetal maintenance during pregnancy. They release IL-10 and TGF-β. T_reg cell numbers increase in blood and other tissues in cancer patients, thus favoring tumor progression. Chemotherapeutic drugs such as dacarbazine prevent tumor proliferation and thus have an immunosuppressive effect on T-lymphocyte populations including the T_reg cells.

Cytotoxic T Lymphocytes

Cytotoxic T cells (CTL), also known as CD8⁺ T cells, provide immunity against viral and intracellular bacterial infections and cancer. They are responsible for graft rejection and are the primary effector cells in tumor immunity (antitumor activity). They recognize antigens through T-cell receptor (TCR) when presented in association with the MHC class I molecule and selectively kill the target cell with the help of cytotoxin with-

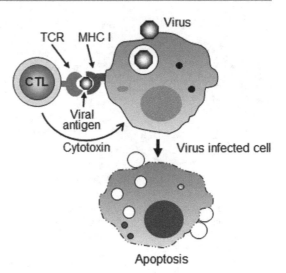

Fig. 3.12 Cytotoxic T-lymphocyte (CTL)-mediated killing of target cells

out affecting the neighboring cells (Fig. 3.12). The cytotoxin or perforin (pore-forming toxin) produced by the CTL induces apoptosis in the target cells. CTL also produce granulysin, which kills host cells infected with intracellular bacteria such as *Listeria monocytogenes*. CTL-mediated killing is antigen-specific and requires cell-to-cell contact; however, the CTL itself is not injured during this interaction. They secrete large amounts of IFN-γ which in turn increases expression of MHC class I molecules by the infected cells for enhanced killing.

Natural Killer Cells

Natural killer (NK) cells are prototypical member of the innate lymphoid cell (ILC) family and classified as group 1 ILC. NK cell function and development are mediated by the transcription factor T-bet, but it is not strictly dependent on T-bet. NK are called large granular lymphocytes (LGL) because they carry numerous large cytoplasmic granules containing cytotoxic perforin and granzymes to cause programmed cell death of targets. They are called non-T and non-B cells because of lack of characteristic T-cell receptor-like CD3 or B-cell receptor-like immunoglobulin. NK cells express CD56 and depending on the stage of the development, some may express

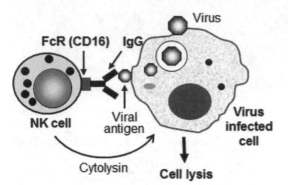

Fig. 3.13 Specific destruction of virus-infected cell by the natural killer (NK) cell with the help of virus-specific antibody. This event is also called antibody-dependent cell cytotoxicity (ADCC)

CD16 and hence are designated CD3⁺/CD56⁺/ CD16⁺ NK cell. They are also known as lymphokine-activated killer (LAK) cells and are activated by IL-12 for enhanced cytotoxic effect. NK cells secrete IFN-γ and IL-2 and are effective against tumor cells or virus-infected cells. The target cell killing is generally nonspecific, i.e., not restricted to antigen–MHC presentation; however, specific target cell destruction is achieved by using immunoglobulin and is termed antibody-dependent cell cytotoxicity (ADCC) (Fig. 3.13).

Natural Killer T Cells

Natural killer T (NKT) cells share properties of both NK and T cells and express variants of TCR, CD3, and CD4. They carry the nonclassical MHC class I molecule CD1d and are activated in the presence of microbial lipids including those of *Mycobacterium* and *Leishmania*. NKT cells secrete IFN-γ, TNF-α, IL-2, IL-4, IL-5, IL-13, and GM-CSF. NKT cells contribute to the innate and preadaptive immune response.

Cytokines

Cytokines are protein hormones produced by immune cells. Cytokines produced by mononuclear phagocytes (monocytes) are called monokines. Cytokines produced by activated T lymphocytes are called lymphokines, and those produced by leukocytes are called interleukins (IL). Cytokines play important roles in both natural immunity and adaptive immunity and regulate lymphocyte activation, growth, and differentiation. Cytokines also activate nonspecific inflammatory cells and stimulate immature leukocyte growth and differentiation.

Properties of Cytokines

1. Cytokines are produced during the effector phase of natural and specific immunity. For example, LPS activates macrophages which produce IL-1 during natural immunity. IL-1 stimulates the thermoregulatory center that increases body temperature (fever) to inhibit bacterial growth.
2. Cytokine secretion is brief and self-limiting because the cytokine mRNA is unstable.
3. Many individual cytokines are produced by multiple diverse cell types. For example, IL-2 is produced by T and NK cells, while IL-6 is produced by B cells, T cells, macrophages, fibroblasts, and endothelial cells.
4. Cytokines act upon different cells. Cytokine action is redundant, and many cytokines have a similar function.
5. Cytokines often influence the synthesis of other cytokines and can have antagonistic or synergistic action.
6. Cytokine action is mediated by binding to its receptor. Cytokine action can be autocrine (the action is on the secreting cell), paracrine (the action is on nearby cell), or endocrine (systemic circulation allows action on distantly located cells) similar to hormones.
7. Cytokines regulate cell division. For example, GM-CSF helps granulocyte–monocyte cell division and growth.

Cytokines in Innate and Adaptive Immunity

Two major cytokines are produced during innate immunity: Type I interferon (IFN) and TNF. During adpative immune response, Type II interferon and interleukins are predominant. Cytokines promote growth, differentiation, and activation of immune cells and can initiate inflammatory reactions that protect against microbial infections.

Interferon

Interferons (IFN) are of two types: type I and type II. The type I interferon is again divided into two groups depending on the molecular weights and the type of cells that produce them: IFN-α and IFN-β. IFN-α is an 18 kDa polypeptide and is called leukocyte interferon since it is produced by the mononuclear phagocytes. IFN-β is a 20 kDa polypeptide and is called fibroblast interferon since it is produced by fibroblast cells. Type I is involved in the natural immunity and inhibits the viral replication by establishing the "antiviral state" in infected cells. Type I IFN stimulates the production of oligoadenylate synthetase, which acts on adenosine triphosphate (ATP) to produce adenine trinucleotide (AT). AT activates RNase L (endoribonuclease) which cleaves the viral mRNA, thus inhibiting viral protein synthesis resulting in "antiviral state." Interferon action is host species-specific but not virus-specific.

An example of type II interferon is IFN-γ which has both immunostimulatory and immunomodulatory effects and plays a key role in adaptive immunity. IFN-γ is a 21–24 kDa polypeptide and is produced by NK and NKT cells during innate immune response and by Th1 CD4$^+$ and CD8$^+$ cells during the adaptive immune response. IFN-γ upregulates both MHC class I and class II molecules. It activates mononuclear phagocytes, increases MHC class II expression, and increases phagocytosis and antigen presentation. It also aids in T- and B-cell differentiation. IFN-γ plays an important role in clearance of intracellular pathogens including bacterial (*Salmonella, Listeria*), viral, and protozoan (*Toxoplasma, Leishmania*) diseases. IFN-γ is also involved in tumor immunity.

Tumor Necrosis Factor

Tumor necrosis factor (TNF) is one of the major cytokines in innate immunity and is responsible for immune cell regulation and cell signaling. Originally, it was discovered for its activity to induce cell death and suppress tumor growth. Two TNF have been described: TNF-α and TNF-β (lymphotoxin). TNF-α consists of three subunits of 17 kDa each with a total molecular mass of 51 kDa. It is produced primarily by macrophages but also by other cell types such as CD4$^+$ T cells, NK cells, eosinophils, neutrophils, mast cells, fibroblasts, epithelial and endothelial cells, and neuron. TNF-α is produced largely in response to bacterial infections primarily due to the release of elevated levels of LPS from Gram-negative bacteria and peptidoglycan and other components of Gram-positive bacteria. TNF induces inflammation and activates NF-κB and MAPK (mitogen-activated protein kinase) pathways to produce inflammatory cytokines for immune cell recruitment, differentiation, and proliferation. TNF-α also stimulates macrophages to produce IL-1. TNF-α, together with IL-1, raises body temperature (fever) through a prostaglandin-mediated pathway to inhibit bacterial growth. TNF-α blocks lipoprotein lipase action; therefore, the fatty acids are not released from lipoprotein for use by the body cells. As a result, the patient loses muscle mass and the body weight leading to cachexia (a wasting condition characterized by loss of body mass). The cachectic condition is also mediated by TNF-induced appetite suppression in the host. Large amounts of TNF-α production affects heart muscle contraction and vascular smooth muscle tone resulting in lowered blood pressure and ultimately septic shock.

Interleukins

Interleukins (ILs) are produced by leukocytes in response to antigens and serve as effector molecules in innate and adaptive immune response by

Table 3.6 Summary of interleukins and their target cells

Name	Produced by	Target cells
IL-1α	Macrophages	T, B
IL-1β	Macrophages	T, B
IL-2	T, NK	T (Th, CTL), B, NK, macrophages
IL-3	T	Hematopoietic stem cells, B, mast, macrophages
IL-4	Th2	Stem cell, T, B, macrophages
IL-5	Th2	T, B, eosinophil
IL-6	T, B, macrophage, fibroblast, endothelial cells	B, T, stem cells, neurons, hepatocytes, macrophages
IL-7	Bone marrow stromal cells	Pre-B, pre-T, and T cells (immature lymphocytes)
IL-8	Macrophages, lymphocytes, hepatocytes, endothelial cells	T, neutrophil, basophil (chemoattractant)
IL-9	CD4$^+$ T	Th cells, none on CTL
IL-10	Macrophages, T$_{reg}$ cells	Macrophages, DC
IL-11	Stromal cell	B cells, platelet production
IL-12	NK, macrophage	Th1, NK
IL-13	CD4+ T	B cell, epithelial, fibroblast, macrophages
IL-15	Macrophage	NK, T cells
IL-16	T, mast, epithelial, eosinophils	CD4 + T, monocytes, eosinophils
IL-17A/B/C/D/F	Th17	Mucosal tissues, epithelial and endothelial cells
IL-18	Monocytes, macrophage, DC	NK, Th1, monocyte, neutrophils
IL-19	Monocytes, macrophages	Keratinocytes, macrophages
IL-20	Monocytes	Keratinocytes
IL-21	Th2, Th17	Th17, B cell, NK, DC
IL-22	Th1, Th17, NK	Fibroblasts, epithelial cells, hepatocytes
IL-23	Macrophage, DC	Th17 differentiation
IL-24	Monocytes, CD4$^+$ T	Keratinocytes
IL-25 (IL-17E)	Th2, mast cells, eosinophils, macrophages	NK?
IL-26	Activated T cells, monocytes	Unknown
IL-27	Activated DC, macrophages	T cells, NK
IL-28/IL-29	DC?	Unknown
IL-31	Activated T cells, Th2	Myeloid progenitor, keratinocytes
IL-33	Smooth muscle cells, keratinocytes, fibroblasts	Th2, mast cells
IL-35	T$_{reg}$	Effector T cells

activating other cells (NK, T, B, and hematopoietic stem cells). Several interleukins are identified and their source and functions are summarized in Table 3.6.

B Lymphocytes and Antibodies

B lymphocytes express surface receptors that interact with antigens. The B-cell receptors are in fact membrane-bound antibody. Antibodies secreted from B cells are present in the γ-globulin fraction of the serum and hence called immunoglobulin (Ig). Immunoglobulins are produced by mammals and avian species, and there are subtle differences among them, which are described below.

Structure of Immunoglobulin

A typical immunoglobulin (Fig. 3.14) consists of two identical heavy chains and two light chains. Two heavy chains (55 or 70 kDa) are joined by two disulfide bridges in the hinge region. Heavy chains have one variable domain called V_H and three constant domains designated C_H1, C_H2, and C_H3. The variable domain (V_H)

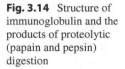

Fig. 3.14 Structure of immunoglobulin and the products of proteolytic (papain and pepsin) digestion

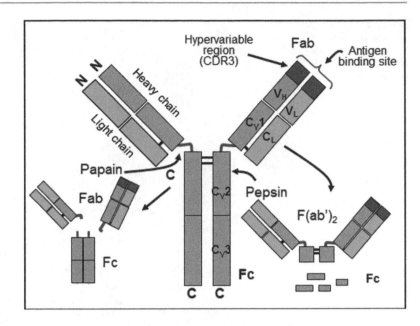

has a hypervariable region, which is also called complementarity-determining region (CDR). The CDR again has three subregions, CDR1, CDR2, and CDR3, of which CDR3 is the most variable and is located close to the N-terminal end. Two light chains κ and λ (24 kDa each) contain one variable (V_L) and one constant (C_L) domain. Light chains are linked to the heavy chain by disulfide bonds. Each structurally distinct domain (such as C_H1, C_H2, and C_H3) in each chain consists of 110 amino acids, and each domain is joined by disulfide bridges. Based on the heavy chain types, immunoglobulins are classified as IgA, IgG, IgM, IgE, and IgD (Fig. 3.15). IgA is made with α-heavy chain, IgG with γ, IgE with ε, IgD with δ, and IgM with μ-heavy chain (Table 3.7).

The antigen-binding end, which is also the N-terminal end, is called Fab, and the C-terminal end is Fc (fragment crystalline). It is the C-terminal end that binds Fc receptors present on various cell types. The Fc region contains carbohydrate molecules and a complement-binding site. Proteolytic enzymes are used to cleave Fab from Fc region. Papain cleaves immunoglobulin molecule in the hinge region resulting in two separate Fab molecules and 1 Fc molecule, while pepsin cleaves the hinge region below the disulfide bond to produce an intact $F(ab)_2$ appearing as the letter "V" and fragmented Fc.

Classes of Immunoglobulins

IgA It consists of a heavy chain type $\alpha1$ or $\alpha2$ and exists either as a monomer or a dimer. In the dimeric form, a joining chain (15 kDa) holds two monomers together. IgA is found in the serum at low concentrations (3 mg ml^{-1}). Dimeric IgA also exists as a secretory form (sIgA) where a secretory molecule of 70 kDa is attached that protects sIgA from proteolytic degradation. sIgA is about 400 kDa and is abundant in saliva, mucus, and other body fluids (bile, synovial fluids, respiratory and intestinal tract secretions).

IgE It is about 190 kDa and contains an additional domain in Fc region. Generally IgE level is very low in serum (0.05 mg ml^{-1}), but its concentration increases during an allergic response or during parasitic infection such as helminth infection.

IgD It is about 180 kDa and remains mostly membrane bound. Thus, it is present in serum in trace amounts. IgD acts as a receptor for antigens when B cells serve as an APC.

IgM It is present in pentameric form (950 kDa) where 5 units are joined by a joining chain (J-chain). This is the predominant antibody in the blood (1.5 mg ml^{-1}) during primary immune

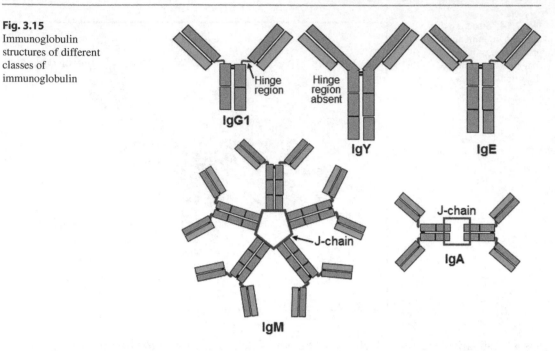

Fig. 3.15 Immunoglobulin structures of different classes of immunoglobulin

Table 3.7 Classes and subclasses of immunoglobulins

Class	Heavy chain type	Subclass	Molecular weight (kDa)
IgA	α1, α2	IgA1, IgA2, sIgA	150, 300, or 400
IgD	δ	None	180
IgE	ε	None	190
IgG	γ	Mouse: IgG1, IgG2a, IgG2b, IgG3	About 150 each
		Human: IgG1, IgG2, IgG3, IgG4	
IgM	μ	None	950 (pentamer)

response when challenged with an antigen. IgM is very unstable and rapidly loses its activity if it is subjected to temperature abuse.

IgG The molecular mass of IgG is 150 kDa. Subclasses of IgG vary between mice and humans depending on the heavy chain type. In mice, subclasses are IgG1, IgG2a, IgG2b, and IgG3; in humans, subclasses are IgG1, IgG2, IgG3, and IgG4. IgG concentration in blood is very high (IgG1, 9 mg ml^{-1}; IgG2, 3 mg ml^{-1}; IgG3, 1 mg ml^{-1}; IgG4, 0.05 mg ml^{-1}) and is the predominant immunoglobulin during secondary immune response. IgGs are very stable.

Avian (chicken) antibodies consist of three immunoglobulin subclasses: IgA, IgM, and IgY. IgA and IgM are similar to mammalian IgA and IgM, while IgY is equivalent to mammalian IgG. These antibodies are found in serum as well as in eggs. In eggs, IgA and IgM are present in albumen in low concentrations (0.15 and 0.7 mg ml^{-1}, respectively), while IgY is found in yolk in a large amount (~25 mg ml^{-1}). Structurally, IgY (180 kDa) is larger than the mammalian IgG (150 kDa), and it can be readily harvested in large quantities from eggs. The H-chain in IgY is 68 kDa and consists of four constant domains (Cv1–Cv4) instead of three for IgG. IgG constant domains Cγ1, Cγ2, and Cγ3 are equivalent to IgY Cv1, Cv3, and Cv4, respectively. IgY lacks a hinge region (Fig. 3.15).

Diversity of Antibodies

A total number of antibody specificities that an individual can produce is called the "antibody repertoire." There are about 10^7–10^9 different antibody molecules with unique amino acid sequences in the antigen-binding site. This generates significant diversity in a host. The variable

Fig. 3.16 B-cell maturation, growth phases, and synthesis of immunoglobulin

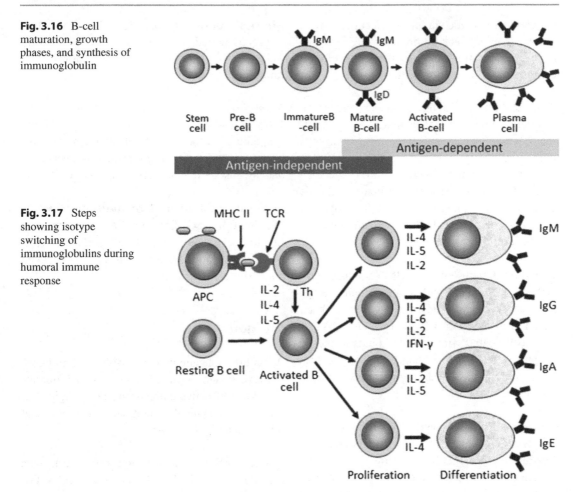

Fig. 3.17 Steps showing isotype switching of immunoglobulins during humoral immune response

region has a hypervariable region or CDR. The unique determinant of the CDR varies from antibody to antibody and is called *idiotope*, while the collection of idiotopes on a particular antibody molecule constitute *idiotype*.

Antibody Production

The sequence of events that takes place for B cells to produce antibody molecules for a specific antigen begins in the stem cell (Fig. 3.16). B cells originate from stem cell and give rise to pre-B, immature B, and mature B cells, sequentially. Immature and mature B cells leave bone marrow and migrate to secondary lymphoid organs such as the lymph nodes and spleen where they transform into antibody-secreting B cells. Immunoglobulin isotype switching takes place at this stage and depending on the antigen type B

cells can produce antibody subclass of IgM, IgG, IgA, or IgE (Fig. 3.17).

Function of Antibody

Acts as B-Cell Receptor Membrane-bound antibody binds antigens which allow B cells to process and present antigenic peptides to T cells via MHC class II molecules. Hence, B cells are efficient antigen-presenting cells.

Neutralization of Antigen Antibody neutralizes toxins, viruses, or bacteria. It binds to the antigenic determinant and prevents the antigen from interacting with the host cell by "steric hindrance."

Activation of Complement IgG or IgM after forming a complex with an antigen activates complement cascade via the classical pathway to produce complement by-products such as C3b and membrane attack complex (MAC) that facilitate lysis of the microorganism.

Opsonization The antibody also serves as an opsonin, a molecule that facilitates the recognition of target microorganisms by phagocytic cells (e.g., macrophages) for elimination. Antibody first forms a complex with microbes, and then the Fc domain of the antibody binds to the Fc receptor on the phagocytic cells (macrophage, neutrophil or eosinophil) for engulfment and destruction.

Antibody-Dependent Cell Cytotoxicity (ADCC) NK, neutrophil, and eosinophil recognize and specifically destroy target cells when coated with antibodies (IgG, IgE, and IgA). A target cell coated with antibody binds to the Fc receptor (CD16) of the phagocytic cell for specific recognition and destruction of virus or tumor infecting a target cell.

Immediate Hypersensitivity Reaction by IgE Basophils or mast cells have Fc receptor (FcεR1) for IgE. IgE forms a complex with the antigen (allergen) and binds to the Fc receptor of the mast cell. This interaction prompts the release of vasoactive amines such as histamine, leukotrienes, or prostaglandins that are responsible for immediate hypersensitivity reaction (Fig. 3.18). Examples are hay fever, asthma, and food allergy.

Mucosal Immunity by IgA Secretory IgA (sIgA) is abundant in mucus and other bodily fluids and provides specific or nonspecific immunity at the mucosal surface.

Neonatal Immunity-Mediated by IgG Colostrum, milk produced by mother immediately after the birth of the fetus, contains a high level of maternal IgG, which protects neonate against pathogens in the early part of its life.

Feedback Inhibition of Immune Response Is Mediated by IgG Binding of excess circulating IgG to Fcγ receptor on the B-cell surface inhibits further activation of B cells. This process is called antibody feedback inhibitions.

Antigen

A foreign molecule that is capable of stimulating immune system is called an antigen or immunogen. Most effective antigens are large, rigid, stable, and chemically complex. Factors that influence antigenicity are:

Molecular Size of the Antigen Large molecules are more antigenic than the small molecules. For example, 14.4 kDa lysozyme is better antigen than 1 kDa angiotensin, the 69 kDa albumin is better antigen than lysozyme, γ-globulin (156 kDa) is better than albumin, fibrinogen (~340 kDa) is better than γ globulin, and IgM (900 kDa) is better than the fibrinogen.

Fig. 3.18 Allergen-mediated activation of basophil and mast cell for secretion of vasoactive amines in the IgE-mediated hypersensitivity response

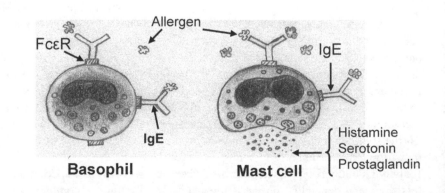

Complexity of Antigen LPS and proteins are complex, whereas lipid, carbohydrate, and nucleic acids as polymers of repetitive units are considered poor antigens.

Structural Stability of Antigen Immune response recognizes shape and stability. Flexible molecules have no shape and thus are considered poor antigens. For example, gelatin is wobbly and is a poor antigen. Similarly, flagella are structurally unstable and thus are considered poor antigens.

Degradability of Antigen Antigens should be degraded relatively easily for processing and presentation by APC. Stainless steel pins and plastic joints are made of large inert organic or inorganic polymers, which the immune system cannot degrade. Thus, they are recognized as poor antigens. Proteins made of D-amino acid isomeric forms are poor antigens because immune cells cannot degrade them. Only L-amino acid isoforms are used by cells.

Foreignness The immune system does not respond to self-antigens, and the immune cells which respond to self-antigens are eliminated or turned off. The degree of foreignness also determines the antigenicity.

Types of Antigens

Microbial and nonmicrobial antigens are diverse. Bacterial antigens include cell wall or somatic antigen (O), capsule (K), pili or fimbriae (F), flagella (H), cell membrane, proteins, ribonucleoprotein, enzymes, and toxins (endotoxins and exotoxins).

Viral antigens include envelope (lipoprotein, glycoprotein), capsomere (a subunit of capsid), and receptor proteins. Cell surface antigens of mammalian cells include (1) blood group antigens, ABO, Rh, MN, Kell, Duffy, Luthern, and Lewis; (2) histocompatibility antigens (MHC class I, class II); (3) cluster differentiation (CD) or lymphocyte surface antigens; and (4) autoantigens such as hormones, myelin, and DNA. Protein antigens are called T-dependent antigen, i.e., these antigens activate T cells for effector function, whereas lipopolysaccharide (LPS) or peptidoglycans (PGN) consisting of carbohydrates are called T-independent antigens which activate B cells. However, some zwitterionic polysaccharides that carry both positive and negative charges such as polysaccharide A from *Bacteroides fragilis* can activate T cells (Th1 cells).

Epitope or Antigenic Determinant

Antigenic determinant, or epitope, is the most immunodominant structure in an antigen and binds to the antibody. The number of epitopes in a molecule depends on the size of the antigen; larger antigens have multiple epitopes. Epitopes can be either linear, conformational, or neoantigenic. Linear antigenic determinants remain intact even after denaturation by the action of physical or chemical treatments. Conformational epitopes are destroyed by denaturation and are unable to bind to the antibody. In this case, an antibody only recognizes the natural conformation of the antigen. Neoantigenic determinants are unraveled only after treatment of the antigen with an enzyme or heat that allows binding to an antibody.

Hapten

Hapten means "to grasp" or "fasten." Small molecules that are unable to induce an immune response by themselves are called haptens. Examples are penicillin, allergens, and peptides. They need carrier molecules to induce antibody production. Carrier molecules such as KLH (keyhole limpet hemocyanin) and BSA (bovine serum albumen) are large and immunogenic and are used as carriers for experimental production of antibodies against haptens. Naturally, in some individuals, penicillin or allergens, though small, can interact with serum proteins in the blood forming a larger complex that is capable of inducing an immune response. Generally, an immune response against penicillin leads to a hypersensitive, or allergenic, response.

Antigen–Antibody Reaction

Antigen–antibody reactions are specific and reversible. Binding forces involved during antigen–antibody reactions are (1) Vander Waal force, where transient dipole interactions allow weak bonding between the antigen and antibody, (2) hydrogen bond [O]–H–[O] or [N]–H–[N] which forms between molecules in an antigen and antibody, (3) electrostatic bond ([+]–[−]) which forms by positive and negative charges contributed by both antigen and antibody, and (4) hydrophobic bond which forms by removing water molecules from the binding sites of antigen and antibody.

Binding strength between an antigen and antibody molecule is expressed as *affinity. Affinity* is the strength of binding between a single complementary site of the antibody and an epitope of an antigen. The *avidity* of an antibody defines the strength of attachment of combining sites of all available epitopes, i.e., the overall strength of binding.

The Major Histocompatibility Complex

The major histocompatibility complex (MHC) molecules, or MHC antigens, are expressed on the surface of a variety of cells. They are essential for antigen presentation to T cells. They are also responsible for graft rejection. During organ or graft transplant, if the host receives tissues or organs that are expressing the same MHC, the graft is accepted, but if the MHC are different, the graft is rejected. There are two classes of MHC antigens: class I and class II. T cell recognizes foreign antigen only when presented within MHC class I or class II. Exogenous peptide fragments always bind to the class II (example, bacterial exotoxins or other surface proteins), whereas endogenously synthesized peptides are presented with class I (example, viral proteins or tumor proteins).

Structure of MHC

MHC Class I

MHC class I proteins consist of a single α-chain (also known as a heavy chain) (Fig. 3.19). The molecular mass of α-chain in human is 44 kDa and in mouse 47 kDa. The α-chain has three domains, α1, α2, and α3, which resemble the constant regions of an immunoglobulin. The β-chain (12 kDa; a β2-microglobulin), a non-MHC coded gene product, noncovalently interacts with the α-chain to provide stability to the MHC class I molecule. An antigen-binding pocket is formed by the α1 and α2 domains, which can accommodate a small peptide (processed antigen) consisting of 8–10 amino acid residues. MHC class I molecules are expressed by most nucleated cells in the body and present antigen to CD8$^+$ T cells (CTL) for effector function (Fig. 3.19). In other words, during infection with intracellular pathogens (bacteria, virus, and protozoa), pathogen-specific antigens are presented on the surface of infected host cells by MHC class I for targeted destruction by CTL cells.

MHC Class II

The MHC class II antigen consists of two noncovalently associated polypeptide chains (Fig. 3.19). The α-chain is about 32–34 kDa, while the β-chain is 29–32 kDa. The peptide-binding pocket is made up of both α1 and β1 domains that hold a peptide of 13–25 amino acids. Both α- and β-chains are encoded by two separate MHC genes and are expressed only in few cells: macrophages, dendritic cells, Langerhans cells, B cells, some T cells in humans and rats but not in mice, and some endothelial or epithelial cells after induction by IFN-γ. CD4$^+$ T cells (T helper cells) respond to the antigen when presented by class II molecules (Fig. 3.20).

Antigen-Presenting Cells

Antigen-presenting cells (APC) are defined as those which after internalization of an antigen (either by phagocytosis or by the active invasion of intracellular microbes) process and present antigens on their surface either in association with MHC class I or in class II molecules for recognition by T or B cells. Most cells in the body when infected by intracellular pathogens (virus, intracellular bacteria, protozoa) or tumors can

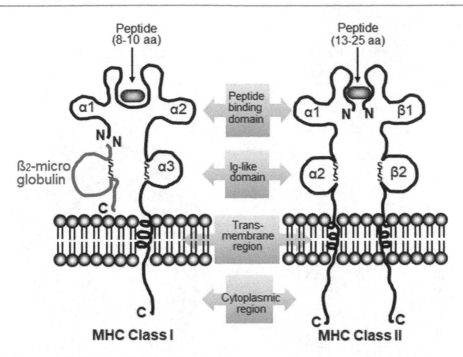

Fig. 3.19 Structure of MHC class I and class II molecules

express class I molecule, while only a limited number of specialized cells can express class II molecules and present the antigen. Cells that use MHC class II for antigen presentation are:

1. *Mononuclear phagocytes*: Macrophages actively phagocytose large particles and infectious organisms such as extracellular bacteria and parasites and degrade them, and the immunodominant peptide is presented in association with class II molecules to the Th or B cells.
2. *Dendritic cells*: DCs are originated from bone marrow. They reside in the spleen, lymph nodes, and submucosa. They process and present antigen to T cells.
3. *Langerhans cells*: These cells are located in the skin. They contain characteristic Birbeck granule (rod shaped or tennis racket shaped) in their cytoplasm, express class II molecules, and serve as APC of the skin.
4. *Venular endothelial cells*: These cells also express class II molecule and present antigen to T cells.
5. *Epithelial cells*: They also can express class II molecules and serve as APC when activated

by IFN-γ. Note: In general, IFN-γ can induce class II expression in any APC cells.

6. *B lymphocytes*: They are considered highly efficient antigen-presenting cells because they carry surface immunoglobulin (IgD or IgM), which serves as a receptor for a protein antigen, and they present antigens to Th cells. B cells bind antigen efficiently even at a low concentration with high affinity. They are important for T-cell-dependent antibody production.

MHC-Restricted Antigen Processing and Presentation

Endogenously synthesized antigens are restricted to class I-mediated presentation, while the exogenously synthesized antigens are restricted to the class II-mediated presentation.

Class II-Restricted Antigen Presentation

Class II-restricted antigen processing takes place with an exogenous protein antigen. The sequence

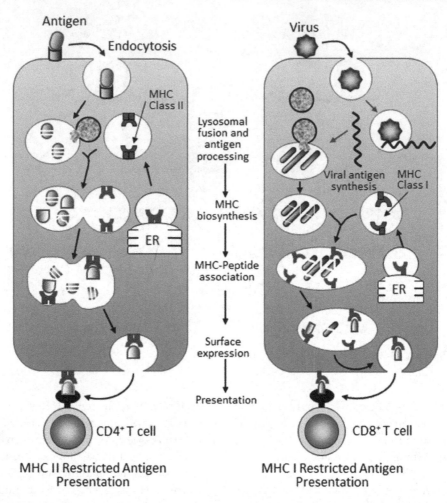

MHC II Restricted Antigen
Presentation

MHC I Restricted Antigen
Presentation

Fig. 3.20 MHC class I- and class II-restricted exogenous and endogenously synthesized antigen processing and presentation

of events that take place for antigen processing and presentations are summarized below (Fig. 3.20):

1. Initially, binding and internalization of protein antigens or native proteins occur by phagocytosis, endocytosis, or pinocytosis, and the antigens are trapped in the endosome or phagosome.
2. In the processing step, a fusion of the phagosome with the lysosome (carries degradative enzymes) results in the proteolytic digestion of proteins in the acidic environment and the generation of immunogenic peptides.
3. Next, the fusion of endosome containing peptides with the MHC class II proteins containing vesicle takes place. In this stage, class II molecule binds to the appropriate immunodominant

peptide fragments (13–25 amino acids). Unbound peptides are further degraded to amino acids and discarded from the cells by exocytosis.
4. In the subsequent step, a fusion of vesicles containing MHC class II with the bound peptide and the cytoplasmic membrane occurs. Through exocytosis (reverse phagocytosis), the MHC with bound peptide is displayed outside for recognition by CD4+ T cells.

Class I-Restricted Antigen Presentation

Intracellular bacteria, viruses, parasites, or tumor antigens are capable of infecting different types of host cells and are processed and presented by MHC class I molecule on the surface of infected cells also known as "target cells." After the invasion, intracellular pathogens escape from the phagosome and multiply inside the host cytoplasm. Thereafter, the sequence of events are somewhat similar to class II-mediated pathway; however, in this case, the pathogen-specific proteins are synthesized inside the host cell through active gene transcription and translation process (Fig. 3.20). (1) In the first step, the bacterial DNA or viral DNA or RNA is transcribed to synthesize proteins. At the same time, DNA for MHC class I is also transcribed to synthesize class I molecules. (2) Pathogen-derived proteins are synthesized in the cytoplasm and then are degraded by host cell proteasome into small peptides. (3) The peptides (eight to ten amino acids) are transported into the lumen of the rough endoplasmic reticulum and bind to MHC class I molecules. The peptide-bound MHC complex is transported to the Golgi for further processing. (4) Vesicles containing MHC class I with bound peptides are fused with cytoplasmic membrane and are transported to the exterior of the cells through exocytosis and presented to the CD8+ T (CTL) cells.

Accessory Molecules Involved during MHC-Restricted T-Cell Activation

During the antigen presentation by MHC class I or class II molecules to T cells, several other molecules are also involved (Fig. 3.21). T-cell receptor (TCR) consisting of αβ- or γδ-chains are the primary component that recognizes the MHC–peptide complex on the surface of APC. The CD3 and CD4/CD8 present on T cells also bind to the MHC–peptide complex and stabilize the structure. In addition, the CD2 molecule on the surface of T cell interacts with the LFA-3 (lymphocyte function-associated antigen) of APC, and LFA-1 of T cell interacts with ICAM-1 (intracellular adhesion molecule) of APC. These accessory molecules increase the strength of the adhesion between T and APC or target cells, serve as surface markers, and transduce the biochemical signal to the interior of the cell.

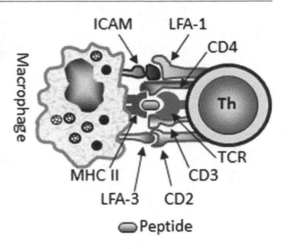

Fig. 3.21 Involvement of accessory molecules during antigen presentation to T cells

The Complement System

The complement proteins are synthesized in the liver and are present in serum. They play important roles during both innate and adaptive immune response. Jules Border in his classical experiment demonstrated that if an antibacterial antibody is mixed with fresh serum at 37°C, it caused lysis of the bacteria; however, when the serum was heated to 56°C or higher prior to mixing, lysis did not occur. Bacterial lysis was reestablished when the fresh serum was added, suggesting the involvement of a heat-labile component, which can complement the antibody function, and hence was given the name "complement." Complement consists of a series of proteins designated C1 to C9, each of which remains in serum in an inactive form. Complement cascade can be activated sequentially by proteolytic enzymes through three mechanisms: (1) classical pathway, (2) alternative pathway, and (3) lectin pathway (Fig. 3.22). The classical pathway is activated by an antigen–antibody (IgG or IgM) complex during the adaptive immune response. An infectious agent and a lectin-like glycoprotein can activate alternative and lectin pathways, respectively, generally during innate immune response. The primary and the most abundant component of the complement cascade is C3, which is converted to C3a and C3b by C3 convertase. Individuals with complement deficiency are susceptible to many infective agents.

Fig. 3.22 Activation of
complement by the
classical, alternative, and
lectin pathways

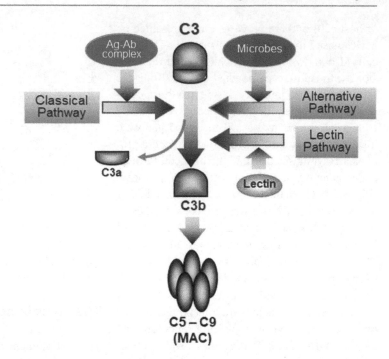

The Classical Pathway

The classical pathway is activated during the specific immune response (Fig. 3.23). Circulating IgG or IgM in the blood form a complex with the specific pathogen or antigen and activates the inactive form of the protein C1. C1 is a large multimeric protein consisting of three subunits: C1q, C1r, and C1s. The active protein complex C1qrs then catalyzes inactive C4 to produce active C4b and a soluble by-product, C4a. C4 can also be activated by mannose-binding lectin (MBL), which is a pattern recognition receptor-specific to bacterial carbohydrates. C2 catalyzes the C4b to form C4b2. C1qrs complex can also activate C4b2 to form the C4b2a, which is also called C3 convertase. The C3 convertase catalyzes the C3 to form C3a and C3b, a major component of the complement activation pathway. C3b also forms a complex with C4b2a forming the C4b2a3b complex, which is known as C5 convertase. The C5 convertase catalyzes C5 to form C5a and C5b. C5b forms a complex, C5b6789, also known as membrane attack complex (MAC) through a sequential reaction of C5b with C6, C7, C8, and C9 components. MAC

forms a donut-shaped hole on the surface of the target pathogen. The complement activation by-products C3a, C4a, and C5a are collectively called anaphylatoxins, which are a chemoattractant for immune cells and are known to induce hypersensitivity response.

Alternative Pathway

During the innate immune response, the complement system plays a major role in protecting the host against invading pathogens in the absence of antibodies. Viral proteins, bacterial LPS, peptidoglycan, teichoic acid (TA), lipoteichoic acid (LTA), and microbial surface polysaccharide can activate complement directly via the alternative pathway. The resulting complement by-products play important roles in the host natural immunity. In the alternative pathway (Fig. 3.24), several serum factors including B, D, H, I, and properdin systems are involved. Initially, microbial factors activate C3 to form C3b. In the presence of factor B, factor D catalyzes C3b to form C3bBb. The C3bBb complex is called C3 convertase and is very unstable. Factor I rapidly degrades this pro-

Fig. 3.23 Complement activation through classical pathway

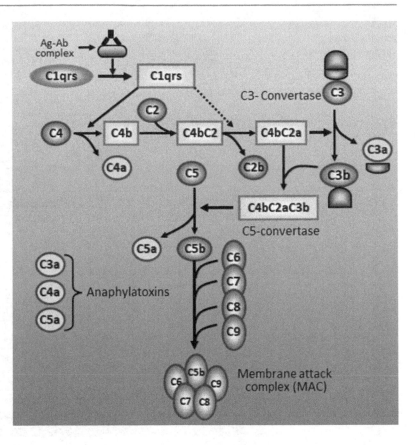

tease. Properdin proteins, however, bind and stabilize C3 convertase. C3 convertase catalyzes C3 to form C3b and a small molecule, C3a. C3b forms a complex with C3bBb to produce C3bBb3b, a C5-convertase, which converts C5 to form C5b and C5a. Sequential catalysis of C6, C7, C8, and C9 results in the formation of C5b6789 complex, called the MAC, similar to a by-product generated in the classical pathway.

Function of Complement

The complement system plays a major part in the defense against microbes in both innate and adaptive immune responses.

1. Complement-mediated bacterial lysis is accomplished by MAC, which is produced in both classical and alternative pathways. MAC also can attack host cells by inserting into the plasma membrane causing collateral damage (Fig. 3.25); however, a majority of the host cells are protected by CD59, a membrane-bound glycoprotein that inhibits MAC formation by blocking the aggregation of C9.

2. Complement protein C3b aids in the opsonization (phagocytosis) process by serving as an opsonin, which binds to microbes. Phagocytes (macrophages and neutrophils) express the C3b receptor (CD35) which binds to the opsonin-coated microbes for engulfment. The opsonization process can be enhanced several-fold when both antibody and C3b coat the target pathogen as an opsonin.

3. Complement protein by-products C3a, C4a, and C5a also serve as anaphylatoxins. They induce the release of soluble inflammatory mediators such as histamine to increase membrane permeability, smooth muscle contraction, and diffusion of inflammatory cells such as neutrophils and macrophages. C5a acts as a chemoattractant for neutrophils at the site of

Fig. 3.24 Complement
activation through
alternative pathway

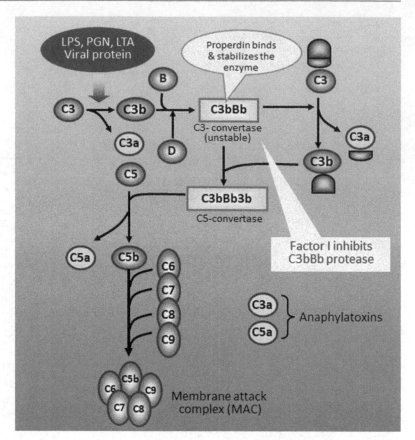

Fig. 3.25 Diagram
showing the insertion of
MAC (membrane attack
complex) into the
plasma membrane

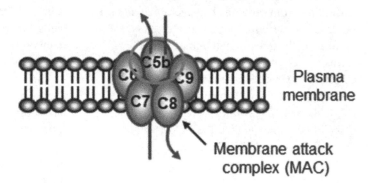

inflammation thus facilitating increased
clearance of the pathogens from the site of
infection.

4. Complement proteins also aid in the phagocytic clearance of immune complexes.
Antigen–antibody complexes formed during
immune response may be trapped in a variety
of organs/tissues including the kidneys resulting in pathological consequences such as
glomerulonephritis. Since the complement
system enhances phagocytic activity, the

antigen–antibody complex is rapidly cleared
from the body by phagocytes, thus preventing
ill effects of the antigen–antibody complex.

Control of Complement Activation

Activation of complement and resulting by-
products are able to eliminate pathogens from the
host; however, activation can also cause collateral
damage to the nearby host cells. If C3b binds to

the host cells, those cells become the target of phagocytic cells causing cell/tissue damage. In addition, the MAC can cause host cell damage. As a protective measure, the host cells carry surface proteins which prevent C3b from binding to host cells. For example, the C3bH formed during the alternative pathway of activation allows degradation of C3b by protein I that in turn prevents the formation of C3bBb. In addition, the host cells contain sialic acids, and C3b does not bind to sialic acid. However, some bacteria take advantage of the situation by producing a capsule that contains sialic acid, thus preventing C3b from binding to self and phagocytic killing. Mannose-binding protein also prevents activation of complement. Some bacteria express mannose on their surface, thus preventing complement activation. Macrophages produce IL-6, which can activate liver cells to produce mannose-binding protein, which binds to mannose on the bacterial surface. This alters the configuration and can activate complement.

Immunity to Microbes

The principal physiological function of the immune system is to protect the host against pathogenic microbes such as extracellular bacteria, intracellular bacteria, viruses, and parasites. Important characteristics of immune response to microbes are as follows. (1) Defense against microbes is mediated by both innate and adaptive immunity. (2) Different types of microbes stimulate distinct populations and subpopulations of lymphocytes. (3) Survival of microbes in the host or progression of the disease depends on their ability to evade the immune system. (4) Tissue injury and disease related to infection are caused primarily by the overt host response rather than the microbe itself.

Extracellular Bacteria

Extracellular bacteria replicate outside the host cells, i.e., in circulation, interstitial space, and the lumen of the respiratory and the intestinal tracts. Examples of extracellular bacteria are *Clostridium*, *Staphylococcus*, *Bacillus*, *Streptococcus*, *Escherichia coli*, and *Vibrio* species. Pathogens cause disease by inducing inflammation or by the direct action of toxins or enzymes that they produce.

Inflammation

Bacterial cells provoke inflammation that results in tissue destruction and an influx of neutrophils and macrophages at the site of infection. During inflammation, complement activation takes place, and cytokines are produced by macrophages. Enzymes and toxins released from the dead neutrophils and macrophages at the site can directly cause cell injury and tissue damage resulting in suppurative infection characterized by pus formation. Three cardinal signs of inflammation are redness, swelling, and pain: (1) increased blood supply to the site of infection causes "redness," (2) increased capillary permeability and fluid accumulation result in swelling, and (3) recruitment of neutrophils and macrophages by chemoattractant C5a results in tissue injury that invokes pain.

Toxins

Exotoxins are protein toxins that are secreted by the microbes and cause diverse pathological effects resulting in cell injury or cell death (see Chap. 4). Shiga toxin and diphtheria toxin inhibit protein synthesis. Botulinum toxin blocks neurotransmitter release. Cholera toxin stimulates cAMP synthesis resulting in Cl$^-$ secretion and H$_2$O loss. Endotoxins are bacterial components, such as LPS and PGN, and are called pyrogens because they induce fever in the host. Pyrogens stimulate the release of cytokines, IL-1 and TNF-α, activate B cells, and act as an adjuvant to stimulate macrophages to produce more cytokines. These cytokines induce fever, decrease smooth muscle contraction, increase membrane permeability, lower blood pressure, and under severe conditions can prompt septic shock and death.

Innate Immunity

The innate immune response against extracellular bacteria is spontaneous and is mediated by immune cells and cytokines. (1) Neutrophils, monocytes, and tissue macrophages actively phagocytose bacteria. (2) Complement is activated via the alternative pathway or mannose lectin pathway by bacterial peptidoglycan or mannose-binding lectin. The complement activation by-product C3b enhances opsonization, while MAC induces cell lysis. (3) LPS or PGN activates macrophages to produce inflammatory cytokines TNF-α, IL-1, and IL-6, which in turn activate neutrophils and macrophages for enhanced bacterial clearance. These cytokines also induce a fever to retard bacterial growth or may induce "septic shock" or "endotoxin shock" leading to fatal consequences.

Adaptive Immune Response

The adaptive or specific immune responses against extracellular bacteria are mediated by cell wall components and the exotoxins, which are summarized below:

1. In the specific immune response, cell components (i.e., LPS and PGN) act as T-independent antigens and activate B cells to produce specific IgM. Antigen-mediated specific T-cell activation results in specific cytokine release, which is responsible for isotype switching to produce specific immunoglobulin class (IgG, IgA, IgE).
2. Exotoxins are T-dependent antigens and are processed by APC (macrophages, B cells) and presented via MHC class II molecules to activate CD4+ T cells and B cells for cytokine and antibody production, respectively. Certain toxins act as superantigens, which induce T cells to produce increased cytokines in the absence of antigen processing and classical MHC presentation leading to toxic shock syndrome (TSS). For example, staphylococcal enterotoxin B (e.g., SEB) binds to MHC class II molecule on the macrophages directly. This

MHC–SEB complex interacts with T-cell receptor and activates CD4+ T cells to produce copious amounts of IFN-γ, which in turn induces and enhances MHC expression on many cell types, antigen (SEB) presentation, and cytokine production by macrophages leading to toxic shock syndrome.
3. Antibodies (IgG and IgM) help in the opsonization of pathogens, neutralization of toxins, and activation of complement via the classical pathway. The activation by-products C3b and MAC cause increased phagocytosis and cell lysis, respectively.

Evasion of Immune System by Extracellular Bacteria

In order to cause a successful infection, bacteria must be able to evade the immune system. Indeed, they have developed various strategies to achieve that.

1. Antigenic variation is an important strategy for some microorganisms such as *Salmonella enterica*, *Haemophilus influenzae*, and *E. coli*. Bacteria use surface molecules such as pili, fimbriae, flagella, or other surface proteins for adhesion and colonization; however, genetic variation in bacterial surface molecules may preclude recognition by the antibodies previously produced against the same organism.
2. Some bacteria (*Bacillus anthracis*, *B. cereus*, and *Pneumococcus* spp.) exhibit antiphagocytic mechanisms. They express capsules made of sialic acid and hyaluronic acid, both of which are also present in host cell membrane and prevent binding of C3b. As a result, these bacterial cells are not recognized by macrophages for destruction.
3. Sialic acid component of the capsule also inhibits complement activation and helps evade pathogen inactivation.
4. Some bacteria (*Staphylococcus aureus*, *Streptococcus pyogenes*) cover themselves with host antibody leaving the immune system unable to recognize them as foreign. *S. aureus* produces protein A, and *S. pyogenes*

produces protein G which binds IgG at the Fc domain masking the bacteria from macrophages even though they carry a receptor for Fc.

5. Some bacteria such as *Streptococcus pneumoniae*, *Haemophilus influenzae*, *Neisseria gonorrhoeae*, and *Neisseria meningitides* produce IgA-specific proteases that degrade these antibodies.

Intracellular Bacteria

Several foodborne pathogens maintain an intracellular lifestyle as part of their infection strategy. After binding to a specific host cell receptor, intracellular bacteria modulate signaling events resulting in cytoskeletal rearrangement and facilitating their own entry (induced phagocytosis). Bacteria trapped inside a phagosome either escape by lysing the vacuolar membrane or produce effector molecules that sustain their intracellular life cycle. They either spread from cell to cell or induce apoptosis leading to host cell lysis as part of their pathogenic mechanism. Examples of intracellular foodborne bacterial pathogens are *Listeria monocytogenes*, *Yersinia enterocolitica*, *Shigella* spp., and *Salmonella enterica*.

Innate Immunity

Innate immunity including gastric acid, bile, antimicrobial peptides, mucins, and natural microflora provide some protection; however, many pathogens are resistant to those. Therefore, the innate immunity is less effective. Furthermore, phagocytes (macrophages and neutrophils) are also less effective since bacteria are resistant to degradation by lysosomal degradative enzymes.

Adaptive Immunity

The humoral immune response has limited contribution toward protection since the bacteria are mostly localized inside the host cell; thus, the cell-mediated immunity provides the most effective protection. However, it is important to point out that despite the lack of protection by humoral immunity, antibodies against bacterial antigens are found in the serum, thus indicating that the antigens are processed and presented by MHC class II pathways for response by CD4$^+$ T cells and B cells. Furthermore, the bacterial antigen–antibody complex can activate complement through the classical pathway and can contribute toward adaptive immunity. The properties of adaptive immunity are as follows:

1. Cell-mediated immunity (CMI) is the major protective immune response against intracellular bacteria, where CD8$^+$ T cell acts as the primary effector cell. These cells recognize the infected target cells expressing antigens on their surface in association with MHC class I molecule.

2. Macrophages activated by T-cell-derived cytokines such as IFN-γ also play an important role by promoting active phagocytosis of bacteria.

3. In some cases, the host induces granuloma (nodule or polyp) formation to contain the infective agent from spreading, which is seen in chronic infections caused by *Mycobacterium*, *Histoplasma*, and some mold species. There is an onset of a delayed type of hypersensitivity reaction (DTH) to these infections. Both CD4$^+$ and CD8$^+$ T cells respond to the soluble protein antigens and intracellular bacterial antigens, respectively. Following activation, these cells secrete TNF and IFN-γ, which activate vascular endothelial cells, which in turn recruit neutrophils, lymphocytes, and monocytes. IFN-γ also activates macrophages, which convert them into epithelioid or giant cells for enhanced elimination of pathogens. If the antigen stimulation continues to persist, macrophages are chronically activated to secrete additional cytokines and growth factors, which help recruit fibroblast cells to encase the bacteria-containing cells to form a protective nodule called a "granuloma." These nodules of various sizes are characteristic for many diseases including tuberculosis and histoplasmosis.

Evasion of Immune System by Intracellular Bacteria

Intracellular bacteria utilize various strategies that allow them to maintain intracellular lifestyle.

1. Some bacteria (example, *Mycobacterium*) inhibit phagolysosomal fusion, thus protecting the bacteria from toxic lysosomal contents.
2. *Escape from phagosome*: *Listeria monocytogenes* secretes a hemolysin that forms pores in the phagosome to help bacteria escape into the cytoplasm before phagolysosomal fusion takes place, thus averting degradation by lysosomal contents. Furthermore, hemolysin also blocks antigen processing, and *Listeria* antigens are not presented on the surface by MHC class I molecule for recognition by T cells.
3. Intracellular bacteria such as *Listeria monocytogenes*, *Salmonella enterica*, *Yersinia* spp., and *Shigella* spp. are capable of scavenging reactive oxygen species to avert toxic effects of oxygen radicals. Superoxide dismutase (SOD) inactivates toxic "O" radicals. Bacterial catalase breaks down toxic H_2O_2.
4. Intracellular bacteria have antiphagocytic activity. For example, *Yersinia enterocolitica* outer membrane proteins such as YopE has antitoxin activity, YopH has tyrosine phosphatase activity, and YpkA blocks signaling pathway that is required for phagocytosis.
5. Intracellular localization also prevents the bacteria from being seen by the immune system.

Immunity to Virus

Viruses are obligatory intracellular pathogens, which upon entry replicate inside the host cell. Viral surface proteins first bind to a host cell receptor leading to entry into the cell. Inside the cell, viral nucleic acid replication, protein synthesis, nucleic acid packaging, and matured virion synthesis take place. Upon host cell lysis, newly developed virions are released and proceed to infect neighboring cells (see Chap. 6). Viral proteins also block host cell protein synthesis and thus may induce host cell death showing characteristic cytopathic effects. Many viruses infect immune cells as their primary target. For example, HIV-I binds to CD4 molecule on helper T cells and then enter the cell eventually leading to acquired immune deficiency syndrome (AIDS) when the infection has resulted in substantial loss to the Th population. Epstein–Barr virus (EBV) binds to type 2 complement receptor (CR2) on B cells and causes infectious mononucleosis. It is also associated with Burkitt lymphoma and other carcinomas. Rhinovirus binds to intercellular adhesion molecule (ICAM) and causes inflammation of the nasal passages. SV40 (simian virus) binds to MHC class I to enter cells and causes cancer in the monkey.

Innate Immunity

Innate immunity against virus infection is multifaceted. (1) Virus infection stimulates the production of type I interferon (both IFN-α and IFN-β) from infected cells, which inhibits viral replication by creating the "antiviral state" discussed earlier. (2) NK cells can kill or lyse a variety of virally infected cells. Moreover, the type I interferon enhances NK activity. (3) Viral protein can activate complement cascade in the alternative pathway and the complement by-product, C3b, can enhance phagocytosis.

Adaptive Immunity

Both humoral and cell-mediated immunity (CMI) are important in protecting the host against viral infection. During the humoral immune response, the antibody is produced against the viral envelope or capsid proteins, which are responsible for binding to host cell receptor. Antibodies are generally very effective in the initial stage of viral infection. Circulating antibodies bind viral surface proteins and block the viral interaction and entry into the host cells. On the mucus membrane, sIgA plays an important role since they are abundant in the mucus of respiratory and the digestive tract, which serves as a primary portal for viral entry. Antibody-dependent immune responses are as follows. (1) Antibody–viral particle complex can help enhance phagocytic clearance by macrophages from the blood circulation by opsonization. (2) Antiviral antibody forming a complex with a viral antigen can activate complement through the classical pathway and promote phagocytosis. (3) Humoral immunity against viral antigens is used as a prophylactic vaccine to protect the host from future infections. Attenuated or killed virus generates circulating antibodies in serum that prevent viral binding to host cells during a later exposure. Current vaccination strategy embraces the induction of mucosal immunity since many foodborne viruses (hepatitis A, norovirus, and rotavirus) use intestinal mucosa as the site of entry to initiate infection.

Though the antibodies are important protective components of immunity to viruses, they may not be fully effective since viruses hide inside the cells. Therefore, CMI provides the most protection. The principal mechanism of specific immunity against viral infection is the generation of cytotoxic T cells (CD8+ T cells). These cells recognize endogenously synthesized viral proteins in association with MHC class I molecules on the surface of target cells and destroy them by producing pore-forming cytolysins (perforin and granulysin).

Evasion of Immune System by Viruses

Viruses use several strategies to overcome immune system. The most important of all is their intracellular life cycle and utilization of host cell machinery for replication and spread, in which immune effector cells are unable to find them. Antigenic variation is also an important characteristic. There are a large number of serologically distinct strains present which show huge antigenic variations in their proteins. Therefore, protective antibodies or vaccine is ineffective. This characteristic is most important with influenza virus (flu virus) in which the genetic variation is very common. In addition, some viruses directly infect immune cells and cause immune suppression. For example, influenza and rhinovirus suppress the immune system, HIV-I diminishes CD4+ T cell populations, and EBV inhibits production of IL-2 and IFN-γ.

Immunity to Parasites

Many of the foodborne protozoan species (see Chap. 7) cause chronic disease where the natural immunity is weak and less effective. Protozoan species such as *Giardia*, *Entamoeba*, *Cryptosporidium*, and *Cyclospora* invade intestinal cells and cause massive damage to the site of infection leading to diarrhea, which may be bloody and mucoid. Helminths such as *Trichinella* and *Taenia* species (tapeworms) invade liver and muscle tissues and cause inflammation resulting in chronic infection. During chronic infection, the clinicopathologic consequences are mostly due to the host response to the parasites.

Innate Immunity

Acid, mucus, bile, antimicrobial inhibitors, resident microbiota in the gastrointestinal tract, complement, and phagocytic cells play important roles in antiparasitic immunity. Complement is activated through the alternative pathway and results in the formation of MAC, which causes lysis of parasites. Many parasites are resistant to phagocytic killing since they are capable of replicating inside the macrophage. Parasites also induce inflammation and recruit polymorphonuclear cells, macrophages, dendritic cells, and neutrophils.

Adaptive Immunity

Adaptive, or specific, immunity to parasites is effective and is mediated by both humoral and CMI. In humoral response, IgE plays an important role. Parasitic antigens are presented on the surface of APC and are recognized by CD4+ T cell, which produces IL-4 and IL-5. IL-4 also promotes IgE production from B cells, and IL-5 recruits eosinophils. The IgE-bound parasite complex is then recognized by eosinophils via receptor FcεR, and consequently, the parasite is destroyed by toxins (peroxidase, cationic protein, reactive oxygen species) produced by eosinophil.

Parasites also induce granuloma formation, which is seen in infection caused by *Trichinella*. During infection, both CD4$^+$ and CD8$^+$ T cells are activated, and they produce cytokines that recruit macrophages and fibroblasts. Fibroblast cells help encase the infective agents resulting in the formation of granulomas. CMI also induces DTH response. *Trichinella* (roundworm) eggs cause liver cirrhosis when deposited in the hepatocytes. *Schistosoma mansoni*, a waterborne parasite (flatworm), form granulomatous polyps in the nasal passage in herbivores and anal polyps in humans. Cytotoxic T cells (CD8$^+$) destroy host cells that carry intracellular protozoa and serve as the primary effector component of the CMI.

Evasion of Immune System by Parasites

Parasites also use numerous strategies to evade the immune system. (i) Protozoa grow intracellularly leaving them inaccessible to immune cells. This strategy is called "anatomic concealment." Some parasites develop cysts (example, *Trichinella*) that are not detected by immune cells. (ii) Some parasites mask themselves with a coat of host cell proteins on the surface. For example, *Schistosoma mansoni* larvae coat themselves with ABO blood group antigens and MHC molecules before they reach the lungs masking them from host immune cells. (iii) Some parasites (e.g., *Entamoeba histolytica*) spontaneously shed antigen coats after binding to the antibody and thus are not recognized by phagocytic cells.

Summary

To understand the pathogenic mechanism of microbial infection, one has to have a clear knowledge of immune response, both innate and adaptive, that are in place to protect the host against infective agents. Microbial dominance results in the disease, while successful host response averts the full-blown infection. Foodborne pathogens affect primarily the digestive system; therefore, the natural immunity of the gastrointestinal tract is the most important protective immune response. Moreover, the onset of symptoms for some diseases (i.e., intoxication) is very fast – appearing within 30 min to an hour. Thus, protection by adaptive immune response would have little impact since it is slow to develop, typically requiring 4–7 days. However, the adaptive immune response is essential for foodborne pathogens that have a prolonged incubation time and are responsible for systemic infection such as *Listeria monocytogenes*, *Shigella* spp., *Campylobacter* spp., *Salmonella enterica*, *Yersinia* spp., hepatitis A virus, *Toxoplasma gondii*, *Trichinella* species, and so forth. In innate immunity, gastric acids, bile, mucus, antimicrobial peptides, natural microflora, macrophages, neutrophils, NK cells, interferons, and complement proteins play a critical role. In the adaptive immune response, T and B lymphocytes produce cytokines and antibodies, respectively. CD4$^+$ T cells are most important for protection against extracellular bacterial infections and their exotoxins, while CD8$^+$ T cells are involved in the elimination of intracellular bacterial, viral, and parasitic infective agents. Antibodies neutralize pathogens or toxins by preventing them from binding to the host cell receptors, so they become the target for elimination by macrophages and neutrophils. An antibody–antigen complex also activates complement for inactivation of pathogens via opsonization. However, the immune system sometimes fails to protect the host. It is because pathogens have developed strategies to overcome immune defense by producing virulence factors that ensure their invasion, survival, replication, and spread inside the tissues. It is important to recognize that the pathogenic action by foodborne exotoxins is very quick and the host has literally no time to mount any immune response; thus, a majority suffers from this form of food poisoning (intoxication) irrespective of their health status or immune response.

Further Readings

1. Abbas, A.K., Lichtman, A.H., and Pillai, S. (2015) *Cellular and Molecular Immunology*. Philadelphia, PA: WB Saunders.
2. Amalaradjou, M.A.R. and Bhunia, A.K. (2012) Modern approaches in probiotics research to control foodborne pathogens. *Adv Food Nutr Res* **67**, 185-239.
3. Bevins, C.L. and Salzman, N.H. (2011) Paneth cells, antimicrobial peptides and maintenance of intestinal homeostasis. *Nat Rev Microbiol* **9**, 356-368.
4. Blaser, M.J. and Falkow, S. (2009) What are the consequences of the disappearing human microbiota? *Nat Rev Microbiol* **7**, 887-894.
5. Brandtzaeg, P. (2003) Role of secretory antibodies in the defence against infections. *Int J Med Microbiol* **293**, 3-15.
6. Hansson, G.C. (2012) Role of mucus layers in gut infection and inflammation. *Curr Opin Microbiol* **15**, 57-62.
7. Heller, F. and Duchmann, R. (2003) Intestinal flora and mucosal immune responses. *Int J Med Microbiol* **293**, 77-86.
8. Janssens, S. and Beyaert, R. (2003) Role of toll-like receptors in pathogen recognition. *Clin Microbiol Rev* **16**, 637-646.
9. Kovacs-Nolan, J., Phillips, M. and Mine, Y. (2005) Advances in the value of eggs and egg components for human health. *J Agric Food Chem* **53**, 8421-8431.
10. Lievin-Le Moal, V. and Servin, A.L. (2006) The front line of enteric host defense against unwelcome intrusion of harmful microorganisms: mucins, antimicrobial peptides, and microbiota. *Clin Microbiol Rev* **19**, 315-337.
11. Round, J.L. and Mazmanian, S.K. (2009) The gut microbiota shapes intestinal immune responses during health and disease. *Nat Rev Immunol* **9**, 313-323.
12. Sakaguchi, S., Miyara, M., Costantino, C.M. and Hafler, D.A. (2010) FOXP3+ regulatory T cells in the human immune system. *Nat Rev Immunol* **10**, 490-500.
13. Santaolalla, R. and Abreu, M.T. (2012) Innate immunity in the small intestine. *Curr Opin Gastroenterol* **28**, 124-129.
14. Sekirov, I., Russell, S.L., Antunes, L.C.M. and Finlay, B.B. (2010) Gut microbiota in health and disease. *Physiol Rev* **90**, 859-904.
15. Shao, L., Serrano, D. and Mayer, L. (2001) The role of epithelial cells in immune regulation in the gut. *Semin Immunol* **13**, 163-175.
16. Spits, H., Bernink, J.H. and Lanier, L. (2016) NK cells and type 1 innate lymphoid cells: partners in host defense. *Nat Immunol* **17**, 758-764.
17. Tscharke, D.C., Croft, N.P., Doherty, P.C. and La Gruta, N.L. (2015) Sizing up the key determinants of the CD8+ T cell response. *Nat Rev Immunol* **15**, 705-716.
18. Turner, J.R. (2009) Intestinal mucosal barrier function in health and disease. *Nat Rev Immunol* **9**, 799-809.
19. Ryan, V. and Bhunia, A.K. (2017) Mitigation of foodborne illnesses by probiotics. In Foodborne Pathogens: Virulence Factors and Host Susceptibility. Edited by Joshua Gurtler, Michael Doyle, and Jeffrey Kornacki. pp 603-634, Springer.

General Mechanism of Pathogenesis

4

Introduction

The diseases caused by foodborne pathogens are classified into *foodborne infection*, *foodborne intoxication*, and *foodborne toxicoinfection*, and each is described in detail below (Fig. 4.1). The primary vehicle of transmission of foodborne pathogens is food and water. The oral route is the main portal, and the primary site of action is the gastrointestinal (GI) tract. The GI mucosa comprises of epithelium, lamina propria, and a thin layer of smooth muscle, and it is the major site of interaction for foodborne pathogens. Most foodborne microorganisms cause localized infection and tissue damage, while others spread to the deeper tissues to induce systemic infection. For a successful enteric infection, several factors must work cooperatively in a host. Due to the presence of multiple protective barriers in the GI tract, generally, a high dosage is required to cause disease compared to the other routes of infection, such as intravenous, intranasal, or intraperitoneal. Besides, pathogens can transfer via direct contact with an animal or a human and from environments (soil, air) or from an arthropod vector. Once inside the host, the pathogen must survive in the changing environment, multiply and propagate, and avoid the host immune defense. Pathogens must find a suitable niche for colonization, which is facilitated by adhesion factors, invasion factors, and chemotaxis. For example, bacterial affinity for iron propels the organism toward the liver, which is rich in iron bound to the transferrin protein. Bacterial cell envelope consisting of the capsule can help bacterial survival in the hostile environment, as the capsule protects the bacterium from being engulfed by professional phagocytic cells. In addition, bacterial toxins and enzymes protect the cells from elimination by the host immune system. The presence of commensal bacteria in the site of infection can also assist the invading bacterium to find a niche. For example, in the case of "wound botulism," aerobic bacteria first enter the wound, grow and multiply, and utilize oxygen to create an anaerobic microenvironment. As the wound closes from the action of blood clotting and fibroblast cell accumulation, *Clostridium botulinum*, an obligate anaerobe transmitted to the wound through sharp object or dust, now has a favorable niche for growth and toxin (botulinum) production. Patients show a typical sign of botulism, a neuroparalytic disorder. Pathogens also damage the host tissues and cells by using exotoxins, endotoxins, or enzymes that induce cell death by apoptosis or necrosis and promote bacterial survival, multiplication, and propagation.

Foodborne Infection

Foodborne infection is committed by intact living microorganisms, which must enter the host to cause infection. Following ingestion of food or

© Springer Science+Business Media, LLC, part of Springer Nature 2018
A. K. Bhunia, *Foodborne Microbial Pathogens*, Food Science Text Series,
https://doi.org/10.1007/978-1-4939-7349-1_4

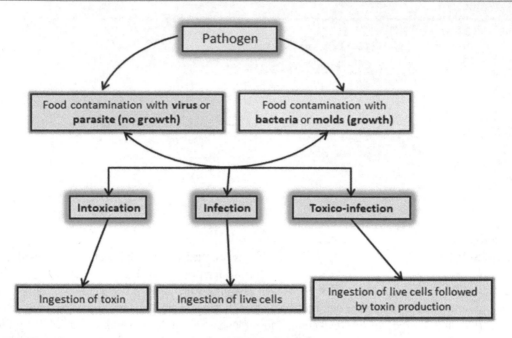

Fig. 4.1 Flow diagram showing the various forms of foodborne diseases

water, microorganisms pass through the acidic stomach environment and move to the intestine, where they attach and colonize using adhesins, fimbriae (pili) or colonization factor antigens (CFA), curli, and flagella. In some pathogens, the ability to form biofilm may aid in pathogen colonization in the intestinal tissues. Invasive pathogens can cross the epithelial barrier through phagocytic M cells located in the lymphoid follicles or Peyer's patches – a passive entry mechanism (Fig. 4.2). Invasive pathogens also can translocate across the intestinal barrier by an active invasion process through intracellular route and/or disrupted epithelial tight junction barrier through paracellular route. Some microorganisms cause local tissue damage and induce inflammation, while others spread to the lymphoid system including mesenteric lymph nodes (MLN) and other extraintestinal sites such as the liver, spleen, gall bladder, kidney, brain, and placenta. Foodborne infection can be *acute* or *chronic*. In acute infection, the onset of a disease is quick and lasts only for a short duration due to a rapid immunological clearance of the microorganism. In chronic infection, the disease is prolonged and the immune clearance is not effective.

Often, the prolonged infection is perpetuated by a strong immune response mounted by the host rather than the infective agent itself, such as seen in chronic shigellosis cases leading to chronic bacillary dysentery. Patients recovering from a foodborne infection may shed the organism for a prolonged period – observed in salmonellosis, in hepatitis A, and in *Norovirus* infection. Some foodborne infections may lead to chronic debilitating sequelae such as Reiter's syndrome, reactive arthritis, Guillain–Barré syndrome and Miller Fisher syndrome.

Infectious Dose

The infectious dose of a pathogen or toxin varies depending on the immunological health status of the host and the natural infectivity of the microorganism (Table 4.1). The infectious dose decreases if the pathogen is consumed with liquid food (milk, soup and beverages) that traverses the stomach rapidly or food (milk, cheese, etc.) that neutralizes the stomach acid. Persons with high gastric pH (achlorhydria) or patients undergoing antibiotic therapy for other ailments are also

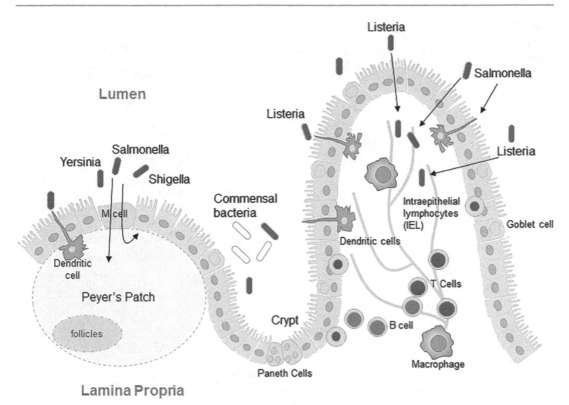

Fig. 4.2 Schematic of intestinal villus structure and bacterial crossing of the epithelial barrier and subcellular translocation (Adapted from Abreu 2010. Nat. Rev. Microbiol. 10, 131–144; and Ribet and Cossart 2015. Microbes Infect. 17, 173–183)

susceptible to foodborne infections because antibiotics reduce the natural microbiota load in the intestine, which renders the host more susceptible to foodborne infections.

Colonization and Adhesion Factors

The gastrointestinal tract, starting with the mouth, esophagus, stomach, small intestine, and large intestine, is very dynamic and active as the epithelial cell surface is constantly washed by mucus and fluids. The GI tract uses peristaltic movement, mucus, and epithelial ciliary sweeping action to expel pathogens. In addition, the bile, proteolytic enzymes, sIgA, and resident microbiota and their metabolic by-products (short-chain fatty acids: acetic, propionic, butyric acids) can also prevent pathogens from colonization. Foodborne pathogens that are involved in either toxicoinfection or infection must colonize the intestine or adhere to the mucus or epithelial surface before exerting their pathogenic actions. Several adhesion or colonization factors are used by pathogens, and general description of each factor is presented below.

Pili, Fimbriae, and Flagella
Pili, also known as fimbriae, are rodlike surface adhesin structures mostly found in Gram-negative bacteria but also present in Gram-positive bacteria such as *Lactobacillus rhamnosus* GG. Pili consist of pilin protein (20 kDa) and have a long helical–cylindrical structure. Pili are located on the surface of bacteria and have a specialized protein tip that helps the bacterium to attach to the host cell receptors composed of glycolipids, glycoproteins (α-D-galactopyranosyl-β-D-galactopyranoside), or other mannose-containing receptors or mucus. Bacteria growing inside the host constantly lose and reform pili. The host immune system produces antibodies against pilus

Table 4.1 Infectious dose and incubation periods of common foodborne pathogens.

Pathogens	Infectious dose	Incubation period
Bacteria		
Escherichia coli O157:H7 and other Shiga toxin-producing *E. coli* (STEC)	10–100 cfu	3–4 days (1–10 days)
Listeria monocytogenes	10^2–10^3 cfu	7–14 days or even longer
Salmonella enterica	1 to 10^9 cfu	6–24 h
Shigella spp.	10–100 cfu	12 h–7 days, but generally in 1–3 days
Vibrio cholerae/parahaemolyticus/vulnificus	10^4–10^{10} cfu	6 h–5 days
Staphylococcus aureus cells	10^5–10^8 cfu	–
Staphylococcal enterotoxin	1 ng/g of food	1–6 h
Bacillus cereus	10^5–10^8 cfu or spores/g	1–6 h (vomiting); 8–12 h (diarrhea)
Bacillus anthracis (inhalation anthrax)	8×10^3–10^4 spores	2–5 days
Clostridium botulinum neurotoxin	0.9–0.15 μg (i.v. or i.m. route) and 70 μg (oral route)	12–36 h; 2 h when large quantities are ingested
Clostridium perfringens	10^7–10^9 cfu	8–12 h
Campylobacter jejuni	5×10^2–10^4 cfu	1–7 days (Avg 24–48 h)
Yersinia enterocolitica	10^7–10^9 cfu	24–30 h and lasts for 2–3 days
Virus		
Norovirus	~10 particles	24–48 h
Hepatitis A	10–100 particles	15–45 days, average 28–30 days
Protozoa		
Entamoeba histolytica	>1000 cysts (as low as 1 cyst)	1–4 weeks
Giardia duodenalis (*lamblia*)	10–100 cysts	1–2 weeks
Cryptosporidium parvum	10 oocysts	2–10 days
Toxoplasma gondii	10 oocysts	2–3 days or longer

Table 4.2 Pili/fimbriae of Gram-negative bacteria.

Types of pili	Classification	Property	Present in
Type I	CFA/II, CFA/IV	Flexible	*Escherichia coli, Haemophilus influenzae*; *Yersinia pestis*
Type IV	–	Rigid	*Pseudomonas, Vibrio,* enteropathogenic *E. coli* (EPEC), and *Neisseria meningitides*
P pili	CFA/I	Rigid	*E. coli, Haemophilus influenzae, Yersinia pestis*
BFP (bundle-forming pili)	CFA/III	Flexible	EPEC
Curli		Flexible	*Salmonella enterica, E. coli*

that can prevent bacterial adhesion to the host cells by physical hindrance or neutralization.

There are four types of pili: type I, type IV, P pili, and bundle-forming pili (BFP) (Table 4.2). Type 1 pili are also known as colonization fimbriae antigen (CFA: CFA/II, CFA/IV), and they are flexible. P pili (CFA/I) and type IV are rigid, whereas BFP (CFA/III) is flexible. Type I and P pili are found in *E. coli, Haemophilus influenzae*, and *Yersinia pestis*. ETEC fimbrial adhesins K88 (F4), K99 (F5), F41, and F17 attach to the host mucus and cause diarrhea in swine (piglet). Type IV pili are found in *Pseudomonas, Vibrio*, enteropathogenic *E. coli* (EPEC), and *Neisseria*. BFP is found in EPEC. Pili are encoded by *pap* gene in *E. coli* and the *pap* operon consists of *papD, papC, papA, papH, papE, papF, papG*, and *papK* genes.

Flagella consist of several thousands of subunits of flagellin protein and interact with mucus to aid in bacterial adhesion. EPEC, enterohemorrhagic *E. coli* (EHEC), *Clostridium difficile*, and *Campylobacter* spp. use flagella to adhere to the mucus layer of the intestine.

Curli

Curli is a thin-coiled fimbriae-like extracellular protein fiber of 6–12 nm wide, produced by bacteria from *Enterobacteriaceae* family. They were first discovered in the 1980s from *E. coli* that were responsible for bovine mastitis. Since then it has isolated from *Salmonella* spp. and other pathogenic *E. coli*, including serovar O157:H7. Curli is considered an amyloid fiber and such fiber formation is typical for many human diseases, such as Alzheimer's, Huntington's, and prion diseases. The major curli subunit is CsgA (15 kDa), which is capable of self-polymerization forming a β-sheet-rich amyloid fiber. In bacteria, curli promote bacterial adhesion to the host cell extracellular matrices such as fibronectin, laminin, and type 1 collagen. It also promotes cell aggregation and biofilm formation. Curli can also induce a strong inflammatory response in the host. It is typically produced under stressful conditions, such as suboptimal growth temperature (below 30 °C), osmolarity and nutrient-limiting environments, and in the stationary phase of growth.

Adhesion Proteins

Adhesion proteins promote tighter binding of pathogens to the host cells and are important for attachment and invasion (Table 4.3). For example, *Yersinia enterocolitica* YadA adhesin binds to fibronectin, laminin, collagen, and β1-integrin on host epithelial cells while *Yersinia* invasin binds to β1-integrin. In EPEC and EHEC, the attachment and effacement (EAE) protein, intimin binds to the translocated intimin receptor (TIR) and aids in the formation of a pedestal. *Campylobacter* spp. use CadF to interact with fibronectin-binding protein on the host cells. In *Listeria monocytogenes*, surface protein internalin A (InlA) binds to E-cadherin located in the adherens junction of the epithelial cell junction; internalin B (InlB) binds to the tyrosine kinase Met receptor, a hepatocyte growth factor; LAP (*Listeria* adhesion protein) binds to the heat-shock protein 60 (Hsp60); and fibronectin-binding protein binds to host fibronectin. Other factors such as autolysin amidase (AMI) and LTA (lipoteichoic acid) are also involved in *Listeria* adhesion. *Vibrio cholerae* uses toxin-coregulated

Table 4.3 Adhesion factors and corresponding host receptors for select pathogens

Pathogens	Adhesion factors	Host receptor
Listeria monocytogenes	Internalin A (80 kDa)	E-cadherin
	Internalin B (63 kDa)	c-Met, gC1q-R/p32
	Vip (virulence invasion protein) (90 kDa)	Gp90
	LAP (*Listeria* adhesion protein) (104 kDa)	Hsp60
Campylobacter spp.	CadF (37 kDa)	Fibronectin
Arcobacter	Hemagglutinin (20 kDa)	Glycan receptor
Enteropathogenic and enterohemorrhagic *E. coli* (EPEC, EHEC)	Intimin (94 kDa)	Translocated intimin receptor (TIR)
Yersinia enterocolitica	YadA (160–240 kDa)	Collagen/fibronectin/laminin/β1-integrin
	Invasin (92 kDa)	β1-integrin
Staphylococcus aureus	Fibronectin-binding protein (FnBP)	Fibronectin
Vibrio cholerae	Toxin-coregulated pili (TCP)	Glycoprotein
	Mannose–fucose-resistant cell-associated hemagglutinin (MSHA), mannose-sensitive hemagglutinin (MSHA)	Glycoprotein?
Influenza virus	Hemagglutinin (HA)	Sialic acids attached to galactose through α2,3-linkage (bird) or α2,6-linkage (human)
Norovirus	Viral capsid protein (VP1)	Histo-blood group antigen (HBGA)
Hepatitis A virus	Capsid protein	Hepatitis A virus cell receptor 1 (HAVCR1)

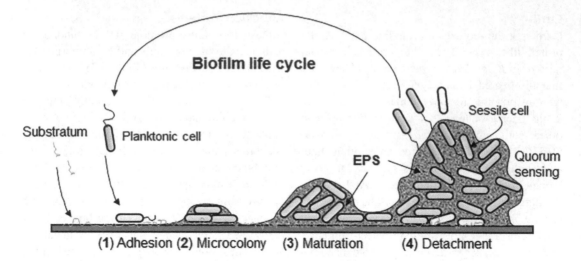

Fig. 4.3 Biofilm life cycle (Adapted from Ray and Bhunia, Fundamental Food Microbiology, 5th edition, 2014, CRC Press)

pili (TCP), mannose-sensitive hemagglutinin (MSHA), fucose-sensitive hemagglutinin (FHA), and/or mannose–fucose-resistant hemagglutinin (MFRHA) for adhesion and colonization in the intestine. Hemagglutinin (HA) expressed on the viral envelope helps avian influenza virus (H5N1) to bind to sialic acids attached to galactose through α2,3-linkage on the host cell surface, while human influenza virus binds to 2,6-linked sialic acid located in mucosal epithelial cells in the respiratory tract.

Histo-blood Group Antigen Adhesins

Campylobacter jejuni, Helicobacter pylori, and Norovirus bind to histo-blood group antigens (HBGA) expressed on the gut mucosa and on epithelial cells. *H. pylori* use blood group-binding adhesin (BabA) and sialic acid-binding adhesin (SabA) to bind to mucus in the stomach, in which the bacterium finds its niche to cause gastric ulcer and cancer. *C. jejuni* is highly motile and uses flagella to colonize the mucosa; however, HBGA lectins are also implicated in adhesion to mucus. Norovirus interacts with HBGA found on the epithelial cells in the gastrointestinal tract and causes severe gastroenteritis. The carboxyl-terminal protruding (P1) domain present in the Norovirus capsid is directly involved in binding to HBGA.

Biofilm Formation

Biofilm formation facilitates bacterial colonization in the host tissues (biotic surface) as well as on inert food processing or medical equipment (abiotic surface). Some bacteria produce extracellular polymeric substances (EPS) consisting of polysaccharides, proteins, phospholipids, teichoic acids, nucleic acids, and other polymeric substances. The polysaccharide is made of poly-β-1,6-N-acetylglucosamine. Bacteria can form dense, multi-organism layers on food contact surfaces, in the mouth, teeth, or intestinal lining. The first layer of the biofilm attaches directly to the host cell surface. Additional layers attach to the basal layer with a polysaccharide matrix. The EPS provides protection against biocides, desiccation, antibiotics, and toxins.

Biofilm formation is comprised of several stages: (1) attachment, (2) microcolony formation, (3) maturation, and (4) detachment or dispersion (Fig. 4.3). Biofilm can be made of a single organism or multi-organisms. In the biofilm, microorganisms express fimbriae, curli, flagella, adhesion proteins, and capsules to attach firmly to the surface. Microorganisms grow in close proximity and can communicate inter- and intraspecies and are called "quorum sensing (QS)", as described below. As the microcolony continues to grow, cells accumulate and form a matured biofilm with a three-dimensional scaffold. Loose cells are then sloughed off from the matured biofilm and convert into planktonic cells, which again attach to a new surface preconditioned by food particles or substrates completing the life cycle of a biofilm. In addition to exopolysaccharides, proteins are found to play a critical role in biofilm formation. A biofilm-associated protein (Bap) and the Bap family of proteins are involved in biofilm formation in many pathogens: *Staphylococcus aureus, S. epidermidis, Enterococcus faecalis, Streptococcus pyogenes, Salmonella enterica* serovar Typhimurium, and *E. coli.* In addition, type IV pili in *Vibrio* promotes biofilm formation while curli in *Salmonella* and *E. coli.*

Biofilms are important in human diseases, including dental plaque caused by *Streptococcus mutans*, mastitis by *S. aureus*, intestinal colonization by *Salmonella*, endocarditis by *Enterococcus faecalis*, and lung infection by *Pseudomonas aeruginosa* in cystic fibrosis patients.

Quorum Sensing

Bacterial quorum sensing (QS) plays a major role during the bacterial response to a changing environment in food or in the host. QS is known as a bacterial cell-to-cell communication system, mediated by small signaling molecules and is identified in both Gram-negative and Gram-positive bacteria. Microbes use both interspecies and intraspecies communication system to develop strategies for the population to adapt to the harsh environment encountered during colonization in the host or in other environments. QS is cell density-dependent: when bacteria reach to a certain cell mass, bacteria secrete small diffusible signaling molecules known as autoinducer (AI). Under stressful conditions such as during starvation and high temperature, in the presence of antimicrobial agents, acids, or other microbes, bacteria secrete AI. AI binds to the receptor on the cell surface and activates gene expression to alter behavior, i.e., adaptation to the harsh environment, tolerance to desiccation, and morphological transformation leading to biofilm formation or production of spores. AI also regulates gene expression to facilitate microbial survival, growth, density, virulence, and resistance to antimicrobials.

There are four major categories of AIs: (1) Autoinducer-1 (AI-1), for example, N-acyl homoserine lactone (AHL), alkyl quinolone, α-hydroxyketones, and a diffusible signal factor, is produced by Gram-negative bacteria and is used for intraspecies communication. (2) Autoinducer-2 (AI-2) is produced by both Gram-positive and Gram-negative bacteria and is used as a universal signaling molecule. (3) Autoinducer-3 (AI-3) is produced by EHEC during infection. (4) Autoinducing peptides (AIPs) are produced and used by Gram-positive bacteria.

Invasion and Intracellular Residence

Some bacteria maintain intracellular lifestyle and they enter cells by two mechanisms: (1) passive entry into the host through uptake by natural phagocytic cells such as M (microfold) cells in the intestinal lining and macrophages and dendritic cells and (2) active invasion via induced phagocytosis in the nonprofessional phagocytic cells such as epithelial, endothelial, and fibroblast cells. M-cell-mediated entrance is used by many pathogens including *Shigella* spp., *Listeria monocytogenes*, and *Salmonella enterica*. Forced or induced phagocytosis is used by *L. monocytogenes, Salmonella, Yersinia*, and *Shigella*, which use invasin proteins to promote bacterial uptake by the host cells. *Listeria* and *Yersinia* also use caveolae and clathrin-coated vesicles to gain entrance. Some pathogens such as uropathogenic and diarrheagenic *E. coli* and viruses such as hepatitis A virus (HAV) and foot-and-mouth disease (FMD) virus use sphingolipid–cholesterol raft that forms invaginations during pinocytosis for host cell entry. *Toxoplasma gondii* and *Cryptosporidium parvum* exhibit gliding motility (1–10 μm/s and counterclockwise), which helps parasite entry into the enterocytes. The gliding motility is an active process for the parasite, but the host cell does not play an active role in the invasion, as invasion does not alter host cell actin cytoskeleton or induce phosphorylation of tyrosine residues.

To enter host cells, the virus first binds to the host cell receptor and is then taken up by a process known as endocytosis (see Chap. 6). The virus may remain trapped inside the endocytic vesicle until a signal is triggered for the nucleic acid release into the cytoplasm. In the case of enveloped virus, membrane fusion is needed to penetrate a cell. Viral envelope (membrane) fuses with the cytoplasmic membrane, and the capsid containing the viral genome is released into the cytoplasm. In non-enveloped virus, viral genome is incorporated into a protein shell in the cytoplasm and then is assembled, and viruses are released by cell lysis.

Natural Phagocytosis

M cells are naturally phagocytic and are present throughout the intestine in follicle-associated lymphoid tissue such as Peyer's patch (Fig. 4.2). Their primary function is to transport intact particles across the epithelial membrane without degrading or processing them. Some pathogens such as *Salmonella*, *Shigella*, and *Listeria* use M cells to reach the subepithelial layer. In contrast, professional phagocytic cells such as macrophages, dendritic cells, and neutrophils spontaneously phagocytose bacteria. Intracellular pathogens may survive inside the phagocytic cells and resist killing by using specialized virulence factors such as hemolysin, superoxide dismutase (SOD), and catalase. Hemolysin forms pore in the phagosome membrane, SOD inactivates toxic oxygen radicals, and the catalase breaks down hydrogen peroxide. *Salmonella*, *Legionella*, *Mycobacterium*, and *Brucella* block or alter the maturation of phagosome, therefore allowing these pathogens to survive inside the phagosome. While *Listeria monocytogenes* and *Shigella* spp. escape the phagosome with the help of hemolysin before phagolysosomal fusion takes place. Bacteria multiply in the cytoplasm, induce actin polymerization for intracellular motility, and infect neighboring cells. Both pathogens also avoid autophagosomal degradation (autophagy) using specialized virulence factors. Pathogens such as *Shigella*, *Yersinia*, and *Salmonella* may induce apoptosis in macrophages/dendritic cells and persist in subepithelial regions in the intestine.

Invasin-Mediated Induced Phagocytosis

Intracellular pathogens may provoke phagocytic ingestion or induced phagocytosis using invasin molecules after interaction with the cellular receptors (Table 4.3). Induced phagocytosis involves coordinated interaction of sophisticated quorum sensing and signals transduction mechanisms. During this process, cytoskeletal proteins undergo rearrangement and accommodate bacterial entry. Induction of actin polymerization is a crucial event, which aids in cytoskeletal rearrangements to accommodate bacterial entry. Two distinct mechanisms are identified for induced phagocytosis: (1) zipper-like mechanism and (2) trigger mechanism. *Yersinia pseudotuberculosis* and *Listeria monocytogenes* use the zipper mechanism, while *Salmonella enterica* and *Shigella* spp. use the trigger mechanism (Fig. 4.4).

In the zipper-like mechanism, a relatively modest cytoskeletal rearrangement and membrane extension occur following binding of bacterial invasin proteins to a host cell receptor. Ligand–receptor interaction initiates a signaling cascade, which in turn promote recruitment of effector molecules such as actin to form a cuplike structure that accommodates the bacterium inside the cup to complete invasion-mediated induced phagocytosis. The zipper-like mechanism is a three-step process: (1) contact and adherence, (2) phagocytic cup formation, and (3) phagocytic cup closure. (1) Contact and adherence happen independent of the actin cytoskeletal rearrangement but involve ligand and receptor interaction. For example, *L. monocytogenes* uses InlA to interact with E-cadherin, located in the adherens junction of the polarized epithelial cell; *L. monocytogenes* also uses InlB to interact with c-Met, a transmembrane receptor tyrosine kinase. *Yersinia* invasin protein interacts with β1-integrin located on the basolateral side of the polarized epithelial cells. (2) The phagocytic cup is formed by membrane extension, which is initiated by the signaling events and induction of actin polymerization through the Arp2/3 pathway. (3) Phagocytic cups are then sealed off trapping the bacterium inside, and this process is facilitated by "membrane retraction" and "actin depolymerization."

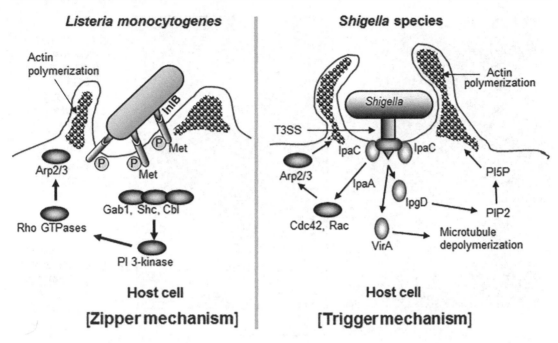

Fig. 4.4 Schematic drawing showing *Listeria monocytogenes*-induced bacterial entry into the cell by zipper mechanism and *Shigella* entry by a trigger mechanism. In zipper mechanism, listerial InlB interacts with Met receptor, which autophosphorylates and recruits protein adapters: Gab1, Cbl, and Shc. These proteins activate phosphatidylinositol 3-kinase (PI 3 kinase) and small Rho GTPase-kinase, Rac, which in turn promotes actin polymerization via Arp2/3. In zipper mechanism, *Shigella* bypass the cell adhesion to the receptor but inject effector proteins by type III secretion (T3SS) apparatus directly into host cell cytosol and induces massive actin polymerization. *Shigella* injects IpaC, VirA, IpgD and more proteins into the host cytosol. IpaC activates small GTPases, Cd42, and Rac and promotes actin polymerization by Arp2/3. VirA inhibits microtubule formation, and IpgD hydrolyzes phosphatidylinositol (4,5) biphosphate (PIP2) to phosphatidylinositol (5) phosphate (PIP) and disconnects actin from the membrane (Adapted and redrawn from Veiga and Cossart 2006. Trends Cell Biol. 16, 499–504)

The trigger mechanism comprises of four steps: (1) pre-interaction stage, (2) interaction stage, (3) macropinocytic pocket formation, and (4) actin depolymerization and closure of the macropinocytic pocket. In the pre-interaction stage, the bacterium synthesizes necessary proteins in preparation for initiating an infection. In the interaction stage, the bacterium injects dedicated bacterial effector proteins through the type III secretion system (T3SS). In *Salmonella*, the effector proteins are SipB and SipC and in *Shigella*, IpaB and IpaC. During formation of the macropinocytic pocket, bacteria initiate signaling events that trigger massive actin polymerization and extension to form the entry foci, known as membrane ruffling. In the final stage, the actin depolymerization occurs and invasion proceeds.

Once the bacterium enters the cell, it employs adaptation strategies to promote intracellular survival, effectively avoiding lysosomal enzymes, antibacterial peptides, low pH, reactive oxygen radicals, and low nutrient concentrations. Some bacteria also induce actin polymerization for inter- and intracellular movement. When the bacterium reaches the plasma membrane of the adjacent cell, it forms a protrusion, which is endocytosed by the neighboring cell, and traps the bacterium inside a vacuole. The bacterium then lyses the double membrane of the vacuole using hemolysin and phospholipase and continues the infection process. Some bacteria use a sugar uptake strategy to garner energy from the host cells. There are several advantages of intracellular residence: avoidance of killing by pha-

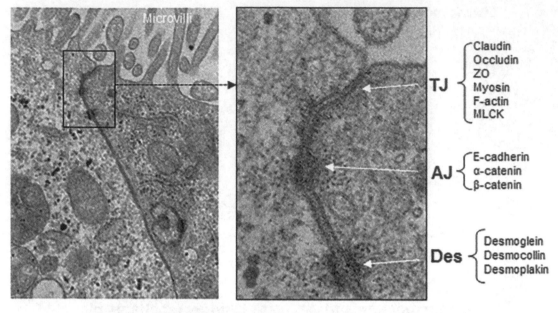

Polarized Caco-2 cells

Fig. 4.5 Transmission electron microscopy picture of epithelial cell–cell junction in polarized Caco-2 cells. TJ, tight junction; AJ, adherens junction; Des, desmososme (Image courtesy of Rishi Drolia and Arun Bhunia)

golysosomal degradative enzymes, access to an abundance of nutrients in the cytoplasm, and protection from antibiotics, antibodies, and complement-mediated cell lysis.

Pathogen Translocation by Epithelial Barrier Disruption

Some pathogens disrupt epithelial tight junction integrity as a strategy to cross the epithelial barrier in the gut. Epithelial barrier architecture consists of four regions: tight junction (TJ), adherens junction (AJ), desmosome (Des), and gap junction (Fig. 4.5). In TJ, claudin and occludin are the major membrane proteins that are stabilized by zona occludin (ZO), F-actin, myosin, and the myosin light-chain kinase (MLCK). In the AJ, E-cadherin is the major protein accompanied by two other structural proteins: α-catenin and β-catenin. In the desmosome, desmoglein, desmocollin, and desmoplakin are the dominant proteins. The cytoskeletal proteins are highly regulated permitting nutrient and small molecule

translocation from the lumen to the submucosal location. However, the enteric pathogens have developed diverse fascinating strategies to disrupt the epithelial barrier for the localized or systemic spread. Pathogens induce the production of inflammatory cytokines: TNF-α and IL-6, through activation of NF-κB, a central regulator of the epithelial innate immune response. The relationship between the cytokine (TNF-α and IL-1β)-induced NF-κB activation and the TJ barrier disruption is well documented (Fig. 4.6). Many enteroinvasive pathogens modulate NF-κB signaling to establish a successful infection. EHEC, EPEC, *Helicobacter pylori*, *Campylobacter jejuni*, and *Yersinia pseudotuberculosis* activate NF-κB, and the resulting inflammatory cytokines dysregulate the junctional proteins to allow bacterial passage across the epithelial barrier. *Yersinia pseudotuberculosis* secretes *Yersinia* outer protein J (YopJ) to induce IL-1β production in the gut to disrupt the intestinal epithelial barrier – both paracellular and transcellular permeabilities through the activation of MLCK. *Helicobacter pylori* also induces the

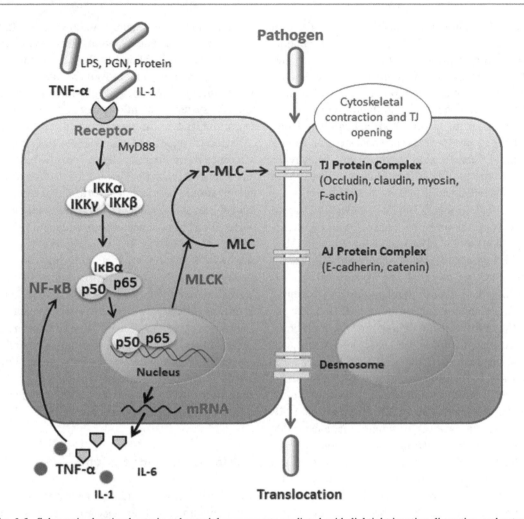

Fig. 4.6 Schematic showing bacteria or bacterial components mediated epithelial tight junction disruption and enteric bacterial paracellular translocation

paracellular permeability by generating cell-signaling events that counteract the normal function of protein kinase C (PKC). *Salmonella enterica* serovar Typhimurium induces increased paracellular permeability associated with the mislocalization of occludin and ZO-1. *Campylobacter jejuni* also increases paracellular permeability by activating phosphoinositide 3-kinase (PI3-K). *Clostridium difficile* toxin B increases PKC-dependent RhoA glycosylation and actin depolymerization to increase paracellular permeability. *Shigella flexneri* also increases paracellular permeability by regulating tight junction-associated proteins. In *Listeria monocytogenes*, LAP disrupts intestinal barrier functions

by activating NF-κB and MLCK to facilitate bacterial paracellular translocation across the epithelium. *Streptococcus pneumonia* and *Haemophilus influenzae* exploit toll-like receptor (TLR)-mediated downregulation of TJ protein to facilitate translocation across the epithelium.

Iron Acquisition

Iron (Fe^{3+}) is essential for metabolic process, survival, and growth in all living organisms. It is required for electron transport, metalloenzyme activity (oxidoreductases, cytochromes, and non-heme oxidases), and oxygen transport (higher animals). Pathogenic microbes must acquire iron from the host during infection. However, a majority of iron is present as insoluble ferric oxide/hydroxide in the environment. In the human body, free iron concentration is very low because iron is bound to proteins in the form of lactoferrin, transferrin, ferritin, and hemoglobin. Three different ways bacteria can sequester iron from a host: (1) use of siderophore (from the Greek: "iron carriers"), (2) direct binding of bacteria to the host cells, and (3) killing of the host cells.

(1) Siderophores are low molecular weight organic compounds that are produced by microorganisms and chelate iron with very high affinity. Examples of siderophores are aerobactin, enterobactin, catechol, hydroxamate, and so forth. After binding of siderophore to the ferric iron, the siderophore–iron complex is taken up by the cell following interaction with the siderophore receptor located on the bacterial cell surface. Inside the cell, the complex is degraded and the iron is released. Siderophores are generally pathogen-specific: *Salmonella* produces enterobactin, a cyclic trimer of dihydroxybenzoic acid; *Yersinia* spp. produce yersiniabactin, a catechol-type siderophore; *Vibrio cholerae* produces vibriobactin; and *E. coli* produces aerobactin. (2) Pathogenic microbes also sequester iron by directly binding to the host cells that carry iron on their surface. (3) Some bacteria produce exotoxin such as hemolysin that lyses the iron (heme or ferritin)-bearing cells, such as the red blood cell (RBC), when the iron level is low. The bacterium scavenges the iron from lysed cells. Many pathogens produce hemolysin such as listeriolysin O (LLO) by *Listeria monocytogenes*, α-lysine by *Staphylococcus aureus*, perfringolysin O (PFO) by *Clostridium perfringens* O, streptolysin O (SLO) by *Streptococcus pneumoniae*, and so forth.

Motility and Chemotaxis

Microbial motility is very important for survival and existence in nature and in a host. For enteric pathogens to colonize and persist in the gut, microbes must overcome many obstacles, since the intestine is constantly washed with fast-moving fluids, the ciliary action of microvilli, and the antimicrobials. Microbes display swimming, swarming, and gliding movement on surfaces. Motile microbes have flagella and have a complex sensory system that allows them to move by swimming and swarming motions in the direction of nutrients (sugars and amino acids) and oxygen. Microbes move directionally toward the mucosal membrane, which provides a greater chance for colonization. For example, *Vibrio* expresses long filamentous pili called TCP (toxin-coregulated pili), which promote bacterial motility and colonization of epithelial cells. *Campylobacter jejuni* also uses flagella to colonize and invade the gut epithelial cells. *Salmonella*, *Clostridium*, *Yersinia*, *Serratia*, *Proteus*, and *Escherichia* use flagella for swarming motility and for surface colonization. Quorum-sensing molecules or chemosensors facilitate bacterial colonization through the action of swarming movement and by forming biofilms. In addition, peptides and amino acids act as a chemotactic factor for *Proteus* sp. while polysaccharides for *Salmonella* Typhimurium and *E. coli*.

Evasion of Immune System

Microbial ability to evade the host immune system is an important strategy for a pathogen to cause disease. Certain bacteria such as *Streptococcus pyogenes*, *S. pneumoniae*, *Bacillus* species, and *Yersinia pestis* express capsules, which exert anti-phagocytic action. The capsule prevents serum protein B from binding to the complement protein, C3b, but helps protein H to bind to C3b forming the C3bH complex, which is easily degraded by protein I. As a result, no C3bBb, also known as C3 convertase, is formed. This prevents the production of C3b and the complement cascade from forming the membrane attack complex (MAC). As a result, the microbes are protected from complement-mediated killing. Sialic acid and the hyaluronic acid, components of the capsule, are also present in the mammalian cell membrane, thus avert binding of the C3b to bacterial capsule preventing macrophage recognition and phagocytosis. Binding of bacterial lipopolysaccharide to C3b also prevents the formation of C3 convertase and prevents the formation of MAC and complement-mediated killing. Intracellular pathogens also evade the immune system and survive phagocytosis by producing hemolysin, superoxide dismutase that destroys "O" radical, and catalase that breaks down H_2O_2.

Some pathogens evade the host antibody response by shifting the antigenic structure, such as structural variations in pili as well as in other surface proteins through mutation in corresponding genes. The antigenic variation disguises the pathogen and fools the immune system from recognition. In addition, some pathogens evade the immune system by shrouding themselves with the host proteins. For example, *Streptococcus* species cover themselves with host fibronectin, and *Staphylococcus aureus* and *Streptococcus pyogenes* coat themselves with host IgG using the surface-expressed protein A and protein G, respectively. Protein A and protein G bind to the Fc part (as opposed to the antigen-binding Fab part) of the IgG, therefore preventing macrophages from recognizing the antigen–antibody complex. Many pathogens such as *Streptococcus pneumoniae*, *Haemophilus influenzae*, *Neisseria gonorrhoeae*, and *Neisseria meningitides* produce immunoglobulin (sIgA)-specific proteases to cleave IgA in the hinge region to make it ineffective for antibody-dependent inactivation.

Intoxication

Ingestion of preformed exotoxins such as staphylococcal enterotoxin, botulinum toxin, *Bacillus cereus* emetic toxin, mycotoxin, and seafood toxins cause food poisoning or intoxication (Table 4.4). Bacteria present in foods grow under a favorable condition and produce toxins. Following ingestion of food, toxins are absorbed through the gastrointestinal epithelial lining and cause local tissue damage, induce inflammation, and promote vomiting and diarrhea. In some cases, toxins are disseminated through the blood or lymphatics to distant organs or tissues such as the liver, kidney, or peripheral or central nervous system where they can cause damage. In most intoxication cases, the microbial cells in the food matrix transit through the digestive system without causing any harm. The general mechanism of pathogenesis of toxins involved in intoxication is described below.

Toxicoinfection

Some bacteria are responsible for causing toxicoinfection, which happens when the ingested bacteria first colonize the mucosal surface and then produce exotoxins in the GI tract. Similar to intoxication, the toxins induce toxic effects on the local cells or tissues, and in some cases, toxins enter the bloodstream and cause disease. Toxicoinfection-causing organisms are *V. cholerae* that produce cholera toxin, enterotoxigenic *E. coli* (ETEC) that produce heat-labile (LT) and heat-stable (ST) toxins, and *Clostridium perfringens* that produce clostridium perfringens enterotoxin (CPE). In some cases, toxins kill polymorphonuclear leukocytes (PMNL) and aid bacterial growth and spread as seen in patients suffering from gas gangrene and myonecrosis caused by *Clostridium* spp. In addition, *Clostridium* spp. produce hydrolytic enzymes such as lecithinase to break down lecithin, hyaluronidase and protease to disrupt extracellular matrix and tissue structure, and DNase to reduce the viscosity of debris from dead host cells, to aid in bacterial propagation.

Toxins

Broadly, two types of toxins are produced by microbes: exotoxin and endotoxin. Exotoxins are usually excreted outside in the extracellular milieu through an active transport system or remained cell associated until they are released from the cell after lysis. Examples of exotoxins are a botulinum toxin, cholera toxin, Shiga toxin, diphtheria toxin, and so forth, while the endotoxins are part of the cell structure, such as lipopolysaccharide (LPS) in Gram-negative bacteria and peptidoglycan (PGN) and lipoteichoic acid (LTA) in Gram-positive bacteria. Some toxins are

Table 4.4 Bacterial toxins and their characteristics

Toxin type	Toxins	Host cell receptor	Producing bacteria	Mode of action	Target
Membrane damaging toxin	Hemolysin	Cholesterol	*E. coli*	Pore formation	Plasma membrane
	Listeriolysin O (LLO)	Cholesterol	*L. monocytogenes*	Pore formation	Cholesterol
	Perfringolysin O (PFO)		*C. perfringens*	Pore formation	Cholesterol
	α-toxin	Cholesterol	*S. aureus*	Pore formation	Plasma membrane
	Streptolysin O	Cholesterol	*S. pyogenes*	Pore formation	Cholesterol
Inhibit protein synthesis (A–B type)	Shiga toxin or Shiga-like toxin	Gb3 (globotriaosylceramide), Gb4 (globotetraosylceramide)	*Shigella* spp. *E. coli*	N-glycosidase	28S rRNA
	Diphtheria toxin	A growth factor receptor	*C. diphtheriae*	ADP-ribosylation	Elongation factor-2
	Anthrax protective antigen (PA)	ATR (anthrax toxin receptor)	*B. anthracis*	Translocates anthrax LF and EF	Clathrin–coated pit
Activate second messenger pathways (A–B type)	Heat-labile toxin (LT)	Ganglioside (GM1)	*E. coli*	ADP-ribosyltransferase	G-proteins
	Heat-stable toxin (ST): STa and STb	STa: Guanylate cyclase C STb: Sulfatide (Glycosphingolipid)	*E. coli*	Stimulates guanylate cyclase	Guanylate cyclase
	Cholera toxin	Ganglioside (GM1)	*V. cholerae*	ADP-ribosyltransferase	G-protein
Activate immune response	Enterotoxins, toxic shock toxins	MHC class II	*S. aureus*	Superantigen	TCR and MHC II
Protease action	Lethal factor (LF)	MAPK	*B. anthracis*	Metalloprotease	MAPK1 MAPK2
	Edema factor (EF)	ATP	*B. anthracis*	Adenylate cyclase	ATP
	Botulinum neurotoxin	Ganglioside	*C. botulinum*	Zinc metalloprotease	Synaptobrevin, SNAP-25, syntaxin
Apoptosis-inducing toxins	IpaB	Membrane	*Shigella* spp.	Apoptosis	
	LLO	Cholesterol	*L. monocytogenes*	Apoptosis	

Adapted and modified from Schmitt et al. 1999. Emerg. Infect. Dis. 5, 224–234

Fig. 4.7 Diagram showing the mechanism of action of an A–B-type toxin

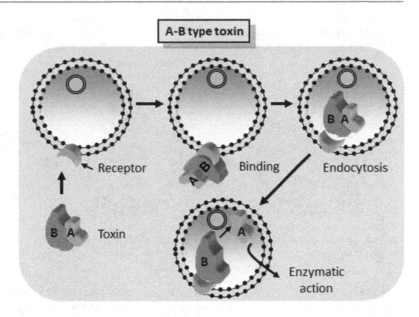

designated enterotoxins such as cholera toxin and *E. coli* LT toxin because they are responsible for gastroenteritis. Some toxins are also called cytotoxins because these are either cell- or tissue-specific: neurotoxin affects nerve cell, leukotoxin attacks leukocyte, hepatotoxin injures liver cell, and cardiotoxin damages cardiac tissue. Toxin designation is occasionally based on the bacterial species that produces them: cholera toxin is produced by *Vibrio cholerae*, Shiga toxin by *Shigella* species or Shiga-toxigenic *E. coli*, diphtheria toxin by *Corynebacterium diphtheriae*, tetanus toxin by *Clostridium tetani*, and botulinum toxin by *Clostridium botulinum*. Sometimes the toxin is named based on the action it exerts such as adenylate cyclase produced by *Bordetella pertussis* and lecithinase by *Clostridium perfringens* and *Listeria monocytogenes*. Toxins may also be designated by a letter such as staphylococcal enterotoxin A (SEA) or SEB or SEC produced by *Staphylococcus aureus*.

Structure and Function of Exotoxins

There are six types of exotoxins, and they are grouped based on the structure and the mode of action they exert on the host (Table 4.4; Fig. 4.7): (1) A–B-type toxin is a single protein or an oligomeric protein complex organized with A and B domain structure–function organization. The A domain has the enzymatic or catalytic activity, while the B domain has the receptor binding function and has the tropism for a specific host cell. The sequence of events for A–B-type toxins is binding to the receptor, internalization, membrane translocation, and enzymatic activity and target modifications (Fig. 4.7). Examples of A–B-type toxins are Shiga toxin, cholera toxin, botulinum toxin, and ST (heat satble) and LT (heat labile) toxins from *E. coli*. (2) Membrane-disrupting toxins either insert into the cell membrane to cause pore formation (e.g., hemolysin) or remove lipid head groups to destabilize the lipid bilayer membrane (e.g., phospholipases). (3) Toxin-mediated activation of secondary messenger pathway: Toxins like *E. coli* LT and ST or cholera toxin interfere with cellular signal transduction pathways. (4) Superantigens activate the immune system to produce excess amounts of proinflammatory cytokines, which exert their deleterious effects on the host cells, provoking toxic shock syndrome or septic shock. The examples of superantigens are staphylococcal enterotoxin (SE), exfoliative toxin, and toxic shock syndrome toxin (TSST). (5) *Proteases*: Some toxins such as botulinum neurotoxin inactivate metalloproteases (zinc metalloprotease) action, thereby interfering with the propagation of nerve impulse. (6) *Toxin-induced apoptosis or necrosis*: Some toxins kill cells by inducing programmed cell death (apoptosis or necrosis).

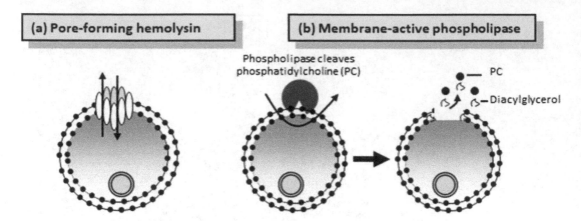

Fig. 4.8 Diagram showing the mechanism of action of pore-forming toxins such as (**a**) hemolysin and (**b**) phospholipase

A–B Toxins

The A–B-type toxin is a single protein or an oligomeric protein complex organized with A and B domain structure–function organization. The A domain has a catalytic activity, while the B domain has receptor binding function and has the tropism for specific host cells. The A and B domains are linked by a disulfide bond or by a noncovalent bond. B domain can be a single monomeric form (B) or an oligomeric form (B5). The B domain binds to a specific receptor such as a glycoprotein or glycolipid on the cell surface and binding is very specific. If the receptor is found only on the neuron, B domain will bind to the neuron only. After interaction of A–B toxin with the receptor, the entire complex is internalized by endocytosis. The B domain also has the capacity to help translocate the A domain across a lipid bilayer, either at the plasma membrane or within the endosome by forming a pore or channel. Through a cellular process, the A domain is separated from the B and exerts its enzymatic or catalytic activity; however, the function of A is not specific for a cell. If A is delivered experimentally into a cell, it will exert its function independent of the B domain. In some toxins, the A domain has ADP-ribosyltransferase activity. This removes ADP-ribosyl group from NAD^+ and attaches it to a host cell protein, such as elongation factor 2 (EF2) in a process called ADP-ribosylation of EF2. Modified EF2 is unable to help in host protein synthesis. For example, diph-

theria toxin (DT) and *Pseudomonas* exotoxin-A cause ADP-ribosylation of EF2 and consequently lead to cell death. In contrast, cholera toxin and *E. coli* LT cause ADP-ribosylation of the G-protein, a signal transduction protein, which controls the adenylate cyclase activity, which in turn increases the cAMP levels. The increased cAMP controls ion (Na^+, K^+, Cl^-) pumps and fluid movement in cells (see Chap. 14 and 18). Stx displays a different type of action – it has adenine glycohydrolase activity. It removes a single adenine residue from the 28S rRNA and blocks protein synthesis (see Chap. 14).

Membrane-Disrupting Toxins

A large group of bacterial pore-forming toxins (PFT) is reported. These are protein toxins and form pores in the membrane of bacteria, mammalian cells, and plant cells resulting in increased membrane permeability and ionic imbalance (Fig. 4.8). As water enters the cell, it swells and ruptures the cell. The soluble PFT toxin produced by bacteria inserts into the target membrane forming a stable multimeric structure. Depending on the type of multimeric structure the PFT forms, it is classified into α-PFT and β-PFT. α-PFT forms α-helix structure when forming a pore in the membrane while the β-PFT forms β-sheet during membrane pore formation. The best-described α-PFT is colicin, a bacteriocin (ColE) produced by *E. coli*, and it causes lysis of other *E. coli* cells.

β-PFTs are hemolysins that oligomerize and form stable multimeric structures in the mammalian cell membrane. The amphipathic β-hairpins on a monomeric subunit help insertion of the toxin into the hydrophobic membrane to form pores, and the size of pores ranges from 2 to 50 nm. The largest group of β-PFTs are cholesterol-dependent cytolysins (CDC) produced by Gram-positive bacteria such as listeriolysin O (LLO) by *L. monocytogenes*, streptolysin O (SLO) by *Streptococcus pyogenes*, pneumolysin O (PLO) by *S. pneumoniae*, and perfringolysin O (PFO) by *C. perfringens*. The CDC pore complex formation occurs before the protein insertion into the membrane. The steps include lateral diffusion of monomers with the membrane, pre-pore monomer oligomerization, pore formation, and insertion of the oligomer into the membrane forming a channel. Phospholipases are also considered hemolysins since they remove the charged phosphate head group $\left(PO_3^+\right)$ from the lipid (diacylglycerol) portion of the phospholipid, which destabilizes the membrane structure to cause cell lysis (Fig. 4.8).

Induction of Second Messenger Pathways

Some toxins alter signal transduction pathways, which are required for various cellular functions. For example, *E. coli* CNF (cytotoxic necrotizing factor) 1 or 2 and LT, *Clostridium botulinum* toxin and C3 toxin, and *Vibrio cholerae* cholera toxin inactivate or modify a small GTP-binding signal transduction protein, Rho, which regulates actin cytoskeleton formation. Destabilized epithelial cell architecture increases membrane permeability and fluid loss.

E. coli STa binds to the membrane guanylate cyclase C receptor on epithelial cells in the small intestine and colon. STa binding to guanylate cyclase C activates protein kinase G (PKG) and protein kinase C (PKC), increasing IP_3-mediated calcium to increase intracellular cyclic GMP (cGMP). Increased cGMP activates calcium ion channel and cystic fibrosis transmembrane regulator (CFTR) and increases the concentration of Cl^- ions in the extracellular space. As a result, electrolyte imbalance in the bowel leads to fluid accumulation within the lumen of the intestine and diarrhea.

Superantigens

Superantigens are a group of protein toxins that show unusual activation of the immune system resulting in septic shock. Superantigens are produced by *Staphylococcus aureus*, *Streptococcus* spp., *Mycoplasma arithiditis*, and *Yersinia pseudotuberculosis*. Staphylococcal enterotoxins (SEA, SEB, SEC, SEE, etc), exfoliative toxins, and toxic shock syndrome toxins (TSST) are well-studied superantigens. Unlike other protein antigens, superantigens are not processed (i.e., proteolytic digestion) by the antigen-presenting cells (APC). Instead, these toxins bind directly to the MHC class II molecules of APC and activate T cells carrying a T-cell receptor (TCR) composed of the Vβ chain (Fig. 4.9). Members of the superantigens possess different structures that have an affinity for MHC class II molecules. Activated T cells then produce a large quantity of IL-2, which in turn activates macrophages for increased production of IL-1, IL-6, TNF-α, and IFN-γ. IFN-γ also induces increased expression of class II molecules on professional phagocytic and nonprofessional phagocytic cells such as epithelial cells and subsequently enhanced presentation of superantigens to T cells. Production of large quantities of IL-2 and TNF-α also induces nausea, vomiting, malaise, fever, erythematosus lesion, and toxic shock.

Protease

Some toxins also possess enzymatic activities and cleave target proteins thus interfere with cellular function. For example, botulinum neurotoxin with zinc metalloprotease activity cleaves SNARE (soluble *N*-ethylmaleimide-sensitive factor attachment protein receptor) proteins, which consists of synaptobrevin, SNAP-25 (synaptosomal-associated protein-25), and syntaxin. The SNARE complex is responsible for acetylcholine release from the synaptic vesicle at the neuromuscular junction. As a result, neurotransmitter release is impaired causing paralysis (see Chap. 12 for details).

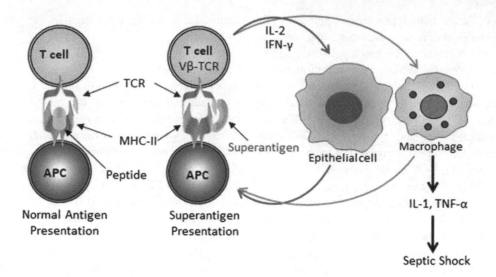

Fig. 4.9 Diagram showing the mechanism of action of a superantigen. APC, antigen-presenting cell, TCR, T-cell receptor, MHC, the major histocompatibility complex

Cells Death

Host cell death is a defensive strategy for the host against microbial infection thus helping maintain immune homeostasis and host defense. In response to an infection, the host cell may undergo programmed cell death for its own benefit: eliminate the pathogen during the early phase of infection without any inflammatory signals and activate dendritic cells to remove apoptotic bodies containing pathogen to induce MHC-mediated antigen presentation and protective immunity. Cell death also benefits the pathogen – successful exit from the infected cell to infect neighboring cells, access to nutrients, evade immune system, and persistence in a host. However, some intracellular pathogens have developed strategies to prevent host cell death during infection for their replication, escape, and dissemination of new host cells. Nevertheless, the cell death is a highly controlled and regulated event that plays a significant role during microbial pathogenesis. There are three types of cell deaths: apoptosis, necrosis, and pyroptosis (Fig. 4.10). Apoptosis is also known as programmed cell death referred to as cell "suicide." It is an active, programmed process of autonomous cellular destruction without evoking an inflammatory response. Caspase-1 is not involved

in this type of cell death. Necrosis is considered a passive event described as "accidental or uncontrolled death" inducing a strong inflammatory response. In recent years, however, programmed cell death by necrosis has been proposed as "programmed necrosis," "regulated necrosis," or "necroptosis." Pyroptosis is considered a non-apoptotic cell death but is dependent on caspase-1 enzyme activity and induction of inflammatory response. The cell death is characterized by the loss of membrane integrity, cytoplasmic content release, and inflammatory response (IL-1β and IL-18).

Apoptosis Microbial toxins induce apoptosis as part of their infection strategy. The dead cells are then removed by phagocytes, thus avoiding the onset of an inflammatory response. The typical signs of apoptosis (Fig. 4.11) are as follows:

(a) Cell shrinkage – condensation of cell cytoplasm and nucleus.
(b) Entrance of cells into the zeiosis stage – cell morphology alters and shrinkage continues.
(c) Chromatin condensation/margination – crescent-shaped nucleus localizes to the nuclear membrane.

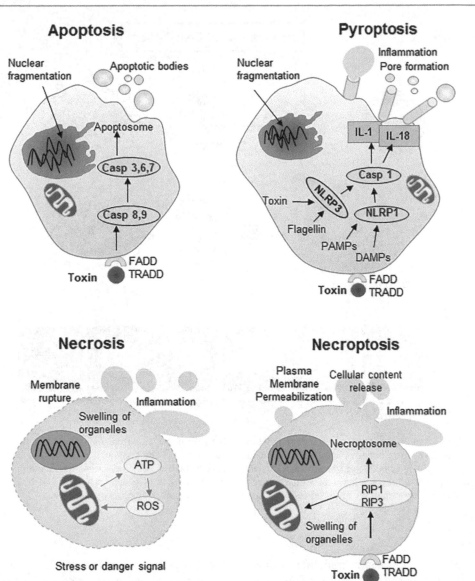

Fig. 4.10 Depiction of four types of cell death: apoptosis, pyroptosis, necrosis, and necroptosis. *Casp*, caspase; *FADD*, Fas-associated death domain; *TRADD*, TNF receptor-associated death domain; *ROS*, reactive oxygen species; *RIP*, receptor-interacting kinase; *PAMPs*, pathogen-associated molecular patterns; *DAMPs*, danger-associated molecular patterns; *NLRP*, NOD-like receptor protein

(d) DNA fragmentation – cysteine proteases (caspases) and nucleases are activated which cleave DNA in the factor of ~168 bp fragments resulting in the laddering of DNA bands, which can be detected by gel electrophoresis.

(e) Membrane blebbing of apoptotic bodies containing DNA fragments occurs which can be seen under a microscope.

(f) Clearance of apoptotic bodies by phagocytic cells – macrophages/dendritic cells engulf apoptotic cells/bodies to prevent inflammation.

The recognition of apoptotic cells by macrophages is mediated by binding to the phosphatidylserine (PS) displayed on the lipid bilayer membrane. Normally, PS is located in the inner

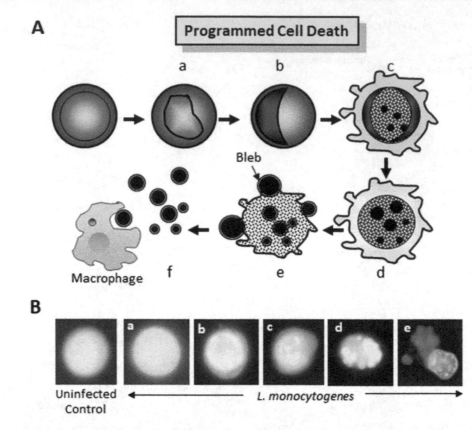

Fig. 4.11 (**A**) Schematics of different stages of apoptotic cell death and (**B**) fluorescence photographs of human hybridoma B cells infected with *Listeria monocytogenes* and stained with ethidium bromide and acridine orange at different stages of apoptotic cell death: (**a**) cell shrinkage; (**b**) zeiosis stage, cell morphology alters and shrinkage continues; (**c**) chromatin condensation/margination, crescent-shaped nucleus; (**d**) DNA fragmentation; (**e**) membrane blebbing of apoptotic bodies; and (**f**) clearance of apoptotic bodies by phagocytic cells

leaflet of the cytoplasmic lipid bilayer membrane; however, when cells are undergoing apoptotic cell death, the PS translocates from the inner leaflet of the membrane to the outer leaflet and becomes a target for macrophage-mediated clearance.

Apoptosis is a highly regulated process of programmed cell death. Apoptotic stimuli such as bacterial toxins activate death ligands FADD (Fas-associated death domain) and TRADD (TNF receptor-associated death domain), which in turn activate procaspase-8. Procaspase-8 phosphorylates Bcl-2-associated death domain (BAD), which consequently activates DNase and procaspase-9 (cysteine protease). Procaspase-8 and procaspase-9 are known as initiator caspase.

Procaspase-9 activates caspase-3, caspase-6, and caspase-7 (known as executioner caspase), which break down nuclear proteins histone, actin/lamins, and poly-ADP-ribose polymerase (PARP) leading to apoptosis. C-Myc and p53 proteins induce programmed cell death, while the Bcl-2 protein family, normally present in the malignant cells, suppresses apoptosis.

Shiga toxin and IpaB protein in *Shigella*, LLO in *L. monocytogenes*, cholera toxin in *V. cholerae*, and *Clostridium perfringens* enterotoxin (CPE) are some examples, which induce programmed cell death. In contrast, some toxins such as LLO and CPE at high concentrations induce oncosis, a form of necrosis by causing physical trauma that leads to cell swelling, lysis, random DNA shearing, and the release of toxic intracellular contents, which induce inflammation.

Necrosis Necrosis is called non-apoptotic or accidental cell death characterized by membrane lysis, organelle swelling, and the release of cellular contents, which induce strong inflammation. Necrosis occurs independently of caspase enzyme involvement in the cell death process. Necrosis can be triggered by reactive oxygen species (ROS) production or danger signals, such as depletion of ATP, lysosomal destabilization resulting from bacterial infection, or physical damage. However, some forms of necrotic deaths are genetically programmed and termed "programmed necrosis." Programmed necrosis is similar to the conventional necrosis process but involves activation through intrinsic and extrinsic factors resulting in the activation of signaling cascades leading to cell death referred to as "programmed necrosis," "regulated necrosis," or "necroptosis." It can be initiated by death ligands, TLR (toll-like receptor), FADD, TRADD, NLR (NOD-like receptor) and microbial infection. In programmed necrosis, the signal from death receptor such as TNF receptor activates the receptor-interacting kinase 1 (RIP1) and 3 (RIP3), which consequently activate downstream targets, leading to calcium and sodium influx, membrane lysis, and programmed necrosis. Programmed necrosis is a strong antiviral and antibacterial defense strategies and seen against vaccinia virus, reovirus, *Mycobacterium*, *Salmonella*, *Yersinia* infection, and *C. perfringens* toxins.

Pyroptosis Pyroptosis is a non-apoptotic cell death, but it depends on the activity of caspase-1 enzyme, which promotes the production of proinflammatory cytokines, IL-1β and IL-18, leading to a strong inflammatory response. In addition, caspase-4, caspase-5, and caspase-11 are also involved in pyroptosis. The pyroptosis executioner molecule, gasdermin D, serves as a substrate of caspase-1, caspase-4, caspase-5, and caspase-11, and the breakdown product gasdermin-N domain forms a pore in the plasma membrane to induce cell swelling and osmotic lysis. Pathogen-associated molecular patterns (PAMPs) or danger-associated molecular patterns (DAMPs) are recognized by inflammasomes (e.g., NLRP, NOD-like receptor protein), which activate caspase-1and trigger pyroptosis. Pyroptosis is characterized by DNA fragmentation, loss of membrane integrity, cytoplasmic content release, and inflammatory response.

Endotoxin

Bacterial structural components, such as LPS in Gram-negative bacteria and PGN, WTA and LTA in Gram-positive bacteria, are called endotoxin or pyrogen. They are released after cellular destruction or lysis. LPS consists of the hydrophobic lipid A component which is highly toxic (endotoxin), a nonrepeating "core" oligosaccharide, and a distal polysaccharide (or O-antigen). A high level of pyrogen in blood circulation is referred to septicemia and is responsible for the rise in body temperature (fever). LPS forms a complex with LPS-binding protein (LBP) and then interacts with the cellular membrane receptor, TLR4, and CD14 on macrophages/monocytes, and through MyD88-dependent signaling pathway, it activates NF-κB to produce inflammatory cytokines, IL-1β and TNF-α, which in turn cause the hypothalamus to release prostaglandins (Fig. 4.12). Increased prostaglandin level causes the body temperature to rise resulting in fever development. In principle, this defense strategy is designed to suppress bacterial growth. Aspirin blocks prostaglandin release and thus lowers the body temperature.

LPS at low concentration stimulates macrophages to produce IL-1 and acts as a polyclonal activator of B cells. LPS at a higher concentration can trigger a massive inflammatory response leading to septicemia and septic shock. LPS activates monocytes and macrophages to release IL-1, IL-6, IL-8, and TNF-α, which induce prostaglandin and leukotriene production, leading to the damage of endothelial cells, tissue injury, and disseminated (widespread) intravascular

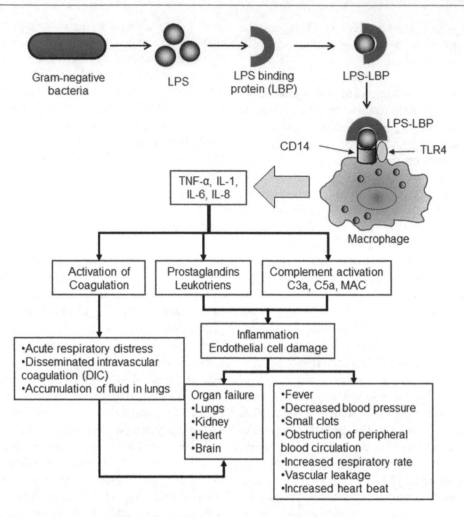

Fig. 4.12 Lipopolysaccharide (LPS)-mediated inflammation and septic shock

coagulation (DIC) allowing neutrophils, lymphocytes, and monocytes to stick to the endothelial surfaces (Fig. 4.12). As a result, blood pressure drops, the patient enters into a shock (referred to as septic shock), and death ensues. During septic shock, the circulatory system collapses, blood pressure drops, fever increases, heart rate increases, multiple organs fail, and death follows. Prolonged exposure to LPS invokes cachexia, characterized by wasting of the muscle and fat cells. LPS is also responsible for appetite suppression resulting in weight loss. Endotoxins also activate the complement pathway (alternative pathway) to form C3a and C5a, which also cause severe endothelial cell damage. A membrane attack complex (MAC) forms and causes cell lysis.

Genetic Regulation and Secretion Systems for Virulence Factors

Pathogenicity Islands

A genomic segment that carries a cluster of virulence genes is called pathogenicity island (PAI), a term that was first introduced by Prof. Jorg Hacker in the late 1980s. It was first discovered in an uropathogenic *E. coli* strain 536, and it was demonstrated that the deletion of the PAI resulted in a nonpathogenic phenotype. The PAI encodes genes for iron acquisition, adhesions, pore-forming toxins, proteins responsible for apoptosis, secondary messenger pathway toxins, superantigens, lipases, proteases, O antigens, and protein secretion systems such as type I, type II, type III, type IV, type V, type VI, and type VII secretion systems.

Specific characteristics of a PAI are as follows. (1) PAI carries one or more virulence genes. (2) It is present in pathogenic species but absent in nonpathogenic species. (3) The size range of PAI is 10–200 kb. (4) The G + C content of the PAI genome (40–60%) is lower than that of the core genome (25–75%). (5) PAI are located close to tRNA genes, indicating that the tRNA gene serves as an anchor point for insertion of foreign genes. Note: some bacteriophages also use tRNA as the insertion points in the host genome. (6) PAI may be unstable and deleted from the genome with a frequency higher than the normal rate of mutation. (7) PAIs are frequently associated with mobile genetic elements and are flanked by the direct repeats (DR). DRs are 16–20 bp long sequence with perfect or near-perfect sequence repetitions, which serve as integration sites for bacteriophages. PAIs also carry genes for integrases or transposes, probably acquired from the bacteriophage. The insertion sequence (IS) elements are also found in PAI. In addition, PAI can represent integrated plasmids, conjugative transposons, and bacteriophages.

Protein Secretion System

Bacteria have developed highly specialized macromolecular structures to secrete a wide range of molecules, including small molecules, proteins, and DNA. These secreted molecules perform a variety of tasks, including pathogenicity, cell envelope assembly, metabolism, and interaction with the host cells. In Gram-negative bacteria, the secretion is mediated by a different mechanism than that of the Gram-positive bacteria because of the presence of an outer membrane (OM) structure (see Chap. 2). In Gram-negative bacteria, the secreted molecules are translocated either to the outer membrane, to the extracellular milieu, or to the eukaryotic cells during host cell interaction. Therefore, different types of secretory machinery are required to achieve the same goal. The secretion systems in Gram-negative bacteria are classified as type I secretion system (T1SS), T2SS, T3SS, T4SS, T5SS, T6SS and T7SS (Fig. 4.13). Structural organization shows that the T1SS, T3SS, T4SS, and T6SS span from the inner cytoplasmic membrane (CM) to the outer membrane (OM) and deliver molecules directly from cytoplasm to the extracellular loca-

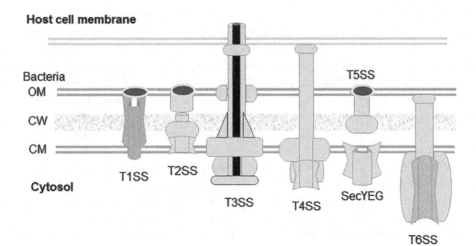

Fig. 4.13 The depiction of protein secretion system in Gram-negative bacteria (Adapted and redrawn from Costa et al. 2012. Nat. Rev. Microbiol. 13, 343–359). *OM*, outer membrane; *CW*, cell wall; *CM*, cytoplasmic membrane

tion, while T2SS and T5SS span only to the OM and protein secretion involves a two-step process, first from cytosol to the periplasmic space facilitated by SecYEG translocon or Tat translocon located in the cytoplasmic membrane and second from periplasmic space to the extracellular location using a dedicated OM spanning secretion system. T7SS is found only in *Mycobacterium* spp., since it has a specialized envelope similar to a Gram-negative bacterium.

Type I Secretion System

The type I secretion system (T1SS) has an assembly of an ATP-binding cassette (ABC) transporter protein located within the inner membrane, a periplasmic membrane, and an outer membrane. T1SS transports a variety of proteins of different sizes and functions from cytoplasm to extracellular milieu including hemolysins. T1SS is equivalent to the RND (resistance– nodulation– division) family of the multidrug efflux pump. RND pump secretes small molecules including antimicrobial compounds out of the cell, thus contributing to the antibiotic resistance.

Type II Secretion System

The type II secretion system (T2SS) consists of 12–15 components that are located in the inner membrane, the periplasm, and the outer membrane. It is responsible for secretion of folded proteins from the periplasm to the extracellular environment across the outer membrane of both pathogenic and nonpathogenic Gram-negative bacterial species that are important for bacterial survival and growth. T2SS transports proteins, such as pseudolysin, a hydrolyzing enzyme of *Pseudomonas aeruginosa*; pullulanase, an amylolytic enzyme of *Klebsiella pneumoniae*; and cholera toxin of *V. cholerae*. The T2SS substrates are transported as unfolded proteins from the cytoplasm to periplasm by using SecYEG translocon located in the cytoplasmic membrane. The equivalent of T2SS in Gram-positive bacteria is called the Sec (secretory) system (see below).

Type III Secretion System

The primary function of the type III secretion system (T3SS) is to translocate effector proteins directly from bacteria to the eukaryotic membrane or cytoplasm during interaction with the host cell. It is called a molecular syringe (injectisome) and its assembly requires the function of several genes. In *Salmonella enterica* serovar Typhimurium, it involves about 25 genes and a multiprotein (25 proteins) complex of 3.5 MDa. A typical T3SS apparatus has four major parts: inner rings, neck, outer rings, and a needle (Fig. 4.14). Assembly of the T3SS is similar to that of the flagellum assembly machinery. The T3SS is involved in pathogenesis. During intimate contact with the host cells, bacteria use the T3SS to inject virulence proteins into the host cell to modulate specific host cell function to promote colonization, invasion, actin-based intra- and intercellular motility, to induce apoptosis, and to interfere with intracellular transport process. The protein translocation is an ATPase-dependent process. T3SSs are present in various pathogenic Gram-negative bacterial genera such as *Salmonella*, *Shigella*, *Yersinia*, *Pseudomonas*, EPEC, and EHEC. Genes required for T3SS formation are encoded in plasmids or in PAI. Genes for T3SS in *Salmonella enterica* are located in SPI-1 and SPI-2, in the locus of enterocyte effacement (LEE) of EPEC and EHEC, and in PAI in a large plasmid of *Shigella*.

Type IV Secretion System

The type IV secretion system (T4SS) is similar to T3SS, but in addition to proteins, it can also translocate DNA or DNA–protein complex into the eukaryotic host. It is the most ubiquitous secretion system in nature and found in both Gram-positive and Gram-negative bacteria and in some Archaea. Since it translocates plasmids, it contributes to the spread of antibiotic resistance. It is also involved in bacterial pathogenesis and helps translocation of toxins and other effector molecules to sustain an intracellular lifestyle. T4SS plays an important function in the pathogenesis of human pathogens, *Bordetella pertussis*, *Legionella pneumophila*, *Brucella* spp., and *Helicobacter pylori*. The best-studied T4SS is a plant pathogen, *Agrobacterium tumefaciens*, where it mediates the translocation of DNA–protein into the plant cells to induce tumor

Fig. 4.14 A schematic model of the type III secretion system (T3SS) of *Salmonella enterica* (Adapted and redrawn from Galan and Wolf-Watz 2006. Nature 444, 567–573; and Galan and Collmer 1999. Science 284, 1322–1328)

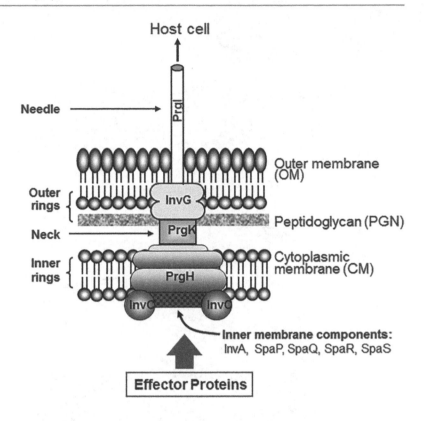

formation. T4SS is a complex structure composed of about 12 proteins and is encoded by Ti plasmid of *A. tumefaciens*.

Type V Secretion System

The type V secretion system (T5SS) is also known as an autotransporter. It is a single membrane-spanning secretion system. The secretion system and the substrates are synthesized as preproteins and are released into the periplasm via the SecYEG translocon. After proteolytic cleavage of the N-terminal leader peptide, the transporter domain of the proprotein is oligomerized to form a β-barrel structure in the outer membrane, which allows the "passenger" domain to be excreted outside. The T5SS secretes proteins involved in pathogenesis, cell-to-cell adhesion, and biofilm formation. In pathogenic *E. coli*, T5SS is encoded by genes located on EspC PAI and in *Salmonella enterica*; it is encoded by SPI-3.

Type VI Secretion System

The T6SS is composed of 13 components and spans from the cytoplasm to the outer membrane and is distributed among *Proteobacteria*. It has two main complexes: the cytoplasmic membrane proteins similar to T4SS and the tail complex related to the contractile bacteriophage tail. T6SS translocates toxic effector proteins into eukaryotic and prokaryotic cells and plays important role in pathogenesis and bacterial competition.

Type VII Secretion System

The type VII secretion system (T7SS) is a specialized secretion system found in *Mycobacterium tuberculosis* and is responsible for secretion of virulence proteins. The cell envelope of *Mycobacterium* consists of a cytoplasmic membrane, a periplasmic space that contains peptidoglycan, and a thick complex waxy outer membrane called mycomembrane consisting of arabinogalactan, mycolic acids, and free lipids.

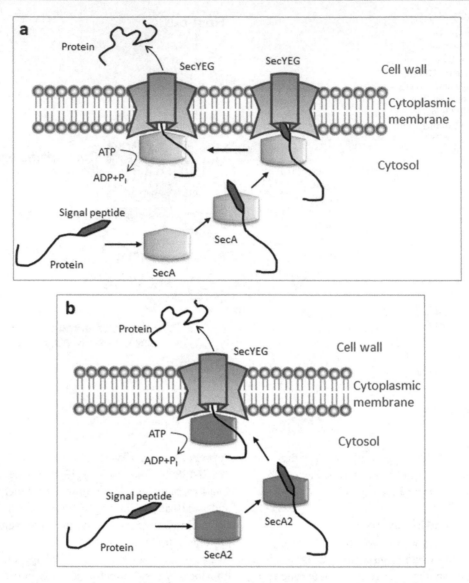

Fig. 4.15 The Sec machinery in Gram-positive bacteria for protein transport. (**a**) Canonical SecA-mediated protein transport system, (**b**) SecA2-dependent system

(Adapted and redrawn from Feltcher and Braunstein 2012. Nat. Rev. Microbiol. 10, 779–789)

T7SS is a 1.5 MDa protein complex and the gene clusters have been identified in *Listeria monocytogenes, Bacillus subtilis*, and *Staphylococcus aureus*.

SecA System

All bacteria possess a conserved general secretion pathway (Sec) to translocate proteins across the cytoplasmic membrane (CM) (Fig. 4.15). The Sec pathway or the SecYEG translocon consists

of three proteins: SecY, SecE, and SecG. It translocates primarily the unfolded precursor proteins across the CM. Most, but not all, Sec-dependent proteins contain a classical Sec signal sequence at the amino acid terminus, except for a few proteins whose translocation and/or secretion are dependent on SecA. SecA, a cytosolic motor protein, binds to preprotein and drives its translocation through the SecYEG channel. During this event, the Sec signal peptide sequence remained associated with SecYEG channel. There are two SecA proteins, SecA1 and SecA2, and both are present in all mycobacteria and a small subset of Gram-positive bacteria including *Staphylococcus*, *Listeria*, and *Streptococcus* spp. SecA1 is responsible for protein transport via a canonical Sec pathway as described above and is essential. SecA2 is present in some species of Gram-positive bacteria but is absent in Gram-negative bacteria and is dispensable. SecA2 contributes to virulence and may have a specialized function for proteins, which cannot be efficiently translocated by the canonical Sec system. For example, SecA2 is required for optimal secretion of superoxide dismutase (SodA) in *L. monocytogenes* and SodA and catalase–peroxidase in *Mycobacterium* and therefore contributes to bacterial virulence by countering the oxidative defenses of the host cells. SecA2 is also important in the pathogenesis of *Streptococcus gordonii*, *Streptococcus parasanguinis*, *Staphylococcus aureus*, *Corynebacterium glutamicum*, *Clostridium difficile*, and *Listeria monocytogenes* by facilitating secretion of many virulence proteins.

Regulation of Virulence Genes

Foodborne pathogens encounter a variety of harsh environments in food products and during transit through the gastrointestinal tract. Food-associated environments may include acids, salts, preservatives, peroxides, antimicrobial chemicals, flavoring agents, sugars, and storage temperatures. While in the GI tract, pathogens encounter gastric acid, bile salts, mucus, lysozyme, and natural microbiota. Although these environments are stressful, the pathogens must be able to express necessary colonization and invasion factors for their survival and multiplication. A complex regulatory element is thought to play important roles during maintenance of saprophytic lifestyle and subsequent interaction with the host system.

Protein synthesis in prokaryotes consists of two major steps: transcription and translation. During transcription, the instructions stored in the DNA are transferred to the mRNA.

Transcription consists of three steps: initiation, elongation, and termination of RNA synthesis, in which the RNA polymerase is involved in the synthesis of mRNA from the DNA. The core RNA polymerase (RNAP) has five subunits: α_1, α_2, β, β', and ω. For the RNAP to bind a specific promoter, another subunit known as sigma factor (σ) is required. The sigma factor greatly increases the specificity of binding of RNAP and helps in the separation of DNA strands during the initiation of the transcription process. The sigma factors detach from the DNA after the initiation process of transcription has begun.

Prokaryotic sigma factors are classified into three structurally unrelated families: σ^N, σ^{54}, and σ^{70}. The σ^N is mostly involved in nitrogen metabolism and is also involved in a variety of metabolic processes. The σ^{70} family is larger and more diverse than σ^{54} and is divided into four groups (group I–IV) based on their primary amino acid sequences and structures. The group I sigma protein/the primary sigma factor (e.g., σ^A of *Bacillus subtilis*) is also known as "housekeeping" sigma factor because it regulates expression of "housekeeping" genes or genes responsible for basic metabolic processes and cell functions. The other groups are known as alternate sigma factors and they regulate specific physiological processes such as survival during stationary phase and during exposure to various stressful environments. Another intriguing aspect is that alternate sigma factors are now shown to regulate virulence genes or virulence-associated genes required for bacterial pathogenesis. Some σ^{70} family members like the σ^S of Gram-negative bacteria (group II), σ^B of Gram-positive bacteria (group III), and extracytoplasmic functioning sigma factors (group IV) contribute to the bacterial stress responses. A few of the alternate sigma factors also play a role in the bacterial virulence as these respond to the host environment, which is critical in the infection process during the intestinal phase of infection.

Synthesis of stress proteins depends on the expression of stress-related genes, some are inducible whereas others are constitutive but expressed at low levels when the cells are not under the stress. As some of the gene systems are global, gene expression by one stress can also help cells to adapt to other related stresses. Expression of stress-related genes is initiated by specific polypeptides or sigma factor (σ) synthesized by specific genes. Some of these, such as σ^B or σ^{37} (encoded by gene *sigB*), helps cope with general stress in Gram-positive bacteria; σ^{32} (encoded by *rpoH* gene) and σ^{24} (encoded by *rpoE* gene) help cope with heat response; and σ^{38} or σ^s (encoded by *rpoS* gene) helps cope with general stress and starvation in Gram-negative bacteria (i.e., *Salmonella*). Under a specific stress (such as heat stress), *rpoH* is turned on and promotes synthesis of RpoH or σ^{32} protein in high amounts. This sigma factor (also called a regulon) then combines with the core RNA polymerase, consisting of four subunits, $\alpha\alpha\beta\beta$, to form the complete RNA polymerase enzyme or holoenzyme. This holoenzyme then binds to the promoter of a heat-shock gene family, leading to the synthesis of heat-shock proteins, which then protect the structural and functional units of stressed cells susceptible to heat damage (e.g., DNA and proteins). Sigma factors can also protect against other stresses, such as cold, heat, low pH, and UV.

Pathogens Survival in Acidic pH

The enteric pathogens, especially the Gram-negative bacteria are susceptible to low pH and die off rapidly in the stomach and high-acid foods (pH ≤ 4.5). Also, at low pH, normal cells are susceptible to other antimicrobial treatments, such as hydrostatic pressure, pasteurization temperature, or preservatives at a much lower level. However, if bacteria are first acid-adapted, they become relatively resistant to low pH and other treatments at minimal levels and can survive well in the food and in the GI tract. Recent foodborne disease outbreaks of *Salmonella enterica*, *E. coli* O157:H7, and *L. monocytogenes* linked to the fruit juice, fermented sausage, and acidified foods helped these pathogens to be acid-adapted. Acid adaptation possibly helped these pathogens to withstand high acidity and thermal processing. To overcome such problem, it is necessary to avoid exposure to the food containing pathogens to mild treatments; instead, hurdle concept consisting of several mild treatments simultaneously could effectively inactivate the pathogens.

Summary

Foodborne pathogens cause three forms of disease: *foodborne infection*, *foodborne intoxication*, and *foodborne toxicoinfection*. The prinicipal route of infection/intoxication for foodborne pathogens is oral and the primary site of action is the gastrointestinal tract. The infectious dose of foodborne pathogens varies and depends on the type of organisms or toxin as well as the type of food (liquid vs. solid) ingested. The pathogens that are responsible for infection colonize the gut by producing various adhesion factors including fimbriae, curli, adhesin proteins, and extracellular matrices that allow biofilm formation. Invading pathogens have developed strategies to cross the epithelial barrier. Some use M cells to reach the subcellular location or some pathogens actively penetrate epithelial cells by rearranging the host cell cytoskeletal structure. Pathogens localized in the subcellular locations multiply, move from cell-to-cell, and induce inflammation and elicit cell damage to induce diarrhea and gastroenteritis. Some intracellular pathogens induce apoptosis or necrosis in macrophages, dendritic cells, neutrophils, and other cells, thus ensuring their survival in host tissues. Pathogens may also translocate to deeper tissues including the liver, lymph nodes, spleen, brain, and placenta. Foodborne intoxication is mediated by exotoxins produced by pathogens in the food, which induces cell damage, fluid and electrolyte losses, and apoptosis or blocks nerve impulse following consumption of the contaminated food. The mechanisms of exotoxin action may vary, and based on the toxin action, the toxins can be classified as A–B type toxins, membrane-acting toxins, superantigens, proteases, protein synthesis inhibitors, and signal transduction modulators. The bacterial cell wall or membrane-associated endotoxins (LPS, PGN) are generally associated with systemic foodborne infection, and these toxins modulate the immune system to induce the release of large quantities of cytokines that promote fever, decrease blood pressure, and induce septic shock. In most pathogens, virulence factors encoded genes are located in pathogenicity islands or islets, which may be found on plasmids, bacteriophage, or the chromosome. Virulence proteins are exported from the microbes by various secretory machinary and those are designated type I–type VII and Sec secretory systems. Finally, bacterial virulence gene expression is a complex process that may be controlled by different regulatory elements in the food system as well as in the host. Alternate sigma factors are found to be crucial in virulence gene expressions in foodborne pathogens.

Further Readings

1. Abreu, M.T. (2010) Toll-like receptor signalling in the intestinal epithelium: how bacterial recognition shapes intestinal function. *Nat Rev Microbiol* **10**, 131–144.
2. Al-Sadi, R., Boivin, M. and Ma, T. (2009) Mechanism of cytokine modulation of epithelial tight junction barrier. *Front Biosci* **14**, 2765–2778.
3. Ashida, H., Mimuro, H., Ogawa, M., Kobayashi, T., Sanada, T., Kim, M. and Sasakawa, C. (2011) Cell death and infection: A double-edged sword for host and pathogen survival. *J Cell Biol* **195**, 931–942.
4. Barnhart, M.M. and Chapman, M.R. (2006) Curli biogenesis and function. *Annu Rev Microbiol* **60**, 131–147.
5. Barreau, F. and Hugot, J.P. (2014) Intestinal barrier dysfunction triggered by invasive bacteria. *Curr Opin Microbiol* **17**, 91–98.
6. Barry, S.M. and Challis, G.L. (2009) Recent advances in siderophore biosynthesis. *Curr Opin Chem Biol* **13**, 205–215.
7. Cossart, P. and Sansonetti, P.J. (2004) Bacterial invasion: The paradigms of enteroinvasive pathogens. *Science* **304**, 242–248.
8. Costa, T.R.D., Felisberto-Rodrigues, C., Meir, A., Prevost, M.S., Redzej, A., Trokter, M. and Waksman, G. (2015) Secretion systems in Gram-negative bacteria: structural and mechanistic insights. *Nat Rev Microbiol* **13**, 343–359.
9. do Vale, A., Cabanes, D. and Sousa, S. (2016) Bacterial toxins as pathogen weapons against phagocytes. *Front Microbiol* **7**.
10. Evans, M.L. and Chapman, M.R. (2014) Curli biogenesis: Order out of disorder. *Biochim Biophys Acta (BBA) - Mol Cell Res* **1843**, 1551–1558.
11. Feltcher, M.E. and Braunstein, M. (2012) Emerging themes in SecA2-mediated protein export. *Nat Rev Microbiol* **10**, 779–789.
12. Fink, S.L. and Cookson, B.T. (2005) Apoptosis, pyroptosis, and necrosis: Mechanistic description of dead and dying eukaryotic cells. *Infect Immun* **73**, 1907–1916.
13. Galan, J.E. and Wolf-Watz, H. (2006) Protein delivery into eukaryotic cells by type III secretion machines. *Nature* **444**, 567–573.
14. Guttman, J.A. and Finlay, B.B. (2009) Tight junctions as targets of infectious agents. *Biochem Biophys Acta* **1788**, 832–841.
15. Harshey, R.M. (2003) Bacterial motility on a surface: Many ways to a common goal. *Annu Rev Microbiol* **57**, 249–273.
16. Hawver, L.A., Jung, S.A. and Ng, W.-L. (2016) Specificity and complexity in bacterial quorum-sensing systems. *FEMS Microbiol Rev.* **40**, 738–752.
17. Henkel, J.S., Baldwin, M.R. and Barbieri, J.T. (2010) Toxins from bacteria. *EXS* **100**, 1–29.
18. Juge, N. (2012) Microbial adhesins to gastrointestinal mucus. *Trends Microbiol* **20**, 30–39.
19. Kazmierczak, M.J., Wiedmann, M. and Boor, K.J. (2005) Alternative sigma factors and their roles in bacterial virulence. *Microbiol Mol Biol Rev.* **69**, 527–543.
20. Lertsethtakarn, P., Ottemann, K.M. and Hendrixson, D.R. (2011) Motility and chemotaxis in *Campylobacter* and *Helicobacter*. *Annu Rev Microbiol* **65**, 389–410.
21. Rajkovic, A. (2014) Microbial toxins and low level of foodborne exposure. *Trends Food Sci Technol* **38**, 149–157.
22. Ray, B. and Bhunia, A. (2014) Microbial Attachments and Biofilm Formation. In *Fundamental Food Microbiology*. pp.73–78. Boca Raton, FL: CRC Press.
23. Ribet, D. and Cossart, P. (2015) How bacterial pathogens colonize their hosts and invade deeper tissues. *Microbes Infect* **17**, 173–183.
24. Salyers, A.A. and Whitt, D. (2002) *Bacterial pathogenesis: A molecular approach*. Washington, D.C.: ASM Press.
25. Schmidt, H. and Hensel, M. (2004) Pathogenicity islands in bacterial pathogenesis. *Clin Microbiol Rev.* **17**, 14–56.
26. Schmitt, C., Meysick, K. and O'Brien, A. (1999) Bacterial toxins: friends or foes? *Emerg Infect Dis* **5**, 224–234.
27. Sibley, L.D. (2004) Intracellular parasite invasion strategies. *Science* **304**, 248–253.
28. Solano, C., Echeverz, M. and Lasa, I. (2014) Biofilm dispersion and quorum sensing. *Curr Opin Microbiol* **18**, 96–104.
29. Sridharan, H. and Upton, J.W. (2014) Programmed necrosis in microbial pathogenesis. *Trends Microbiol* **22**, 199–207.
30. Veiga, E. and Cossart, P. (2006) The role of clathrin-dependent endocytosis in bacterial internalization. *Trends Cell Biol* **16**, 499–504.
31. Wree, A., Broderick, L., Canbay, A., Hoffman, H.M. and Feldstein, A.E. (2013) From NAFLD to NASH to cirrhosis - new insights into disease mechanisms. *Nat Rev Gastroenterol Hepatol* **10**, 627–636.

5

Animal and Cell Culture Models to Study Foodborne Pathogens

Introduction

Our knowledge of the pathogenic mechanism of foodborne pathogens has stemmed largely from the use of various mammalian cultured cell lines and animal models. In the early years, animal models were often used to confirm the pathogenic attributes of a microbial isolate that was involved in a disease outbreak. In recent years, animals are still used as a model to study not only the pathogenic mechanism and the immune response but also the efficacy of a vaccine. Cultured cell lines are still considered an indispensable powerful tool in studying the molecular and cellular mechanisms of pathogenesis. The knowledge gained from the cell culture model is then substantiated on an animal model, a necessary step toward understanding the pathogenic mechanism, and associated pathology and clinical signs. Thus, one must be familiar with various cell culture and animal models and their applications while studying microbial pathogenesis.

While studying pathogen–host interaction, some guidelines must be followed: (1) it is important to use the same strain of the microbe that caused the disease because of the clonal nature of the pathogen. (2) Pathogens may also lose virulence traits during subculture; therefore, one has to avoid multiple subculturing (passage) before performing the pathogenicity experiment. (3) During pathogenicity testing, some strains may not exhibit pathognomonic symptoms in either cell culture or in an animal model. Therefore, finding an alternative sensitive cell culture system or an animal model is crucial for studying the host–pathogen interaction. (4) In some cases, organ culture such as the ligated intestinal loop or an embryonated egg model can be used.

Animal Model

Humans are the ideal model to study human pathogens; however, it is not possible to use humans for safety, bioethics, and expense-related concerns. However, in certain nonfatal diseases, human volunteers have been used. Animal models are used frequently as a substitute. Among them, nonhuman primates (monkey, baboon, and chimpanzee) are ideal for mimicking many diseases because of a higher genetic relatedness between primates and humans. However, the pain and suffering inflicted on animals during experimentation and expense-related considerations may limit their widespread application. Thus, animal use in research is highly regulated and an alternative to the animal models is highly desirable. Rodents are the most commonly used animal models that include mice, rats, rabbits, hamsters, gerbils, guinea pigs, and ferrets (Table 5.1). Pigs or piglets, bovine calves, sheep, goat, dogs, cats, chicken, and zebrafish have been used. In addition, insects, such as fruit fly (*Drosophila melanogaster*) and larvae of the

A. K. Bhunia, *Foodborne Microbial Pathogens*, Food Science Text Series, https://doi.org/10.1007/978-1-4939-7349-1_5

Table 5.1 Common laboratory rodents' breeding data sheet

Species	Strains	Life span (years)	Adult body weight	Sexual maturity	Gestation period (days)	Average litter size	Age at weaning (weeks)
Mouse	Balb/C, A/J, C57BL/6	1.5–2.5	20–25 g	6–8 weeks	19–21	6–8 pups (2–12)	3 weeks
Gerbil (desert rats)	Great gerbil, Mongolian gerbil	3–4	70 g	8–12 weeks	24–28	4 (1–8) pups	3–4 weeks
Rat	Wistar, Sprague Dawley, Fischer 344 Zucker	2	250–500 g	5 weeks	21–23	10–13 pups	4 weeks
Guinea pig	Dunkin-Hartley	4–5	700–1200 g	4 weeks	63–68 (59–72)	3 (1–6) pups	3 weeks (pups born fully developed)
Rabbit	New Zealand White, Baladi	7–8	2–6 kg	5–7 months	31–32	7–9 kits	5–8 weeks

greater wax moth (*Galleria mellonella*), and nematodes, in particular, *Caenorhabditis elegans* (*C. elegans*), have been used to study microbial pathogenesis. Some animals such as horse, donkey, sheep, goat, camels, llama, and alpacas are used for the production of large quantities of polyclonal antibodies. Therefore, careful considerations should be given while choosing an animal model for pathogen–host interaction study: pathogens must infect animals by the same route as humans and exhibit a similar colonization pattern, a similar tissue distribution pattern, and the same degree of virulence as in humans.

Advantages of Animal Models

The major advantages of using an animal model, especially the rodent model, are the availability of an inbred line; therefore, the infection could be reproduced and the variability between the animals could be minimized. Rodents are small, hence, require less housing space compared to the primate or other animals, and are less expensive to house or maintain. Rodents breed rapidly and have a big litter size; thus, a large number of the same inbred line can be obtained in a short period, for use in an experiment to obtain a statistically significant result (Table 5.1). In addition, genetically engineered rodents such as transgenic mice or rats with targeted gene knockout can elucidate the role of a specific gene product in microbial pathogenesis.

Limitations and Cautions of Animal Models

Animal models present some limitations thus requiring careful considerations before use. Animal use requires prior institutional approval, and there are strict guidelines one has to follow to use animals for research purposes. Animals must be procured from authentic vendors or sources. Proper housing facility, breeding, handling, and veterinary care must be provided. Caregivers or animal handlers must take precautions to avoid accidental exposure to the infective agents.

Researchers must use appropriate personal protective equipment (PPE) that include a lab coat, mask, hairnet, and gloves before conducting an animal experiment using pathogens. Animals also can attack, scratch, and bite during handling. Furthermore, there are some fundamental differences in physiology between the rodents and humans. Rodents are coprophagy; thus, reingestion of the same bacteria may compromise the experimental data especially for an experiment conducted with the foodborne pathogens. Resident microflora population and composition in rodents may be different from that of the human; thus, rodents may respond differently to the human pathogens and the outcome of the disease may be different. Most importantly, some animals are not sensitive to pathogens and thus may not show the same human-type symptoms or tissue distribution. For example, *Salmonella enterica* serovar Typhimurium do not show gastroenteritis or same tissue distribution as humans when administered orally into mice. The gastroenteritis symptom (diarrhea) is achieved only when the mice are pretreated with antibiotic, streptomycin, to inhibit resident gut microbiota.

Yet, rodents are being used as prevailing models. Several strategies are employed to overcome shortcomings of rodent models. (1) Related bacterial strains, which are infective, are used. (2) Different routes of administration are employed, i.e., ingestion vs. intraperitoneal route, intranasal route, or intravenous route. (3) Neonatal animals are used since their immune system has not fully developed thus exhibiting high susceptibility to a pathogen. (4) Germ-free gnotobiotic animals without any resident microbiota living in or on them may show enhanced susceptibility to a pathogen. (5) Humanized mouse models that are expressing human cell and tissue transplants or a specific gene for a receptor protein or cytokine can show increased interaction to a pathogen. For example, *Salmonella enterica* serovar Typhi (causative agent for typhoid fever) does not show typhoid-like symptom in mice, but a humanized mouse can show the symptom. (6) Immunocompromised or immunodeficient animals with impaired immune system are highly susceptible to foodborne pathogens.

Trangenic Animal Models

Immunocompromised conditions can be induced by exposing animals to gamma irradiation, which destroys immune cells (T cells, B cells, NK cells, macrophages) since the immunological stem cells are susceptible to irradiation. Immunodeficient mice are genetically engineered (transgenic mice) to have impaired immune system, and these include "nude" mice, "SCID" (severe combined immunodeficient) mice, and "Rag"-deficient mice. Nude mice have a homozygous mutation in *Foxn1*, which encodes a transcription factor for thymus development and hair follicle growth thus making the mouse athymic and hairless. The absence of the thymus prevents CD4$^+$ and CD8$^+$ T cells from differentiation and maturation, making the nude mice (*Foxn1$^{-/-}$*) deficient in functional T cells. The SCID mice have a homozygous mutation in *Prkdc*, which encodes DNA-dependent protein kinase required for DNA repair. DNA damage occurs during somatic recombination of TCR (T-cell receptor) and Ig (immunoglobulin) gene; thus, *Prkdc$^{-/-}$* mice lack both functional T and B cells. Likewise, *Rag$^{-/-}$*-deficient mice do not express *Rag1* and *Rag2*, which are required for somatic recombination of TCR and Ig genes resulting in functional T- and B-cell deficiency. The immunocompromised and immunodeficient animals are expensive to procure and require greater care in their handling and maintenance. They need pathogen-free environment since they are highly susceptible to infective agents and even the commensal microbes.

Suckling mouse model has been used to study the pathogenesis of diarrheagenic microbes including *Vibrio cholerae* and enterotoxigenic *E. coli* (ETEC). In this model, the test microorganisms or the toxin preparations are administered orally in 2–4-day-old suckling mice, and mice are sacrificed and examined for fluid accumulation in the gut after 4–24 h. The ratio of gut weight to the body weight has been used to determine the enterotoxic effect of the toxins. Suckling mice have also been used to study the translocation of protozoan parasites including *Cryptosporidium parvum* in the gastrointestinal tract and their oocyst counts in the intestinal contents.

Organ Culture

The animal organ can be used as an alternative to the whole animal and is ideal for investigating certain pathogen–host interaction. Ligated intestinal loop and the embryonated eggs are examples of organs that are routinely used to study infectious agents. The advantages of organs are as follows: they are genetically intact, multiple cell types are represented, and the cells retain their original shape and configurations.

Ligated Intestinal Loop Assay

Ligated intestinal loop (LIL) assay has been used to study intestinal inflammation and fluid secretion by diarrheagenic microorganisms and their toxins such as the species of *Salmonella*, *Clostridium*, *Bacillus*, *Vibrio*, *Shigella*, and *Escherichia*. In addition, LIL has been used to investigate microbial translocation across the intestinal barrier including the involvement of M cells in pathogenesis. The LIL from mice, rats, guinea pigs, rabbits, chicken, calves, and piglets is used. Animals are deprived of food and water for about 12 h and anesthetized before initiating the experiment. A small incision is given in the abdomen, and a portion of the small intestine is pulled out from the abdominal cavity. The knots are placed in every 2–3 cm intervals using strings to create loops. Test materials along with proper controls are injected into the loops and the intestine is placed back inside, and the skin opening is closed by suturing. After a period of 18–24 h, the intestine is examined for fluid accumulation, typical "ballooning" due to the action of the diarrheagenic toxin.

Embryonated Chicken Egg Assay

The embryonated chicken egg (Fig. 5.1) offers an alternative to investigate microbial pathogenesis. The immunity of the egg is similar to that of higher mammals. Twelve- to fourteen-days-old embryonated hen's eggs are injected aseptically with test organisms in the chorioallantoic

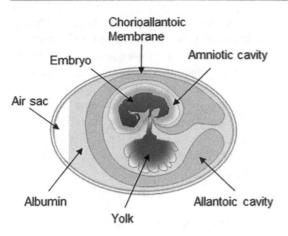

Fig. 5.1 Schematic of an embryonated hen egg

membrane (CAM) or directly in the embryos of eggs and incubated for 3–4 days. The death of embryos or characteristic cytopathic effects (CPE) is indicative of the infective nature of the test organism. This model is used for various bacterial pathogens including *Listeria monocytogenes*, *Yersinia enterocolitica*, and *Klebsiella pneumoniae*; viruses such as influenza virus, herpes simplex virus, Rous sarcoma virus, and Newcastle disease virus; and molds such as *Aspergillus fumigatus* and *Candida albicans*.

Zebrafish Embryo

Zebrafish (*Danio rerio*) embryo is used as a model host to study the function of the vertebrate immune system (both innate and acquired immunities) during microbial infection. The acquired immunity in the embryo is developed only after day 30. The embryo is used in the first 7 days after the eggs are deposited. The pathogens are injected in different areas by using a microinjection system, and primarily the host innate immunity is studied during the pathogen–host interaction. Since the embryo is transparent, every organ and tissue are easily visible under a microscope for manipulation and imaging. This model has been used to study intracellular pathogens *Salmonella enterica* serovar Typhimurium and *Mycobacterium marinum*. The zebrafish model requires specific facilities to maintain the fish and the expertise for handling and microinjection.

Cultured Cell Lines

Cells derived from human, animal, or insect tissues are attractive models for studying the pathogenesis of infective agents. Cultured cells are also used for cancer research, drug screening and toxicity testing, viral growth, vaccine development, tissue or organ transplant, stem cell therapy, and genetic engineering to produce insulin, hormones, and antibodies. Microorganisms or toxins have tropism for different cell types originating from different organs or tissues; therefore, the use of a specific cell type can provide insights of the mechanism of infection for the test organism in vitro. The predominant cell type is epithelial and endothelial cells; however, fibroblast, lymphocyte, and neuronal cells are also used.

There are two types of cells, primary and secondary. Primary cells are harvested directly from the animal tissues or organs. Tissues are homogenized, processed by enzymatic digestion, and maintained using appropriate cell culture growth media. The primary cells consist of mixed cell types and are short-lived with a maximum lifespan of 12–14 days. Secondary cells are immortal and are derived from a single lineage of the cell and have a uniform genetic composition, thus referred to as a "cell line" (monoculture). The secondary cells are originated either from tumor cells, from transformed cells, or from hybridoma cells. The transformed cells are generated by introducing viral oncogenes such as those from simian virus 40 (SV40) or Epstein–Barr virus (EBV). This technique is routinely used to convert a primary cell line to a transformed immortal cell line. The hybridoma cells are made from the fusion of primary cells and cancer cells such as myeloma cells, and they share properties of both partner cell lines. Hybridoma B cells generated in this process are immortal and produce monoclonal antibodies. Both primary and secondary cells can grow either in suspension or as adherent cells forming monolayers in tissue culture flasks. Cells are typically grown at 37 °C in a humidified incubator with a supply of CO_2 (5–7%) (see below). Examples of commonly used secondary or transformed cultured cell lines are Caco-2 and HT-29 (colon adenocarcinoma cells), HepG2 (liver cells), HEp-2 (larynx cell), Henle-407 (small intestine–jejunal), HeLa (cervix), J-774 (macrophage), Vero (monkey kidney), and CHO (Chinese hamster ovary) (Table 5.2).

Advantages and Limitations of Cultured Cell Models

There are several advantages of cell culture model for studying microbial pathogenesis. Cultured cell line is a simple and controlled model to study host–microbe interaction and easy to run experiments with radioactive materials, toxins, or transfection with foreign DNA; cells multiply rapidly; thus, experiment can be conducted rapidly; the secondary cells are immortal, provided nutrients and proper culturing conditions are maintained, and relatively inexpensive compared to the animal models.

However, there are several limitations. Cultured mammalian cells are generally derived from tumor cells; therefore, genetic aberrations have occurred in these cells, and the cells may not be genetically identical to the parent cells. Continuous culturing of cells produces mutations and genetic rearrangements in the chromosome; therefore, these cells may lose traits of original tissue and may also lose tissue-specific receptors for bacterial adhesins or unmasking low-affinity receptors. Fundamentally, immortalized cells are different from that of the normal cells in the body. Microbes that normally do not interact with certain host cells, in fact, may adhere and invade cultured cell lines. Sometimes the cell shape and distribution of surface antigens may be affected. Cultured cells are not polarized, while the primary cells in the body are polarized in the intact animal host, i.e., different parts of the cells are exposed to different environments, such as lumen, adjacent cells, underlying blood vessels, and tissues. The polarized cells contain different sets of proteins important for their functions. Physical disruption of polarized cells exposes some surface proteins, which may not be exposed previously. Hormones can help secondary transformed cells to convert into polarized cells. Furthermore, the host tissue consists of multiple cell types such as epithelial cells, fibroblasts, muscle cells, and neurons, but the cultured cells consist of only one type of cell; therefore, bacterial interaction with concerted host cell cannot be studied with a monoculture model. In cultured cells, mucus and other secretory components are absent, which normally interact with pathogens. Monoculture system is also unable to provide intercellular communication, such as communication between the antigen-presenting cell and T cell for immune response, and cytokine or growth factor production.

Co-cultured Cell Model

Though the pathogen interaction with a monoculture can give a glimpse of information about pathogens behavior, it is difficult to extrapolate the functional information for an animal model since multiple different cells in a tissue are involved during pathogen interaction in vivo. Co-culturing of two or more cell types in vitro can overcome some of the limitations of monoculture system. For example, alveolar epithelial cells have been co-cultured with macrophages and dendritic cells to develop a model for alveolar epithelial barrier system for respiratory tract infective agents. Likewise, enterocyte-like Caco-2 cells have been co-cultured with mucus-secreting HT-29 cells to represent mucoid environment encountered by enteric pathogens in the intestine. Co-cultured cells grown in a three-dimensional (3D) scaffold in a gel matrix can provide an opportunity to study pathogen interaction in the 3D organotypic model. Though the cultured cell model can provide in-depth knowledge of specific molecular and cellular mechanism of a pathogen, it is necessary to use animal models to substantiate the findings from the cell culture model and to understand the tissue distribution, associated pathology, immune response, and clinical signs.

Growth and Maintenance of Cultured Cell Lines

To grow cultured cell lines in vitro, appropriate growth media and environment must be provided. The basal growth medium is composed of

Table 5.2 List of human, animal, and insect cell lines used to study the interaction of foodborne pathogens

Bacteria	Cell line	Cell type	Source	Interaction type
Salmonella	CHO	Epithelial	Chinese hamster ovary	Elongation, detachment
	Vero	Epithelial	Monkey kidney	Lysis, protein synthesis inhibition
	HEp-2	Epithelial	Human laryngeal	Invasion
	Henle-407	Epithelial	Human jejunal	Intracellular growth
	J774	Macrophage	Mouse	Intracellular growth
	HeLa	Epithelial	Human cervix	Toxicity, actin polymerization
	HT-29	Epithelial	Human colon	Apoptosis
E. coli	Vero	Epithelial	Monkey kidney	Lysis, protein synthesis inhibition
	CHO	Epithelial	Chinese hamster ovary	Lysis, toxicity
	Henle-407	Epithelial	Human jejunal	Adhesion
	HEp-2	Epithelial	Human laryngeal	Adhesion, toxicity
	MDBK	Epithelial	Madin–Darby bovine kidney	Invasion
	HeLa	Epithelial	Human cervix	Toxicity, apoptosis
	T84	Epithelial	Human colon	Apoptosis
	Y1	Epithelial	Human adrenal gland?	Rounding, detachment, cAMP
	J774	Macrophage	Mouse	Apoptosis
Shigella	HeLa	Epithelial	Human cervix	Cell death, protein synthesis inhibition
	Vero	Epithelial	Monkey kidney	Lysis, protein synthesis inhibition
	3 T3	Fibroblast	Mouse	Invasion, actin polymerization
	U937	Monocyte	Human	Apoptosis
	Mφ	Macrophage	Mouse	Invasion, apoptosis
Campylobacter	HeLa	Epithelial	Human cervix	Distended cells (CDT effect)
	Vero	Epithelial	Monkey kidney	Distended cells (CDT effect)
	AZ-521	Epithelial	Human stomach	Vacuolation
Yersinia	J774	Macrophage	Mouse	Exocytosis, apoptosis
	HEp-2	Epithelial	Human laryngeal	Invasion, lysis
Vibrio	CHO	Epithelial	Chinese hamster ovary	Elongation, cAMP accumulation
Clostridium perfringens	Vero	Epithelial	Monkey kidney	Membrane permeability
	Caco-2	Epithelial	Human colon	Necrosis, apoptosis, pore formation
	MDDK	Epithelial	Madin–Darby dog kidney	Pore formation
Listeria monocytogenes	Caco-2	Epithelial	Human colon	Adhesion, invasion, apoptosis
	CHO	Epithelial	Chinese hamster ovary	Detachment, lysis
	Henle-407	Epithelial	Human jejunal	Intracellular growth, death
	Vero	Epithelial	Monkey kidney	Toxicity, adhesion, invasion
	HEp-2	Epithelial	Human laryngeal	Invasion
	J774	Macrophage	Mouse	Intracellular growth
	RAW	Macrophage	Mouse	Intracellular growth
	HeLa	Epithelial	Human cervix	Toxicity
	HUVEC	Endothelial	Human umbilical vein endothelial cell	Intracellular growth
	Hep-G	Epithelial	Human liver	Intracellular growth, apoptosis
	3 T3	Fibroblast	Mouse	Invasion, plaque formation, lysis
	Ped-2E9	B cell	Mouse hybridoma	Lysis, apoptosis
	RI-37	B cell	Human–mouse hybridoma	Lysis, apoptosis
	S2	Epithelial	Drosophila	Invasion and infection model
Avian flu virus (H5N1)	MDCK	Epithelial cells	Madin–Darby canine kidney	Cytopathic effects, virus multiplication
Hepatitis A virus	Vero	Epithelial	African green monkey kidney	Virus replication and cytopathic effects

Adapted and modified from Bhunia, A.K. and Wampler, J.L. 2005. *Foodborne Pathogens: Microbiology and Molecular Biology*, Caister Academic

Table 5.3 Media used for growing common mammalian cell lines

Cell line	Source	Morphology	Medium used
Caco-2	Human	Epithelial cell (colorectal adenocarcinoma)	Eagle's Minimum Essential Medium (EMEM or Dulbecco's Modified Eagle's Medium (DMEM) + serum (10%)
HT-29	Human	Epithelial (colorectal adenocarcinoma)	McCoy's 5a + serum (10%)
HEK293	Human	Epithelial	Eagle's Minimum Essential Medium (EMEM) + serum (10%)
HUVEC	Human	Endothelial	Hams's F-12 + serum (10%) + heparin (100 μg/ml)
3 T3	Mouse	Fibroblast	Dulbecco's Modified Eagle's Medium (DMEM) + serum (10%)
CHO	Hamster	Epithelial	Hams's F-12 + serum (10%)
Jurkat	Human	Lymphoblast	Roswell Park Memorial Institute 1640 (RPMI-1640) + serum (10%)

glucose, amino acids, fatty acids and lipids, vitamins, trace elements, salts, and phenol red as pH indicator. Several commercial media such as Eagle's Minimum Essential Medium (EMEM), Dulbecco's Modified Eagle's Medium (DMEM), Iscove's Modified Dulbecco's Medium (IMDM), Roswell Park Memorial Institute 1640 (RPMI-1640), McCoy's 5a, and Hams's F-10 and F-12 media are used depending on the cell type (Table 5.3). The basal growth medium is then supplemented with serum (1–10%) to provide proteins, growth factors, cytokines, iron chelators (transferrin, ferritin), and vitamins. Serum source is generally from adult bovine or calf. The medium is also supplemented with 2 mM glutamine since this amino acid tends to degrade over time. Antibiotic supplement (i.e., gentamicin, ampicillin, streptomycin) is also used to prevent microbial contamination, but antibiotic may slow down mammalian cell growth. Mammalian cells are seeded in appropriate flasks and dishes and placed in a specialized cell culture incubator at 37 °C under humidified condition, with a constant supply of CO_2 gas (5–7%). During cell growth, metabolic activity yields acidic by-product, which is toxic to cells. Therefore, maintaining the buffering capacity of the growth media is critical, which is naturally adjusted by the gaseous CO_2 and the CO_3/HCO_3 content in the media. HEPES, a zwitterion, has buffering capacity (in the pH range of 7.2–7.4) and has been used in the medium in the absence of CO_2 atmosphere. The serum also contributes to the media buffering. The standard pH of the growth medium is 7.4 and appears red due to the presence of phenol red indicator. When the medium pH drops below 7.4 due to the buildup of acid, the medium color changes to yellow. Sometimes phenol red may interfere with cellular growth; thus, it may be omitted from the medium formulation.

Measurement of Virulence

Animal Model

In an animal model, infectivity or lethality of a pathogen or toxin may vary depending on the age and immunological health status of the animal and the route of administration of the infective agent. Generally, for experimental purposes, different routes are used to administer pathogens, such as oral, intragastric (i.g.), intravenous (i.v.), intraperitoneal (i.p.), intramuscular (i.m.), subcutaneous (s.c.), and intradermal (i.d.). To assess pathogenicity of an infective agent, the natural route of infection is recommended: an oral or intragastric route for foodborne pathogens or enteric pathogens, the intranasal route for respiratory tract infective agents, and so on. Generally, the virulence measurement is expressed as the infectious dose (ID) or as the lethal dose (LD). The infectious dose 50 (ID_{50}) is defined by the number or concentration of a pathogen required to infect 50% of the animals. The lethal dose 50 (LD_{50}) is defined by the number or concentration required to kill 50% of the animals. Kaplan–Meier survival curve is generated to determine the LD_{50} value (Fig. 5.2). Infectious dose or lethal

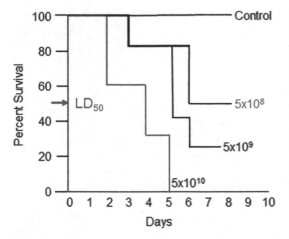

Fig. 5.2 Estimation of LD_{50} values for pathogens extrapolated from a Kaplan–Meier survival curve from a hypothetical animal study. LD_{50} value is 5×10^8 cfu

dose data provide the crude measurement of infection that includes cumulative effects of colonization, invasion, tissue distribution, and symptoms; therefore, ID_{50} cannot be compared with the LD_{50} value. Moreover, these data provide a relative measure of virulence and cannot compare two different diseases such as bacillary dysentery vs. cholera.

Cell Culture Model

Using cell culture models, pathogen or toxin interaction with different cell lines can be studied (Table 5.2). Cytotoxicity or cytopathogenicity assays are used to determine the cytotoxic potential of pathogens or toxins. Similarly, pathogens' ability to adhere, to invade, and to move from cell-to-cell can be assessed using specific cell lines. Cytotoxicity assay is generally used as a confirmatory test for a pathogen, which has already been identified or detected by other methods. This in vitro assay also allows one to study the mechanism of pathogenesis of various microorganisms or toxins, which can be assessed by different methods.

Microscopy

The degree of cell damage can be assessed under a phase-contrast light microscope, fluorescence

microscope, or electron microscope. The typical cytopathic effect (CPE) consists of cell detachment, flocculation, rounding or elongation, cell lysis, cytoplasmic granulation, and cell death (Fig. 5.3). Highest dilutions of a toxin or pathogen showing CPE are considered the cytotoxic titer for that agent. Furthermore, the nature of cell death (necrosis vs. apoptosis) could be detected by using a fluorescence microscope after staining cells with acridine orange, ethidium bromide, or propidium iodide. These dyes can permeate only through the damaged cell membrane and bind to the DNA emitting green or red fluorescence indicative of cell death. The healthy cells with intact cell membrane prevent dye entrance. Scanning electron microscope is often used to examine the membrane damage, pore formations, and apoptotic bodies in the eukaryotic cells (Fig. 5.4).

One of the most commonly used cytotoxicity assays is Vero cell (monkey kidney epithelial cell line) assay, which is performed to determine the cytotoxin (also called verotoxin) production from *Shigella* species, *E. coli*, and *Clostridium difficile*. The cytotoxic or cytopathic effects (CPE) are characterized by cell detachment, flocculation, rounding, and cell death (Fig. 5.3).

Trypan Blue Exclusion Test

Trypan blue is a diazo dye, which is routinely used in the cell culture laboratories to assess the viability of cultured mammalian cells during subculturing. In this assay, the cell suspension is mixed with an equal volume of trypan blue (0.4% v/v) for a few seconds, and a small volume is placed on a hemocytometer, and viable or dead cells are counted under a microscope. Viable cells with intact membrane exclude stain (hence called trypan blue exclusion test) and appear as bright translucent, while the dead or dying cells with damaged membrane allow stain permeation and appear blue. Confluent adherent cell monolayers growing on a flask or Petri dishes also can be stained with the trypan blue to assess the degree of cell damage induced by toxins or pathogens. The cell monolayers are then examined under a light microscope. Trypan blue staining cannot differentiate cell death by necrosis or

Fig. 5.3 Effect of
Escherichia coli
O157:H7 (EDL933)
toxin on Vero cells after
24 h exposure.
Toxin-exposed cells
(panel B) show
cytopathic effects
characterized by cell
rounding, granulation,
and detachment

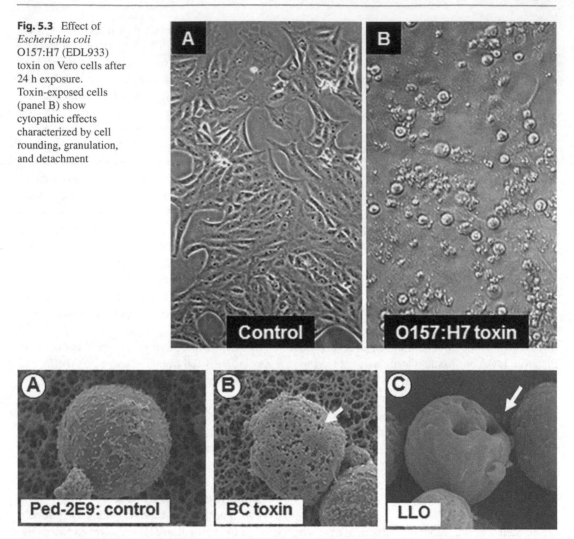

Fig. 5.4 Scanning electron microscopic photographs of lymphocyte cell line, Ped-2E9 (**A**) control, treated with (**B**) *Bacillus cereus* toxin, and (**C**) *Listeria monocytogenes* listeriolysin O (LLO, a hemolysin). Toxin-induced membrane pore formation is marked by arrows

apoptosis, but it can interpret the membrane integrity of a cell.

Alkaline Phosphatase Assay

Alkaline phosphatase enzyme is expressed by all living organisms; however, its expression and function may be tissue-specific. Certain cells, especially those of lymphocyte origin, express endogenous membrane-anchored alkaline phosphatase (ALP) enzyme (100–165 kDa), which is released from cells if the cell membrane is severely damaged. Membrane-active toxins, such as hemolysins or phospholipase enzymes produced by species of *Listeria*, *Bacillus*, *Clostridium*, and *Staphylococcus*, cause membrane pore formation and allow the release of ALP from cells. The ALP enzyme is analyzed by the addition of a substrate, such as *p*-nitrophenylphosphate (PNPP) or methyl umbelliferyl phosphate (MUP) yielding colored or fluorescence product for measurement by a spectrophotometer or a spectrofluorometer. The drawback of this assay is that not all cell types may have adequate amounts of this enzyme and thus

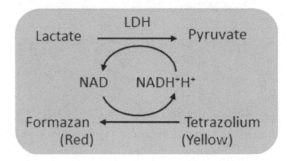

Fig. 5.5 Lactate dehydrogenase (LDH)-mediated color change in the reaction mixture

may not be a good indicator of cell cytotoxicity. In general, epithelial cells carry very low levels, while the cells of B lymphocyte origin carry high amounts. Therefore, pathogen-induced damage in epithelial cells cannot be measured accurately with the ALP assay.

Lactate Dehydrogenase Assay

All living cells including the mammalian cells possess lactate dehydrogenase enzyme (LDH). LDH is a low molecular weight enzyme (35 kDa) and may exist in the tetrameric form. The enzyme is released from the cells even due to a minor perturbation in membrane integrity, thus making the assay highly sensitive. LDH release assay has been widely used for studying cell cytotoxicity induced by varieties of microbial or nonmicrobial interactions with mammalian cells and immune effector cells (NK cells and cytotoxic T cells). The enzyme activity can be assayed by using an appropriate substrate. LDH converts lactate to pyruvate and the NAD receives an electron (H$^+$) from NADH, and during this process, yellow tetrazolium is converted to red formazan, which can be measured by a spectrophotometer for a quantitative cytotoxic response (Fig. 5.5). Commercial LDH assay kits are available for performing cytotoxicity assays.

MTT- and WST-1-Based Cytotoxicity Assays

The eukaryotic cell viability or proliferation could be assayed by a metabolic staining method. Pathogens or toxins that interfere with the metabolic activity of cells can be measured by this assay. The yellow tetrazolium, MTT (3-[4,5-dimethyl thiazolyl-2]-2,5-diphenyltetrazolium bromide), is a water-soluble compound, which is reduced by metabolically active cells. MTT is converted into water-insoluble purple formazan by the action of dehydrogenase enzyme that generates NADH and NADPH. The purple formazan is then quantified by a spectrophotometer at 570 nm. In the MTT assay, pathogen/toxin is added for 1–2 h before the MTT reagent is added. The assay generally takes 24–48 h to complete. The assay has been used to measure cytotoxic effects of bacterial and viral pathogens.

The WST-1-based assay is similar to that of the MTT assay, except that the WST-1 (4-[3-(4-iodophenyl)-2-(4-nitrophenyl)-2H-5-tetrazolio]-1,3-benzene disulfonate) reagent is used and the products are measured at 450/655 nm. The assay takes about 3 h to complete.

Cell Death Analysis

Pathogen interaction with host cells often results in cell death, which could be either necrotic, necroptotic, apoptotic, or pyroptotic (see Chap. 4). Necrosis, a nonspecific or accidental cell death, usually leads to cell swelling with a loss of membrane integrity, lysis, and initiation of a host inflammatory response. In contrast, apoptosis is a tightly regulated method for removal of damaged or unneeded "self"-cellular debris, especially during growth, development, and cell homeostasis. Apoptosis is characterized by cell shrinkage, chromatin condensation and margination, DNA fragmentation, and membrane blebbing (Fig. 5.6). There is no inflammatory response during apoptosis. Many cytokines and cellular proteins regulate cells death via two main pathways (1) by activating tumor necrosis receptor-associated death domain (TRADD) and (2) by activating CD95 receptor (also known as Fas) or Fas-associated death domain (FADD). These receptors bind effector molecules and set an intercellular apoptotic cascade in motion by recruiting, binding, and activating caspase (Cas 3, Cas 6, Cas 7, Cas 8, Cas 9) enzymes or downregulating apoptosis by interacting with inhibitor proteins, such as proteins in the Bcl-2 family (see Chap. 4).

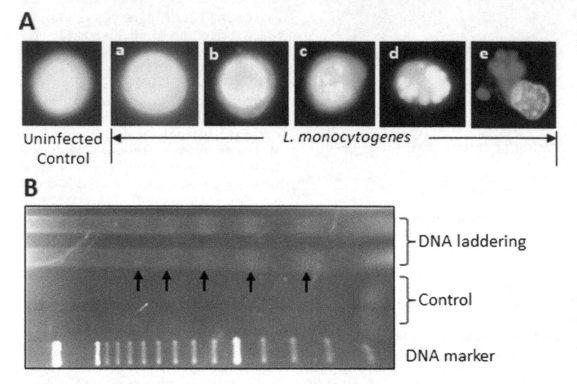

Fig. 5.6 (**A**) Different stages of apoptotic cell death in human B lymphoma cell line by *Listeria monocytogenes*: (a) cell shrinkage; (b) zeiosis stage, cell morphology alters and shrinkage continues; (c) chromatin condensation/margination, crescent-shaped nucleus; (d) DNA fragmentation; and (e) membrane blebbing of apoptotic bodies. (**B**) DNA fragmentation in hybridoma B cells by *L. monocytogenes* (Adapted from Bhunia and Feng 1999. *J Microbiol Biotechnol* 9, 398–403; Menon et al. 2003. *Comp Immunol Microbiol Infect Dis* 26, 157–174)

Apoptosis is regarded as a noninflammatory response that targets specific cell without full-scale tissue damage. However, in some cases, pathogens such as *Shigella* and *Salmonella* induce massive macrophage apoptosis, and this process is often proinflammatory. Consequently, speculation surrounds the question of whether host cells initiate apoptosis in an attempt to reduce widespread infection or if pathogens themselves may trigger apoptosis through activation of the apoptotic protein cascade as an infection strategy without alarming the immune system.

Apoptosis and necrotic cell deaths are differentiated by using a DNA fragmentation assay, characterized by DNA laddering, by flow cytometry analysis of annexin V binding to outer leaflet of the apoptotic cell membrane, or through a specific cell staining method such as acridine orange, ethidium bromide, or propidium iodide staining which intercalate host cell DNA (Fig. 5.6).

Measurement of Specific Steps in Colonization and Invasion

Animal Model

Microbial Enumeration and Staining

To study foodborne pathogen colonization, adhesion, invasion, and translocation to the extraintestinal sites or to study localized tissue damage, pathogens are administered orally or intragastrically. Sometimes animals are starved up to 12 h and/or fed with antacids to neutralize gastric pH before administration to allow safe microbial passage to the intestinal tract. To determine pathogen colonization and localized cell damage, intestinal sections are collected and intestinal contents removed, homogenized, and analyzed for bacterial counts in the mucus membrane by a plating method. In some cases, intestinal sections are treated with a broad-spectrum antibiotic (gentamicin) to kill extracellular bacteria and analyzed for bacterial invasion into the intestinal epithelial cells or submucosal layers. Histological sections of intestine can be stained with hematoxylin and eosin and examined under a microscope to assess recruitment of inflammatory cells (neutrophils and lymphocytes). Bacteria-specific immunostaining can be performed to localize bacterial cells in these sections. To determine bacterial invasion and translocation across the epithelial barrier to the extraintestinal sites, such as the liver, spleen, lymph nodes, brain, kidney, and placenta (pregnant model), each organ/tissue is collected, homogenized, and serially diluted, and bacterial counts are determined by a plating method.

Intravenous or intraperitoneal route of administration of foodborne pathogens has been used to investigate the pathogenic potential, tissue distribution, immune response, and clinical signs when the gastrointestinal route is bypassed.

In the ligated intestinal loop model, pathogens or toxin preparations are injected directly into the loop to determine fluid accumulation or to study bacterial adhesion and translocation to deeper tissues in a controlled environment. Staining of histological sections of the loop can reveal the bacterial adhesion and invasion patterns, the involvement of the M cells in pathogenesis, and the infiltration of inflammatory cells.

Whole-Body Bioluminescence Imaging

Whole-body bioluminescence imaging is also a powerful tool for real-time visualization of pathogen localization and tissue distribution during infection in live animals. In this case, pathogens are engineered to express luciferase enzyme (lux operon from *Photorhabdus luminescens*) that produces bioluminescence. After the animals are infected with the bioluminescent microbes, the entire animal body can be imaged at time intervals to visualize real-time progression of infection and tissue distribution in live animals. This method complements conventional microbial enumeration method from harvested tissues. This method can also reveal unexpected spread of infection in different tissues, which may otherwise be missed if the tissues were not harvested from that site. Whole-

body bioluminescence imaging has been used for studying bacterial pathogens, such as *Salmonella*, *Listeria*, *Escherichia*, *Bacillus*, *Haemophilus*, *Staphylococcus*, *Streptococcus*, and *Mycobacterium*, and parasitic (*Entamoeba*, *Toxoplasma*, *Leishmania*, *Plasmodium*), fungal (*Aspergillus fumigatus* and *Candida albicans*), and viral (*Vaccinia*) diseases.

Cell Culture Model

Adhesion Assay

Microbial adhesion is generally performed on confluent cell monolayers formed in the wells of cell culture plates, small flaskets, chambered slides, and Petri dishes or on coverslips. Depending on the cell type, some cells must be grown for a specified time to allow for cell differentiation and polarization before the adhesion assay can be performed. Cell monolayers are inoculated with the test organisms at MOI (multiplicity of infection) of 0.1, 1, 10, or 100 and incubated for 30 min to 1 h at 37°C in a cell culture incubator. The unbound organisms are removed by washing the cell monolayers at least for three times using cell culture media or a buffer. MOI needs to be optimized for each pathogen – generally, a lower MOI of 1 or 10 is recommended. Bacterial attachment is assessed by performing specific immunostaining, fluorescence staining, or Giemsa staining. Differential fluorescence staining is also used to distinguish intracellular bacteria from surface-attached bacteria. Quantitative bacterial counts are obtained by a plating method. To achieve bacterial counts, cell monolayers are first disrupted by treating with a mild detergent solution such as Triton X-100 (0.1–1%), and then cell suspensions are serially diluted and plated on agar plates.

Invasion Assay

To determine bacterial invasion and intracellular localization, polarized and differentiated cell monolayers are infected with the test microbes at MOI of 0.1, 1, 10, or 100 for 1–2 h to allow pathogen entry into the cell. Cell monolayers are then washed (three times) with cell culture media or buffer to remove loosely attached or unbound microbes and treated with an antibiotic (gentamicin at 10–100 μg ml^{-1}) for 1–2 h to kill extracellular bacteria. Gentamicin at low concentrations cannot permeate through the cell membrane; thus, intracellular bacteria are protected. Subsequently, the cell monolayers are washed as before and treated with mild detergent (Triton X-100) or cold buffer to lyse the cells, serially diluted, and plated to determine the intracellular bacterial counts. Immunofluorescence staining of cells whose membranes are permeabilized before staining can reveal intracellular bacterial localization.

Cell-to-Cell Spread or Plaque Assay

This assay is performed to determine the microbial ability to move from cell to cell. Generally, fibroblast (L2) or epithelial (Caco-2 or HT-29) cells are used for this purpose. Differentiated or polarized cell monolayers are first inoculated with the test organisms for 1–2 h, washed, and treated with gentamicin (100 μg ml^{-1}) for another 1.5 h. After washing the monolayers with a buffer solution, the cell monolayers are overlaid with tempered agarose (0.7%) containing gentamicin (10 μg ml^{-1}) and incubated for about 72 h in a cell culture incubator. The monolayers are stained with 0.1% neutral red and examined for the formation of clear plaques. Clear plaques indicate the capacity of the pathogen to move from one cell to another. This technique has been used to assess the infectivity of intracellular pathogens, such as *Shigella* and *Listeria* species.

Paracellular Translocation Assay

Many pathogens disrupt epithelial barrier to gain entrance into the subcellular space and for the systemic spread. Bacterial ability to translocate through the paracellular route (in between cells) can be examined in an in vitro transwell setup (Fig. 5.7). Confluent cell monolayers are first established on transwell filter inserts with pore sizes of about 3–4 μm. Often, transepithelial electrical resistance (TEER) of the polarized monolayer is measured using a voltmeter to ensure monolayer integrity before conducting the translocation experiment. Bacteria (MOI ~10) to be tested are added to the apical well of the

transwell setup and incubated in a humidified CO_2 incubator at 37 °C for 1–2 h. The liquid from the basal compartment is collected and the translocated bacteria are enumerated by plating. Often, fluorescein isothiocyanate (FITC)-conjugated dextran (4–40 kDa) is used to assess tight junction integrity. FITC-dextran is loaded into the apical well, and after a period of incubation, translocated FITC-dextran level in the liquid in the basal compartment is measured by a spectrofluorometer. The higher the fluorescence reading, the greater is the epithelial permeability. Epithelial tight junction integrity or disruption can also be visualized under a confocal microscopy following immunostaining with the tight junction and adherens junction-specific proteins: ZO-1, occludin and claudin, and E-cadherin. Enterohemorrhagic *E. coli* (EHEC), enteropathogenic *E. coli* (EPEC), *Helicobacter pylori*, *Campylobacter jejuni*, *Yersinia pseudotuberculosis*, *Listeria monocytogenes*, and *Shigella* species are known to disrupt epithelial barrier for paracellular translocation.

Summary

Much of our knowledge and understanding of the pathogenic mechanism of foodborne pathogens are gained from using animal and cell culture models. In modern day, these models are indispensable research tools; thus, one must have apprehension for the availability of different models, their usage, advantages, and limitations. In addition, one must choose an appropriate model for a pathogen in order to learn the microorganism's behavior in that environment. Cell culture models, especially the secondary cell lines, are the most valuable tools that allow determining virulence potential of pathogens. Cytotoxicity assays are measured by microscopic analysis and enzyme release assays (lactate dehydrogenase, alkaline phosphatase, or metabolic staining assay). Cell culture models also allow studying bacteria-induced cell death such as apoptosis or necrosis. In addition, these models help us to study the molecular mechanism of pathogen–host interaction such as microbial adhesion, invasion, paracellular translocation and

Fig. 5.7 Transwell setup for measuring paracellular transepithelial translocation of bacteria

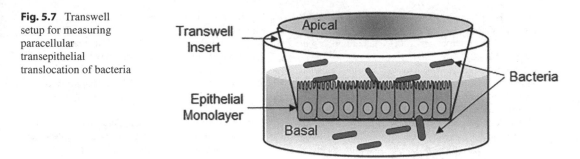

cell-to-cell movement, and host cell signaling events. Animal models are often used to substantiate the pathogenic mechanism that has been established in an in vitro cell culture model. Whole animals provide a better picture of dissemination of bacterial pathogens in different organs and tissues, and some models show clinical symptoms similar to humans. Sometimes animals are insensitive to human pathogens. In such situations, immunocompromised or immunodeficient animals are employed to study microbial pathogenesis. In addition, transgenic mice with targeted gene knockout can elucidate the role of a specific gene product in microbial pathogenesis. Pathogen interactions with animals are measured by determining the lethal dose (LD_{50}) or infective dose 50 (ID_{50}). Tissue distribution of pathogens can be analyzed by whole-body imaging using bioluminescence reporter microbes, conventional microbiological analysis from harvested tissues, and microscopic analysis of histological sections. Animal models also provide the nature of the immune response and the efficacy of a vaccine against a pathogen. Animal use in research is extremely regulated and alternative to animal models is highly desirable. Zebrafish, fruit fly, and nematodes (*C. elegans*) are some of the uncommon models which are gaining popularity as a substitute for laboratory animal models.

Further Readings

1. Andersson, C., Gripenland, J. and Johansson, J. (2015) Using the chicken embryo to assess virulence of *Listeria monocytogenes* and to model other microbial infections. *Nat Protocols* **10**, 1155–1164

2. Andreu, N., Zelmer, A. and Wiles, S. (2011) Noninvasive biophotonic imaging for studies of infectious disease. *FEMS Microbiol Rev* **35**, 360–394

3. Banerjee, P. and Bhunia, A.K. (2009) Mammalian cell-based biosensors for pathogens and toxins. *Trends Biotechnol* **27**, 179–188

4. Banerjee, P., Franz, B. and Bhunia, A. (2010) Mammalian Cell-Based Sensor System. In *Whole Cell Sensing Systems I* eds. Belkin, S. and Gu, M.B. pp.21–55: Springer

5. Bhunia, A.K. and Feng, X. (1999) Examination of cytopathic effect and apoptosis in *Listeria monocytogenes*-infected hybridoma B-lymphocyte (Ped-2E9) line *in vitro*. *J Microbiol Biotechnol* **9**, 398–403

6. Bhunia, A.K. and Wampler, J.L. (2005) Animal and cell culture models for foodborne bacterial pathogens. In *Foodborne Pathogens: Microbiology and Molecular Biology* eds. Fratamico, P., Bhunia, A.K. and Smith, J.L. pp.15–32. Norfolk, UK: Caister Academic Press

7. Decker, T. and Lohmann-Matthes, M.-L. (1988) A quick and simple method for the quantitation of lactate dehydrogenase release in measurements of cellular cytotoxicity and tumor necrosis factor (TNF) activity. *J Immunol Methods* **115**, 61–69

8. Duell, B.L., Cripps, A.W., Schembri, M.A. and Ulett, G.C. (2011) Epithelial cell coculture models for studying infectious diseases: Benefits and Limitations. *J Biomed Biotechnol* **2011**, 852419

9. Glavis-Bloom, J., Muhammed, M. and Mylonakis, E. (2012) Of model hosts and man: Using *Caenorhabditis elegans*, *Drosophila melanogaster* and *Galleria mellonella* as model hosts for infectious disease research. In *Recent Advances on Model Hosts* eds. Mylonakis, E., Ausubel, F.M., Gilmore, M. and Casadevall, A. pp.11–17. New York, NY: Springer New York

10. Hoelzer, K., Pouillot, R. and Dennis, S. (2012) Animal models of listeriosis: a comparative review of the current state of the art and lessons learned. *Vet Res* **43**, 18

11. Hutchens, M. and Luker, G.D. (2007) Applications of bioluminescence imaging to the study of infectious diseases. *Cell Microbiol* **9**, 2315–2322

12. Ito, R., Takahashi, T., Katano, I. and Ito, M. (2012) Current advances in humanized mouse models. *Cell Mol Immunol* **9**, 208–214

13. Martín, R., Bermúdez-Humarán, L.G. and Langella, P. (2016) Gnotobiotic rodents: An in vivo model for the study of microbe–microbe interactions. *Front Microbiol* **7**, 409

14. McCormick, B.A. (2003) The use of transepithelial models to examine host–pathogen interactions. *Curr Opin Microbiol* **6**, 77–81

15. Menon, A., Shroyer, M.L., Wampler, J.L., Chawan, C.B. and Bhunia, A.K. (2003) *In vitro* study of *Listeria monocytogenes* infection to murine primary and human transformed B cells. *Comp Immunol Microbiol Infect Dis* **26**, 157–174

16. Ngamwongsatit, P., Banada, P.P., Panbangred, W. and Bhunia, A.K. (2008) WST-1-based cell cytotoxicity assay as a substitute for MTT-based assay for rapid detection of toxigenic *Bacillus* species using CHO cell line. *J Microbiol Methods* **73**, 211–215

17. Salyers, A.A. and Whitt, D. (2002) *Bacterial pathogenesis: A molecular approach.* Washington, D.C.: ASM Press

18. Trevijano-Contador, N. and Zaragoza, O. (2014) Expanding the use of alternative models to investigate novel aspects of immunity to microbial pathogens. *Virulence* **5**, 454–456

Foodborne Viral Pathogens and Infective Protein

6

Introduction

Foodborne viral diseases account for about 5.5 million illnesses, 15,284 hospitalizations, and 156 deaths each year in the United States of America. Estimated economic burden for foodborne viruses is about 3 billion dollars. Viruses are obligate intracellular parasites, requiring a live host for replication. They cannot grow outside their specific host or within foods. Viruses are generally "host-specific," i.e., infecting specific plants, animal, human, or bacteria and usually do not establish cross-species infections. However, zoonotic viruses are able to infect humans, and other viruses can occasionally undergo genetic modifications to adapt themselves to different hosts.

The majority of foodborne viruses are considered enteric due to their fecal–oral mode of transmission. Enteric viruses are generally non-enveloped virus and are stable in the environment. There are four acute gastroenteritis-causing viruses: *Calicivirus*, *Rotavirus*, *Astrovirus*, and *Adenovirus*. *Norovirus* (*Calicivirus*) causes about 5.4 million illnesses in the USA per annum, while *Rotavirus*, *Astrovirus*, and *Sapovirus* cause approximately 15,000 cases, and hepatitis A virus causes approximately 1500 cases.

Enteric viruses (Table 6.1) are highly infectious. A low dose of virus consisting of as few as 10–100 particles is sufficient to cause foodborne infection. Virus life cycle consists of several steps including: (1) attachment to host receptor, (2) penetration of the host cells, (3) uncoating of RNA/DNA, (4) transcription and/or translation, (5) RNA/DNA replication, (6) assembly and packaging of nucleic acid with viral proteins, and (7) release of matured virus particles (see Chap. 2). RNA virus encodes genes for RNA-dependent RNA polymerase, a nonstructural enzyme (replicase), which is needed for RNA replication inside the host, while the DNA-dependent RNA polymerase (called RNA transcriptase) is required for transcription of RNA from DNA and protein synthesis such as capsid protein for viral packaging. Viruses do not carry any genes for metabolism and hence rely on host cells for propagation.

Sources and Transmission

Contaminated foods readily transmit the virus. Primary contamination occurs before harvesting. Examples include oyster and clams (which concentrate virus particles) and vegetables that are irrigated with contaminated or polluted water. Secondary transmission occurs during processing or handling of products where fecally contaminated hands are in contact with foods. Secondary transmission can also occur when vomitus, containing virus particles, is aerosolized, contaminating foods and food contact surfaces. Uncooked foods that receive no heat treatment are common sources of virus infection.

© Springer Science+Business Media, LLC, part of Springer Nature 2018
A. K. Bhunia, *Foodborne Microbial Pathogens*, Food Science Text Series,
https://doi.org/10.1007/978-1-4939-7349-1_6

Table 6.1 Foodborne viruses

Name	Family (Genus)	Size (genome)	Foodborne	Incubation period in days (median)
Poliovirus, Coxsackievirus, Echovirus, Enterovirus	*Picornaviridae* (*Enterovirus*)	28 nm (ssRNA)	Yes, mainly water, present in shellfish	1–5 (3)
Astrovirus	*Astroviridae*	28 nm (ssRNA)	Yes, shellfish	2–3
Hepatitis A virus	*Picornaviridae* (*Hepatovirus*)	28 nm (ssRNA)	Yes	15–50 (28)
Hepatitis E virus	*Hepeviridae* (*Hepevirus*)	34 nm (ssRNA)	Mainly water	14–60
Rotavirus	*Reoviridae* (*Rotavirus*)	70 nm (dsRNA)	Rare often water	1–4 (2)
Sapovirus	*Caliciviridae* (*Sapovirus*)	34 nm (ssRNA)	Yes (rare), mainly shellfish	1–3 (2)
Adenovirus group F, serotypes 40 and 41	*Adenoviridae* (*Mastadenovirus*)	100 nm (dsDNA)	Water, shellfish	3–10 (5)
Norovirus	*Caliciviridae* (*Norovirus*)	34 nm (ssRNA)	Yes	1–2 (1)
Nipah virus	*Paramyxoviridae* (*Henipavirus*)	500 nm (ssRNA)	Fruit bat, fruits, and sap	7–14
Ebola virus	*Filoviridae* (*Ebolavirus*)	1200 nm (ssRNA)	Yes, Fruit bat and other animals	2–21 (7–10)

These foods include salads, bakery products, and raw shellfish. Ice made from virus-contaminated water can also be a source of transmission.

Virus Classification/Taxonomy

Enteric viruses are generally nonenveloped viruses and belong to the family of *Adenoviridae*, *Caliciviridae*, *Hepeviridae*, *Picornaviridae*, and *Reoviridae*. Viruses are classified based on the size, shape, structure, and nucleic acid content (Fig. 6.1). Viruses contain either DNA or RNA with single- or double-stranded nucleic acid molecules. Adenovirus has double-stranded DNA, and RNA viruses generally have single-stranded RNA molecules. Based on the genetic elements, viruses can be classified into seven types: dsDNA, ssDNA, dsRNA, (+) sense ssRNA, (−) sense ssRNA, RNA reverse-transcribing viruses, and DNA reverse-transcribing viruses.

Foodborne Viral Pathogens

Adenovirus

Adenovirus has a large icosahedral structure containing a double-stranded DNA (genome 28–45 kb long). Adenovirus is a nonenveloped virus. It is a member of *Adenoviridae* family and genus *Mastadenovirus*, which includes >20 known viruses (5 human, 3 bovine and 3 porcine, and 9 other species). There are 51 serotypes of human adenovirus. While many adenoviruses can infect the intestinal tract, serotypes 40 and 41 cause the majority of human adeno-gastroenteritis and are shed in feces in large numbers. The incubation period is about 3–10 days and the illness lasts for about a week. Immunocompetent adults are more resistant to these viruses, while children below 2 years of age are most susceptible. The clinical symptoms are associated with gastrointestinal complications involving watery diarrhea. Waterborne outbreak resulting in conjunctivitis has been reported. Failure in proper chlorination of water may lead to an outbreak. Adenovirus is also frequently found in shellfish.

Astrovirus

Astroviruses are RNA viruses and are small (28 nm diameter) with star-like surface projections. Human astroviruses have eight serotypes, with serotypes 1 and 2 being predominant in children. It causes diarrhea in children and the illness is generally mild. The incubation period is

Fig. 6.1 Schematics of a model virus showing various structural components

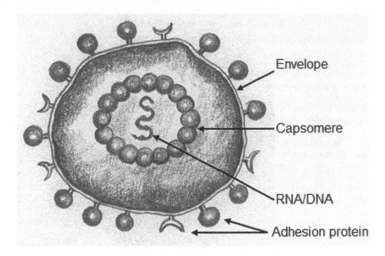

2–3 days and the disease lasts for about 3–4 days. A major foodborne outbreak was reported in Japan in 1994 affecting 1500 school children and the teachers. Serotype 4 was responsible for that outbreak. The virus can be cultured using mammalian cells.

Rotavirus

Rotavirus looks like a wheel (rota means a "wheel") in which capsid proteins are arranged like spokes of a wheel. The capsid structure consists of an inner core, an intermediate capsid, and an outer capsid with short radiating spikes (Fig. 6.2). It is a large dsRNA nonenveloped virus belonging to *Reoviridae* family. The segmented genome encodes six structural viral proteins (VP1–VP4, VP6, and VP7) and six nonstructural proteins (NSP1–NSP4, NSP5/NSP6). Among the VP proteins, VP1 is the RNA-dependent RNA polymerase, VP2 is core protein, VP3 is methyltransferase, VP6 is inner capsid, and VP4 and VP7 are outer capsids. Among the NSPs, NSP1 is an interferon antagonist, and NSP4 is an enterotoxin. Rotavirus is grouped in eight (A–H) groups. Groups A, B, C, and H are known to infect humans. Of which, group A is responsible for more than 90% of all infections in humans.

Rotavirus primarily infects children of less than 5 years of age and causes acute gastroenteritis.

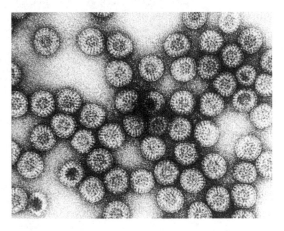

Fig. 6.2 Electron microscopic picture of rotaviruses (70 nm particles). Bar = 100 nm (Courtesy of Dr. Erskine L. Palmer, CDC, Atlanta, GA)

Foodborne rotavirus is responsible for an estimated 15,000 cases per year in the USA worldwide; rotavirus causes about 215,000 deaths annually in children under age 5. The incubation period is about 1–4 days, and the illness is manifested by diarrhea, vomiting, and fever, lasting for a week. Severe dehydration and electrolyte imbalance are responsible for fatalities. A cell culture model (MA104, monkey kidney epithelial cell line) is available to study rotavirus pathogenesis.

Rotavirus infection starts with viral attachment to matured enterocytes using VP4 to the host cell glycan (i.e., sialic acid). The virus then enters the enterocytes at the tip of the small intestinal villi. Histo-blood group antigens

(HBGAs) may serve as a receptor for some rota-viral strains. Virus induces structural changes such as villous atrophy and infiltration of mono-nuclear inflammatory cells in the lamina propria. Rotaviruses are released from infected epithelial cells without destroying them or causing cell death. Thus, maldigestion and maladsorption of nutrients and a consequent inhibition of reab-sorption of water lead to diarrhea. Nonstructural viral protein, NSP4, has been shown to act as an enterotoxin. It promotes chloride secretion and fluid loss. Chloride secretion response is regu-lated by phospholipase C-dependent calcium sig-naling pathway. NSP4 may disrupt tight junction barrier function and may induce paracellular per-meability. Viruses are released in the stool in high numbers (about 10^9 particles per gram) and can contribute to the fecal–oral transmission. Diagnosis is relatively simple which is accom-plished by electron microscopy, agglutination assay or enzyme-linked immunosorbent assay (ELISA), and reverse transcriptase PCR with stool specimen. Rotavirus infection can induce a long-lasting immunity. Two live, attenuated oral vaccines, Rotarix® and RotaTeq® are available to prevent rotavirus-associated gastroenteritis and mortality.

Hepatitis Viruses

Hepatitis viruses are a major public health con-cern worldwide. They are grouped into hepatitis A, B, C, D, and E (Table 6.2). Hepatitis A virus (HAV) and hepatitis E virus (HEV) are known food–/waterborne pathogens, while hepatitis B

(HBV), C (HCV), and D (HDV) viruses are blood-borne and are responsible for acute and chronic liver diseases. A brief description of HBV, HCV, and HDV is included to have some understanding of non-foodborne hepatitis viruses, while HAV and HEV as food–/water-borne pathogens are discussed in detail below.

Hepatitis B Virus

Hepatitis B virus (HBV) is responsible for about 786,000 deaths per year globally. It is a double-stranded DNA virus (3.2 kb) and is transmitted through contact with infected blood and semen. It causes sexually transmitted disease with a high rate of infection seen in homosexuals or heterosexuals with multiple sexual partners, injection drug users, and healthcare personnel. Perinatal infection from mother to neonates is common in the high endemic area. It causes chronic infection. Patients suffer from liver cir-rhosis or hepatic carcinoma. A prophylactic vac-cine made from recombinant DNA that expresses hepatitis B virus surface antigen (HBsAg) is highly effective.

Hepatitis C Virus

Hepatitis C virus (HCV) is a single-stranded RNA virus with 9.6 kb genome. It exhibits a high rate of mutation due to the lack of proofreading activities of RNA-dependent RNA polymerase. It belongs to *Flaviviridae* family. It infects about 130–170 million people worldwide, and these patients are in the high risk of developing hepat-osteatosis (accumulation of lipids in the liver), liver fibrosis (scarring), liver cirrhosis, and hepatic cancer. Transmission occurs predomi-

Table 6.2 Characteristics of hepatitis viruses

Hepatitis virus	Genome	Properties	Clinical significance
HAV	ssRNA(−); 7.5 kb; 27–32 nm	Fecal–oral transmission	Hepatitis, jaundice
HBV	dsDNA; 3.2 kb	Blood-borne	Acute and chronic liver disease
HCV	ssRNA(+); 9.6 kb	Blood-borne	Acute and chronic liver disease; hepatosteatosis (accumulation of lipids in the liver), liver fibrosis (scarring), liver cirrhosis, and hepatic cancer
HDV	ssRNA; 1.7 kb	Blood-borne	Coinfection with HBV
HEV	ssRNA(+); 7.5 kb; 27–34 nm	Fecal–oral transmission; water	Acute hepatitis; increased mortality during pregnancy

nantly through contaminated blood due to blood transfusion, intravenous drugs, organ transplants, and vertical transmission from mother to child.

Hepatitis D Virus

Hepatitis D Virus (HDV) is also known as "delta hepatitis" infection and is often associated with HBV infection. HDV contains a small RNA genome (1.7 kb) with single ORF (open reading frame) and is unable to synthesize its own envelope protein to form a fully functional virion. Thus, it is called defective virus particle or satellite virus and depends on HBV to supply the envelope protein or surface antigens (HBsAgs) for packaging during coinfection with HBV. HDV infection is blood-borne, and infection can spread during coinfection with HBV. HDV infection is involved in acute or chronic liver disease and infection is uncommon in the USA. Globally more than 15–20 million people are infected by HDV.

Hepatitis A Virus

Introduction

The hepatitis A virus (HAV) was identified in 1972 when an immunoelectron microscopy was used for diagnosis. The most common infection routes are person-to-person (fecal–oral route) via household contact, or in homosexual men, intravenous drug users sharing the same needles, and exposure to contaminated food and water. Foods implicated in outbreaks including clams, mussels, raw oysters, lettuce, ice slush beverages, frozen strawberries, blueberries, raspberries, and green onions. The largest outbreak occurred in 1988 in Shanghai (China) due to consumption of contaminated clams, and about 300,000 people showed the symptom of acute hepatitis with 47 deaths. Generally, the HAV infection is asymptomatic in children under 6 years of age, while it is symptomatic in older children and adults with jaundice occurring in greater than 70% patients. Annually, approximately 1.4 million people suffer from HAV infection worldwide.

Biology

HAV particle is 27–32 nm in diameter and has icosahedral symmetry. It is a nonenveloped single-stranded RNA virus (7.5 kb) and belongs to *Picornaviridae* family and genus, *Hepatovirus*. HAV has seven genotypes and four of them (IA, IB, II, III) are associated with human infection and three others (IV, V, and VI) are associated with nonhuman primates. The viral genome encodes a single open reading frame for a polyprotein of 250 kDa. The polyprotein has three regions: P1 represents structural protein, i.e., the capsid protein, and P2 and P3 represent nonstructural proteins required for RNA synthesis and virion assembly.

Pathogenesis

The incubation period of the disease is about 15–50 days (median 28 days). From the intestine, HAV reaches to the liver after a systemic circulation. It interacts with the host cell receptor protein called HAVCR1 (hepatitis A virus cell receptor 1) before entry into hepatocytes. It replicates inside hepatocytes, moves to the gall bladder and released into bile, and eventually is shed in the stool. Infection results in inflammation in the liver. Viruses impair liver function and as a result, bilirubin accumulates in blood and jaundice develops. Two to three weeks after the infection, the immune response to virus develops. Consequently, activated immune cells, CD8[+] T, and NK cells attack virus-infected hepatocytes to eliminate the virally infected cells. As a result, hepatocytes are severely damaged manifesting characteristic viral pathogenesis.

The major symptom of hepatitis is jaundice, characterized by yellow discoloration of the skin and the white part of the eye. In jaundice patient, the stool is pale colored and the urine becomes dark. Anorexia, vomiting, malaise, and fever are manifested in the hepatitis patients, and virus particles are shed in large numbers in the stool (about 10^9 particles per gram). Viruses are also shed through saliva. Liver failure may occur in patients with the underlying chronic liver disease. Children may exhibit asymptomatic HAV infection and shed viruses longer than the adults do, while older children and adults show symptoms. The long-lasting immunity primarily humoral (antibody) response, after primary infection, is seen in patients.

Immune Response

The innate immunity involves the production of interferons (IFNα, IFNβ) and antiviral state of viral recognition through pathogen recognition receptors (PRRs) or pathogen-associated molecular patterns (PAMPs). The adaptive response includes the production of IFNγ and activation of CD4+ regulatory T cells (Treg) that express CD25+ and FoxP3 and suppress the immune response to the pathogen. Cytotoxic CD8+ T cells respond to viral antigens through MHC class I pathway. HAV also induces a long-term humoral antibody response.

Prevention and Control

Detection of HAV is achieved by immunological methods such as immunofluorescence assay, dot blot assay, immunoblotting, and ELISA. In the cell culture using African green monkey kidney cell line (Vero) and the fetal rhesus monkey kidney cell, virus replication could be detected in 2–4 weeks by immunoassays. Reverse transcriptase-polymerase chain reaction (RT-PCR) and real-time PCR assays have been used to detect the virus. More recently, nucleic acid sequencing of the PCR-amplified products has been done to confirm and to determine the genetic relatedness among the HAV isolates.

It is difficult to trace the food source because of the long incubation period of the disease. HAV may survive about a month in the environment. It is resistant to chlorine and requires 1 min exposure to 1:100 dilutions of household bleach (sodium hypochlorite). Inactivation by heating requires >85 °C for 1 min. Immunization has been effective in reducing the hepatitis cases, especially in children. The reduction in children hepatitis cases possibly affects the hepatitis infection cycle thus probably reduces the number of adult hepatitis cases in recent years. However, total numbers of sporadic hepatitis cases have not been reduced. Vaccination of food handlers may reduce HAV cases but may not be cost-effective. Personal hygiene and hand washing after toilet visit are the most effective practice in preventing the transmission of HAV.

Hepatitis E Virus

Introduction

Originally, identified as an "atypical hepatitis A virus" that caused an outbreak in New Delhi (India) and infected about 29,000 people in 1955, hepatitis E virus (HEV) was originally classified as hepatitis non-A, non-B virus. Now it has been renamed hepatitis E virus. This virus causes acute hepatitis, annually infecting about 20 million people worldwide. HEV is a member of family *Hepeviridae*, genus *Hepevirus*, and there are four genotypes. Genotype 1 and 2 are associated with person-to-person human infection, and genotype 3 and 4 are associated with zoonotic transmission to humans, from pigs and other species.

HEV is primarily a waterborne virus and caused several outbreaks in developing countries in Asia, Latin America, and Africa. Swine serves as the most important reservoir for HEV. HEV infection can also transfer through organ transplants. The high-risk group includes patients suffering from chronic liver disease and pregnant women and young children. Adults and young adults are also susceptible. Though it is a waterborne virus, several recent outbreaks were associated with consumption of deer meat and raw or undercooked swine liver in Japan. HEV has been routinely isolated from swine from Canada, South Korea, Japan, Spain, the Netherlands, New Zealand, and the USA implying potential future outbreak of HEV with the consumption of undercooked pork meat.

Biology

HEV is a nonenveloped icosahedral-shaped spherical particle (27–34 nm) containing a single-stranded RNA with genome size, 7.2 kb. The genome contains three ORFs: ORF1 encodes nonstructural proteins, RNA-dependent RNA polymerase (RdRP), protease, methyltransferase, and helicase; ORF2 encodes a structural protein, capsid that is required for viral entry; and ORF3 encodes phosphorylated protein required for viral release from the host cells.

Pathogenesis

The incubation period of HEV varies from 2 weeks to 2 months. Propagation of human HEV is currently challenging but hepatic cell lines, Huh7, HepG2, lung cell line A549, and colon cell line Caco-2, have been used as cell culture models to study pathogenesis. Pathogenesis involves several steps that require viral interaction with the host cell receptor, heparin sulfate proteoglycans, and to another uncharacterized receptor molecule. Upon internalization, viral uncoating allows RNA release followed by RNA replication, packaging, and release of mature virions from the host cells.

The disease is dose dependent: higher dose shows clinical symptoms while lower dose exhibits subclinical infection. It is responsible for acute viral hepatitis similar to HAV and manifests mild jaundice, anorexia, and hepatomegaly. Some patients suffer from abdominal pain, nausea, vomiting, and fever.

Norovirus

Introduction

Norovirus (formerly Norwalk virus) is the leading cause of gastroenteritis and a major public health concern worldwide. The first outbreak occurred in 1968 in children and the adults in Norwalk, OH (USA), hence the name Norwalk, but the virus was not identified until 1972. Dr. Albert Z. Kapikian (1930–2014) at the National Institutes of Health was the first person to identify the virus using an electron microscopy (Fig. 6.3). The Norwalk virus is now called

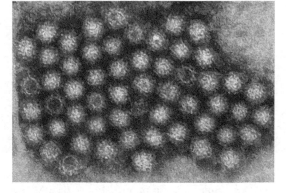

Fig. 6.3 Electron microscopic image of norovirus (Courtesy of Charles D. Humphrey, CDC, Atlanta, GA)

Norovirus (NoV). NoV is highly contagious and the transmission routes are food, water, person, and environment (Fig. 6.4). About 20 million Americans are affected every year resulting in 56,000–71,000 hospitalizations and 570–800 deaths. Of the total illnesses, about 5.5 million illnesses are attributed to food.

NoV has been responsible for numerous outbreaks in various establishments: the cruise ships, restaurants, swimming pool, schools, nursing homes, and hospitals. Primary transmission to humans can happen through food, and then secondary transmission occurs from fecal–oral or person-to-person and from the environment. Secondary transmissions are expedited when people are in close contact in settings like hospitals, cruise ships, hotels, restaurants, nursing homes, day-care centers, prisons, military installments, and sports stadiums. Fresh produce such as lettuce, tomato, melons, green onions, strawberries, raspberries, peppers, fresh-cut fruits, salads, and food handlers, processors, and irrigation water are also involved in the transmission. Transmission can occur through filter-feeding bivalves including muscles, oysters, and clams, which collect viruses in their tissues.

Biology

Norovirus is also known as a small round structured virus (SRSV) of 27–38 nm diameters with icosahedral shape (Fig. 6.3). It is a nonenveloped virus carrying a plus-sense single-stranded RNA, and the 7.4–7.7 kb genome is comprised of three ORFs. ORF1 encodes a nonstructural polypeptide consisting of p48, NTPase, p22, VPg (viral genome-linked protein), viral protease, and RNA-dependent RNA polymerase (RdRP). ORF2 encodes a major capsid protein, VP1, and ORF3 encodes a minor capsid protein, VP2. VP1 binds to the host cell receptor, histo-blood group antigens (HBGAs), promotes viral entry, and determines antigenicity and strain specificity. VP1 also elicits host protective antibody, cellular and humoral immunity; while the VP2 helps in RNA packaging, regulates the synthesis of VP1, and stabilizes the VP1 structure.

NoV belongs to the family of *Caliciviridae* that has six genera: *Norovirus*, *Sapovirus*,

Fig. 6.4 Transmission pathway for norovirus (Schematics based on Ushijima et al. 2014. Food Safety, 2, 37–54)

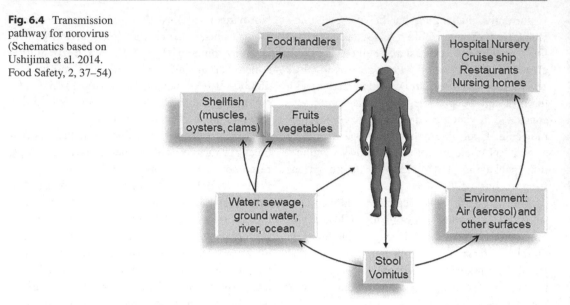

Lagovirus, Vesivirus, Recovirus, and *Becovirus.* Human infections are generally associated with viruses from the genus *Norovirus.* NoV is now grouped into six genogroups (G1-GVI): GI is isolated from humans; GII from both humans and swine; GIII from cattle; GIV from human, feline, and canine; GV from murine; and GVI from the canine. Genogroups are again subdivided into genotypes. Genogroup GI has >9 while GII has >22 genotypes. GII.4 is currently the most common epidemic strain.

Pathogenesis

There is no reliable cell culture or small animal model, which can be used to study human norovirus pathogenesis or viral growth. However, gnotobiotic pigs, gnotobiotic calves, rhesus macaques, and chimpanzees have shown to respond to human norovirus infection. Feline calicivirus (FCV) and murine norovirus (MNV) have been used as surrogates. FCV is the member of family *Caliciviridae* and genus *Vesivirus.* Likewise, MNV is the member of *Caliciviridae* family and genus *Norovirus.* Both possess some attributes similar to the human norovirus, and they can be cultured in cell culture models thus, have been used widely as NoV surrogates. The infectious dose of NoV is very low: about 10–100 virus particles. Severity and the onset of disease may depend on a number of virus parti-

cles ingested. The HBGAs distributed in blood cells, vascular endothelial cells, and mucosal gastrointestinal epithelial cells serve as the receptor for viral interaction. Norovirus infects mature enterocytes covering the small intestinal villi leading to massive cell damage and malabsorption. Damaged cells are rapidly replaced by undifferentiated immature enterocytes originated from the crypt, which are not susceptible to the virus infection. These immature cells cannot function properly; thus, malabsorption continues until the cells mature. The virus causes gastroenteritis, characterized by explosive projectile vomiting, nausea, cramps, diarrhea, dehydration, anorexia, headache, chills, fever, and myalgia. Adults are more susceptible to norovirus than children are, and the incubation period is 24–48 h.

Symptoms appear within 12–24 h after ingestion and last for 2–3 days. The disease is self-limiting, and the viruses are excreted in the vomitus and feces of infected persons at the rate of 10^4–10^5 virus particles per gram of vomitus and 10^8–10^{10} particles per gram of feces. Recovering patients shed virus for an extended period even up to a month or longer. Immunocompromised patients, children, and the elderly may shed for a prolonged duration. Adults may encounter recurrent infections despite the presence of norovirus-specific antibody in the serum. Cell-mediated immunity is Th1-dependent, and IFNγ is the dominant cytokine.

Prevention and Control

Effective sanitization and disinfection are needed to prevent and control the spread of NoV. However, it is resistant to standard sanitizers and disinfectants at the permissible levels during food processing or food production. Since it is resistant to standard chlorination treatment, so a concentration >10 mg L^{-1} is needed to disinfect water. As an industry practice for leafy green sanitization, chlorine is used at 0.2 mg L^{-1}, which is not effective against NoV; however, 15–30 s treatment with 2–5% trisodium phosphate solution (TPS) can effectively reduce viral loads on produce. Human milk, oligosaccharide, milk glycoprotein, and milk glycolipid, contain the same epitope as HBGA and can block viral binding to the epithelial cells and provide a strategic approach to preventing viral interaction with enterocytes. Currently several vaccines are under development based on viral proteins; however, due to the lack of a cell culture model to grow the virus, live attenuated viral vaccine development is not possible. Noroviruses are genetically and antigenically diverse; thus, vaccine efficacy would be limited. The virus is routinely detected by ELISA and by RT-PCR assay targeting the RNA poly-merase gene. Electron microscopy is used as a confirmatory test.

Zoonotic Viral Pathogens

Avian Flu Virus

Introduction

Avian flu is also known as avian influenza and the predominant strain is H5N1. Avian influenza virus infects birds including poultry; however, it is capable of adapting and causing infection in humans, especially the poultry handlers, and thus it is considered a zoonotic pathogen (Fig. 6.5). Avian influenza virus has been isolated from wild birds such as geese, ducks, and waterfowls and mammals including pig, horse, dog, and sea mammals. The migratory birds can readily transmit the virus to different continents. Even though the avian influenza virus has been isolated from poultry meat, there is no evidence for foodborne transmission. Though human-to-human transmission has not been confirmed, scientists predict that avian influenza can possibly cause a pandemic, which is now responsible for the epidemic in Asia, Africa, and Eastern Europe. According to the World Health Organization

Fig. 6.5 Interspecies transmission pathways for avian flu virus (avian influenza H5N1 virus)

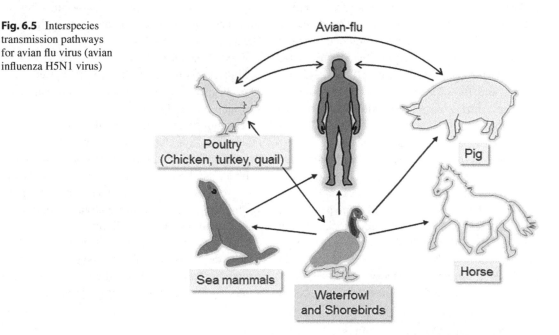

(WHO), between 2003 and 2015, avian influenza virus strain H5N1 infected 840 and killed 447 worldwide.

There are three types of influenza viruses: A, B, and C. Influenza virus type A is responsible for human epidemic every year, and it was responsible for past pandemics. In the human history, three major pandemics of influenza A virus have been reported: 1918 Spanish flu (H1N1), 1957 Asian flu (H2N2), and 1968 Hong Kong flu (H3N2). These three pandemics killed millions of people. The avian flu virus has an avian origin or the virus results from avian-human virus reassortment.

Biology

Avian influenza virus belongs to the family *Orthomyxoviridae* and genus *Influenzavirus A*. Influenza A virus is an enveloped virus and the size is about 80–120 nm in diameter. It is a negative-sense single-stranded RNA virus with eight different segments and encodes ten proteins including surface glycoproteins, hemagglutinin (HA), and neuraminidase (NA) and matrix proteins M2 and M1, nonstructural proteins NS1 and NS2, the nucleocapsid, and the three polymerase enzymes, PB1 (polymerase basic 1), PB2 (polymerase basic 2), and PA (polymerase acidic).

HA and NA are the surface antigens and are a determinant of the pathogenicity, transmission, and adaptation of the virus to other species. HA is the most important determinant and binds to the host epithelial cell receptor for viral entrance and replication, while NA is responsible for the release of newly formed viruses. Both HA and NA are responsible for antigenic variation resulting in antigenic drift and shift and allow the virus to evade the host immune system. The virus lacks a proofreading and correction mechanism during viral replication; hence, the error introduced due to nucleotide substitution, deletion, or point mutation results in antigenic variation. There are 16 subtypes of HA (H1–H16) and 9 subtypes of NA (N1–N9), and many of these are associated with various animals including humans, dogs, pigs, horses, and birds. The subtypes H1, H2, and H3 and N1 and N2 are associated with human infection; H1 and H3 and N1 and N2 are associated

with pigs and H3 and N8 in dogs. A majority of avian influenza viruses are H5 and H7 subtypes, and H5, H7, and H9 have caused sporadic infections in humans. There are two pathotypes, highly pathogenic avian influenza (HPAI) and low pathogenic avian influenza (LPAI). Several different avian influenza virus types have been isolated from different regions or countries: H1N1 (Asia), H4N6 (Canada), H9N2 (China), and H5N1 (Asia). Avian flu virus is currently named by type/place isolated/culture number/year of isolation. For example, the strain isolated from Shanghai, China, is designated as B/Shanghai/361/2002 (H5N1).

Pathogenesis

Typically, the influenza virus affects the respiratory tract. The HA of influenza virus binds to the epithelial sialic acid-containing receptor before initiating an infection. In human, this receptor consists of sialic acid–galactose, and the cross-link between these two molecules consists of α-2,6 linkage (SA α-2,6), while in birds it is α-2,3 linkage (SA α-2,3). This difference possibly prevents the avian influenza virus from readily infecting humans. Moreover, the receptors with the α-2,3 linkages are distributed in the lower part of the respiratory tract of humans thus also reduces the access of the virus. In addition, human isolates of avian H5N1 have shown to display antigenic variation to show binding to human cellular receptor containing α-2,6 linkage. The virus infects and multiplies in nasopharyngeal and alveolar epithelial cells. The virus also exhibits tropism for the liver, renal system, and other tissues showing signs of diarrhea, renal dysfunction, and lymphopenia.

The symptoms of H5N1 infection in humans appear 2–8 days after exposure, and clinical signs include flu-like symptom, fever, cough, shortness of breath, and pneumonia affecting primarily the lower respiratory tract. The disease may progress to vomiting, diarrhea, and abdominal pain. The patient requires mechanical ventilation and dies within 9 days from the onset of symptoms.

In birds, the low infective pathotype exhibits mild symptoms characterized by ruffled feathers and mild respiratory symptoms lasting approximately for

10 days. The high infective pathotype shows severe respiratory and neurological disorders, organ failure, and death within 2–3 days.

Prevention and Control

The avian flu virus can be cultured using Madin–Darby canine kidney (MDCK) cell line or embryonated eggs (see Chap. 5). Viral antigens can be detected from a clinical specimen by enzyme immunoassay or immunofluorescence assay, and viral RNA can be detected by using a reverse transcriptase PCR (RT-PCR) assay that targets genes for HA and NA synthesis. Antiviral drugs, amantadine, rimantadine, oseltamivir and zanamivir, show serotype-specific effectiveness. NA inhibitors (oseltamivir and zanamivir) are shown to be effective against H5N1 in vitro and in a mouse model. Several vaccines based on killed subunit vaccines are under development, but those must be able to protect against different strains currently causing infections in humans globally.

Waterfowl is the natural reservoir for avian influenza, and the virus can be transferred to domestic birds by respiratory and fecal–oral routes through contaminated water, feed, environment, and feces. Infiltration of wild birds should be prevented from poultry farms or premises.

The avian flu virus has the potential for causing a pandemic; thus, precaution should be taken to prevent the spread. Contact with infected domestic or wild birds should be avoided. Routine surveillance of migratory birds (dead or alive) or birds in a poultry farm for the presence of influenza virus should be performed. Vaccination of human populations may be needed to control the spread; however, concerns of antigenic variation may challenge its efficacy and effectiveness.

Can the avian flu virus be a food safety concern? It has been demonstrated that conventional cooking temperature (70 °C or more) can readily inactivate the H5N1 virus; however, the virus may not be killed by refrigeration or freezing, and in fact, the virus has been isolated from frozen duck meat. If the poultry eggs and meats are properly cooked, they can eliminate the virus.

The greatest risk of exposure is through handling and slaughter of live infected poultry.

Nipah Virus

Introduction

The first Nipah virus (NiV) outbreak was reported in Malaysia in 1998–1999, which affected 276 people and 39% fatality, and the infection was originally transmitted through exposure to infected swine. The NiV was amplified at large numbers in the respiratory tracts and facilitated the spread of infection in farm workers. NiV was originally isolated from a patient from Sungai Nipah village in Malaysia, and fruit bat (genus, *Pteropus*), also called flying fox, acts as a natural reservoir. Interaction of fruit bats with swine and humans led to increased numbers of outbreaks in Malaysia. Bats transmit the virus through saliva or urine to the fruits. Swine from a farm located near the bat habitat acquires the organism and aid in the zoonotic transmission of the disease (Fig. 6.6). Domestic animals foraging may eat virus-laden partially eaten fruits and may transmit the disease to humans. NiV is also transmitted through sap (juice) of date palm tree when fruit bats feed on sap at night, and this leads to numerous epidemic and sporadic outbreaks in Bangladesh and the eastern part of India. Date palm tree sap is used for making molasses. The virus survives well in the sap, and unheated sap can transmit the virus to humans. Person-to-person transmission also occurs. Nipah virus outbreak was also reported in Singapore, Cambodia, and Thailand, and virus has been circulating in the natural reservoir in Southeast Asia, including Malaysia, Cambodia, Indonesia, East Timor, Vietnam, Thailand, Bangladesh, India, and Papua New Guinea.

Biology

NiV is a member of the family of *Paramyxoviridae* and genus *Henipavirus*. NiV virus is about 500 nm in diameter and is larger than the typical paramyxoviruses (150–400 nm). The Nipah virus size may vary from 180 nm to 1900 nm. It is an enveloped negative-sense single-stranded RNA virus with genome size 18.246 kb. Six structural proteins are encoded in the genome: two envelope glycoproteins F (fusion) and G (receptor binding), the nucleoprotein N, phosphoprotein P, matrix protein M, and the RNA-dependent RNA polymerase L. The G and F glycoproteins are required for viral attachment and entry into the host cells. The G protein binds to the receptor molecule, Ephrin-B2, which is expressed on neurons, smooth muscles, and endothelial cells. The F protein (546 amino acid residues) is a type I transmembrane protein and facilitates the fusion of virus and the host cell membrane during the infection. Both F and G proteins induce neutralizing antibodies. The M protein (352 amino acid residues) provides rigidity and the structural stability of the virion through its interactions with the F protein, the ribonucleoprotein (RNP) complex, and the inner surface of the virion envelope. The N protein (532 amino acid residues) helps

encapsidation of the viral genome and interacts with P protein. The L protein possesses all the enzymatic activities responsible for initiation, elongation, and termination of both mRNA transcription and genome replication. P protein serves as a scaffold between the L and the encapsidated genome.

Pathogenesis

The incubation period of NiV infection is 1–2 weeks. The virus binds to the cellular receptor Ephrin B2 present on the neuron and endothelium, enters host cells, replicates, and causes cell damage. Systemic vasculitis with extensive thrombosis is seen in patients since virus attacks endothelium in the blood vessels and the CNS. The virus causes high fever, headache, myalgia, dizziness, confusion and lack of consciousness, and encephalitis. In addition, the virus causes acute respiratory tract infection, pulmonary edema, coma, and death.

Fig. 6.6 Transmission pathways for Nipah virus

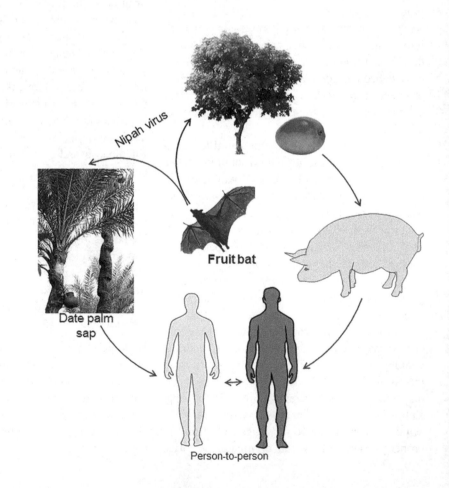

Kidneys are also affected showing signs of glomerular fibrinoid necrosis. The mortality rate in human is about 75%.

Infected bats do not show clinical signs, but serve as the carrier, whereas pigs are highly susceptible to NiV showing signs of meningitis and encephalitis, bronchointerstitial pneumonia, systemic vasculitis, and focal necrosis in the spleen and lymph nodes. Viral antigen is detected in the endothelial and smooth muscle cells of the brain, lungs, and lymphoid system. Virus antigen is present in neurons, glial cells, and epithelial cells of the upper and lower respiratory tracts.

Prevention and Control
Culling of infected swine helped reduce NiV cases in Malaysia. Heat treatment of sap or avoiding unprocessed sap consumption will also help prevent viral infection. Serologic testing for antibody titer in human sera and reverse transcriptase PCR assay have been used for diagnosis and detection from human urine, cerebrospinal fluid, and oral swabs.

Ebola Virus

Introduction
In the past several years, the Ebola virus caused major outbreaks in West and Central Africa with a case fatality rate of 25–90%. The Ebola virus was first identified in 1976 in the Democratic Republic of the Congo, and it was named after the Ebola River located near the epicenter of the first outbreak. The fruit bats are considered the reservoir. Animals or humans have acquired the disease by consuming or coming in contact with the infected bats or animals (gorillas, chimpanzees, and monkeys). Thus, there is a strong evidence for the infection to be of foodborne zoonotic disease, and bush meat may be an important link. The virus can pass through bodily fluids and spread from human-to-human.

Biology
The Ebola virus is an enveloped single-stranded RNA virus (19 kb) of the family *Filoviridae* and genus *Ebolavirus*. It is a filamentous and pleomorphic virus with about 1200 nm in length. The viral genome encodes for a nucleoprotein (NP), glycoprotein (GP), RNA-dependent RNA polymerase (L), and four structural proteins: VP24, VP30, VP35, and VP40. In addition, it expresses a truncated soluble form of GP (sGP) through RNA editing. Five strains are reported: Zaire ebolavirus (EBOV), Sudan ebolavirus (SUDV), Tai Forest ebolavirus (TAFV), Bundibugyo ebolavirus (BDBV), and Reston ebolavirus (RESTV). All are pathogenic to humans except RESTV, which is thought to be pathogenic to nonhuman primates.

Pathogenesis
The incubation period of the disease is 2–21 days (average 7–10 days). The pathogenesis is not fully understood, but the virus is thought to suppress both innate and adaptive (cellular and humoral) immune responses. The virus replicates in monocytes, macrophages, and dendritic cells. The virus is found inside endothelial cells, fibroblasts, hepatocytes, and adrenal cells and disseminates to the lymph nodes, liver, and spleen. Massive production of proinflammatory cytokines (IL-1, IL-6, IL-8, IL-15, IL-16) and several chemokines lead to shock and multi-organ failure and death. The symptoms of infection include lack of appetite, fever, headache, malaise, joint and muscle aches, abdominal pain, diarrhea and vomiting, and internal and external bleeding at the later stage of the disease. Reverse transcriptase PCR is used for diagnosis of the disease. Vaccines are being developed for humans using inactivated virus and DNA vaccine using replication-defective recombinant adenovirus type 5 expressing glycoprotein (GP) and nucleocapsid protein (NP).

Prevention and Control of Foodborne Viruses

Enteric viruses are shed in large numbers from the host through feces and vomitus, and they could be airborne or waterborne. The infectious dose is very low; thus, effective sanitization and control measures need to be employed to prevent contamination and spread. Person-to-person transmission occurs readily when people are in close contacts, especially in cruise ships, in restaurants, and in hospitals.

Depuration helps remove the virus from shellfish; however, proper water temperature should be maintained. During depuration, harvested shellfish are kept in clean fresh water for 24–48 h where viruses are escaped into the water. Food preservatives (chlorine compounds, detergents, etc.), freeze-drying, ultraviolet light, freezing, and heating at 100 °C can inactivate foodborne viruses. In general, viruses are highly stable, because the virus coat proteins most likely provide the protection against processing treatments. Viruses are remarkably stable at high temperatures such as 90–100 °C, possibly because the virus particles may remain aggregated or protected by food particles. UV treatment can inactivate the virus. During the farming of fruits and salad vegetables, clean virus-free water should be used for irrigation. Food handlers serve as a source; thus, workers' health and hygienic practices should receive the greatest attention. Shellfish are a potential source of norovirus and hepatitis A virus, and these animals should not be harvested from water that may have been polluted with sewage.

Infective Proteins

Bovine Spongiform Encephalopathy

Introduction

A group of neurodegenerative infective agents (prions) capable of transmission to various hosts is termed transmissible spongiform encephalopathies (TSEs). Several TSE agents are described in the literature: bovine spongiform encephalopathy (BSE) or "mad cow disease" in cattle, "scrapie" in sheep and goat, chronic wasting disease (CWD) in cervids, transmissible mink encephalopathy (TME) in minks, and Creutzfeldt–Jakob disease (CJD), variant CJD (vCJD), Gerstmann–Straussler syndrome, and "kuru" in humans. In the early 1950s, in the eastern highlands of Papua New Guinea, Kuru was prevalent among the islanders due to a cannibalistic practice of consumption of infected brain tissues of relatives. The disease was characterized by the degenerative brain with spongy appearance, and the victims suffered from rapid physical and mental abnormalities, culminating in paralysis, coma, and death. It was called slow virus because the incubation period is about 2–10 years.

It became a major concern in the early 1990s when the disease was detected in cattle, and the wasting of brain tissue resulted in abnormal behavior in cattle; hence, it was called "mad cow disease." Animals exhibit symptoms of abnormal gait, hyper-responsiveness to stimuli, tremors, aggressive behavior, nervousness or apprehension, changes in temperament, and even frenzy. Cattle over 24–30 months of age are susceptible to this infection. The incubation period of classical BSE is about 2–8 years. Though there is no human case directly linking the consumption of contaminated beef, finding the organism in the late 1990s and early 2000 (2003) in Canada and the USA caused a major beef embargo among developed countries with huge economic impacts costing both countries over 4–6 billion dollars.

WHO reported that from October 1996 to March 2011, 175 cases of vCJD have been reported in the UK; 25 in France; 5 in Spain; 4 in Ireland; 3 each in the Netherlands and the USA; 2 each in Canada, Italy, and Portugal; and 1 each in Japan, Saudi Arabia, and Taiwan. BSE is endemic in the UK with reported 176 cases of vCJD as of April 2012. The incubation period for vCJD in humans is 11–12 years. Before 1980, vCJD occurred due to the use of (1) cadaveric human growth hormone, (2) contaminated surgical instruments, (3) infected dura mater graft, and (4) corneal transplant. In recent years, however, consumption of contaminated animal products with a brain, lymph nodes, or neurons is thought to be responsible for transmission. One suspected source of BSE in beef is presumably due to the feeding of beef cattle with contaminated meat and bone meal (MBM) preparation, for fast growth and increased body weight gain. MBM is often prepared from sheep offal and/or condemned bovines, which are not fit for human food. In 1997, the US-FDA banned the use of proteins derived from mammalian tissues in feeding to ruminants in an effort to prevent transmission of TSE to food animals. Conversely, the UK delayed imposing such a ban and about 100 persons developed fatal cases of vCJD between roughly 1996 and 2005.

Biology

BSE-causing agent was originally thought to be a virus, but later in 1982, Dr. Prusiner discovered that it is a proteinaceous infectious particle called prion protein (PrP). He received the Nobel Prize for his work in 1997. Prion protein has aberrant protein folding, and its accumulation in nervous tissues leads to neurodegeneration. It is resistant to most treatments including heat, chemicals, and proteases. The prion is found primarily in the central nervous systems (CNS) including the brain and neurons and in lymphatic system in the gut. Amino acid sequences of PrP from normal and infected brains are identical but show differences in biochemical and biophysical behaviors. Monomeric form of the PrP protein contains 253 amino acids with a molecular mass of 22–36 kDa, while the abnormal or infective molecule is a macromolecular aggregate with a molecular mass greater than 400 kDa. The normal cellular version of PrP is called PrP^c and is encoded by a single chromosomal gene, *PRNP* located on the chromosome 20 in humans. PrP^C is sensitive (PrP^{sen}) to proteases such as proteinase K and trypsin. PrP-mRNA is 2.1 kb long and is detected primarily in the brain (neurons) and small amounts in the lungs, spleen, and heart. The infective form is resistant to protease and has a drastically different secondary structure and referred to as PrP^{Sc} (PrP from scrapie). The α helical structures are predominant in PrP^C, while a misfolding of the prion protein results in the formation of β-sheet, which is abundant in PrP^{Sc}. Normal PrP^c has α-helix of 40% and β-sheet 3% while in the disease-causing prion (PrP^{Sc}) has α-helix 30% and β-sheet 40%. The prions are highly hydrophobic and form aggregates easily. Aggregates are highly resistant to cellular digestion and accumulate in the lymphoid and nerve tissues and cause a spongiform change in the brain.

Prion (PrP^{Sc}) is highly resistant to heat, certain chemicals, and proteases (proteinase K, trypsin,

etc). It can withstand dry heat treatments of 160 °C for 24 h, or at 360 °C for 1 h, and saturated steam autoclaving at 121 °C for 1 h. The prion protein is resistant to chemical treatments such as 0.5% sodium hypochlorite for 1 h, 3% hydrogen peroxide for 1 h, and ethanol. However, complete inactivation is possible by autoclaving at 132 °C for 1.5 h, and treatment with 1 M sodium hydroxide at 20 °C for 1 h, or sodium hypochlorite (2% chloride) for 1 h at 20 °C.

Pathogenesis

Transmission of prion through digestive tract has been the subject of much investigation in recent years. Prion possibly passes through the M cells overlying the Peyer's patches, and it is then transported by the dendritic cells to the central nervous system and the brain. In another study, it is proposed that the prion bypasses the lymphoid system altogether and is directly transmitted via the peripheral nervous system to reach the CNS. PrP accumulates in the neural cells and disrupts normal neurological function, causing vacuolation (spongy appearance) and cell death.

In humans, the first signs are psychiatric, such as anxiety, depression, insomnia, withdrawal, paranoid delusions, head and neck pain, and progressive dementia. Mean duration of suffering is about 14 months. The neurologic symptom is accompanied by cerebral ataxia (defective muscular coordination) and dementia. In the terminal stage, the patient becomes bedbound, akinetic, and mute, a state in which the person is not able or will not move or make sounds.

Prevention and Control

There is no laboratory test available to use in the live animals or humans for testing of abnormal PrP^{Sc}. Postmortem analysis of brain tissues shows characteristics amyloid plaque, which is a waxy translucent substance, composed of complex protein fibers and polysaccharides that are formed in body tissues in some degenerative diseases, such as Alzheimer's disease, and spongy appearance. Immunoassays (Western blot or ELISA) are used to detect PrP^{Sc} antigens in cattle after slaughter. In humans, magnetic resonance imaging (MRI) has been used as a tentative diagnosis to detect cortical atrophy in the brain, coupled with clinical signs. There is no treatment available for the infectious prion.

PrP^{Sc} is concentrated in certain tissues in infected animals and referred to as SRM (specific risk material). These include the brain, spinal cord, skull, vertebral column, eyes, tonsil, and ileum. BSE can be prevented by several ways: (1) routine surveillance for BSE-infected cattle, (2) prevent entry of BSE agent in cattle population, (3) stop feeding the beef cattle with animal proteins derived from other animals, and (4) identify and condemn the infected cattle before entering into the human food chain. Currently, the European Union and the USA have banned the feeding of animal proteins from other animals. BSE suspect carcasses should not be used for food, and carcasses should be destroyed at 133 °C (under pressure) for 20 min.

foodborne source, but the virus can spread from human-to-human through bodily fluids. Transmissible spongiform encephalopathy (TSE) diseases such as bovine spongiform encephalopathy (BSE) and varient Creutzfeldt-Jakob Disease (vCJD) are caused by misfolded neurodegenerative infectious prion proteins (PrP^{Sc}), which are highly resistant to heat and protease enzymes and can be transmitted by consuming contaminated meat. Preventing the use of meat–bone meal (MBM) or specific risk materials (SRM) can prevent the spread of prions among the meat-producing animals and to humans.

Further Readings

1. Ansari, A.A. (2014) Clinical features and pathobiology of Ebolavirus infection. *J Autoimmun* **55**, 1–9.
2. Arias, C.F., Silva-Ayala, D. and López, S. (2015) Rotavirus Entry: a deep journey into the cell with several exits. *J Virol* **89**, 890–893.
3. Benova, L., Mohamoud, Y.A., Calvert, C. and Abu-Raddad, L.J. (2014) Vertical transmission of hepatitis C virus: Systematic review and meta-analysis. *Clin Infect Dis* **59**, 765–773.
4. Carter, M.J. (2005) Enterically infecting viruses: pathogenicity, transmission and significance for food and waterborne infection. *J Appl Microbiol* **98**, 1354–1380.
5. Chmielewski, R. and Swayne, D.E. (2011) Avian Influenza: Public health and food safety concerns. *Annu Rev Food Sci Technol* **2**, 37–57.
6. Cook, N., Knight, A. and Richards, G.P. (2016) Persistence and elimination of human norovirus in food and on food contact surfaces: a critical review. *J Food Prot* **79**, 1273–1294.
7. Croser, E.L. and Marsh, G.A. (2013) The changing face of the henipaviruses. *Vet Microbiol* **167**, 151–158.
8. de Wit, E. and Munster, V.J. (2015) Animal models of disease shed light on Nipah virus pathogenesis and transmission. *J Pathol* **235**, 196–205.
9. Dormont, D. (2002) Prions, BSE and food. *Int J Food Microbiol* **78**, 181–189.
10. Echeverria, N., Moratorio, G., Cristina, J. and Moreno, P. (2015) Hepatitis C virus genetic variability and evolution. *World J Hepatol* **7**, 831–845.
11. Esona, M.D. and Gautam, R. (2015) Rotavirus. *Clin Lab Med* **35**, 363–391.
12. Fausther-Bovendo, H., Mulangu, S. and Sullivan, N.J. (2012) Ebolavirus vaccines for humans and apes. *Curr Opin Virol* **2**, 324–329.
13. Greenlee, J.J. and Greenlee, M.H.W. (2015) The transmissible spongiform encephalopathies of livestock. *ILAR Journal* **56**, 7–25.

Summary

Foodborne viruses such as rotavirus, norovirus, and hepatitis A virus cause enteric disease characterized by gastroenteritis and other complications and affect a large number of people every year. Foodborne enteric viruses are generally RNA virus, and they are shed in large numbers (about 10^9 particles per gram) from infected patients through vomitus and feces. Person-to-person or fecal–oral transmission is a common mechanism for viral infection. Since viruses are highly infectious, only a small dose of 10–100 particles is sufficient to cause an infection. Zoonotic viral pathogens are transmitted from animals and avian species to humans, sometimes through direct contact with the animal or the meat. Avian influenza virus is transmitted primarily through contact or aerosol to the bird handlers and is not considered a foodborne pathogen, but it has the potential to cause human pandemics. Avian flu virus infection is fatal and affects lower respiratory tract resulting in pneumonia. The Nipah virus is transmitted by fruit bat while feeding on fruits or palm sap, and it is responsible for viral transmission to swine or directly to humans. It has a very high mortality rate. Likewise, Ebola virus is a zoonotic viral pathogen with 25–90% mortality rate. The infection may be the

14. Han, H.-J., Wen, H.-l., Zhou, C.-M., Chen, F.-F., Luo, L.-M., Liu, J.-w. and Yu, X.-J. (2015) Bats as reservoirs of severe emerging infectious diseases. *Virus Res* **205**, 1–6.

15. Huang, C.-R. and Lo, S.J. (2014) Hepatitis D virus infection, replication and cross-talk with the hepatitis B virus. *World J Gastroenterol* **20**, 14589–14597.

16. Kalthoff, D., Globig, A. and Beer, M. (2010) (Highly pathogenic) avian influenza as a zoonotic agent. *Vet Microbiol* **140**, 237–245.

17. Kapikian, A.Z. (2000) The discovery of the 27-nm Norwalk virus: An historic perspective. *J Infect Dis* **181**, S295–S302.

18. Kingsley, D.H. (2016) Emerging foodborne and agriculture-related viruses. *Microbiol Spectrum* **4**.

19. Ksiazek, T.G., Rota, P.A. and Rollin, P.E. (2011) A review of Nipah and Hendra viruses with an historical aside. *Virus Res* **162**, 173–183.

20. Lee, J., Kim, S.Y., Hwang, K.J., Ju, Y.R. and Woo, H.-J. (2013) Prion diseases as transmissible zoonotic diseases. *Osong Public Health Res Perspect* **4**, 57–66.

21. Li, J., Predmore, A., Divers, E. and Fangfei, L. (2012) New interventions against human norovirus: progress, opportunities, and challenges. *Annu Rev Food Sci Technol* **3**, 331–352.

22. Lorrot, M. and Vasseur, M. (2007) How do the rotavirus NSP4 and bacterial enterotoxins lead differently to diarrhea? *Virol J* **4**, 31.

23. Marsh, G.A. and Wang, L.-F. (2012) Hendra and Nipah viruses: why are they so deadly? *Curr Opin Virol* **2**, 242–247.

24. Meng, X.J. (2010) Hepatitis E virus: Animal reservoirs and zoonotic risk. *Vet Microbiol* **140**, 256–265.

25. Meyers, L., Frawley, T., Goss, S. and Kang, C. (2015) Ebola virus outbreak 2014: Clinical review for emergency physicians. *Ann Emerg Med* **65**, 101–108.

26. Moore, M.D., Goulter, R.M. and Jaykus, L.-A. (2015) Human norovirus as a foodborne pathogen: challenges and developments. *Annu Rev Food Sci technol* **6**, 411–433.

27. Nainan, O.V., Xia, G., Vaughan, G. and Margolis, H.S. (2006) Diagnosis of Hepatitis A virus infection: a molecular approach. *Clin Microbiol Rev* **19**, 63–79.

28. Pabbaraju, K., Tellier, R., Wong, S., Li, Y., Bastien, N., Tang, J.W., Drews, S.J., Jang, Y., Davis, C.T., Fonseca, K. and Tipples, G.A. (2014) Full-genome analysis of avian influenza A(H5N1) virus from a human, North America, 2013. *Emerg Infect Dis* **20**, 887–891.

29. Parashar, U.D., Bresee, J.S., Gentsch, J.R. and Glass, R.I. (1998) Rotavirus. *Emerg Infect Dis* **4**, 561.

30. Patel, M.M., Hall, A.J., Vinjé, J. and Parashar, U.D. (2009) Noroviruses: A comprehensive review. *J Clin Virol* **44**, 1–8.

31. Ray, B. and Bhunia, A. (2014) *Fundamental Food Microbiology. Fifth edition*. Boca Raton, FL: CRC Press, Taylor and Francis Group.

32. Sridhar, S., Lau, S.K.P. and Woo, P.C.Y. (2015) Hepatitis E: A disease of reemerging importance. *J Formosan Med Asso* **114**, 681–690.

33. Torres, H.A. and Davila, M. (2012) Reactivation of hepatitis B virus and hepatitis C virus in patients with cancer. *Nat Rev Clin Oncol* **9**, 156–166.

34. Trépo, C., Chan, H.L.Y. and Lok, A. (2014) Hepatitis B virus infection. *The Lancet* **384**, 2053–2063.

35. Ushijima, H., Fujimoto, T., Müller, W.E.G. and Hayakawa, S. (2014) Norovirus and foodborne disease: A review. *Food Safety* **2**, 37–54.

36. Van Kerkhove, M.D., Mumford, E., Mounts, A.W., Bresee, J., Ly, S., Bridges, C.B. and Otte, J. (2011) Highly pathogenic avian influenza (H5N1): Pathways of exposure at the animal-human interface, a systematic review. *PLoS One* **6**.

37. Vaughan, G., Goncalves Rossi, L.M., Forbi, J.C., de Paula, V.S., Purdy, M.A., Xia, G. and Khudyakov, Y.E. (2014) Hepatitis A virus: Host interactions, molecular epidemiology and evolution. *Infect Gen Evol* **21**, 227–243.

38. Walker, C.M., Feng, Z. and Lemon, S.M. (2015) Reassessing immune control of hepatitis A virus. *Curr Opin Virol* **11**, 7–13.

Foodborne Parasites

7

Introduction

Parasites are those organisms that derive nourishment and protection from other living creatures. Some parasites are unicellular organisms, such as protozoa, or multicellular, such as metazoa. Metazoa includes helminths, nematodes, trematodes, and arthropods (flies, mosquitos, ticks). Some of the parasites can be transmitted by the arthropod vectors or food. Parasites are classified into ectoparasites that harbor outside the body, such as ticks, lice, and mites; endoparasites that reside inside the host such as *Toxoplasma*, *Taenia*, *Trichinella*; and epiparasites that feed on another parasite such as endoparasite of a flea living on a dog. Foodborne parasites are endoparasites, and the major groups are protozoa, cestodes (tapeworms), nematodes (roundworm), and trematodes (flatworms or flukes) (Table 7.1). Parasites are significantly larger than the bacteria and vary in their sizes. The general life cycle consists of a stage, in which the parasite is as an oocyst (egg) and can contaminate food and water (Fig. 7.1). The oocysts are morphed into larvae (designated trophozoites, bradyzoites, or merozoites) and then become adults. Larva stage is most infective. Most parasites require more than one animal host, i.e., an intermediate host to complete their life cycle. Generally, the protozoa complete their life cycle in a single host and are called homoxenous, while the metazoa require more than one host and are known as heteroxenous. Foodborne parasites do not proliferate in food but require live hosts for their growth. Parasites are detected by direct means using microscopy, genetic, or antibody-based assays.

Contamination of foods with parasites often indicates inadequate hygienic practices employed during production and processing. Water, fruits and vegetables, and raw or undercooked meat or fish are known to be the major source of parasitic pathogens. The occurrence of parasitic diseases is high in tropical countries, but global warming, the climactic shift in weather patterns, and globalization of food supply may cause increased foodborne parasitic diseases in countries that may not have experienced in the past.

Protozoa

Giardia duodenalis

Introduction
Giardia duodenalis also known as *Giardia lamblia* or *G. intestinalis* is responsible for giardiasis. The disease is characterized by watery diarrhea referred to as traveler's diarrhea, which occurs in people during travel to countries where sanitary practices are inadequate. Giardia is also sometimes responsible for chronic illness. The common source for giardiasis is contaminated water and vegetables. Cysts (eggs) are also found

© Springer Science+Business Media, LLC, part of Springer Nature 2018
A. K. Bhunia, *Foodborne Microbial Pathogens*, Food Science Text Series,
https://doi.org/10.1007/978-1-4939-7349-1_7

in sewage affluent. Wild animals, such as beavers, can carry the protozoa and contaminate water.

Pathogenesis

Giardia is an extracellular parasite and resides in the intestinal lumen. Giardia life cycle has two stages: cyst and trophozoite. Cysts are infective while trophozoites are not. The infective dose is 10–100 cysts, and the incubation period is 1–2 weeks. Once ingested, the cyst (ovoid, 9–12 μm long) arrives in the duodenum and is dissolved by digestive enzymes, and two trophozoites (9–21 μm long and 5–15 μm wide) are released. The trophozoites are teardrop shaped

Table 7.1 Foodborne parasites

Group	Genus/species
Protozoa	*Giardia lamblia; Entamoeba histolytica*
	Cyclospora; Cryptosporidium; Cystoisospora
	Toxoplasma gondii; Trypanosoma cruzi
Cestodes (tapeworms)	*Taenia saginata; T. solium*
	Echinococcus granulosus
Nematode (roundworms)	*Ascaris lumbricoides*
	Trichinella spiralis
Trematodes (flatworms or flukes)	*Clonorchis sinensis*
	Paragonimus spp
	Fasciola hepatica

and contain two nuclei and four flagella, and they display tumbling motility (Fig. 7.2). Trophozoites use surface protein, alpha-giardins to attach to the intestinal epithelium for colonization and infection. They multiply rapidly through asexual reproduction and cause damage to the mucus membrane and disrupt epithelial tight junction barrier to increase intestinal permeability in the upper small intestine. During chronic infection, giardia downregulates tight junction protein, claudin 1, and increases epithelial apoptosis. Some trophozoites mature into the cyst, before being released into the feces completing its life cycle. Giardiasis is characterized by malabsorption, severe watery diarrhea, bloating, and flatulence. Treatment with metronidazole or tinidazole can eliminate the infection. *Giardia* is resistant to chlorine treatment that is applied to municipal water supplies.

Entamoeba histolytica

Introduction

The genus *Entamoeba* consists of several species: *Entamoeba histolytica, E. dispar, E. mosh-kovski, E. coli, E. hartmanni,* and *E. polecki. Entamoeba histolytica* is pathogenic and the

Fig. 7.1 General life cycle of parasites

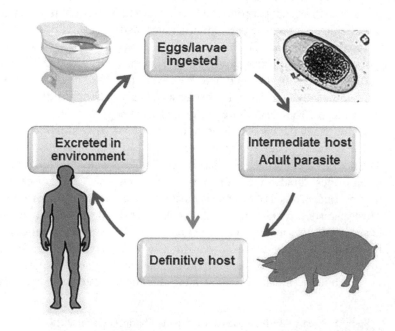

Fig. 7.2 (a) Illustration of *Giardia duodenalis* protozoa (Courtesy of Jennifer Oosthuizen and James Archer, CDC, Atlanta, GA), (b) *Giardia duodenalis* life cycle and tissue damage in the intestine

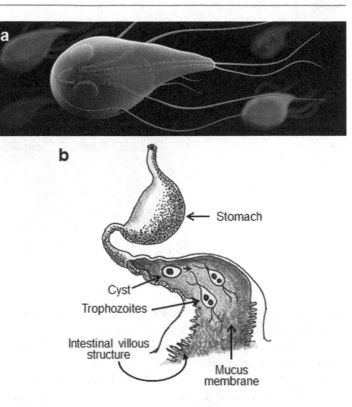

most invasive member of the genus and is responsible for amebic dysentery. Other species of *Entamoeba* exist as commensal in the human intestine. Worldwide, *E. histolytica* causes 50 million cases and 100,000 deaths, and the most amebiasis cases are associated with the developing countries including Africa, Central and South America, and the Indian subcontinent. Amebic cases are also predominant in tropical and subtropical countries. Humans serve as the major reservoir of the organism and can contaminate water and foods. Unsanitary living conditions and unhygienic food preparation practices are the major risk factors for the spread of amebic infection. Transmission may occur through contaminated water, fresh foods, sewage, insects, and fecal–oral route or in homosexuals through oral–anal contacts.

Pathogenesis

Entamoeba histolytica produces cysts (10–15 µm in diameter), which are infective. The infective dose is above 1000 cysts, but a single cyst has the potential to cause disease. The incubation period is 1–4 weeks. Ingested organisms reach to the lower small intestine (terminal ileum) and upper large intestine, where each cyst gives rise to eight daughter trophozoites. The trophozoites (10–60 µm in diameter) are motile (Fig. 7.3) and adhere and invade intestinal epithelial cells, majorly in the large intestine. They produce several virulence factors, which promote adhesion and tissue damage. The parasite uses Gal/GalNAc-inhibitable lectin to interact with the host glycoprotein for adherence, to block complement activation, and to promote cytotoxicity of neutrophils and macrophages. The parasite produces proteolytic enzymes (collagenase and neutral proteases) and cysteine proteases, which promote parasite invasion and tissue damage. It produces amoebapore that forms ion channels in the host cell membrane causing cytolysis. Trophozoites can cause localized infection by invading mucosal membrane in the intestine or pass through the bloodstream to reach to extraintestinal sites such as the liver, brain, and lungs. The growth of trophozoites results in inflammation and ulceration in the mucosal layer. The trophozoites multiply by

Fig. 7.3
Photomicrograph of
Entamoeba histolytica
trophozoite carrying a
number of erythrocytes
(arrow) as inclusion
bodies (Courtesy of Dr.
Mae Melvin; Dr.
Greene, CDC, Atlanta,
GA)

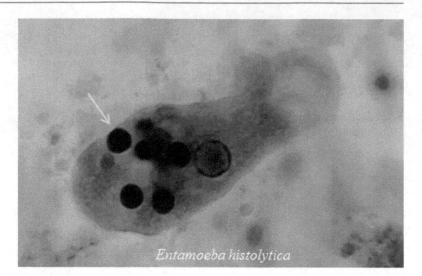

binary fission and produce cysts, and both stages are found in the feces. Each trophozoite carries single nucleus and can convert into precyst and matures into tetranucleated cysts, which are released with feces. Cysts can survive outside for several weeks to months but are sensitive to temperature under −5 °C or over 40 °C.

The symptoms of amebiasis include abdominal cramp and diarrhea which is watery, mucoid, or bloody resembling bacillary dysentery caused by *Shigella dysenteriae* type 1 and *S. flexneri* (see Chap. 19) and enteroinvasive *Escherichia coli* (see Chap. 14). Diarrhea may occur with ten or more bowel movements per day, and some patients show high fever. The patients show signs of anorexia and continue to loose body weight. The disease becomes chronic in some patients, characterized by intermittent diarrhea, flatulence, and ulcerative colitis. The extraintestinal disease condition may include liver abscess, pulmonary disease, peritonitis, pericarditis, and brain abscess. Treatment with metronidazole (tinidazole) is effective against amebiasis.

Toxoplasma gondii

Introduction
The disease caused by *Toxoplasma gondii* is called toxoplasmosis, which normally is transmitted by the domestic cat. *T. gondii* is consid-

ered a zoonotic pathogen and is one of the leading causes of foodborne illnesses and deaths. It causes about 87,000 illnesses, 4428 hospitalizations, and 327 deaths annually in the USA. Drinking water contaminated with cat feces is thought to be responsible for several outbreaks worldwide including a large outbreak in Brazil in 2002. Unwashed hands after contact with pet cats, vegetables washed in contaminated water, and eating of undercooked pork are considered potential sources for this parasite.

Pathogenesis
Toxoplasma gondii is an obligate intracellular parasite and it has very low host specificity. Its life cycle has two phases: intestinal (enteroepithelial) phase, which is seen on the primary host, cats, and extraintestinal phase, which is seen in the secondary host, i.e., all infected animals (Fig. 7.4). Cats can be infected by eating infected prey. Cysts or oocysts in the meat are dissolved in the lumen of the digestive tract and the bradyzoites are released, which penetrate intestinal epithelial cells and undergo asexual multiplication. Later, sexual multiplication follows. Male and female gametes form zygotes, which develop into oocysts and released in the feces. Oocysts mature in the feces and infect warm-blooded animals (intermediate host) including rodents, farm animals, and humans. Oocyst dissolves in the gut and sporozoites are released which then penetrate

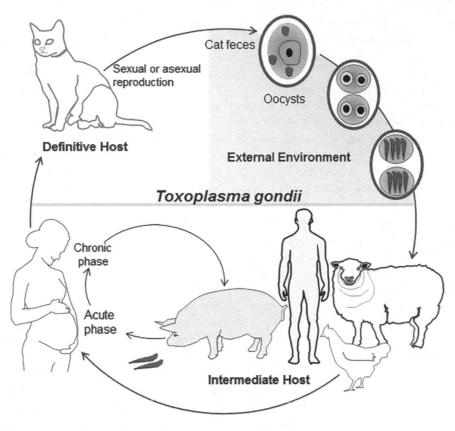

Fig. 7.4 *Toxoplasma gondii* transmission pathway

the intestinal epithelial cells, reach to blood circulation, and invade muscle tissues. Sporozoites bind to host cell receptor, sialic acid, and exhibit gliding motility that helps parasite invasion into enterocytes. Invasion is an active process for the parasite, and the host cell does not play an active role in invasion, i.e., invasion does not alter host cell actin cytoskeleton or phosphorylation of tyrosine residues. Inside the host cell, *Toxoplasma* is trapped in a specially modified vacuole that is primarily derived from the host cell plasma membrane. The parasite avoids fusion of vacuole with normal host endocytic or exocytic vesicle and begins replication. *Toxoplasma* then migrates to the subepithelial region in the basement membrane and penetrates deep into the submucosa and disseminates into central nervous system (CNS), retina, and placenta.

A human can also acquire by eating raw or undercooked meats, water, or food contaminated with cat feces. In healthy individuals, the symptoms appear flu-like, swollen lymph nodes, fatigue, joint and muscle pain, rash, and headache, and the disease is usually self-limiting. In immunocompromised host, the disease could be fatal. In pregnant women, there is a significant risk of transmission of the parasite to the fetus, resulting in spontaneous abortion. Infection during the later stage of pregnancy may be less severe and may not affect the fetus.

An experimental infection induced in mice shares morphologic and histologic characteristics with human inflammatory bowel disease (IBD), which is characterized by the loss of intestinal epithelial architecture, shortened villi, the influx of inflammatory cells (macrophages, monocytes, neutrophils, and lymphocytes), and scattered necrotic patches. The parasite is found mostly in the muscle and CNS. When the infection is unregulated, the inflammatory process results in early mortality. Female animals are more susceptible than the males, suggesting that gender and sex hormones play an important role in determining the susceptibility of the small intestine to *T. gondii*.

Immune Response to Toxoplasmosis

Toxoplasma gondii infection exhibits a significant increase in chemokine secretion and macrophage inflammatory proteins and recruitment of polymorphonuclear cells (PMNs) such as neutrophils, macrophages, monocytes, dendritic cells (DCs), and T lymphocytes. Infiltration of neutrophils is essential to remove *T. gondii* during the first few days of infection. Neutrophil-depleted mice exhibit lesions and greater parasite burden. Activated CD4+ T cells produce increased IFN-γ and TNF-α that are responsible for recruitment of inflammatory cells including macrophages for clearance of parasite.

Cryptosporidium parvum

Introduction

Cryptosporidium parvum is a coccidian protozoa and it causes cryptosporidiosis. The major source of this pathogen is water including agricultural runoff and the sewage effluent. Vegetables or foods exposed to contaminated water serve as the carrier for this pathogen. The first human case was reported in 1976. There are several species in the genus, but *Cryptosporidium parvum* is the most common. Two types of *C. parvum* exist, type I and type II. Type I is exclusively from humans and type II from cattle or humans exposed to infected cattle. *C. parvum* completes its life cycle in one host. The oocyst is 4–6 μm in diameter and carries four sporozoites inside.

Pathogenesis

The infective dose is thought to be very low, about 10 oocysts, and the incubation period is 2–10 days. After consumption of contaminated food or water, one oocyst releases four sporozoites, which undergo asexual reproduction followed by sexual reproduction, and the zygotes are formed. The sporozoites adhere to the epithelial cells and invade apical surface of the epithelial cells by

gliding motility using parasite-driven process. *Cryptosporidium* induces host cytoskeletal structure to create a platform made of host actin filaments composed of α-actinin, ezrin, talin, and vinculin causing severe cellular damage resulting in gastroenteritis.

The disease is characterized by watery diarrhea or cholera-like illness with abdominal cramp, fever, and muscle ache. Anorexia, weight loss, dehydration, and abdominal discomforts are generally associated with this disease. In immunocompromised patients (e.g., the AIDS patients), the organism is highly invasive and can infect lungs and bile ducts which are often life threatening. Infected cattle or human shed large numbers of oocysts in the feces, and the stool samples are used for diagnosis. Oocysts are resistant to chlorine treatment; thus, contaminated water supply becomes a threat for infection. A large outbreak of *Cryptosporidium* occurred among HIV patients in Milwaukee (Wisconsin, USA) in the mid-1990s due to the consumption of contaminated water from the municipal water supply. It was determined that the water was contaminated with cattle manure runoff and the parasite survived the water treatment system.

Cyclospora cayetanensis

There are 17 species of *Cyclospora*. *Cyclospora cayetanensis* is the newest member of the coccidian family that is responsible for food and waterborne diseases. Consumption of oocyst-contaminated food or water leads to the onset of cyclosporiasis disease. The first human case was reported in 1977. Since then, outbreaks of *Cyclospora* have been associated with raspberries, basil, and lettuce. In the late 1990s, contaminated raspberries from Guatemala caused widespread outbreaks in 20 states in the USA, and again in 2004, Guatemalan snow pea caused an outbreak in Pennsylvania (USA).

Infected people shed unsporulated oocyst, which is non-infective and immature. The oocysts are 8–10 μm in diameter and oval shaped, and each differentiates or matures into sporocyst under favorable condition. Each sporocyst carries two infective crescent-shaped sporozoites. The infective dose is unknown but the incubation period is typically 1 week. The disease usually lasts for 2 weeks. Oocysts excyst in the small intestine usually in the jejunum and sporozoites infect the epithelial cells. The detailed mechanism of infection is not clear; however, histopathology shows villous atrophy, crypt hyperplasia with inflammation in the small intestine.

Cyclospora cayetanensis causes self-limiting watery diarrhea, fatigue, nausea, vomiting, myalgia, anorexia, and weight loss, and sometimes it may cause explosive diarrhea. Cyclosporiasis sometimes may lead to chronic sequelae like Guillain–Barré syndrome, reactive arthritis, and acalculous cholecystitis. Trimethoprim–sulfamethoxazole treatment is effective in controlling the infection.

Cystoisospora belli

Cystoisospora belli, formerly known as *Isospora belli*, is a coccidian protozoan parasite and causes cystoisosporiasis, which is manifested by diarrhea. It is distributed in tropical and subtropical countries including Latin America, Caribbean countries, Africa, Southeast Asia, and Australia. It is generally transmitted through vegetables and contaminated water. *Cystoisospora* oocysts (10–19 × 20–30 µm) are much larger than *Cyclospora* or *Cryptosporidium*. Oocysts are elliptical, and one oocyst forms two sporocysts, and each carries four sporozoites. Like other coccidian protozoa, the infective sporozoites invade intestinal epithelial cells, undergo sexual and asexual reproduction, and cause tissue damage. Villous atrophy, hypertrophic crypts, and inflammation and infiltration of epithelial cells with eosinophils are characteristics of cystoisosporiasis infection. Watery diarrhea, abdominal pain, anorexia, and weight loss are symptoms of this infection, and the disease can persist for months to years. It could be fatal in immunocompromised hosts, and the AIDS patients are highly susceptible to this infection.

Trypanosoma cruzi

Introduction

Trypanosoma cruzi causes Chagas disease, also known as American trypanosomiasis, and is a serious tropical disease in Latin America. *Trypanosoma cruzi* was first identified by Carlos Chagas, and he named it *Trypanosoma cruzi* after his mentor, Oswaldo Cruz. It belongs to *Trypanosomatidae* family. *T. cruzi* infects both vertebrates and invertebrates, hence called heteroxenic. This protozoon is transmitted by blood-sucking *Triatomine* insect, also called kissing bug that lives in dwellings in rural poor communities. Contaminated fruit juice laden with feces of Triatominae or crushed bugs during juice preparation is a major source of *T. cruzi* foodborne infection in Brazil and Venezuela. Besides transmission of *T. cruzi* through blood meal feeding during *Triatomine* bite to humans, it can also transmit through blood transfusions, organ transplantation, and from mother to fetus in the womb. Dogs have been recognized as an animal reservoir.

Pathogenesis

Trypanosome generally enters through the skin after an insect bite. During a blood meal, *Triatomine* releases *Trypanosome* through feces on the bite wound. However, as a foodborne pathogen, *T. cruzi* enters gastrointestinal tract through contaminated food and interacts with intestinal mucosal epithelial cells. *T. cruzi* is an intracellular pathogen. It uses surface glycoprotein, gp82, to interact with host sialic acid and other unknown host cell receptors to invade cells. Parasite interaction initiates signaling cascade and promotes cytoskeletal rearrangement facilitating parasite invasion. During this process, the host cell adenylate cyclase is activated promoting an enhancement of cAMP that contributes to the Ca_2^+ release from the endoplasmic reticulum, which promotes lysosome recruitment forming the parasitophorous vacuole where parasite rests and transform into trypomastigote. Ca_2^+ can also facilitate disorganization of the actin cytoskeleton, aiding parasite invasion by disrupting microfilaments. *T. cruzi* produces a toxin (TcTox), which forms pores in the vacuole, and the trypomastigotes escape into the cytoplasm. Trypomastigotes transform into amastigotes, multiply by binary fission inside the cytoplasm, burst out of the cell, and enter the bloodstream. *Triatomine* insect acquires *T. cruzi* from humans during feeding on human or animal blood.

The symptoms of trypanosomiasis (Chagas disease) may include fever, fatigue, body aches, headache, skin rash, loss of appetite, diarrhea, and vomiting. The acute Chagas disease is characterized by swelling of the eyelids if the bug feces are deposited near the eyes during the bite or accidentally rubbed into the eye. Acute phase symptoms disappear within a few weeks or months. Infection persists if untreated. During chronic infection, enlarged heart or cardiomyopathy, altered heart rhythm, and cardiac arrest may occur. Intestinal complications include enlarged esophagus (megaesophagus) or colon (megacolon), and disease becomes fatal, if untreated. Vector control is the key to controlling Chagas disease in the endemic areas.

Cestodes (Tapeworms)

Taenia solium and *Taenia saginata*

Introduction

Cestodes are also known as tapeworms. Two species of tapeworms are of major concerns to humans: *Taenia saginata*, known as "beef tape," associated with beef, lamb, and fish, and *Taenia solium*, associated with pork hence "pork tape." *T. asiatica* also occurs in cattle and pigs and is another major pathogen of concern for humans. Several other *Taenia* species infect dogs, cats, wolves, and foxes and are of less public health concern (Table 7.2). Raw or undercooked pork or beef are the major sources of transmission to humans. Asia and perhaps Africa and Latin America are considered endemic zones. The disease is considered a highly neglected disease, and the parasite is maintained in the rural impoverished parts of the communities where humans, pigs, and dogs coexist. Pigs and dogs feed on human feces or feces-contaminated foodstuff, and the humans consume both pigs and dogs (in

some parts of Asia), and parasite life cycle is sustained (Fig. 7.5).

The length of an adult tapeworm varies between 4 m and 12 m. *T. solium* is much smaller and thinner than the *T. saginata*. Suckers in the scolex (head) help the worm to remain attached to the intestinal wall. The segments or proglottids could give rise to several new progenies. Gravid proglottids may carry up to 80,000 eggs and each egg represents a single unit. Before eggs are released, the scolex evaginates exposing the suckers outside the body, which can attach to the mucus membrane when ingested.

Pathogenesis

Taenia solium causes taeniasis or cysticercosis in humans. Taeniasis is caused by adult worms while cysticercosis is caused by ingestion of cysts (eggs). The lethal form of cysticercosis is neurocysticercosis (NCC). Other forms include ocular (OCC) and subcutaneous cysticercosis (SCC). For completion of the life cycle, the dogs and pigs serve as the intermediate host and humans the definitive host. In humans, after ingestion of contaminated meat, adult tapeworm matures in the intestine, and eggs are released from the gravid proglottids. The gravid proglottids carrying thousands of eggs can be separated from the adult parasites and are discharged with the feces. Eggs also are released with the feces. Pigs and dogs become infected by eating contaminated food and water or direct eating of human feces. A human can also be infected by *Taenia* eggs through food and water. Eggs reach CNS and infect the brain to cause neurocysticercosis which results in seizures, epilepsy, meningitis, hydrocephalus, increased intracranial pressure, and death. Subcutaneous cysticercosis is characterized by the formation of the firm and mobile nodules in the skin. Muscular cysticercosis is characterized by myositis, muscular pseudohypertrophy, fibrosis, and eosinophilia; and ocular cysticercosis presents visual difficulties, retinal edema, hemorrhage, and vision loss. During the intestinal phase of infection (taeniasis), the tapeworms cause severe abdominal pain resembling colic-like sharp pain, headache, and diarrhea. Treatment regimen includes praziquantel and corticosteroid.

Table 7.2 *Taenia* species and their host range

Species	Intermediate host	Definitive host
Taenia solium	Pigs, dogs, humans	Humans
Taenia saginata	Cattle	Humans
Taenia asiatica	Pigs, cattle	Humans
Taenia crassiceps	Rodents, humans (rare)	Dogs, foxes
Taenia ovis	Sheep, humans (rare)	Dogs
Taenia hydatigena	Pigs, sheep, goats, humans (rare)	Dogs
Taenia pisiformis	Rabbits, humans (rare)	Dogs, foxes
Taenia taeniaeformis	Rodents, humans (rare)	Cats
Taenia coenurus	Sheep, rabbits, humans (rare)	Dogs, cats, wolves
Taenia serialis	Rabbits, rodents, humans (rare)	Dogs, wolves, foxes

Adapted from Ito et al. (2016). Infect. Gen. Evol. 40, 357–367

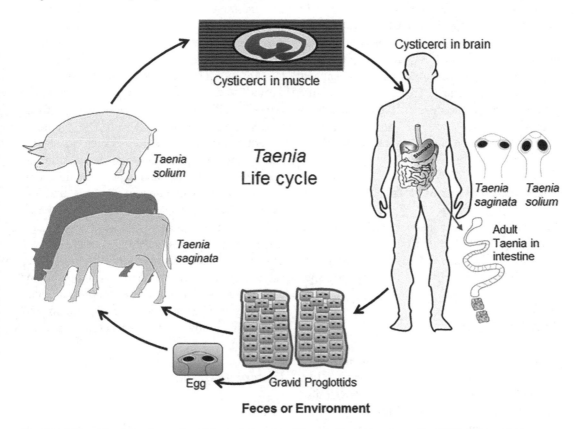

Fig. 7.5 Life cycle of *Taenia* species (Schematic based on lifecycle illustration created by CDC, Atlanta, GA)

Echinococcus granulosus

Echinococcus is a very small tapeworm (3–6 mm) found in carnivores. Dogs and other carnivores are the definitive host, and ungulates (horses, cattle, goats, sheep, pigs, giraffes, camels, deer, and hippopotamuses) are the intermediate host.

Humans are the accidental intermediate host and acquire parasite eggs through contaminated fruits and vegetables. The disease is called echinococcosis or hydatidosis or hydatid disease. Four species of *Echinococcus* are of public health concerns: *Echinococcus granulosus*, *Echinococcus multilocularis*, *Echinococcus oligarthus*, and

Echinococcus vogeli. E. granulosus causes cystic echinococcosis, *E. multilocularis* causes alveolar echinococcosis, and *E. oligarthus* and *E. vogeli* cause polycystic echinococcosis.

E. granulosus is the most important species of concern, and the genotype sheep G1 is responsible for the majority of echinococcosis infection in humans. After ingestion, the larvae reach to the liver, which is the most common extraintestinal site, and the parasite forms polyp (cyst). The lungs, spleen, kidney, heart, and CNS are also affected. The polyp grows very slowly in the organs at the rate of 1–5 cm per year, eventually causing organ dysfunction, and the onset of clinical symptoms. A sudden burst of polyp results in severe hypersensitivity (allergic reaction) response due to the release of polyp content and fatal anaphylactic shock. Treatment involves surgical removal of the cysts (polyps), chemotherapy with benzimidazoles, and percutaneous aspiration, injection, and re-aspiration (PAIR).

Nematodes (Roundworm)

The major foodborne nematodes include *Anisakis* and *Trichinella*. In addition, other nematodes that are also involved in foodborne infections are *Ascaris*, *Gnathostoma*, and *Angiostrongylus* spp.

Anisakis simplex

Anisakis simplex is a nematode responsible for fish-borne disease called "anisakiasis," first reported in the 1960s in the Netherlands. It is also referred to as "worm-herring disease" resulting from the consumption of lightly salted herrings. Consumption of raw or undercooked fish is responsible for this disease. The symptoms associated with anisakiasis are described as gastric, intestinal, extraintestinal, and allergic. Worldwide, about 20,000 people suffer from anisakiasis and about 90% of which are from Japan.

Anisakis life cycle is complex and is maintained in marine mammals and fish. Adult *A. simplex* harbors in the stomach of marine mammals and lays eggs, which pass through the feces. The first-stage larvae are called L1 and are released from the eggs. The larvae molt into L2 form, which is then eaten by crustaceans, where they mature into L3 form. Through predation, fish acquires L3 form, which is infective to fish, and migrates from the intestine to muscle tissues.

Once contaminated fish is ingested by humans, the *Anisakis* larvae reach the stomach and attach to the mucosa using their projections in the mouth. They release proteolytic enzymes, which cause erosion of mucus and submucus layers leading to the hemorrhagic lesion. The acute gastric symptoms appear within 2 h of infection characterized by severe abdominal pain, vomiting, diarrhea, and mild fever. Untreated disease may lead to chronic ulcer-like infection. The onset of the intestinal form of anisakiasis appears within 5–7 days after ingestion of larvae, which primarily affect terminal ileum causing inflammation and granulomatous lesion. The symptoms include constant or intermittent abdominal pain and in rare cases, ascites, small bowel obstruction, ileal stenosis, and pneumoperitoneum. *Anisakis* can also migrate to the peritoneal cavity, liver, pancreas, and ovary.

The immunological response involves adaptive immune response leading to both Th1 and Th2 induction and production of IL-2, IL-4, IL-5, and IFN-γ and proliferation of B lymphocytes for the production of IgE. The cellular immune response involves activation of eosinophils and mast cells leading to a typical allergic response.

Trichinella spiralis

Introduction
Trichinella species are transmitted through ingestion of muscle tissues containing the encysted *Trichinella* larvae. *Trichinella spiralis* causes trichinellosis and is mostly associated with the consumption of raw or undercooked pork products. Both wild and domestic swine are the major source for *Trichinella*; however, other animals such as horses, reptiles, birds, and wild carnivores (fox, hyena, and bear) serve as the reservoir for *Trichinella* spp. *Trichinella spiralis* is the major species associated with swine maintaining

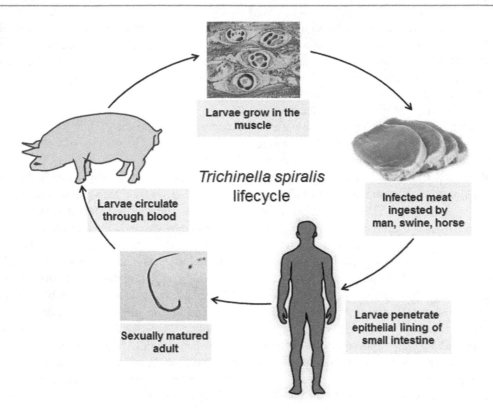

Fig. 7.6 Life cycle of *Trichinella spiralis*

a bioburden of greater than 8000 larvae per gram of tissue without showing clinical symptoms and is a major public health concern. Other species of *Trichinella* include *T. britovi* and *T. nelsoni*, and both can infect swine in lesser frequency than *T. spiralis*. Other *Trichinella* spp. such as *T. nativa*, *T. murrelli*, *T. patagoniensis*, etc., rarely infect pigs are of less public health concerns.

Pathogenesis

Trichinella species completes its life cycle in one host (Fig. 7.6). A gravid female worm living in the mucosal layer in the host intestine releases the newborn larvae, which migrate to lymphatics and blood and then to muscle tissues forming cysts (polyps). Larvae grow in the striated skeletal muscle. When the undercooked infected meat is ingested by man, swine, and horse, the disease propagates. Parasite larvae are released from the meat tissues in the intestine after proteolytic enzymatic digestion, and the larvae penetrate the epithelial lining of the small intestine. Larvae

undergo molting to develop into sexually matured adults within 2 days. The male and female copulate, and the female releases newborn larvae within 5–7 days of post-infection. The larvae reach to muscle tissue through blood circulation and are encapsulated in the muscle. Some larvae may be found in the intestinal mucosa (Fig. 7.6). Calcification of the encapsulated larvae in muscle tissue is seen after 6 months of infection, but the larvae can survive in muscle tissues in humans for up to 40 years. Within months, due to immunological action in the intestine, the adult parasites are continuously expelled in the feces.

The trichinellosis is manifested in two phases: the intestinal phase and the muscular phase. The intestinal phase of the disease is characterized by gastroenteritis seen 2 days after infection: nausea, abdominal pain, diarrhea, and vomiting. The muscular phase is characterized by muscle pain. Trichinellosis is again categorized as acute – or chronic – stage disease. Acute stage is characterized by a headache, fever (lasting for 1–3 weeks), chills, myalgia, pyrexia, eyelid swelling, occasional gastrointestinal disorders, myocarditis, thromboembolic disease, and encephalitis. In some patients infection may be severe, and clinical pathology includes conjunctivitis with ocular edema and vascular lesion (bleeding) in the conjunctiva, uvea, and retina. Some patients experience intense pain in the eyeballs due to the invasion of larvae into the ocular muscle. Patients may also suffer from myocarditis including pericardial pain and tachycardia. In the chronic stage of disease (usually after 3–4 weeks after infection), the patients exhibit signs of encephalitis, neurological disorders, and bronchopneumonia due to secondary infection by bacteria. Symptoms may include persistent formication (insect walking sensation on skin), numbness, excessive sweating, impaired muscle strength, and conjunctivitis.

The immunological response involves inflammation characterized by infiltration of eosinophils, mast cells, monocytes, and lymphocytes. The immune response is primarily mediated by Th2, and cytokine production leads to proliferation and activation of eosinophils as the major effector cells. Eosinophilia (>1000 cells/ml) is common in trichinellosis patients; however, IgE response is inconsistent.

Diagnosis, Prevention and Control
The disease can be diagnosed by clinical symptoms: fever, muscle ache, gastrointestinal symptoms, conjunctivitis, retinal hemorrhage, and eosinophilia. Laboratory diagnosis involves PCR, ELISA, and Western blot (for seroconversion). Trichinellosis can be prevented by routine postmortem inspection of meats from pigs, horses, wild boars, and bear for the presence of larvae or encapsulated calcified nodules in meat. The predilection sites for inspection include the diaphragm, tongue, and masseter muscle (facial jaw muscle). Microscopic examination of enzyme digested muscle tissues, and magnetic stirrer methods have been used for detection of larvae from meats. The cyst could be easily killed by cooking meat at 71 °C or 159.8 °F (internal temperature) at least for 1 min, freezing meat at least at −18 °C for 30 days, and irradiation (0.3 kGy).

Curing and smoking are ineffective in controlling *Trichinella* larvae in meat. For treatment of trichinellosis in humans, anthelmintic drugs such as albendazole and mebendazole are used.

Ascaris lumbricoids

Introduction
Ascaris lumbricoids, a nematode, is associated with food when prepared under poor sanitary conditions. The disease is called ascariasis and is manifested by enlarged liver and malnutrition. It is a soil-transmitted helminth (STH); hence, fruits and vegetables from contaminated soil may carry the worm. The disease affects adults and children primarily in economically poor countries. Children in tropical and subtropical areas carry this parasite for a long period. The disease is prevalent in Asia, Africa, and Europe. Although there have not been any major outbreaks in the USA, the high occurrence of the eggs in sewage municipal waters does present a potential high risk for infection.

Pathogenesis
Ascaris lumbricoides is mainly a human pathogen while *Ascaris suum* infects swine. Report of *A. suum* infection in humans is also documented. Both are prevalent in soil, and soil moisture, atmospheric humidity, and warm temperature help maintain eggs and larvae and facilitate faster development of ova. Fruits and vegetables contaminated with fecal matter or soil containing the eggs is the source of infection in humans. Larvae hatched from eggs, invade the intestinal mucosa, and are transported to lungs via lymphatics and blood circulation. The matured larvae invade the alveolar wall, ascend the bronchial tree to the throat, and are swallowed by the host. In the small intestine, the larvae develop into adult worms and engage in sexual reproduction. The adult *Ascaris* survives in the intestine for 1–2 years. The females are larger (30 cm × 5 mm) than the males. A female may lay 200,000 eggs per day, and eggs are released through feces. Only fertilized eggs are infective. The whole cycle of infection requires 2–3 months for completion.

Ascariasis affects adults but is more common in children from 5–15 years of age. The disease is manifested 4–16 days after infection. Malnourished children are highly susceptible to ascariasis. Infected children exhibit slower growth, poor cognitive development, and slow weight gain since the parasite feed on nutrients of the partially digested food in the intestine. Patients infected heavily with parasite show symptoms of abdominal pain and obstruction of the intestines accompanied with fever, inflammation, and diarrhea. Other complications include pancreatitis, acute cholecystitis, and liver abscess. The diagnosis involves identification of eggs (40–70 μm × 35–50 μm) or the worm from the stool sample. In addition, the larvae can be found in pulmonary secretions (sputum). It is also common for the patients to cough the worm when they are migrating through the throat.

Immune Response

The migration of the larvae to the pulmonary tract results in hemorrhagic lesions in the lungs with eosinophil infiltration and inflammation. The migration of the larvae to the intestine triggers an immunological response with mucosal mast cell recruitment and elevated levels of histamine. The immune response consists of Th2-derived cytokines (IL-4 and IL-13) that stimulate intestinal smooth muscle contractibility, decrease glucose intake, and enhance fluid accumulation leading to the expulsion of the parasite. The larvae, however, can overcome host defense by upregulating gene expression to enhance motility and metabolic activity.

Trematodes (Flatworm or Fluke)

Foodborne trematodes pose a significant public health concern and globally about 750 million people are at risk. Trematodes comprised of liver flukes (*Clonorchis sinensis*, *Fasciola gigantica*, *Fasciola hepatica*, *Opisthorchis felineus*, and *Opisthorchis viverrini*), lung flukes (*Paragonimus* spp.), and intestinal flukes (*Echinostoma* spp.). Transmission to humans occurs through consumption of contaminated raw or undercooked

aquatic foods such as freshwater fish, frogs, shellfish, snails, tadpoles, snakes, and water plants. The life cycle of trematodes starts with the production of a large number of eggs by adult worms following sexual reproduction in the final host (humans or domestic or wild animals). For most foodborne trematodes, eggs are released through feces or sputum (for *Paragonimus* spp.). Embryonated eggs hatch under appropriate environmental conditions including temperature, moisture, and oxygen tension. Hatched eggs release "saclike" larvae called miracidium, which begins swimming, and snails (intermediate host) can ingest the miracidium. Alternatively, miracidium larvae can penetrate the molluscan tissue for life cycle completion. Two representative trematodes are described below.

Clonorchis sinensis

Clonorchis sinensis is also known as "Oriental liver fluke," a major trematode of concern in parts of the Asia including China, Taiwan, Japan, Korea, and Vietnam. Snails serve as the first intermediate host, which transfers the flukes to fish. Freshwater or brackish water fish is the second intermediate host, and human acquires infection upon consumption of raw or undercooked fish. Globally about 35 million people are infected with *C. sinensis*, of which 15 million infections are reported in China. The parasite affects the liver and can obstruct the biliary duct. The symptoms include jaundice, ascites, gastrointestinal bleeding, pancreatitis, gallstone, and fatal cholangiocarcinoma.

Paragonimus Species

Paragonimus species is also known as lung fluke (about 10 mm long) and is found in Southeast Asia, America, and parts of Africa. *P. westermanni* is distributed in Asia while *P. kellicotti* in North America. Freshwater snail is the first intermediate host, and crustaceans (crab and crayfish) are the second intermediate host. Humans and other mammals (cats and dogs) are infected after consumption of raw, undercooked, or pickled crustaceans. Humans release eggs through sputum. Symptoms for paragonimiasis include fatigue, malaise, myalgias, fevers, night sweats, vomiting, swollen lymph nodes, gastroenteritis, hepatosplenomegaly, eosinophilia, and tuberculosis (TB)-like symptoms such as a cough, brown tinged sputum, and chest pain. Paragonimiasis could be easily misdiagnosed. High numbers in lungs can result in serious complications. Patients showing symptoms of coughing, fever, hemoptysis (coughing of blood), and eosinophilia should be examined for *Paragonimus*. Avoiding eating raw crayfish is a good preventive strategy. Praziquantel is the drug of choice for treatment.

Schistosoma Species

Schistosoma species is a blood fluke and causes schistosomiasis. It has not been implicated in any foodborne infection. *Schistosoma* species are considered zoonotic pathogens. There are several species distributed around the globe: *Schistosoma haematobium*, *S. mansoni*, *S. japonicum*, *S. mekongi*, and *S. intercalatum*. *S. haematobium* is widely distributed over the African continent with smaller foci in the Middle East, Turkey, and India. *S. mansoni* is widespread in Africa and Middle East and the only species seen in the Western Hemisphere in parts of South America and some Caribbean islands. *S. japonicum* is predominant in Asia, mainly in China, the Philippines, Thailand, and Indonesia. *S. mekongi* is predominantly found in Southeast Asia. *S. intercalatum* is found in Central and West Africa. *Schistosoma* infects both humans and bovine species (water buffalo, cattle, goats).

Schistosoma life cycle consists of a single intermediate host, a freshwater snail. *Schistosoma* larvae enter through the host skin, while all other trematodes enter by ingestion. *Schistosoma* eggs are released with feces or urine and the eggs hatch and release miracidia, which penetrate tissues of the snail. In the snail, miracidia undergo transformation to form sporocysts and then cercariae. Cercariae (infective) are released into the water, swim freely, and penetrate human skin. In the human host, they become schistosomula

and migrate through several tissues and reside in the veins. Adult worms reside in the mesenteric veins, bladder, and rectal venules depending on the species. Symptoms and pathology include fever, hepatic granulomas, fibrosis, and occasional embolic egg granulomas in the brain or spinal cord, hematuria, scarring, calcification, and squamous cell carcinoma.

Schistosoma spp. also infect ruminants (cattle) causing intestinal, hepatic, and nasal schistosomiasis. The intestinal disease is characterized by diarrhea, weight loss, anemia, hypoalbuminemia, hyperglobulinemia, and severe eosinophilia. In severe cases, the animal dies within a few months of infection. In the hepatic form, granulomas develop in the liver and in the intestine. Lesions are also found in the lungs, pancreas, and bladder of heavily infected animals. In the nasal form of the disease, a cauliflower-like granuloma growth is seen on the nasal mucosa. Partial obstruction of the nasal cavity leads to snoring sounds during breathing and grazing. A blockage also causes dyspnea. Nasal discharge is hemorrhagic and/or mucopurulent, and adult flukes are found in the blood vessels of the nasal mucosa; however, the pathogenesis is associated with the eggs. Rupturing of the abscess releases eggs, which eventually leads to extensive fibrosis.

Prevention and Control of Parasitic Diseases

Water treatment, hygienic practices, and proper washing and cooking of foods are necessary to prevent parasite infection. Proper disposal of fecal material is necessary to prevent the spread of parasitic diseases in economically poor countries since the eggs are present in the stools of an infected person. Arthropod vectors should be controlled to prevent parasite transmission among intermediate and definitive hosts. Food should be stored below 10 °C or frozen since parasites are susceptible to cold. Wash fresh fruits and vegetables thoroughly before consumption. Meat should be cooked thoroughly. For trematodes control in fish, the FDA recommend freezing at −20 °C or below for 7 days or at −35 °C or below for 15 h intended for raw consumption.

Chemotherapy should be used that can affect cysts, larvae, and adult parasites. The drug should kill or paralyze parasites. Many parasites cause chronic diseases so treatments are often for the long term. Thus, drugs present serious side effects. Anti-protozoan drugs include tinidazole, which is effective against trichomoniasis, giardiasis, and amebiasis. Antinematode drug is mebendazole and is effective against pinworms, roundworms, and hookworms. Anthelminthic such as praziquantel is effective against schistosomiasis, cysticercosis, and paragonimiasis. *Anisakis simplex* larvae are removed by endoscopy from infected patients, and granulomas are removed by surgical method.

Common side effects of antiparasitic drugs are flu-like symptoms, stomach pain, vomiting and diarrhea, hypersensitivity reaction, and blurred vision. Serious side effects include drug interactions. These drugs are extremely toxic when taken at high doses leading to liver damage, excessive bleeding, and blood clotting and internal bleeding.

Summary

Protozoa (unicellular) and metazoan (multicellular) parasites are increasingly becoming public health concerns due to their spread through vegetables and fruits since they are often consumed raw. Raw or undercooked fish and meats are also the major sources of infection. Contamination of foods with parasites often indicates inadequate hygienic practices employed during production and processing. Parasitic disease occurrence is high in tropical countries, but global warming, the climatic shift in weather patterns, and globalization of food supply may cause increased foodborne parasitic diseases in countries that have not experienced in the past. The major protozoan parasites include *Giardia*, *Entamoeba*, *Cyclospora*, *Cryptosporidium*, and *Toxoplasma*. *Trypanosoma cruzi* is becoming an emerging foodborne protozoon due to its accidental transmission through fruit juices. The major metazoan parasites of concerns are *Echinococcus*, *Anisakis*, *Taenia*, and *Trichinella* species. The general life cycle consists of a stage, in which the parasite is maintained as oocyst (egg) and can contaminate food and water. Oocysts or eggs hatch to produce larvae and then mature into adults. Larvae stage is the most infective, and some parasites require an intermediate host and a definitive host to complete their life cycles. Immunocompromised people are highly susceptible to the intracellular protozoan parasites. The hygienic practices during food production and processing are essential for controlling and preventing parasitic diseases. Antiparasitic drugs are effective but often present serious side effects to the patients. Vector control is also an important step for parasite control.

Further Readings

1. Conlan, J.V., Sripa, B., Attwood, S. and Newton, P.N. (2011) A review of parasitic zoonoses in a changing Southeast Asia. *Vet Parasitol* **182**, 22–40.
2. Dawson, D. (2005) Foodborne protozoan parasites. *Int J Food Microbiol* **103**, 207-227.
3. Dorny, P., Praet, N., Deckers, N. and Gabriel, S. (2009) Emerging food-borne parasites. *Vet Parasitol* **163**, 196–206.
4. Gottstein, B., Pozio, E. and Noeckler, K. (2009) Epidemiology, diagnosis, treatment, and control of *Trichinellosis*. *Clin Microbiol Rev* **22**, 127–145.
5. Ito, A., Yanagida, T. and Nakao, M. (2016) Recent advances and perspectives in molecular epidemiology of *Taenia solium* cysticercosis. *Infect Genet Evol* **40**, 357–367.
6. Jones, J.L. and Dubey, J.P. (2012) Foodborne Toxoplasmosis. *Clin Infect Dis* **55**, 845–851.
7. Mascarini-Serra, L. (2011) Prevention of soil-transmitted helminth infection. *J Global Infect Dis* **3**, 175–182.
8. Ortega, Y.R. (2006) *Foodborne parasites*: Springer, NY.
9. Pozio, E. (2014) Searching for *Trichinella*: not all pigs are created equal. *Trends Parasitol* **30**, 4–11.
10. Robertson, L.J., Sprong, H., Ortega, Y.R., van der Giessen, J.W.B. and Fayer, R. (2014) Impacts of globalisation on foodborne parasites. *Trends Parasitol* **30**, 37–52.
11. Sullivan, J.W.J. and Jeffers, V. (2012) Mechanisms of *Toxoplasma gondii* persistence and latency. *FEMS Microbiol Rev* **36**, 717–733.
12. Torgerson, P.R., de Silva, N.R., Fèvre, E.M., Kasuga, F., Rokni, M.B., Zhou, X.-N., Sripa, B., Gargouri, N., Willingham, A.L. and Stein, C. (2014) The global burden of foodborne parasitic diseases: an update. *Trends Parasitol* **30**, 20–26.
13. Yoshida, N., Tyler, K.M. and Llewellyn, M.S. (2011) Invasion mechanisms among emerging food-borne protozoan parasites. *Trends Parasitol* **27**, 459–466.

Molds and Mycotoxins

8

Introduction

Fungi are also called molds or microfungi. Molds are ubiquitous in nature and cause diseases in humans, animals, and plants. Fungi infection in mammals is referred to as mycosis, which is seen in the form of skin infections: ringworm and athlete's foot. Invasive mycoses are caused by inhalation of spores, and systemic spread of the infection has been seen in healthy as well as in immunocompromised hosts, which can be fatal. Mycoses are caused by two categories of pathogens: primary pathogens such as *Coccidioides immitis* and *Histoplasma capsulatum* and the opportunistic pathogens such as *Aspergillus fumigatus* and *Candida albicans*. Some filamentous molds produce mycotoxin, a secondary metabolite with no apparent function in the normal metabolism of fungi. It is, however, toxic to humans, animals, and birds.

Mold contamination can occur at various stages of plant-based food production: in the field, during harvest, processing, transportation, and storage. Mold infection in crops causes reduced yield, and feeding of meat animals with the contaminated feed results in poor animal health, death, and substantial economic losses. Molds also cause spoilage of foods. The major crops that are affected include wheat, maize, peanuts, tree nuts, cottonseed, and coffee. Food and Agriculture Organization (FAO) of the United Nations estimates 25% of the world's crops are affected each year with annual losses account for about 1 billion metric tons of foods. Economic losses are attributed to (1) yield loss due to diseases by toxigenic fungi, (2) reduced crop value resulting from mycotoxin contamination, (3) losses in animal productivity from mycotoxin-related health problems, and (4) human health costs. Additional costs associated with mycotoxins include the cost of management at all levels: prevention, sampling, mitigation, litigation, and research.

Filamentous toxigenic molds produce mycotoxins, and both molds and mycotoxins affect food and feed supply chains including crop producers, animal producers, grain handlers and distributors, processors, consumers, and society as a whole. Twenty-five to 50% of the foodstuffs are presumed contaminated with mycotoxins. Generally, the cereal foods and nuts are the major sources of mycotoxins. In addition, spices, fruits, and their by-products may be a source. The disease caused by mycotoxin is called mycotoxicosis, and it is nontransmissible and does not respond to antibiotics, and the associated outbreak is usually seasonal.

© Springer Science+Business Media, LLC, part of Springer Nature 2018
A. K. Bhunia, *Foodborne Microbial Pathogens*, Food Science Text Series,
https://doi.org/10.1007/978-1-4939-7349-1_8

Biology

Molds are multicellular microorganisms, and a typical mold possesses hyphae, conidiophore – consisting of stalk, vesicle, sterigmata, and conidia (spores) (Fig. 8.1). Often the identification of mold species is done by examining the morphological shape and size of the spores under a light microscope. The cell wall consists of a polymer of hexose chitin and N-acetyl glucosamine (NAG). Mold spores or hyphae are allergenic to humans, and frequent exposure may occur in damp buildings with high humidity with poor air circulations.

Toxigenic filamentous molds produce mycotoxins. There are about 400 different mycotoxins, and only a few have been studied in depth. The same mycotoxin can be produced by different species of mold, or many species can produce the same mycotoxin. Mycotoxins are low molecular weight secondary metabolites. The molecular weight may vary from 50 Da to >500 Da and generally non-immunogenic. The structure may vary from a simple heterocyclic ring to 6–8 irregularly arranged heterocyclic rings. They are highly stable, hence difficult to destroy by traditional food processing conditions. Under certain wavelength of UV light, they emit fluorescence thus can be used for screening of contaminated food or feed stuff. The optimal condition for mycotoxin production is at a water activity (Aw) of 0.85–0.99 and at a temperature range of 10–30 °C (Table 8.1).

The toxigenic molds are a major problem in agriculture products such as grains, cereals, nuts, and fruits. Mycotoxins (Fig. 8.1) produced in these products can cause severe health problems in both humans and animals. When mycotoxins are consumed at high dosage or over a long period, they may cause severe disease. Mycotoxins can cause acute disease manifested by kidney or liver failure or chronic diseases such as carcinoma, birth defects, skin irritation, neurotoxicity, and death. Three general mechanisms of mycotoxin action are described: mutagenic, teratogenic, and carcinogenic. During the mutagenic action, the toxin binds to DNA, especially the liver mitochondrial DNA resulting in point muta-

tion or frameshift mutation due to deletion, addition, or substitution of nucleotide bases in DNA, and can affect liver function (hence hepatotoxic). Teratogenic action leads to birth defect, and the carcinogenesis causes irreversible defects in cell physiology resulting in abnormal cell growth and metastasis.

Three major genera of molds, *Aspergillus*, *Penicillium*, and *Fusarium*, are of significant interest in food safety for the production of mycotoxins (Table 8.1); however, several other mold species are also recognized in recent years for production of a variety of mycotoxins.

Aflatoxin

Aflatoxin (*Aspergillus flavus* **toxin**) is produced primarily by *Aspergillus flavus* and *A. parasiticus* (Fig. 8.2). Aflatoxins (AF) are difuranocoumarins and occur in different chemical forms: AFB1, AFB2, AFG1, AFG2, and AFM1. The B and G forms stand for the blue fluorescence and green fluorescence, respectively, which are evident when the samples containing those mycotoxins are exposed to UV light. Aflatoxins are found in nuts, spices, and figs and are produced during storage under hot and humid conditions. The allowable toxin limits are 20 ppb in nuts such as in Brazil nuts, peanuts, and pistachio. Aflatoxin-contaminated feed causes high mortality in farm animals. Cows fed with aflatoxin-contaminated feed convert the aflatoxin B1 into hydroxylated form called aflatoxin M1 (milk), and the toxin is released through milk, and the allowable limit in milk is 0.5 ppb. The allowable limit in meats, corn, and wheat is also 0.5 ppb. The 50% lethal dose (LD_{50}) for rat is 1.2 mg/kg. The acute lethal dose for an adult human is thought to be 10–20 mg. The primary target organ for aflatoxin is the liver. Mitochondrial cytochrome P450 enzyme in the liver converts aflatoxin into reactive 8,9-epoxide form, which binds to DNA and results in GC to TA transversions, leading to carcinogenesis. Aflatoxin causes gross liver damage, resulting in liver cancer (hepatocarcinogen). It can also cause colon and lung cancer. The International Agency for Research on Cancer

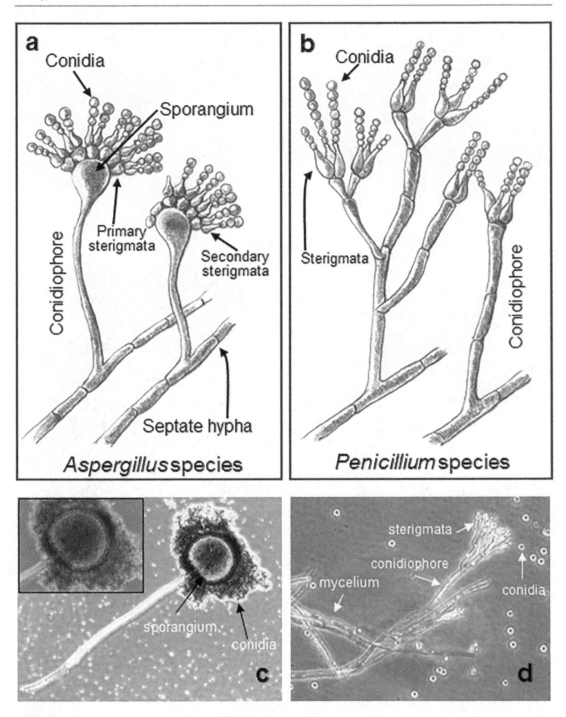

Fig. 8.1 Schematics of (**a**) *Aspergillus* species and (**b**) *Penicillium* species. Panels (**c**) and (**d**) are the light microscopic photograph of (**c**) *Aspergillus niger* and (**d**) *Penicillium citrinum* (magnification 400×)

Table 8.1 Optimal condition for mycotoxin production from molds

Mold	Mycotoxin	Optimum temperature (°C)	Optimum water activity (Aw)
Aspergillus flavus A. parasiticus	Aflatoxin B1, B2, G1, G2	33	0.99
Aspergillus ochraceus	Ochratoxin	30	0.98
Penicillium verrucosum	Ochratoxin	25	0.90–0.98
Aspergillus carbonarius	Ochratoxin	15–20	0.85–0.90
Fusarium verticillioides	Fumonisin	10–30	0.93
Fusarium verticillioides	Deoxynivalenol (DON)	11	0.90
Fusarium graminearum	Zearalenone	25–30	0.98
Penicillium expansum	Patulin	0–25	0.95–0.99

Adapted and modified from Murphy et al. (2006). J Food Sci 71, R51–R65

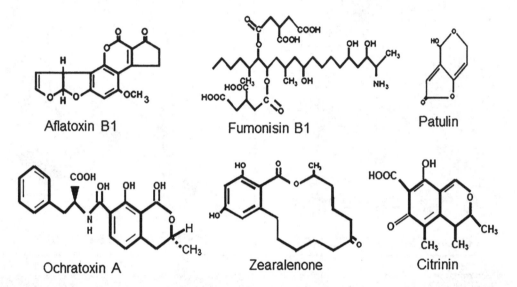

Fig. 8.2 Chemical structures of selected mycotoxins

(IARC) has classified aflatoxin B1 as a group I carcinogen. AFB1 suppresses the immune system, promotes inflammation, and retards the growth of humans and animals.

Aspergillus flavus also causes "Aspergillus ear rot" in corn. Climate change can affect Aspergillus flavus growth and aflatoxin B1 production. Hot and dry climates have the propensity for high risk of aflatoxin production.

Ochratoxin

Aspergillus ochraceus and several other species including Penicillium spp. produce seven structurally related secondary metabolites called ochratoxin (Fig. 8.2). Ochratoxin is a phenylalanyl derivative of a substituted isocoumarin and is found in a large variety of foods including wheat, corn, soybeans, oats, barley, coffee beans, meats, and cheese. Barley is thought to be the predominant source. The toxin is heat stable requiring exposure to 250 °C for several minutes to reduce activity. Ochratoxin is hepatotoxic and nephrotoxic and a potent teratogen and a carcinogen. Nephropathy and renal pathology are predominant consequences of ochratoxin poisoning. The toxin is structurally similar to phenylalanine (Phe), thus inhibits enzymes that use Phe as a substrate. Specifically, it inhibits phenylalanine–tRNA synthetase to synthesize phenylalanine required for protein synthesis. It causes mitochondrial damage, oxidative burst, and lipid peroxidation and inhibits cellular function and ATP production. It also causes apoptosis in some mammalian cells. The LD_{50} value in rats is 20–22 mg/kg^{-1}. The IARC considers ochratoxin as the category 2B carcinogen. The provisional tolerable weekly intake (PTWI) is 120 ng kg^{-1} body weight. The toxin is analyzed by using the high-performance liquid chromatography (HPLC) technique and a mass spectrometry.

Fumonisins

Fumonisins are synthesized by the condensation of amino acid alanine into acetate-derived precursor, and the most abundant form is fumonisin B1 (Fig. 8.2). Other forms of the toxins are FB2 and FB3. The FB1 has structural similarity to sphinganine, the backbone precursor of sphingolipids. Fumonisins are produced by *Fusarium verticillioides*, *F. proliferatum*, and *F. nygamai*. *Fusarium verticillioides* under ideal conditions can infect corn causing seedling blight, stalk rot, and ear rot and are present virtually in all matured corns. Corns, tomatoes, asparagus, and garlic are the major source of fumonisin. Fusarium ear rot of corn is also caused by *Fusarium verticillioides*. Fumonisins are highly water-soluble, and they do not have any aromatic structure or unique chromophore for easy analytical detection; however, HPLC with a fluorescence detector has been used for identification. Fumonisins are highly stable to a variety of heat (150 °C) and chemical processing treatments and are not degraded during fermentation.

Fumonisin B1 strongly inhibits ceramide (CER) synthase enzyme that catalyzes the acylation of sphinganine and recycling of sphingosine. As a result, the intracellular sphinganine and other sphingoid levels increase, which are highly cytotoxic. This cellular imbalance of sphinganine and sphingosine may be responsible for toxicity and possibly carcinogenicity. In animals, fumonisins cause a variety of diseases including leukoencephalomalacia, pulmonary edema, and hydrothorax. The toxins are reported to cause esophageal cancers in humans.

Trichothecenes

Trichothecenes are a group of structurally related compounds. Over 180 trichothecenes are reported, and they are produced by a number of fungal genera including *Fusarium*, *Trichoderma*, *Myrothecium*, *Stachybotrys*, *Trichothecium*, and others. The trichothecenes (T) are divided into four groups (A–D). Group A consists of the most common trichothecenes: 3-acetyl deoxynivalenol (DON), HT-2, and T-2 toxin, and group B are represented by DON. Groups C and D trichothecenes are of less importance. T-2 and HT-2 are produced by *Fusarium sporotrichioides*, *F. langsethiae*, *F. acuminatum*, and *F. poae*, while DON is produced by *F. graminearum*, *F. culmorum*, and *F. cerealis*. These toxins are associated with several different cereal products, meat, and dairy products. The toxins can be detected by HPLC and the thin layer chromatography.

Trichothecenes inhibit eukaryotic protein synthesis by binding to the 60S ribosomal subunit and by interacting with the peptidyl transferase enzyme thus preventing peptide bond formation. The T-2 toxin can also inhibit DNA and RNA syntheses and cell death. Toxins cause hemorrhage in the gastrointestinal tract, diarrhea, and vomiting. Toxins are cytotoxic, immunotoxic, and hepatotoxic, and direct contact may cause dermatitis. For trichothecenes (T-2 and HT-2), the tolerable daily intake (TDI) is 0.1 µg per kg body weight, and provisional maximum tolerable daily intake (PMTDI) is 0.06 µg per kg body weight per day. For DON, both TDI and PMTDI are 1 µg per kg body weight per day, and acute reference dose (for 1-day exposure) (ARfD) is 8 µg per kg body weight per day.

Patulin

Patulin (Fig. 8.2) is known as toxic lactones produced by *Penicillium clariform*, *P. expansum*, *P. patulum*, *Aspergillus* spp., and *Byssochlamys* spp. *Penicillium expansum* is a common contaminant of damaged fruits. Therefore, fruits (apricots, grapes, peaches, pears, and apples) or juices are the major sources. Patulin does not survive the apple cider making process; hence, it is sensitive to some processing treatments. In addition, patulin is also found in bread and sausage. Patulin is needed in high dosage to cause diseases such as gastrointestinal ulceration, immunotoxicity, and neurotoxicity. It has a strong affinity for sulfhydryl group in enzymes thus inhibits enzyme activity. Patulin can affect immune response and suppress macrophage function. The LD_{50} value in rats is 15–25 mg per kg. It is a carcinogenic toxin and is reported to be responsible for subcutaneous sarcoma. The provisional maximum tolerable daily intake limit is 0.4 mg kg^{-1} body weight.

Zearalenone

Fusarium graminearum and other *Fusarium* species (*F. culmorum*, *F. cerealis*, *F. equiseti*, *F. verticillioides*, and *F. incarnatum*) produce zearalenone (Fig. 8.2). Zearalenone has two derivatives: α-zearalenone and β-zearalenone. All zearalenones have strong estrogenic properties and resemble the 17β-estradiol, the principal hormone produced by the human ovary. The estrogenic potential of α-zearalenol is higher than that of β-zearalenol due to a greater affinity toward the estrogen receptor. Zearalenone is classified as a nonsteroidal estrogen or a mycoestrogen. Thus, the toxin designation does not appear to be appropriate for zearalenone. Occasional outbreak of mycotoxin in livestock results in infertility. Zearalenone concentrations of 1.0 ppm can cause the hyperestrogenic syndrome in pigs, and even higher concentrations can cause abortion and other fertility-related problems. Reproductive problems are also reported in cattle and sheep. The major concern is that zearalenone can disrupt sex steroid function in humans. It has been used to treat postmenopausal problems in women and has been patented as oral contraceptives. Zearalenone promotes estrus in mice, and LD_{50} in the rat is reported to be 10,000 mg per kg. Corn, wheat, oats, and barley are known sources for this toxin. The provisional maximum tolerable daily intake in human is 0.5 μg per kg body weight per day, and the tolerable daily intake is 0.25 μg per kg body weight. Zearalenone is sensitive to heat exceeding 150 °C.

Citrinin

Citrinin (Fig. 8.2) is produced by *Penicillium citrinum*, *Penicillium expansum*, and *Penicillium viridicatum* including some strains of *Penicillium camemberti*, used in cheese production. Citrinin is also produced by several species of *Aspergillus* including *Aspergillus oryzae*, used in the production of sake, miso, and soy sauce. The major source of this toxin is rice (also called yellow rice), bread, ham, wheat, oats, rye, and barley. Citrinin is a nephrotoxin and causes nephropathy in animals. The LD_{50} for citrinin in chicken is 95 mg per kg; in rabbits, 134 mg per kg, and its significance in human health is unknown.

Alternaria Toxin

There are more than 70 *Alternaria* mycotoxins produced by several species of *Alternaria*: *Alternaria alternata*, *A. citri*, *A. solani*, and *A. tenuissima*. *Alternaria* toxins are fetotoxic and have teratogenic and mutagenic effects. Other known *Alternaria* mycotoxins are alternariol, altenuene, and tenuazonic acid. *A. alternata* is the most important mycotoxin-producing species. This toxin is generally associated with apples, tomatoes, and blueberries.

Emerging Mycotoxins

There are several other mycotoxins reported in recent years, for example, fusaproliferin, enniatins, beauvericin, and moniliformin produced by different species of *Fusarium*. These mycotoxins are teratogenic and cytotoxic, and detailed mechanism of action is yet to be determined.

Ergot Alkaloids

Claviceps purpurea produces a toxic cocktail of alkaloid, ergot, which is not considered a typical mycotoxin. *Claviceps purpurea* grows on the heads of grasses such as wheat and ryes, and the ergot is transmitted after ingestion of bread or other products prepared with contaminated rye or wheat. The alkaloids are relatively thermolabile, and some may not survive during the bread making process. In addition, other *Claviceps* species including *Claviceps paspali* (forage grass), *Claviceps fusiformis*, *Claviceps gigantea*, and *Sphacelia sorghi* (an anamorphic form of *Claviceps*) produce alkaloids called ergometrine, ergotamine, and ergotoxine.

The disease caused by ergot is called ergotism. It is also known as "holy fire" or "St. Anthony's fire" because of severe burning sensations in the limbs and extremities of the victim. Two forms of ergotism are reported: gangrenous and convulsive. In the gangrenous form, the blood supply is affected causing tissue damage. In the convulsive form, the toxin affects the central nervous system. With advanced food processing techniques, the ergotism may have been eliminated from humans, but it remains a serious problem in animals including cattle, sheep, pigs, and chicken resulting in gangrene, convulsions, abortion, hypersensitivity, and ataxia. In cattle, ergotism spreads around the hooves, and the animal may lose hooves and is unable to walk and die by starvation.

Prevention and Control of Mycotoxins

Good agricultural practices (GAP) and good manufacturing practices (GMP) to control molds in preharvest and postharvest crops should be employed. Those include soil testing, crop rotation, irrigation, antifungal treatments, appropriate harvesting conditions, drying, and storage. Traditionally mold was controlled by adjusting the temperature, pH, and moister levels of the stored grains, cereals, and fruits. Proper food processing approaches such as physical removal of the moldy grains, nuts, and fruits can lower the mycotoxin levels. Washing, flotation and density segregation, dehulling, milling, steeping, and extrusion can also reduce mycotoxin content. High temperature, acids, and alkaline conditions in food can also chemically modify the mycotoxins. Natural enzymes in the food or enzymes derived from fermentation or deliberately intro-

duced into the food can also be used for detoxification. Other modern food processing technologies, such as cold plasma and irradiations both ionizing (gamma) and non-ionizing (UV, microwave), can inactivate molds and reduce mycotoxins.

In modern day, HACCP (hazard analysis critical control points) is employed to reduce mold and mycotoxins in products. Implementation of HACCP is aided by improved analytical techniques for sensitive detection of mycotoxins and stringent regulatory standards to exclude products for human consumption that contain mycotoxin levels over the allowable limits. In addition, development of transgenic plants that are able to increase the insect and mold resistance may aid in reduced levels of mycotoxins in products.

Except for supportive therapy, there is no treatment currently available for foodborne mycotoxin poisoning.

Summary

Molds (fungi) or mycotoxins infect humans, animals and plants. The common mold genera of importance in food safety are *Aspergillus*, *Penicillium*, *Fusarium*, and *Alternaria*. Mold causes mycosis, while the mycotoxin causes mycotoxicosis in humans and animals. Molds also cause food spoilage, and the major crops that are affected include wheat, maize, peanuts and other nut crops, cottonseed, and coffee. The Food and Agriculture Organization (FAO) of the WHO estimates that about 25% of the world's crops are affected by molds. The toxigenic molds produce a variety of mycotoxins, and some are studied in detail. A majority of mycotoxins such as aflatoxin, fumonisin, trichothecene, deoxynivalenol (DON), patulin, ochratoxin, and citrinin are carcinogenic, mutagenic, teratogenic, or cytotoxic and are acquired by consuming mycotoxin-contaminated cereal foods, nuts, milk, and meat. Some mycotoxins are hepatotoxic and nephrotoxic and are life-threatening. *Claviceps purpurea* produces a toxic cocktail of alkaloid, ergot, which is responsible for ergotism characterized by gangrene and neurological disorder. A variety of mycotoxins may be present in a food, but sensitive analytical tools are needed to monitor their presence. Avoiding the use of mold-contaminated raw products for food/feed production can reduce mycotoxicosis in humans and animals. There is no treatment currently available for foodborne mycotoxin poisoning.

Further Readings

1. Bennett, J.W. and Klich, M. (2003) Mycotoxins. *Clin Microbiol Rev* **16**, 497–516.
2. Cousin, M.A., Riley, R.T. and Pestka, J.J. (2005) Foodborne mycotoxins:chemistry, biology, ecology, and toxicology. In *Foodborne Pathogens: Microbiology and Molecular Biology* eds. Fratamico, P., Bhunia, A.K. and Smith, J.L. pp.163–226. Norfolk: Caister Academic Press.
3. Edite Bezerra da Rocha, M., Freire, F.d.C.O., Erlan Feitosa Maia, F., Izabel Florindo Guedes, M. and Rondina, D. (2014) Mycotoxins and their effects on human and animal health. *Food Control* **36**, 159–165.
4. Karlovsky, P., Suman, M., Berthiller, F., De Meester, J., Eisenbrand, G., Perrin, I., Oswald, I.P., Speijers, G., Chiodini, A., Recker, T. and Dussort, P. (2016) Impact of food processing and detoxification treatments on mycotoxin contamination. *Mycotox Res*, 1–27.
5. Marin, S., Ramos, A.J., Cano-Sancho, G. and Sanchis, V. (2013) Mycotoxins: Occurrence, toxicology, and exposure assessment. *Food Chem Toxicol* **60**, 218–237.
6. Marroquín-Cardona, A.G., Johnson, N.M., Phillips, T.D. and Hayes, A.W. (2014) Mycotoxins in a changing global environment – A review. *Food Chem Toxicol* **69**, 220–230.
7. Medina, A., Rodriguez, A. and Magan, N. (2014) Effect of climate change on *Aspergillus flavus* and aflatoxin B1 production. *Front Microbiol* **5**.
8. Murphy, P.A., Hendrich, S., Landgren, C. and Bryant, C.M. (2006) Food mycotoxins: An update. *J Food Sci* **71**, R51–R65.
9. Zheng, M.Z., Richard, J.L. and Binder, J. (2006) A review of rapid methods for the analysis of mycotoxins. *Mycopathologia* **161**, 261–273.

Fish and Shellfish Toxins

9

Introduction

Fish and shellfish harvested from marine, brackish (a mixture of fresh and seawater), and freshwater environments may carry marine biotoxins and when consumed can cause foodborne intoxication. The toxins are generally derived from harmful single-cell marine plant (phytoplankton) algae, in which fish or shellfish feeds on. The toxins could also be originated from microbial breakdown of fish proteins. Worldwide, about 60–80 different species of toxic microalgae are reported, of which dinoflagellates represent about 75% of all marine species. Marine toxins can kill fish, shellfish, or marine animals. Fish and shellfish also accumulate toxins in their tissues upon feeding of toxic algae, and humans suffer from fish and shellfish poisoning after consuming raw, cooked, or partially cooked products, because a majority of toxins are heat stable. The toxins are responsible for two types of symptoms: diarrheagenic and neurotoxic. Diarrheagenic toxins affect intestinal epithelial barrier and cause fluid loss, while the neurotoxins block Na-channel and interfere with nerve impulse.

There is a direct correlation between algae bloom and the toxins in seafood, which generally occur in the late spring to summer months. Global warming and the climate change can have a significant impact on marine organisms leading to toxic algal bloom. Sea level rise and flash flood can cause increased release of nutrients, nitrogen, and phosphorous to the coastal and marine waters that can alter phytoplankton community and algal bloom. The highest incidence of seafood-related outbreaks is associated with coastal inhabitants. Algal toxins are originated from dinoflagellates, while bacterial toxins are from cyanobacteria (blue-green algae). Dinoflagellates produce several polyether ladder toxins which include ciguatoxin (CTX), brevetoxin A (BTX A), brevetoxin B (BTX B), maitotoxin (MTX), and yessotoxin (YTX).

Diseases Caused by Fish and Shellfish Toxins

The majority of fish and shellfish toxins are derived from single-cell marine algae. There are six classes of algal toxins that are responsible for food poisoning: (i) paralytic shellfish poisoning (PSP), (ii) diarrhetic shellfish poisoning (DSP), (iii) neurotoxic shellfish poisoning (NSP), (iv) ciguatera fish poisoning (CFP), (v) azaspiracid shellfish poisoning (AZP), and (vi) amnesic shellfish poisoning (ASP). In addition, the scombroid toxin is an important class of toxin representing the biogenic amines, histamine, cadaverine, putrescine, and saurine, produced by bacterial decarboxylation of fish proteins. Pufferfish poisoning is associated with tetrodotoxin produced by marine bacteria (Table 9.1).

© Springer Science+Business Media, LLC, part of Springer Nature 2018
A. K. Bhunia, *Foodborne Microbial Pathogens*, Food Science Text Series,
https://doi.org/10.1007/978-1-4939-7349-1_9

Table 9.1 Fish and shellfish toxins and their regulatory limits

Type of poisoning	Toxin(s) involved [microbial source]	FDA regulatory limit	Fish and shellfish source	Symptoms	Symptom onset time
Scombroid	Histamine [multiple bacterial species]	USA: 50 ppm Europe: 200 mg/kg in fresh fish and 400 mg/kg in fishery products	Tuna, mahi-mahi, bonito, marlin, bluefish, wahoo, mackerel, and salmon	Severe headache, dizziness, nausea, vomiting, allergic response (flushed skin, urticaria, and wheezing)	Minutes to 1 h
Ciguatera fish poisoning (*CFP*)	Ciguatoxin (CTX) [algae: *Gambierdiscus toxicus*]	0.01 ppb (Pacific ciguatoxin) 0.1 ppb (Caribbean ciguatoxin)	Coral reef fish: amberjack, snappers, grouper, goat fish, barracuda, sea bass, surgeonfish, ulua, and papio	Abdominal pain, diarrhea, vomiting, paresthesias (tingling, burning, numbness), cold-to-hot sensory reversal, weakness, and myalgias	30 min–4 h
Paralytic shellfish poisoning (*PSP*)	Saxitoxins [algae: *Alexandrium acatenella*]	0.8 ppm (80 µg/100 g)	Molluscan shellfish: mussels, clams, and oysters	Vomiting, diarrhea, facial paresthesias, and respiratory paralysis leading to death	5–30 min
Neurotoxic shellfish poisoning (*NSP*)	Brevetoxins (BTX) [algae: *Karenia brevis*]	0.8 ppm	Mussels and clams	Diarrhea, vomiting, abdominal pain, myalgias, paresthesias, and ataxia	30 min–3 h
Amnesic shellfish poisoning (*ASP*)	Domoic acid (neurotoxin) [algae: *Pseudo-nitzschia pungens*]	20 ppm	Mussels, clams, crabs, and anchovies	Vomiting, diarrhea, headache, myoclonus, loss of short-term memory, confusion, disorientation, seizures, coma, and hemiparesis	15 min–> 35 h
Diarrhetic shellfish poisoning (*DSP*)	Okadaic acid, dinophysistoxins, pectenotoxins, yessotoxin [algae: *Dinophysis fortii*]	0.16 ppm	Mussels, clams, and scallops	Diarrhea, nausea, vomiting, and abdominal pain	30 min–3 h
Azaspiracids shellfish poisoning (*AZP*)	Azaspiracid-1 [algae: *Azadinium spinosum*]	0.16 ppm	Mussels	Diarrhea, nausea, vomiting, and abdominal pain	30 min–6 h
Pufferfish poisoning (*PFP*)	Tetrodotoxin (TTX)	–	Ocean sunfishes, porcupine fishes, and fugu (pufferfish)	Paresthesias, headache, vomiting, diaphoresis, respiratory paralysis, circulatory collapse, and death	10–45 min
Cyanotoxin	Nodularin and cylindrospermopsin [cyanobacterial blooms]	–	Crayfish	Hepatotoxicity, gastroenteritis	

Adapted from Kalaitzis et al. (2010). Toxicon 56, 244–258; Tsabouri et al. (2012). Ped. Allergy Immunol. 23, 608–615; FDA (http://www.fda.gov/downloads/Food/GuidanceRegulation/UCM252395.pdf)

Scombroid Toxin

Scombroid toxin means "mackerel-like," i.e., the toxin production is associated with scombroid fish (dark meat fish). These fish include tuna, mackerel, skipjack, bonito, marlin, sardine, yellow tail (amberjack), and dolphin (mahi-mahi) fish. The toxin is also found in bluefish, herring, and anchovies. Most of these fish have a high level of free L-histidine amino acid in their tissues. The scombroid toxin is produced due to the breakdown of L-histidine of flesh by bacterial histidine decarboxylase within 3–4 h. Bacterial decarboxylation of histidine produces small amines such as histamine, saurine, putrescine, and cadaverine, which produce typical allergic symptoms for the toxin when consumed.

The most important bacterial species associated with histamine production in fish is *Morganella morganii*. Several other species are also involved, namely, *Klebsiella pneumoniae*, *Proteus* spp., *Enterobacter aerogenes*, *Hafnia alvei*, and *Vibrio alginolyticus*, which are present in the gill and the gastrointestinal tract. In fermented fish products, *Lactobacillus* spp. can also produce histamine. Toxin level increases with increasing fish decomposition (spoilage).

The scombroid toxin is highly heat-stable. Cooking, canning, or freezing cannot reduce the toxicity of the fish products. Symptom severity depends on the amount of toxin ingested. The symptoms are generally manifested by an allergic reaction such as flush, body rash, headache, shortness of breath, dizziness, and tachycardia, which appear within hours of consumption of toxin. Dilatation of the peripheral blood vessels results in hypotension, flushing, and headache, while the increased capillary permeability causes urticaria, hemoconcentration, and eyelids edema. The gastrointestinal symptoms are nausea, vomiting, cramp, and diarrhea. Antihistamine drugs can be used to treat scombroid fish poisoning.

Paralytic Shellfish Poisoning

Saxitoxin is responsible for causing paralytic shellfish poisonings (PSP). The toxin was originally isolated from a clam, *Saxidomus giganteus*. However, the toxin is now found in mackerel, snapper, grouper, sea bass, barracudas, mollusks (moon snails and whelk), scallops, and oysters that feed on the toxic dinoflagellates (algae). The toxigenic dinoflagellate species are *Alexandrium acatenella*, *A. andersonii*, *A. catenella*, *A. tamarense*, *A. fundyense*, *A. hiranoi*, *A. monilatum*, *A. minutum*, *A. lusitanicum*, *A. tamiyavanichii*, *A. taylori*, *A. peruvianum*, *Gymnodinium catenatum*, and *Pyrodinium bahamense*. Many cyanobacterial species also produce saxitoxin: *Aphanizomenon flosaquae*, *Anabaena circinalis*, *Lyngbya wollei*, *Planktothrix*, and *Cylindrospermopsis raciborskii*. When dinoflagellates bloom, the water appears red, and it is referred to as "red tide."

Saxitoxin is a heat-stable potent neurotoxin and blocks the production of action potentials in neuronal cells. The guanidinium moiety of the saxitoxin binds to a region of the sodium channel known as "site1" and blocks the opening of the sodium channel. Saxitoxin is responsible for paralytic and neurotoxic shellfish poisoning. The symptoms appear within 30 min–2 h after consumption and include paresthesia (tingling and pricking sensation) of the mouth and extremities, burning sensation, drowsiness, incoherent speech, numbness, rash, and paralysis. In severe cases, saxitoxin poisoning may cause respiratory paralysis leading to death if respiratory support is not provided.

Diarrhetic Shellfish Poisoning

Diarrhetic shellfish poisoning (DSP) is a severe form of gastrointestinal illness. Two major toxins, okadaic acid (OA) and dinophysistoxins (DTXs) are responsible for diarrhetic shellfish poisoning. These toxins are produced by dinoflagellates: *Dinophysis acuta*, *D. acuminate*, *D. caudata*, *D. fortii*, and other *Dinophysis* species. They are also produced by *Prorocentrum arenarium*, *P. lima*, *P. hoffmannianum*, *P. maculosum*, and other species. OA also acts as a tumor promoter, induces lipid peroxidation, cytotoxicity, and apoptosis in cultured mammalian cells. DSP is associated with the consumption of scallops, mussels, and clams that feed on the toxic algae. The major symptoms are diarrhea, nausea, vomiting, and abdominal pain, which appear within 30 min–3 h and may last up to 4 days. The toxins are heat-stable; thus, cooking cannot inactivate them. The incidence of the disease can be reduced by removing the digestive organs of the shellfish, which tend to accumulate the majority of toxins.

Neurotoxic Shellfish Poisoning

Neurotoxic shellfish poisoning (NSP) is caused by brevetoxin (brevetoxin A, BTX A; brevetoxin B, BTX B) produced by dinoflagellate *Karenia brevis*. Algae growth in water produces red tide, and the molluscan shellfish feeds on the algae and accumulates toxin in their muscle. Humans consuming the contaminated shellfish exhibit neurological symptoms similar to the paralytic shellfish poisoning, such as tingling and numbness of the lips, tongue, and throat, muscular aches, and dizziness, but less severe. Gastrointestinal symptoms include diarrhea and vomiting. The symptoms appear very quickly within 30 min–3 h and generally subside in a few hours. *Karenia brevis* bloom is a recurring problem in the Gulf of Mexico.

Ciguatera Fish Poisoning

Ciguatera fish poisoning (CFP) is associated with the consumption of ciguatera toxin (CTX), which is derived from the single-celled marine macroalgae, *Gambierdiscus toxicus*. CTX is a lipid-soluble and heat-stable toxin and is produced and released into the water by *G. toxicus* bloom. Toxins enter the marine food chain and ultimately affect humans. Fish such as snapper, amberjack, grouper, barracuda, and sea bass live in reefs or shallow water which is the potential source of this toxin. CTX is concentrated in herbivorous fish consuming *G. toxicus* in the reefs of tropical and subtropical waters. The carnivore fish eats herbivore and acquires the toxin. CTX toxin occurrence is seasonal and found in fish harvested from the coastal waters of Florida, Hawaii, Puerto Rico, and the Virgin Islands during late spring through summer months.

CTX is a neurotoxin and causes paralysis. CTX has the similar mode of action as brevetoxin, which selectively targets the common binding "site 5" on the α-subunit of the neuronal sodium channel. The symptoms are tingling sensation of the lips, tongue, and throat, irregular heartbeat, and reduced blood pressure. It also causes a headache, muscle pain, and progressive weakness leading to paralysis. The gastrointestinal symptoms that develop within 2 h are nausea, vomiting, cramps, and diarrhea and last for a short time. Neurological and cardiovascular symptoms usually emerge within 6 h.

Azaspiracid Shellfish Poisoning

Azaspiracid shellfish poisoning (AZP) is associated with the consumption of molluscan mussels, which accumulate azaspiracid (AZA) toxin in tissues. This toxin is produced by a dinoflagellate, *Azadinium spinosum*. The symptoms that develop within minutes to hours after consumption of the contaminated shellfish last for several days. Symptoms are similar to the diarrhetic shellfish poisoning and are manifested as diarrhea, nausea, vomiting, and abdominal pain.

Amnesic Shellfish Poisoning

Amnesic shellfish poisoning (ASP) results from consumption of the molluscan shellfish (mussels, clams) and crabs due to a neurotoxin, domoic acid, which is produced by an alga, *Pseudonitzschia pungens*. The neurological and gastrointestinal symptoms associated with ASP consumption include vomiting, diarrhea, headache, disorientation, seizures, and hemiparesis (lack of voluntary movement of limbs and fingers similar to patients suffering from a stroke), which appear within 24–48 h after consumption. In addition, the patient may suffer from a short-term memory loss and confusion, respiratory difficulty, and coma.

Pufferfish Poisoning

Pufferfish poisoning (PFP) is associated with consumption of pufferfish (also known as fugu), ocean sunfish, and porcupine fish. Pufferfish is a culinary delicacy in Japan. The active agent is tetrodotoxin (TTX), which is produced by marine bacteria, and the toxin may accumulate in the skin and viscera (liver and ovary) of the pufferfish. Tetrodotoxin is a neurotoxin, which blocks sodium channel in the neuron. The toxin is heat-stable and heating may enhance toxicity. The symptoms appear within 10–45 min of consumption and manifest as tingling, burning, numbness, headache, vomiting, diaphoresis, respiratory paralysis, circulatory collapse, and death. The lethal potency of TTX is 5000–6000 MU (mouse unit) mg^{-1}. One MU is defined as the amount of toxin needed to kill a 20 g male mouse within 30 min after intraperitoneal injection. In humans, the minimum lethal dose is about 10,000 MU (~2 mg). Pufferfish poisoning is most prevalent in Asian countries with an estimated two to three deaths happening annually in Japan. Cases have also been reported from China, Taiwan, Singapore, Bangladesh, and Europe. There is no antidote for TTX.

Prevention and Control

To prevent fish- and shellfish-related food poisoning, fish and shellfish should be harvested only from officially approved bodies of water. Growth (bloom) of toxic algae is influenced by water temperature, salinity, runoff, and the presence of nutrients in the water. Therefore, harvesting of fish should be avoided when there is an evidence of algal bloom or red tide. Satellite is used to monitor ocean color including the red tide. Shellfish growing in the shallow water tend to accumulate more toxins. Periodic monitoring of shellfish for toxins can also help make a decision on closing contaminated shellfish beds for harvesting. Furthermore, shellfish should be purchased from a certified or licensed angler or fishmonger.

Toxins are heat-stable; therefore, cooking may not be able to inactivate the toxins. A severe heating process such as retorting may be able to reduce, but may not completely eliminate the toxicity.

Good manufacturing practice (GMP) and the hazard analysis critical control points (HACCP) should be applied to reduce the scombroid toxin production on fish. Preventing temperature abuse is essential. Low-temperature storage of fish can slow down the bacterial growth and the scombroid toxin production. Once histidine decarboxylase has been formed, it continues to produce histamine even at the refrigeration temperature. In frozen fish, histamine remains stable and can be reactivated after thawing. Storage of fish at −18 °C or below can stop the bacterial growth and prevent preformed histidine decarboxylase from producing histamine. Cooking can inactivate both histidine decarboxylase enzyme and the microbes, but not histamine. Antihistamine drug can be used to treat scombroid poisoning. For other toxin poisoning cases, supportive therapy is recommended.

Summary

Fish- and shellfish-associated toxins are generally derived from toxic microalgae or cyanobacteria growing in the water from which the fishes are harvested with the exception of the scombroid toxin, which is associated with bacterial decarboxylation of fish protein. Six types of algal toxins that are involved in fish and shellfish poisoning include (i) paralytic shellfish poisoning, (ii) diarrhetic shellfish poisoning, (iii) neurotoxic shellfish poisoning, (iv) ciguatera fish poisoning, (v) azaspiracid shellfish poisoning, and (vi) amnesic shellfish poisoning. Scombroid toxin shows allergenic response, while the majority of the algal toxins (Na-channel blocker) affect nerve impulse propagation. Fish and shellfish poisoning are manifested by the diarrhetic, neurological, or anaphylactic response and are prevalent in people living in the coastal areas because of the increased consumption of seafood. Global warming and climate change have been attributed to increased fish-related food poisoning due to increased toxic algal bloom. Toxins are heat-stable; thus, cooking or heating may not be able to inactivate them. To prevent food poisoning, fish and shellfish should be harvested from clean water or purchased from a certified or licensed source before consumption.

Further Readings

1. Bane, V., Lehane, M., Dikshit, M., Riordan, A. and Furey, A. (2014) Tetrodotoxin: Chemistry, toxicity, source, distribution and detection. *Toxins* **6**, 693–755.
2. Furey, A., O'Doherty, S., O'Callaghan, K., Lehane, M. and James, K.J. (2010) Azaspiracid poisoning (AZP) toxins in shellfish: Toxicological and health considerations. *Toxicon* **56**, 173–190.
3. Hinder, S.L., Hays, G.C., Brooks, C.J., Davies, A.P., Edwards, M., Walne, A.W. and Gravenor, M.B. (2011) Toxic marine microalgae and shellfish poisoning in the British isles: history, review of epidemiology, and future implications. *Environ Health* **10**, 54.
4. Hungerford, J.M. (2010) Scombroid poisoning: A review. *Toxicon* **56**, 231–243.
5. Kalaitzis, J.A., Chau, R., Kohli, G.S., Murray, S.A. and Neilan, B.A. (2010) Biosynthesis of toxic naturally-occurring seafood contaminants. *Toxicon* **56**, 244–258.
6. Noguchi, T., Onuki, K. and Arakawa, O. (2011) Tetrodotoxin poisoning due to pufferfish and gastropods, and their intoxication mechanism. *ISRN Toxicol* **2011**, 10.
7. Ray, B. and Bhunia, A. (2014) Opportunistic Bacterial Pathogens, Molds and Mycotoxins, Viruse, Parasites, and Fish and Shellfish Toxins. In *Fundamental Food Microbiology*. pp.387–406. CRC Press, Taylor and Francis Group.
8. Tirado, M.C., Clarke, R., Jaykus, L.A., McQuatters-Gollop, A. and Frank, J.M. (2010) Climate change and food safety: A review. *Food Res Int* **43**, 1745–1765.
9. Tsabouri, S., Triga, M., Makris, M., Kalogeromitros, D., Church, M.K. and Priftis, K.N. (2012) Fish and shellfish allergy in children: Review of a persistent food allergy. *Ped Allergy Immunol* **23**, 608–615.

Staphylococcus aureus

10

Introduction

There are more than 30 species in the genus *Staphylococcus*, and the species with the greatest impact on human health is *Staphylococcus aureus*. While *S. aureus* is a natural inhabitant (commensal) of human and animal skin, nares, and respiratory and genital tracts, as an opportunistic pathogen, it can cause invasive and fatal infections that affect many organs. Of particular public health concern is the emergence of drug-resistant strains of *S. aureus*, which are now one of the most frequently isolated pathogens from hospital-associated (nosocomial) infections. In the USA, about half a million people annually acquire skin and soft tissue infections by *S. aureus*. The bacterium is also involved in foodborne outbreaks associated with food poisoning, leading to approximately 241,000 annual illnesses, an estimate, which probably does not account for sporadic cases.

Staphylococci form clusters when grown in liquid or solid media, a characteristic, which led to the name staphylococcus (*staphyle* means a bunch of grapes and *kokkos* means a grain or a berry in Greek) (Fig. 10.1). In 1871, Von Recklinghausen, a German scientist, observed cocci in a diseased kidney and called them "micrococci." Later, based on cell arrangements, Billroth (1874) classified them as "monococcus," "diplococcus," "streptococcus," and "gliacoccus."

In 1880, Sir Alexander Ogston, a Scottish surgeon, and Louis Pasteur, a French scientist, confirmed that cocci-forming organisms are capable of causing disease. Later, Ogston coined the name "Staphylococcus," and he was given the credit for the discovery of the pathogen. In 1914, Barber discovered that a toxin produced by staphylococci was responsible for staphylococcal food poisoning.

Staphylococci are mostly associated with community-acquired and nosocomial infections and may be life-threatening in immunodeficient conditions. *Staphylococcus aureus* infections are traditionally treated with the β-lactam antibiotic penicillin, but bacteria frequently develop resistance by producing penicillinase (β-lactamase). To overcome resistance to penicillin, the β-lactamase-resistant drug methicillin was synthesized; however, some strains developed resistance to methicillin and are called methicillin-resistant *S. aureus* (MRSA). Several strains are also resistant to vancomycin (VRSA) and multiple other antibiotics and are routinely isolated from hospital settings. Antibiotic resistance and high virulence potential make this organism a very dangerous pathogen, and infection may be fatal because of lack of alternative antibiotics. MRSA may be either hospital associated (HA-MRSA) or community associated (CA-MRSA); however, in recent years MRSA has been associated with livestock (LA-MRSA). This later group may be responsible for the

© Springer Science+Business Media, LLC, part of Springer Nature 2018
A. K. Bhunia, *Foodborne Microbial Pathogens*, Food Science Text Series,
https://doi.org/10.1007/978-1-4939-7349-1_10

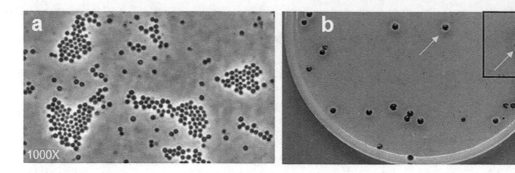

Fig. 10.1 (a) Phase contrast microscopic photograph of *Staphylococcus aureus* cells. Cells appear as clusters (magnification 1000×); (b) typical *S. aureus* colonies on Baird-Parker agar showing the characteristic black appearance surrounded by halo

transfer of the pathogen from animal to animal or animal to human and thus may have serious implication as a zoonotic pathogen. Most staphylococci are responsible for skin infections such as abscesses (boil, carbuncle, and furuncle), but some may cause life-threatening endocarditis, toxic shock syndrome, sepsis, and pneumonia. *S. aureus* also causes food poisoning, resulting in severe vomiting and cramping with or without diarrhea. Staphylococci also cause mastitis in cows and joint infection in humans, in animals, and in poultry, leading to edema and arthritis.

Classification

In the *Bergey's Manual of Determinative Bacteriology*, *Staphylococcus* has been placed in the family of *Micrococcaceae*. DNA–ribosomal RNA hybridization and comparative oligonucleotide analysis of 16S rRNA gene have demonstrated that staphylococci form a coherent group at the genus level. Staphylococci are differentiated from other close members of the family by their low G + C content of DNA, ranging from 30 to 40 mol%. The genus *Staphylococcus* has been further classified into more than 30 species and subspecies by biochemical analysis and by DNA–DNA hybridization. *Staphylococcus aureus* is the primary species in the genus *Staphylococcus* and is responsible for food poisoning and nosocomial and hospital-acquired infections. Other species which belong to this genus include *S. intermedius*, *S. chromogenes*, *S. cohnii*, *S. caprae*, *S. caseolyticus*, *S. delphini*, *S. epidermidis*, *S. felis*, *S. gallinarum*, *S. haemolyticus*, *S. hyicus*, *S. lentus*, *S. saprophyticus*, *S. sciuri*, *S. simulans*, *S. succinus*, *S. warneri*, and *S. xylosus*. A majority of them produce enterotoxins.

Morphology

Staphylococcus aureus is a Gram-positive coccus (1 μm in diameter) appearing microscopically as grape-like clusters due to three incomplete planar divisions (Fig. 10.1). They are nonsporeforming, are nonmotile, and produce golden yellow-pigmented colonies. This pigmentation is alluded to in the microbe's name, as aureus means golden, i.e., a gold coin of Rome. The cell wall of *S. aureus* contains three main components: the peptidoglycan comprising repeating units of *N*-acetylglucosamine β-1,4 linked to *N*-acetylmuramic acid; a ribitol teichoic acid bound via *N*-acetyl mannosaminyl-β-1,4-*N*-acetylglucosamine to a muramyl-6-phosphate; and protein A, which is covalently linked to the peptidoglycan. Protein A is characterized by its ability to bind to the Fc component of mammalian immunoglobulin molecules, which results in autoagglutination of mammalian plasma. Most of the other species of staphylococci lack protein A in their cell wall and therefore do not exhibit autoagglutination properties.

Cultural and Biochemical Characteristics

Staphylococcus aureus (Table 10.1) is a catalase-positive, facultative anaerobe, and grows abundantly under aerobic conditions, except for *S. saccharolyticus*, which is a true anaerobe. Under aerobic condition, it produces acetoin as the end product of glucose metabolism. It ferments mannitol, causes coagulation of rabbit plasma (coagulase positive), produces thermonuclease, and is sensitive to lysostaphin, a metalloendopeptidase. *S. aureus* produces hemolysins and causes an α-, β-, and α + β (double) hemolysis on blood agar plates. *S. aureus* can grow in a wide range of temperatures (7–48 °C; optimum 30–37 °C), pH (4–10; optimum 6–7), and water activity (A_w 0.83–0.99, optimum 0.98). *S. aureus* is highly salt-tolerant (up to 20% NaCl) and relatively resistant to drying and heat. The enterotoxins are produced at temperature 10–46 °C (optimum, 37–45 °C), pH 4–9.6 (optimum pH 7–8), A_w 0.85–0.99 (optimum 0.98), and NaCl 0–10% (optimum 0%). *S. aureus* coagulates rabbit plasma relatively quickly, while *S. intermedius* and *S. hyicus* subsp. *hyicus* cause delayed coagulation. *S. epidermidis* is coagulase negative and does not ferment mannitol. Many growth media have been developed for selective isolation of *Staphylococcus* species: mannitol salt agar (MSA) contains mannitol, a phenol red indicator, and 7.5% sodium chloride producing small colonies surrounded by yellow zones indicating mannitol fermentation. Baird-Parker media containing tellurite and egg yolk is suitable for isolation of coagulase-positive *S. aureus*, which produce black, shiny colonies surrounded by clear zones (Fig. 10.1).

Virulence Factors

Staphylococcus aureus produces a family of virulence factors such as adhesion proteins, enterotoxins, superantigens, toxic shock syndrome toxins (TSST), exfoliative toxins (ET), pore-forming hemolysins, ADP-ribosylating toxins, and proteases (Table 10.2).

Adhesion Proteins

Staphylococci possess multiple adhesion molecules, which are collectively known as MSCRAMM (microbial surface components recognizing adhesive matrix molecules). Internalization of the organism by host cells is triggered by MSCRAMM. Adhesion proteins include Bap (biofilm-associated proteins), which is responsible for biofilm formation and colonization in the mammary gland during mastitis. The C-terminus of Bap contains typical cell wall anchoring domain comprising of LPXTG motif, a transmembrane sequence, and a positively charged C-terminus. The fibronectin-binding protein (Fbp) binds to host fibronectin. Bacterial binding to fibronectin also facilitates internalization into nonprofessional phagocytes, such as

Table 10.1 Typical characteristics of *S. aureus*, *S. epidermidis*, *S. intermedius*, and *S. hyicus*

Characteristic	*S. aureus*	*S. epidermidis*	*S. intermedius*	*S. hyicus*
Catalase activity	+	+	+	+
Hemolysis	+	+/−	+	−
Coagulase production	+	−	+/−	+/−
Thermonuclease production	+	−	+	+/−
Hyaluronidase production	+	−	−	+
Lysostaphin sensitivity	+	+/−	+	+
Anaerobic utilization of				
Glucose	+	+	−	−
Mannitol	+	−	−	−

+, Most (90% or more) strains are positive
−, most (90% or more) strains are negative

Table 10.2 Virulence factors and enzymes produced by *Staphylococcus aureus*

Virulence factors	Receptors
Adhesin proteins	
Spa (protein A)	Fc part of IgG
Bap (biofilm-associated proteins)	Unknown
Fbp (fibronectin-binding protein)	Fibronectin, fibrinogen, elastin
ClfA (fibrinogen-binding protein)	Fibrinogen
Cna (collagen adhesin)	Collagen
IsdA, IsdB, IsdC, IsdH (iron-regulated surface proteins)	Hemoglobin, transferrin, hemin
Pls (plasmin-sensitive cell wall protein)	Cellular lipid called ganglioside GM_3
Atl (autolysin amidase)-bacteriolytic action	Fibronectin, fibrinogen, vitronectin
Enolase	Laminin
Teichoic acid	Unknown – binds epithelial cells
Enterotoxins (24)	
Staphylococcal enterotoxin SEA–SElX, except SEF	Glycosphingolipid
Pore-forming hemolysins	
Hemolysins α, β, γ, δ	Cholesterol
Superantigens	
TSST (toxic shock syndrome toxin)	MHC class II
Enterotoxins	Glycosphingolipid, MHC class II
Exfoliative toxins (ETA, ETB)	Desmoglein-1
ADP-ribosylating toxins	
Panton-Valentine Leukocidin, pyrogenic exotoxin	Complement receptor, C5aR
Proteases	
Metalloprotease, collagenase, hyaluronidase, endopeptidase, elastase	
Others	
Nuclease, lysozyme, phospholipases, coagulase	

keratinocyte, epithelial cell, endothelial cell, and osteoblast. Staphylococci also produce other adhesion factors including ClfA, a fibrinogen-binding protein that activates platelets aggregation and plays a role in staphylococcal arthritis; Pls, a plasmin-sensitive cell wall protein that binds to ganglioside GM_3 of host cells and promotes adhesion to nasal epithelial cells and; Cna, a collagen adhesin binds to collagenous tissues, i.e., cartilages.

Toxic Shock Syndrome Toxin-1 (TSST-1)

TSST-1 is a 22 kDa protein and acts as a superantigen, which generates a strong immune response in the host. The toxin stimulates the release of IL-2, TNF-α, and other proinflammatory cytokines (IL-1, IFN-γ, IL-12). In the1980s, TSST-1 was mostly recognized for its role in the outbreaks of toxic shock syndrome associated with

tampon use. Only a small number (<10%) of *S. aureus* strains produce TSST-1, and toxic shock syndrome (TSS) is a rare but severe and potentially fatal disease. TSST is responsible for acute illness, high fever, erythematous lesion (rash), hypotension, and septic shock. It can cause organ failure and DIC (disseminated intravascular coagulation). Unlike the enterotoxin superantigens, TSST does not cause emesis.

Exfoliative Toxin

The exfoliative toxin (ET) is a serine protease of 30 kDa and has two serotypes: ETA and ETB. Strains can produce one or the both and dependent on the isolates of different geographical origins. ET is responsible for a severe skin disease that primarily affects infants, called staphylococcal scalded skin syndrome (SSSS), which is characterized by a bright red rash, blisters, and severe skin lesions. The ET binds to receptor desmoglein-1, a desmosomal glycoprotein and selectively hydrolyzes the desmosomal cadherins of the superficial skin layer (stratum granulosum). This results in the destruction of the superficial skin layers causing dehydration and increased susceptibility to secondary skin infections.

Miscellaneous Enzymes and Toxins

Staphylococcus aureus produces coagulase, which contributes to the fibrin clot formation and accumulation on the bacterial cell surface, and aids in bacterial evasion of phagocytosis. The bacterium also secretes four types of hemolysins (α, β, γ, and δ); membrane active lipase such as phospholipase C (PLC); Panton-Valentine leukocidin, which destroys leukocytes; collagenase, which hydrolyzes collagen; hyaluronidase, which hydrolyzes hyaluronic acid component of the cellular basement membrane; metalloprotease; pyrogenic exotoxin; and staphylokinase, which degrades fibrin clots.

Enterotoxins

Staphylococcus aureus produces a large number of extracellular proteins and toxins. The most important toxins are called staphylococcal enterotoxins (SEs) and SE-like toxins (SEls) (Table 10.3), which share four common properties: (i) structural similarity, (ii) resistance to heat and proteolytic enzymes, (iii) superantigenicity, and (iv) emetic activity. Twenty-four major serologically distinct SEs are reported: SEA through SElX with no SEF. The SEF is similar to other SEs, but rather than inducing emesis, it causes toxic shock syndrome (TSS), hence designated TSST-1. One of the characteristics of SEs is the induction of emesis. If a SE is structurally similar to other SEs, but has not been tested for induction of emesis or negative for emesis in a primate model (monkey), the International Nomenclature Committee for Staphylococcal Superantigens (INCSS) designates them as SE-like toxins (SEls). The first five well-characterized SEs are SEA to SEE, and they all cause emesis. The SEC has three antigenically distinct subtypes: SEC1, SEC2, and SEC3. The SEG–SEI and SER–SET also have strong emetic activity, while SElL and SElQ are not emetic. The other SEls including SElJ, SElK, SElM, SElN, SElO, and SElP have been tested in primate models and have shown emetic activity but much lower than the SEA or SEB. SElU–SElX have not been tested in a primate model. SEs are responsible for food poisoning, acute illness, fever, erythematous lesions, and hypotension.

SEs are a heterogeneous group of water-soluble, single-chain globular proteins of 168–257 amino acids with molecular weight of about 19.3–29 kDa. The SE polypeptide chain contains relatively a large number of lysine, aspartic acid, glutamic acid, and tyrosine residues. The SEs are generally heat-resistant (121 °C for 10 min), and a heat-denatured enterotoxin can be renatured by prolonged storage or in the presence of urea. Toxins remain active even after boiling for 30 min. In food, such as in mushrooms, they are stable at 121 °C for 28 min. The SEs also are resistant to proteolytic digestion such as trypsin and pepsin thus retain activity in the gastrointestinal tract to cause food poisoning.

SEA is the most common serotype found in *S. aureus*. It is the most common SE responsible for about 78% of food poisoning outbreak followed by SED and SEB. SEA is a 27.1 kDa toxin, and its production is not regulated by *agr* (accessory gene regulator). SEB is a 28.4 kDa toxin and is the most heat-resistant (stable at 60 °C for 16 h) among all the toxins. SEB also is resistant to gastrointestinal proteolytic enzymes such as trypsin and pepsin. SECs are a group of highly conserved proteins, and there are three antigenically distinct subtypes: SEC1, SEC2, and SEC3. Staphylococcal isolates from different animal species produce host-specific SECs. SED (24 kDa) is the second most serotypes responsible for food poisoning. SED has the ability to form a homodimer in the presence of Zn^{2+} which facilitates its binding to MHC class II molecule on antigen-presenting cells and enables it to serve as a superantigen (see below).

Table 10.3 Biological characteristics of staphylococcal enterotoxins and staphylococcal enterotoxin-like toxins

Enterotoxin	Genetic element	Molecular weight (kDa)	Superantigenic activity	Emetic activity in monkey[a]
1. SEA	Prophage	27.1	+	25
2. SEB	Chromosome, plasmid	28.4	+	100
3. SEC1	SaPI	27.5	+	5
4. SEC2	SaPI	27.6	+	NT
5. SEC3	SaPI	27.6	+	<50
6. SED	Plasmid	26.9	+	NT
7. SEE	Prophage	26.4	+	NT
8. SEG	Egc, chromosome	27	+	160–320
9. SEH	Transposon	25.1	+	30
10. SEI	Egc, chromosome	24.9	+	300–600
11. SElJ	Plasmid	28.6	+	NT
12. SElK	SaPI	25.3	+	NT
13. SElL	SaPI	24.7	+	Not emetic
14. SElM	Egc, chromosome	24.8	+	NT
15. SElN	Egc, chromosome	26.1	+	NT
16. SElO	Egc, chromosome	26.8	+	NT
17. SElP	Prophage	26.7	+	NT
18. SElQ	SaPI	25.2	+	Not emetic
19. SER	Plasmid	27	+	<100
20. SES	Plasmid	26.2	+	<100
21. SET	Plasmid	22.6	+	<100
22. SElU	Egc, chromosome	27.2	+	NT
23. SElV	Egc, chromosome	27.6	+	NT
24. SElX	Chromosome	19.3	+	NT

Adapted from Hu and Nakane (2014)
Egc enterotoxin gene cluster, *SaPI S. aureus* pathogenicity island, *NT* not tested
[a]µg/animal after oral administration

The genes for enterotoxin production are located either in a bacteriophage, chromosome, transposon, or in plasmids. The *sea* and *sep* are located in a bacteriophage; *seb*, *seh*, and family of *sec* are in the chromosome; and *sed*, *sej*, and *ser* are located in a plasmid. The *sed* and *sej* genes are colocalized and the same strain always produces these two toxins (SED and SEJ) together. Enterotoxin gene cluster (*egc*) can encode for several SEs such as SEG, SEI, SEM, SEN, and SEO. Certain enterotoxin genes are also located in the pathogenicity islands (PAI). There are five staphylococcal PAI: SaPI-1, SaPI-2, SaPI-3, SaPI-4, and SaPI-bov. SaPI-1 contains genes for TSST-1 and SEK and SEQ. SE genes located on mobile elements can result in horizontal gene transfer. Production of enterotoxins is not restricted to *S. aureus* alone, as other non-*aureus* staphylococci are reported to produce

enterotoxin. The accessory gene regulator system (*agr*) is a main regulatory mechanism controlling the expression of virulence factors in *S. aureus*; however, not all SEs or SEls are regulated by *agr*.

Enterotoxins are expressed differentially, and the toxin production depends on the growth phase of bacteria, bacterial density, pH, and CO_2 levels. SEA and SEJ are synthesized mostly during the exponential phase, while SEB, SEC, and SED are produced during the transition from exponential to the stationary phase of growth.

Molecular Regulation of Virulence Gene Expression

The pathogenic potential of staphylococci in humans can be attributed to the expression of a wide array of virulence factors, most of which are governed by Staphylococcal virulence regulators. Four loci have been implicated in the regulation of expression of virulence factors: accessory gene regulator, *agr*; the staphylococcal accessory regulator, *sar*; *S. aureus* exoprotein expression, *sae*; and exoprotein regulator, *xpr*.

Enterotoxin production is regulated by the two-component regulatory system, *agrAC*. In the two-component system, one protein serves as a sensor and transfers signal by phosphorylating the intracellular activator, while another regulates genes to provide response called response regulator. The *agr* locus comprises the global regulatory system, which regulates virulence gene expression in the post-exponential phase of growth. It consists of two divergent units driven by promoters P_2 and P_3. The P_2 transcript includes four open reading frames (ORFs) referred to as *agrA*, *agrB*, *agrC*, and *agrD*. The P_3 transcript RNA III is the actual effector molecule and activates secretion of enterotoxins and other exoproteins, primarily at the transcriptional level. In the *agr* locus, P2 operon includes the genes for response regulator, *agrA*; histidine kinase, *agrC*; autoinducing peptide, *agrD*; and autoinducing ligand inducer (AIP), *agrB*. The strains that produce high amounts of enterotoxin have a high concentration of RNA III. An intact *agr* locus is necessary for maximum SEC expression. The *agr* locus regulates SEC expression at the posttranscriptional level, presumably at the level of translation or secretion. RNA III is produced at lower levels under alkaline pH and SEC production decreases at low pH. In addition, Sar family of proteins also regulates virulence gene expression. SarA is a DNA-binding protein and binds to *agr* promoter stimulating the transcription of RNA II. Activation of RNA II and subsequently RNA III leads to alternating target gene expression. Expression of SEB is positively controlled by *sarA*. Xpr represents an additional genetic element involved in regulation of some *agr*-regulated proteins. Low levels of SEs production were observed in *xpr* mutants. Reduced levels of RNA II and III were also observed in these mutants.

Food Association and Enterotoxin Production

Staphylococcal food poisoning is one of the most common foodborne illnesses reported worldwide. Nearly one-third of all the food poisoning cases in the USA were caused by staphylococci during the 1970s and 1980s, which in general has decreased over the past two decades. However, it remains the main reported cause of food poisoning in a number of countries including Brazil, Egypt, Taiwan, Japan, and most of the other developing countries. About 50–80% of *S. aureus* isolates are positive for at least one superantigen gene. MRSA strains isolated from hospital patients tend to harbor more superantigen genes than the methicillin-sensitive strains. Intoxication occurs due to the ingestion of one or more preformed SEs or SE- SEls in contaminated food.

Staphylococcal foodborne illness is often associated with creamy food prepared with milk and milk products, cheese, custard (pudding), cream-filled pastries, cakes, salad dressing, shellfish, fish, meat, and hams, as well as foods that require greater hand preparation such as pasta and chicken salads, deli foods, and sandwiches. Staphylococci can be transmitted to food through meat grinder's knives and food handlers. The bacteria replicate in foods subject to temperature abuse, such as foods left at room temperature for a long period.

Generally, staphylococcal enterotoxins are produced at temperatures between 10 °C and 46 °C (optimum, 37–45 °C), pH 4–9.6 (optimum pH 7–8), A_w 0.85–0.99 (optimum 0.98), and NaCl 0–10% (optimum 0%). At temperature 60 °C or higher, the organism will not grow; however, below 60 °C, the organism will grow and produce toxins.

Mechanism of Pathogenesis

Emesis and Diarrhea

The infectious dose of *S. aureus* is 10^5–10^8 cfu g^{-1} of food and a toxin concentration of 1 ng g^{-1} of food. The emetic dose (ED_{50}) of SEA toxin in

Rhesus monkey is 5–20 µg kg^{-1}, while 200 ng kg^{-1} may be needed to show intoxication syndrome in human. SEB is highly toxic and a dose of 400 ng kg^{-1} is required for humans. Following consumption of *S. aureus* enterotoxin contaminated food, toxins are absorbed and cause typical gastroenteritis, while bacteria pass through the intestine without causing any adverse effects on the host. Studies conducted with SEA and SEB have shown that SE binds to submucosal mast cells and induces degranulation and release of 5-HT (5-hydroxytryptamine), a neurotransmitter (also known as serotonin). 5-HT interacts with the 5-HT$_3$ receptor on adjacent vagal afferent neurons in the stomach lining and stimulates medullary vomiting center (medulla oblongata) to induce a violent emetic reflex (Fig. 10.2). 5-HT synthesis inhibitors or 5-HT$_3$ receptor antagonist (ondansetron hydrochloride (Zofran)) drugs can prevent SE-induced vomiting. Although the mast cell receptor for SE is unknown, studies in kidney cells indicate that the putative receptor for SE is a glycosphingolipid. SE also activates Ca^{2+} signaling pathway in the intestinal epithelial cells. Enterotoxins elicit damage to the intestinal epithelial cells resulting in villus destruction, villi distension, crypt elongation, and lymphoid hyperplasia.

Superantigen Activity

The superantigenic property of staphylococcal enterotoxins distinguishes them from other bacterial toxins. Superantigens are the molecules that have the ability to stimulate an exceptionally high percentage of T cells (CD4$^+$ and CD8$^+$ cells), massive cytokine release, and systemic shock (Fig. 10.3). Enterotoxins cross the epithelial barrier, enter blood circulation, and bind to the α-chain of the MHC class II molecules on the surface of macrophages. The toxin is presented to the T cells that carry TCR (T-cell receptor) made with β-chain, also called Vβ carrying T cells. T cells proliferate and produce large quantities of IL-2 and IFN-γ. Elevated levels of IFN-γ also induce increased MHC class II expression in macrophages and other cells, which in turn bind more superantigens, and activate more T cells. Inflammatory cytokines such as IL-1 and TNF-α are produced by activated macrophages, and the cytokines initiate typical toxic shock syndrome with disseminated intravascular coagulation (DIC), high fever, low blood pressure, massive shock, and death (Fig. 10.3).

Superantigens exert immunopathological response in the GI mucosa. The intestinal mucosa harbors all of the major Th-cell subsets (Th1,

Fig. 10.2 The pathogenic mechanism of intoxication with enterotoxin from *Staphylococcus aureus*. 5-HT, 5-hydroxytryptomine

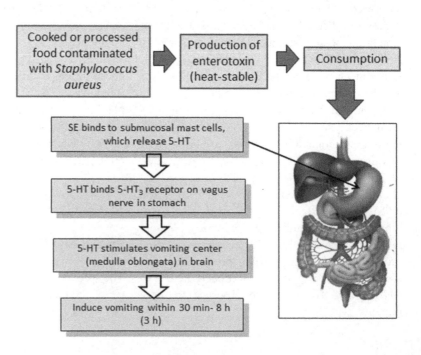

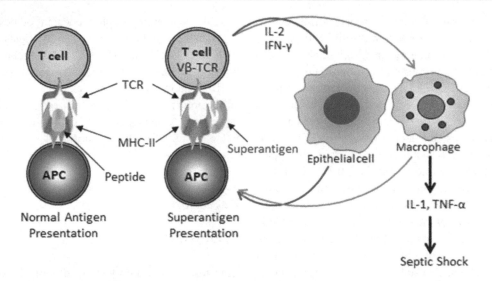

Fig. 10.3 Mechanism of superantigen action of staphylococcal enterotoxin

Th2, Th17). Superantigens, unlike others, do not stimulate T cells indiscriminately, as only specific Vβ sequences in the TCR are recognized. SE increases epithelial permeability and intestinal secretion by reducing TJ and AJ protein expression in epithelial cell–cell junctions. Epithelial permeability is mediated by enhanced secretion of IFN-γ and TNF-α from lymphocytes (*see* Chap. 4). SEs activate gastrointestinal T cells and provoke a "cytokine storm," in which Th1 releases IL-2 and IFN-γ, Th2 releases IL-4, and Th17 secrets IL-17, and activated macrophages and other cell types secrete TNFα, IL-1β, IL-6, and IL-8. These cytokines may act as chemoattractants and induce expression of adhesion molecules, favoring localization of diverse immune cells responding to SE. Cytokines also affect epithelial cell functions primarily the ion and water transport.

been used as a small animal model which shows an emetic response after oral and intraperitoneal administration.

Staphylococcal enterotoxins act as superantigens. Examination of these toxins for the ability to cause mammalian cell damage provides a means to assay these toxins. A bioassay for superantigen on a T-lymphocyte cell line (Raji) has been developed. In this assay, the SEA-induced cytotoxic action was detected colorimetrically using the CytoTox 96-well lysis detection kit. This system can detect SEA at picomolar concentrations. *Staphylococcus aureus* enterotoxins are also detected in Madin-Darby bovine kidney (MDBK), bovine embryo lung (PEB), and dog carcinoma cell line (A-72) for possible cytopathic effects. The PEB cell line is the most susceptible, and the cytopathic effect is observed after 2 h of incubation.

Animal and Cell Culture Models

Animal models have been developed to study emetic effect after oral administration. Nonhuman primate (monkey, i.e., *Macaca mulatta*), pigs, piglets, and dogs are sensitive to *S. aureus* emetic effects, while rodents (mice, rats, and rabbits) are less susceptible. The house musk shrew (*Suncus murinus*), a rodent, has

Symptoms

Staphylococcal infection may cause skin infections such as abscesses (boil, carbuncle, and furuncle), staphylococcal scalded skin syndrome, impetigo, and cellulitis. Severe infections may cause life-threatening endocarditis, osteomyelitis, toxic shock syndrome, sepsis, high fever, pneumonia, and sudden infant death syndrome.

Symptoms of staphylococcal gastrointestinal intoxication appear within 30 min–8 h (average of 3 h) and include hypersalivation, nausea, violent vomiting in spurts, abdominal cramping with or without diarrhea, headache, dizziness, shivering, and general weakness. The significant fluid loss will cause dehydration and hypotension. In severe cases, headache, prostration, low blood pressure, and anaphylactic shock may happen. The mortality rate is very low, 0.02% occurring in the most susceptible persons, infants, and the elderly. The disease is self-limiting and may resolve within 24–48 h, but infants and elderly may require hospitalization. In the case of aerosol exposure of enterotoxin, sudden onset of fever, chills, headache, and cough occur. Fever may last for several days and the cough can last for 10–14 days.

Prevention and Control

For treatment of skin infection or systemic infection, antibiotics and other supportive therapies are needed. The staphylococcal food intoxication is mostly self-limiting; therefore, only fluid therapy and bed rest are recommended without any antibiotic therapy.

To prevent *S. aureus*-related food poisoning, cooking food thoroughly is important, but preventing contamination and cross-contamination is critical. After cooking, food should not be left at room temperature for longer than 2 h, because the permissive temperature for bacterial growth and toxin production is between 10 °C and 46 °C. Thus, food should be held at above 60 °C or cooled rapidly to below 5 °C or stored refrigerated to prevent toxin production. Since one of the major sources of *S. aureus* is human skin, handwashing and the use of protective gloves, masks, and hairnets before food handling should reduce the chance of food contamination. Maintaining cold chain during food preparation and processing will prevent bacterial growth. Ensuring quality of raw ingredients, processing methods, adequate cleaning, and disinfection of equipment should be a routine part of food production practices to prevent pathogen contamination and growth in food. Strict hygienic practices are

crucial in preventing staphylococcal food poisoning. Implementation of Hazard Analysis and Critical Control Points (HACCP), good manufacturing practices (GMP), good hygienic practices (GHPs), and rapid microbiological analysis will aid in preventing pathogens in food manufacturing facility. Ondansetron hydrochloride (Zofran) is the choice of drug for controlling nausea and vomiting.

Detection

Culture Methods

Conventional culture methods allow isolation of bacteria on mannitol salt agar or Baird-Parker agar plates. The colonies on Baird-Parker appear black, shiny, circular, convex, and smooth with the entire margin forming a clear zone with an opaque zone (lecithinase halo) around the colonies (Fig. 10.1).

Nucleic Acid-Based Methods

Nucleic acid-based detection systems offer a very good alternative to conventional culture methods for detection of *Staphylococcus* in food. Nucleic acid probe-based methods have been developed for the detection and enumeration of staphylococci. PCR-based detection of enterotoxin genes including *egc* (enterotoxin gene cluster: SEA to SEE; SEG, SEH, SEI, SEM, SEJ, SEN, and SEO), TSST-1, exfoliative toxins A and B (*etA* and *etB*), methicillin-resistant (*mecA*) gene, and 16S rRNA from *S. aureus* has been reported. Fluorescence-based real-time PCR (TaqMan-PCR) has been demonstrated for enterotoxins A to D and *mecA* for rapid analysis of a large number of samples. A DNA microarray was developed for detection and identification of 17 staphylococcal enterotoxin (*ent*) genes simultaneously. The assay is based on PCR amplification of the target region of the *ent* genes with degenerate primers, followed by characterization of the PCR products by microchip hybridization with oligonucleotide probes specific for each *ent* gene. The use of degenerate primers allowed the simultaneous amplification and identification of as many as nine different *ent* genes in one *S. aureus* strain.

Quantitative detection of staphylococci or their toxin genes was achieved through the use of quantitative real-time PCR (qRT-PCR). Commercial rapid assay kits that detect *S. aureus* 23S rRNA is available. A commercial array chip called Staphychips also is available for identification of five different staphylococci in an array format.

Immunoassays

Immunoassays based on ELISA are widely used for detection of enterotoxins. Automated commercial detection systems are available. The detection limit is <0.5–1 ng g^{-1} of enterotoxin. Enzyme immunoassay is available for the detection of superantigens: SEA, SEB, SEC, TSST-1, and streptococcal pyrogenic exotoxin A (SPEA) in the blood serum. A fluorescence immunoassay has been developed for SEB with a detection limit of 100 pg per well and is more sensitive than the conventional ELISA assay. Magnetic bead-based immunoassay has been used to detect SEs in a sandwich format. Rapid latex agglutination tests for SEA to SEE with a detection limit of 0.5 ng/ml are available.

Other Rapid Methods

Although above methods are widely applied for staphylococcal detection, some of the other methods which directly detect the whole cell or their metabolites include, direct epifluorescence technique (DEFT), flow cytometry, impedimetry, ATP-bioluminescence, are commonly used for the routine analysis of milk. Toxins are also detected by mass spectrometry.

Summary

Staphylococcus aureus, a natural inhabitant of the human and animal body, is mostly associated with community-acquired and nosocomial infections, which can be fatal in immunodeficient patients. Methicillin and vancomycin-resistant *S. aureus* can cause serious nosocomial infections in humans. *S. aureus* also causes mastitis in the cows, and joint infection in humans, animals and poultry. Staphylococci are also responsible for food poisoning characterized by severe vomiting and cramping with or without diarrhea. *S. aureus* produces a large number of toxins and enzymes, of which the enterotoxins (24 serotypes of toxins are identified) are most important in the production of gastroenteritis (vomiting and diarrhea) and superantigen-associated illness. Enterotoxins are heat-stable and are produced when the temperature of food is at or above 46 °C. Consumption of preformed toxins induces vomiting with or without diarrhea within 30 min–8 h (average 3 h). The enterotoxin induces the release of 5-HT (5-hydroxytryptamine) from mast cells, which stimulates vagal nerves in the stomach lining and induces vomiting. Enterotoxins are also called superantigens, because they form a complex with MHC class II molecules on the surface of antigen-presenting cells, activating and proliferating T cells to produce massive amounts of cytokines (IL-2, IFNγ, IL-1, TNF-α) that contribute to fatal toxic shock syndrome. The genes for enterotoxin production are present in pathogenicity islands in the chromosome, in plasmids, in transposons, and in temperate bacteriophages. Toxin production is regulated by a two-component regulatory system called *agrAC* (accessory gene regulator). Strict hygienic practices are crucial in preventing staphylococcal food poisoning. Skin infection or systemic infection requires antibiotic therapy, while the foodborne intoxication does not require antibiotic therapy since the disease is caused by the toxin and it is mostly self-limiting.

Further Readings

1. Argudín, M.Á., Mendoza, M.C. and Rodicio, M.R. (2010) Food poisoning and *Staphylococcus aureus* enterotoxins. *Toxins* **2**, 1751–1773.
2. Balaban, N. and Rasooly, A. (2000) Staphylococcal enterotoxins. *Int J Food Microbiol* **61**, 1–10.
3. Bronner, S., Monteil, H. and Prevost, G. (2004) Regulation of virulence determinants in *Staphylococcus aureus*: complexity and applications. *FEMS Microbiol Rev* **28**, 183–200.

4. Bukowski, M., Wladyka, B. and Dubin, G. (2010) Exfoliative toxins of *Staphylococcus aureus*. *Toxins* **2**, 1148.

5. Clarke, S.R. and Foster, S.J. (2006) Surface adhesins of *Staphylococcus aureus*. *Adv Microb Physiol* **51**, 187–225.

6. Heilmann, C. (2011) Adhesion Mechanisms of Staphylococci. In *Bacterial Adhesion: Chemistry, Biology and Physics* eds. Linke, D. and Goldman, A. pp.105–123. Dordrecht: Springer Netherlands.

7. Hennekinne, J.-A., De Buyser, M.-L. and Dragacci, S. (2012) *Staphylococcus aureus* and its food poisoning toxins: characterization and outbreak investigation. *FEMS Microbiol Rev* **36**, 815–836.

8. Hu, D.-L. and Nakane, A. (2014) Mechanisms of staphylococcal enterotoxin-induced emesis. *Eur J Pharmacol* **722**, 95–107.

9. Kadariya, J., Smith, T.C. and Thapaliya, D. (2014) *Staphylococcus aureus* and staphylococcal food-borne disease: An ongoing challenge in public health. *Biomed Res Int* **2014**, 9.

10. Omoe, K., Hu, D.-L., Ono, H.K., Shimizu, S., Takahashi-Omoe, H., Nakane, A., Uchiyama, T., Shinagawa, K. and Imanishi, K.I. (2013) Emetic potentials of newly identified staphylococcal enterotoxin-like toxins. *Infect Immun* **81**, 3627–3631.

11. Otto, M. (2010) Basis of virulence in community-associated methicillin-resistant *Staphylococcus aureus*. *Annu Rev Microbiol* **64**, 143–162.

12. Otto, M. (2014) *Staphylococcus aureus* toxins. *Curr Opin Microbiol* **17**, 32–37.

13. Pinchuk, I.V., Beswick, E.J. and Reyes, V.E. (2010) Staphylococcal enterotoxins. *Toxins* **2**, 2177–2197.

14. Principato, M. and Qian, B.-F. (2014) Staphylococcal enterotoxins in the etiopathogenesis of mucosal auto-immunity within the gastrointestinal tract. *Toxins* **6**, 1471–1489.

15. Silversides, J.A., Lappin, E. and Ferguson, A.J. (2010) Staphylococcal toxic shock syndrome: Mechanisms and management. *Curr Infect Dis Rep* **12**, 392–400.

16. Wendlandt, S., Schwarz, S. and Silley, P. (2013) Methicillin-resistant *Staphylococcus aureus*: A food-borne pathogen? *Annu Rev Food Sci Technol* **4**, 117–139.

Bacillus cereus and Bacillus anthracis

11

Introduction

Christian Gottfried Ehrenberg, a German scientist, in 1835, described a soil-borne organism, *Vibrio subtilis*. Later, in 1872, Ferdinand Julius Cohn renamed the organism, *Bacillus subtilis*. This organism was also called hay bacillus, grass bacillus, or *Bacillus globigii*. *B. subtilis* is a member of the *Bacillaceae* family and the genus *Bacillus*, and it has been used as a model organism to study genetics and physiology of Gram-positive bacteria. The prototype strain is "*Bacillus subtilis* (Ehrenberg) Cohn." Robert Koch, a German physician and a microbiologist in 1876, discovered that the *Bacillus anthracis* is the causative agent for anthrax when investigating the disease in farm animals.

Bacillus species are aerobic endospore former and are widely distributed in nature including air, dust, soil, water, plants, animals, and humans. A majority of bacilli have significant industrial applications including the production of antibiotics and fermented foods; development of flavor in cocoa, coffee, and vanilla; and use as probiotic and plant growth promoter through nitrogen fixation in the plant rhizosphere, phytohormone production, root nodulation, and increasing nutrient availability. Bacilli also cause food spoilage, produce off-flavors, damage texture, and unwanted growth in commercially sterile products. However, some are pathogenic and produce multiple toxins. Of these, three species have significant importance in public health and in agriculture because of their ability to cause diseases in humans, animals, and insects.

Bacillus cereus is responsible for emetic and diarrheal food poisoning leading to about 63,000 illnesses annually in the USA. In addition, it can cause systemic fatal infections in immunocompromised patients and neonates, irreversible eye infection from post-traumatic injury, and metastatic endophthalmitis, i.e., the infection spread from other sites to the eye. *Bacillus anthracis* causes anthrax, which is characterized by septicemia and toxemia in animals and humans with a very high mortality rate. The *B. anthracis* spores are used as bioterrorism agents. *B. thuringiensis* is pathogenic to insects and has been associated with foodborne disease. Other *Bacillus* species including *B. licheniformis*, *B. subtilis*, and *B. pumilus* have occasionally been associated with foodborne illnesses. *B. licheniformis* is also implicated in sporadic abortion in cattle and sheep.

Classification

The *Bacillaceae* family consists of at least 56 genera and 545 species. The new species have been delineated based on the 16S rRNA gene sequence and DNA-DNA relatedness. The genus *Bacillus* contains about 136 species. The

© Springer Science+Business Media, LLC, part of Springer Nature 2018
A. K. Bhunia, *Foodborne Microbial Pathogens*, Food Science Text Series,
https://doi.org/10.1007/978-1-4939-7349-1_11

major pathogenic species of *Bacillus* belong to the "*Bacillus cereus* group" which consists of seven species: *B. cereus*, *B. anthracis*, *B. thuringiensis*, *B. mycoides*, *B. pseudomycoides*, *B. weihenstephanensis*, and *B. cytotoxicus*. A majority of these species produce enterotoxin and cause gastroenteritis. Some strains of other *Bacillus* species, such as *B. circulans*, *B. lentus*, *B. megaterium*, *B. oleronius*, *B. licheniformis*, *B. thuringiensis*, *B. pumilus*, *B. subtilis*, *B. sphaericus*, *B. carotarum*, *B. pasteurize*, and *Paenibacillus polymyxa*, may produce enterotoxin.

Biology

Bacillus species are Gram-positive, aerobic, endospore-forming rods (Fig. 11.1) with an exception of *B. saliphilus*, which is an endospore-forming coccus. The size of the vegetative cells is in the range of 0.5×1.2–2.5×10 μm and occurs singly or in chains. Most bacilli are motile and express peritrichous flagella. The colonies formed by the bacilli are rough with irregularly shaped perimeter and exhibit swarming patterns, while *B. polymyxa* forms a round smooth circular colony (Fig. 11.2). The members of the genus *Bacillus* carry a crystalline surface layer of protein or glycoproteins called S-layers, which vary in molecular weights (40–200 kDa) among strains. Some bacilli express capsule consisting of homo-polypeptide of D- or L-glutamic acid or heteropolysaccharides (carbohydrates). The cell wall of *Bacillus* species is composed of peptidoglycan containing *meso*-diaminopimelic acid (*m*-DAP), teichoic acid, or teichuronic acid (uronic acid-based polymer). *Bacillus* carries plasmids which encode for toxins, bacteriocins, or capsule biosynthesis, and/or antibiotic resistance.

Growth temperature varies between 4 °C and 50°C with an optimum temperature from 25 to 37°C. Psychrotolerant bacilli are found in milk and dairy products. Some thermophilic strains can grow at 75°C. *B. cereus* usually does not grow below pH 4.5. Biophysical and biochemical properties including motility, capsule production, spore structure, hemolysis pattern, urea utilization, nitrate reduction, and sensitivity to penicillin can be used to differentiate the major pathogenic *Bacillus* species (Table 11.1). Illustration of hemolytic and phospholipase activity on blood agar and egg yolk agar plates is shown in Fig. 11.3.

Bacilli produce round, oval, ellipsoid, or cylindrical endospores located centrally to the terminal end (Fig. 11.1). The spore contains a signature molecule, dipicolinic acid

a

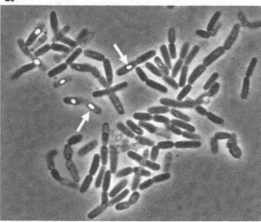

b

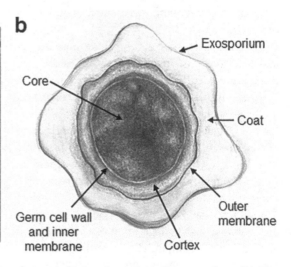

Fig. 11.1 (**a**) Phase-contrast light microscopic picture of cells of *Bacillus cereus* showing ellipsoid or cylindrical spores (see *arrows*), (**b**) schematics of a transmission electron microscopic picture of a spore from *Bacillus anthracis*

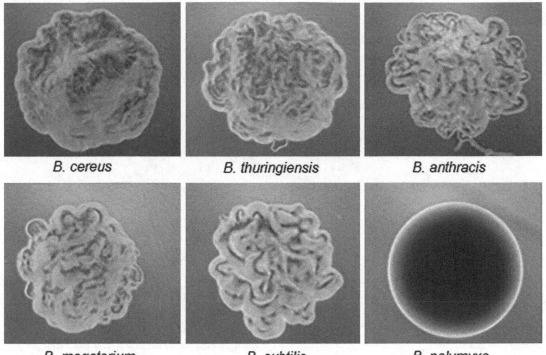

B. cereus B. thuringiensis B. anthracis

B. megaterium B. subtilis B. polymyxa

Fig. 11.2 Colony morphology of *Bacillus* species grown on modified mannitol egg yolk polymyxin (mMYP) agar plate without the egg yolk. Colony diameter is about 1 mm (Images, courtesy of Atul Singh and Arun Bhunia)

Table 11.1 Summary of biochemical properties of common *Bacillus* species

Bacillus species	Hemolysis	Motility	Capsule	Urea hydrolysis	Nitrate reduction	Sensitivity to penicillin
B. cereus	+	+	v	v	v	−
B. anthracis	−	−	+	−	+	+
B. thuringiensis	+	+	−	+	+	−
B. mycoides	−	−	−	v	v	−
B. megaterium	−	+	−	−	−	−
B. subtilis	−	+	−	v	v	−

+ Positive, − Negative, *v* Variable

(2,6-pyridinedicarboxylic acid, DPA), which is essential for sporulation, spore germination, and spore structure. Bacilli are naturally found in soil, dust, and water and remain viable in nature under harsh conditions because of their ability to form endospores, which are difficult to inactivate. The endospores are resistant to heat, radiation, disinfectants, and desiccation. They are also adhesive and can stick to food processing and clinical equipment and resistant to cleaning and sanitization. Foods especially the dried products such as spices, milk powder, and cereal products can be contaminated with spores.

Bacillus species that are responsible for food poisoning may produce multiple toxins including emetic toxin (cereulide) and diarrheal enterotoxins. Three diarrheal enterotoxins have been reported – hemolysin BL (HBL), nonhemolytic enterotoxin (Nhe), and cytotoxin K (CytK) (Table 11.2). HBL and Nhe productions are linked to the members of "*Bacillus cereus* group" only. In addition, *B. cereus* also produces a cholesterol-binding cytolysin (cereolysin O or hemolysin I), hemolysin II, hemolysin III, and hemolysin IV. The latter two hemolysins are the same as cytotoxin K. The enzymes include

Fig. 11.3 Hemolytic and phospholipase activity of *Bacillus cereus* on (**a**) blood agar plate and on (**b**) BACARA™ plate (bioMerieux), respectively (Photo courtesy of Atul Singh and Arun Bhunia)

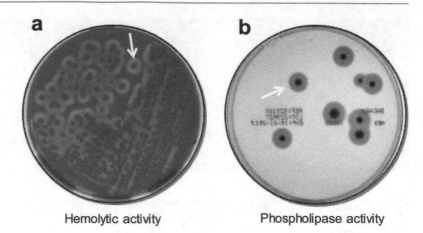

Hemolytic activity   Phospholipase activity

phospholipases, lecithinase, sphingomyelinase, collagenase, protease, amylase, and β-lactamase. β-lactamase inactivates penicillin and thus makes this organism resistant to penicillin. *Bacillus* species also produce many antibiotics such as subtilin, gramicidin, bacitracin, polymyxin, colistin, and so forth.

Bacillus cereus Group and Other Pathogens

Bacillus cereus is the major pathogen in the "*Bacillus cereus* group." It causes a variety of diseases, including food poisoning, fulminant bacteremia, meningitis and brain abscesses, endophthalmitis, pneumonia, and gas gangrene-like cutaneous infections. Food poisoning is characterized by two types of symptoms, vomiting and diarrhea. Vomiting is induced by an emetic toxin, cereulide, and diarrhea by diarrheagenic toxins, HBL, Nhe, and CytK (Table 11.2). The mechanism of action is described in detail below. Only a small percentage of strain produce emetic toxin. The same strain can cause both forms of disease, or separate strains in the same food can cause two types of diseases. Foods contaminated with endospores will survive cooking and germinate, multiply, and produce toxins under favorable conditions. Intoxication or diarrhea depends on the amount of toxin or the vegetative cells ingested.

Bacillus anthracis causes "anthrax," a serious life-threatening systemic disease primarily in the herbivore animals and humans (*see the section on B. anthracis*). The organism is closely related to *B. cereus*, but instead of enterotoxin, it produces anthrax toxin. A *B. cereus* strain has been identified that possess virulence factors common to *B. anthracis* implicating a close relationship exists between these two pathogens.

Bacillus thuringiensis is a well-known insect pathogen. It produces a wide variety of crystal toxins (133 kDa) known as delta endotoxins (bioinsecticide), which are toxic to insects in an Order-specific manner. Since 1905, these endotoxins have been applied commercially to biologically control pestiferous insects in the Orders *Coleoptera*, *Lepidoptera*, and *Diptera*, such as beetles, mosquitoes, butterfly, and moths. The gene for crystal toxin production (*cryA*) is plasmid (pXO12) encoded. Transgenic crops have been developed that carry insecticidal *cry* genes to make the plants' insect resistant. A plasmid-cured *B. thuringiensis* strain is biologically indistinguishable from *B. cereus*. *B. thuringiensis* carries enterotoxin genes encoding HBL, Nhe, and CytK, and it is implicated in foodborne outbreaks. The gene for the emetic toxin, cereulide, is not found in *B. thuringiensis* strains.

Bacillus weihenstephanensis is psychrotolerant and phylogenetically distinct from mesophilic strains of *B. cereus*. It may produce cereulide at 8 °C but does not produce CytK. No diarrheal outbreaks are associated with this organism.

Table 11.2 Toxins produced by *Bacillus cereus*

Toxins	Genes	Molecular weight (kDa)	Activity
Emetic toxin	*ces*	1.2	Emesis (vomiting)
Diarrheal toxin		38–43	Diarrhea
Hemolysin – HBL			Hemolytic, enterotoxic, dermonecrotic
B-component	*hblA*	37.8	
L-component (L1)	*hblD*	38.5	
L-component (L2)	*hblC*	43.5	
Nonhemolytic enterotoxin (Nhe)			Enterotoxic
NheA	*nheA*	41	
NheB	*nheB*	39.8	
NheC	*nheC*	36.5	
Enterotoxin T (BcET)	*bceT*	41	Unknown
Enterotoxin FM (EntFM)	*entFM*	45	Enterotoxic
Cytotoxin K (CytK) (Hemolysin IV)	*cytK*	34	β-barrel pore-forming toxin; necrotic dermatitis
Hemolysin I (cereolysin O; CLO)	*hlyI*	55	Cholesterol-binding pore-forming toxin
Hemolysin II (HlyII)	*hlyII*	382	β-barrel channel-forming toxin
Hemolysin III (HlyIII)	*hlyIII*	24.4	Pore formation, hemolysis

Adapted and modified from Gray et al. (2005). J. Clin. Microbiol. 43, 5865–5872

Bacillus licheniformis has occasionally been associated with foodborne illnesses showing symptoms of nausea, vomiting, stomach cramp, and diarrhea. It produces a cereulide-like toxin called lichenysin A, but its mechanism of action is different from that of cereulide. *B. licheniformis* is also implicated in sporadic abortion in cattle and sheep.

Bacillus subtilis is also associated with foodborne illness primarily exhibiting vomiting symptom and occasional diarrhea. The onset of symptoms may be within 10 min–14 h, and the disease is associated with food prepared with fish or meat and cereal-based components such as bread, pastry, or rice.

Bacillus cereus

Foods Involved

Bacillus cereus-related outbreaks are generally associated with cereal foods, pudding, and soups. The emetic syndrome is generally associated with pasta, rice dishes, beef, poultry, milk pudding, vanilla sauce, and infant formulas, while diarrheal syndrome is associated primarily with meat, fish, soups, dairy products, vegetables such as corn, cornstarch, and mashed potatoes. *Bacillus* contamination can deteriorate the quality of milk and can produce "bitty cream" by forming a creamy layer of the milk due to the action of lecithinase. *B. cereus* can also form sweet curdle without lowering the pH of the milk. Raw milk is generally contaminated in the farm from soiled udders; thermoduric spores may then survive subsequent pasteurization treatments.

Virulence Factors

Emetic Toxin

The molecular mass of emetic toxin is 1.2 kDa, and it consists of three repeats of four modified amino acids [D-O-Leu-D-Ala-L-O-Val-L-Val]$_3$ forming a ring structure (dodecadepsipeptide) and is known as cereulide. The gene, *ces*, encoding cereulide synthesis is located in a mega plasmid, similar to pXO1 virulence plasmid in *B. anthracis* (see section on *Bacillus anthracis*). The cereulide is produced non-ribosomally by a large multidomain enzyme complex. The toxin is highly heat-stable (121 °C for 90 min) and resistant to frying, roasting, and microwave cooking. It is hydrophobic and active over a broad pH range (pH 2–11) and is not digested by pepsin or trypsin. The toxin is nonantigenic. Only a narrow range of *B. cereus* strains produces cereulide toxin. Foods contaminated with 10^5–10^8 cell g^{-1} are sufficient to produce toxins to induce emetic syndrome. Maximum cereulide production occurs in cultures incubated between 12 °C and 22 °C during the stationary phase of growth under aerobic and microaerobic environments. No cereulide production has been seen under 12 °C, but *B. weihenstephanensis*, a psychrotolerant, may produce detectable cereulide at 8 °C. *Bacillus* spores usually survive during cooking or pasteurization and germinate into vegetative cells when food is temperature abused. The emetic toxin is produced during prolonged food storage, but the toxin production is not associated with sporulation. Reheating of food kills vegetative cells, but not the toxin.

Diarrheal Toxins

Three major toxins are implicated in causing diarrheal disease: hemolysin BL (HBL), nonhemolytic enterotoxin (Nhe), and cytotoxin K (CytK). In addition, other toxins such as enterotoxin T (BcET), enterotoxin FM (EntFM), and other hemolysins may contribute to the disease (Table 11.2). Preformed diarrheal toxins in food probably do not contribute to the disease since these toxins are sensitive to heating (55 °C for 20 min), acids (pH~3.0), and gastric enzymes (trypsin, pepsin, and chymotrypsin). Ingested bacteria produce toxins in the intestine to cause diarrheal disease.

Hemolysin BL

The best-characterized *Bacillus* spp. enterotoxins are 38–43 kDa heat-labile proteins, which include HBL, Nhe, and CytK. HBL consists of a single B-component (37.8 kDa) and two L-components: L1 (38.5 kDa) and L2 (43.5 kDa). All three subunits are highly heterogeneous in *B. cereus* strains showing variable molecular weights, and all three subunits are required for producing transmembrane pore and for maximal activity. HBL produces unique discontinuous hemolysis pattern characterized by the presence of a clear zone and a zone of incomplete hemolysis on blood agar plate. The highest hemolytic activity

is seen in guinea pig blood followed by swine, calf, sheep, goat, rabbit, human, and horse. HBL is considered an important virulence factor for *B. cereus* because it is hemolytic, cytotoxic, and dermonecrotic and induces vascular permeability as demonstrated in a rabbit ileal loop assay. About 50–66% of *B. cereus* strains carry genes for HBL.

Nonhemolytic Enterotoxin

Another three-component pore-forming enterotoxin produced by *B. cereus* is the nonhemolytic enterotoxin, comprised of NheA (41 kDa), NheB (39.8 kDa), and NheC (36.5 kDa) subunits. The genes encoding these three subunit proteins are located in an operon, and transcription of this operon is regulated by a pleiotropic regulator, PlcR (phospholipase C regulator), that also controls phospholipase C expression. The Nhe carries a high degree of sequence homology with HBL. Much like the HBL complex, all three components are necessary for full activity. NheB was shown to interact with host cell receptor; however, its specific role in gastroenteritis has not been clearly established. Most, if not all, *B. cereus* strains carry the gene for Nhe.

Cytotoxin K

CytK is a single component of 34 kDa toxin and belongs to the family of β-barrel pore-forming toxin similar to a prototype toxin α-hemolysin from *Staphylococcus aureus* and β-toxin from *Clostridium perfringens*. CytK is now designated as hemolysin IV. The toxin is dermonecrotic, cytotoxic, and hemolytic. The toxin is present in about 90% of all *B. cereus* and *B. thuringiensis* strains. CytK has been implicated in a *B. cereus* outbreak resulting in bloody diarrhea, necrotic enteritis, and death in the elderly patients.

Enterotoxin BcET

Originally *B. cereus* enterotoxin T (BcET) was isolated from the B-4 ac strain. The molecular weight of BcET is 41 kDa and encoded in the *bceT* gene (2.9 kb). The function of BcET in pathogenesis is controversial; some studies suggest it has an enterotoxin activity, while others dispute its existence and consider it an artifact.

Enterotoxin FM

Enterotoxin FM (EntFM) was originally isolated from *B. cereus* FM-1 strain that was responsible for foodborne intoxication. EntFM is a 45 kDa polypeptide, and it is rich in beta structure and contains some unusual sequence arrangements such as repeating asparagine (N) amino acid around the residue number 280. EntFM has not been implicated in the food-poisoning outbreak; however, the *entFM* gene is present in most outbreak-associated strains and is actually the most prevalent enterotoxin gene in all *B. cereus* strains tested. In the laboratory experiment, purified EntFM caused increased vascular permeability and fluid accumulation in rabbit/mouse ileal loop models and was lethal to mice.

Hemolysin I

Hemolysin I, known as cereolysin O (CLO), is a 55 kDa cholesterol-dependent cytolysin (CDC), which forms pores in the eukaryotic membrane (Fig. 11.3). CLO has high sequence similarities (57–68%) with other CDCs such as alveolysin O (AVO), perfringolysin O (PFO), listeriolysin O (LLO), and streptolysin O (SLO) from *Paenibacillus alvei*, *Clostridium perfringens*, *Listeria monocytogenes*, and *Streptococcus pyogenes*, respectively. Hemolysin I from *B. thuringiensis* is called thuringiolysin O (TLO) and from *B. anthracis*, anthrolysin O (ALO). CLO and TLO are regulated by a transcriptional activator, PlcR, a 34 kDa pleiotropic regulatory protein. CLO is heat-labile, and the hemolytic activity can be inhibited by cholesterol. The contribution of CLO in pathogenesis is not fully understood but proposed to cause necrotic cell death during endophthalmitis, while ALO may decrease epithelial barrier function by rearranging junctional occludin protein.

Hemolysin II

Hemolysin II (HlyII) is 42.6 kDa heat-labile toxin, and the hemolytic activity is not inhibited by cholesterol. HlyII is present in *B. cereus* group including *B. thuringiensis*. HlyII belongs to the family of β-barrel channel-forming toxins including the α-toxin, leukocidin, and γ-toxin from *S. aureus*, γ-toxin from *C. perfringens*, and CytK from *B. cereus*. HlyII produces pores with a diameter of 7 Å in a planar lipid bilayer, and it has been demonstrated to cause lysis of phagocytic cells, such as insect hemocytes, mouse macrophages, human monocytes, and dendritic cells, but not epithelial cells. HlyII is responsible for immune cell death by apoptosis but is unlikely to be involved in gastroenteritis. Toxin production is not regulated by PlcR.

Hemolysin III

Hemolysin III (HlyIII) is a 24.4 kDa heat-labile toxin, and it forms a pore on the membrane with a pore diameter of 3–3.5 nm. However, its role in pathogenesis is not known.

Phospholipases

Bacillus cereus produces two phospholipase C enzymes which are specific for phosphatidylcholine (PC-PLC) also known as lecithinase and phosphatidylinositol (PI-PLC). PC-PLC is a 29.9 kDa enzyme and possesses three zinc atoms in its active site. The zinc atoms contribute to the structural and catalytic properties of the enzyme. PI-PLC is a 34.6 kDa enzyme. Both enzymes may be important during respiratory tract infections resulting in necrosis and tissue hemorrhage.

Sphingomyelinase

Sphingomyelinase (SMase) is a 34.2 kDa enzyme, and it has two magnesium atoms at its active site. SMase hydrolyzes sphingomyelin to ceramide and phosphorylcholine (ChoP) located in the eukaryotic membranes.

Regulation of Toxin Biosynthesis

The cereulide expression is regulated by transition state and sporulation regulators Spo0A and AbrB, and the maximum expression is observed during early to the mid-exponential growth phase. The optimum temperature for emetic toxin production is 20–30 °C under aerobic but not under anaerobic environment. Diarrheagenic toxins are produced during the late exponential to early stationary phase of growth or in the intestine. Glucose acts as a catabolic repressor of HBL, while sucrose enhances expression when grown in a modified defined medium. Expression of *hbl* and *nhe* and 13 other virulence genes is regulated by PlcR (34 kDa), which is present in *B. cereus* and *B. thuringiensis*. A truncated PlcR polypeptide is also present in *B. anthracis* due to a nonsense mutation in the *plcR* gene, and it is probably nonfunctional in this pathogen.

Mechanism of Pathogenesis

Consumption of food contaminated with preformed *B. cereus* emetic toxin results in food-associated intoxication, which is characterized by nausea and vomiting, while consumption of bacteria followed by subsequent toxin production de novo in the intestinal tract results in gastroenteritis and signs of toxicoinfection. The infectious dose of *B. cereus* is highly variable, ranging between 10^5 and 10^8 viable cells or spores per gram to cause intoxication or toxicoinfection.

Emetic Toxin

The emetic syndrome is associated with consumption of contaminated rice dishes, pasta, and noodles. Occasionally, cream, milk pudding, and reconstituted infant formulas can cause the emetic symptom. The emetic dose of toxin is estimated to be 0.02–1.83 μg kg^{-1} body weight in humans. *B. cereus* strain produces 0.004–0.13 μg of cereulide per 10^6 cells in the exponential growth phase; thus, a cell population of 10^5–10^8 cfu g^{-1} food is sufficient to cause emetic disease. The toxin binds to vagus afferent nerve receptors, 5-HT$_3$, in the stomach and subsequently induces vomiting by stimulating the medulla oblongata in the brain. A 5-HT$_3$ receptor antagonist, ondansetron hydrochloride (Zofran), can abolish cereulide-mediated vomiting. It is not clear whether the toxin directly interacts with the vagus nerve or releases serotonin, 5-HT (5-hydroxytryptamine), from enterochromaffin cells or mast cells in the stomach, which binds to the 5-HT$_3$ receptor on the nerve. Cereulide can also cause liver failure and brain edema. Cereulide has structural similarity to valinomycin, a potassium-specific transporter (K$^+$ ionophore), which facilitates K$^+$ movement through lipid membrane following the electrochemical potential gradient. Cereulide inhibits fatty acid oxidation affecting mitochondrial activity in hepatocytes, which results in massive degeneration of hepatocytes.

The typical symptoms are nausea and vomiting, which appear within 30 min–6 h following ingestion, and the disease resembles intoxication caused by staphylococcal food poisoning. The disease subsides within 24 h but can cause a serious problem in infants with liver failure and brain edema.

Diarrheagenic Toxin

Ingestion of spores or vegetative cells is required for diarrheal illness. The infective dose is 10^5–10^8 cells g^{-1} of food. Three major toxins are believed to be responsible for diarrheal symptoms: HBL, NHE, and CytK. The toxins are produced after colonization of bacteria in the small intestine. Though the mode of action of toxins is not fully understood, it is speculated that the enterotoxins form transmembrane pores and osmotic cell lysis resulting in the fluid loss. In addition, the toxins may induce diarrhea by stimulating conversion of intracellular ATP to cyclic adenosine monophosphate (cAMP). As a result, Na$^+$, Cl$^-$, and H$_2$O are lost from the epithelial cells resulting in electrolyte imbalance.

For HBL-mediated pore formation, it has been hypothesized that each member of HBL binds to the membrane forming transmembrane pores by oligomerizing all three components of hemolysin (Fig. 11.4). HBL causes fluid accumulation in experimental rabbit ileal loop model resulting in necrosis of villi, edema in submucosal layer, infiltration of lymphocytes in the interstitial spaces, and accumulation of blood. Similarly, CytK also forms pores with a predicted pore size of approximately 7 Å. The CytK toxin spontaneously oligomerizes within the lipid bilayer membrane and becomes resistant to sodium dodecyl sulfate (SDS).

Diarrheagenic symptoms appear within 8–16 h after consumption of bacteria and begin with mild diarrhea, nausea, abdominal cramping, followed by watery stools. The disease is self-limiting and

Fig. 11.4 A schematic diagram showing the possible mechanism of pore formation by *B. cereus* HBL complex on epithelial cells. Each component is believed to participate in the oligomerization process (The model is based on the description of Schoeni and Wong (2005). J. Food Prot. 68, 636–648)

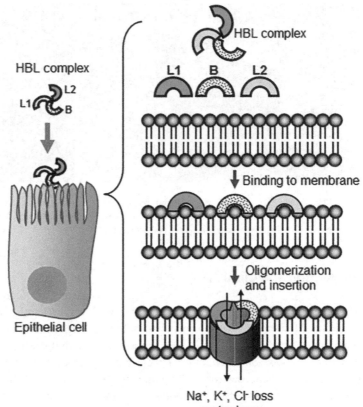

Animal and Cell Culture Models

resolves within 24–48 h. In a rare occasion, the disease can be severe in the elderly patients causing bloody diarrhea, necrotic enteritis, and death. Vomiting is occasionally induced in patients exposed to the diarrheagenic toxin.

Rhesus monkey and house musk shrew have been used to determine the emetic dose of toxin after oral administration. Rabbit ileal loop assays and suckling mice models are used to determine the action of diarrheagenic toxin (see Chap. 5).

The biological activity of *B. cereus* toxins has been assessed on a number of eukaryotic cell lines. The emetic toxin is assayed colorimetrically by using the HEp-2 cells by using MTT (3-(4,5-dimethylthiazolyl-2)-2,5-diphenyltetrazolium bromide)-based metabolic staining or by microscopic analysis of vacuole formation in HEp-2 cells. Another bioassay based on the loss of motility of boar spermatozoa has been developed. In this assay, *B. cereus* emetic toxin affects the mitochondrial function and oxidative phosphorylation process and inhibits the motility of boar spermatozoa, which can be assessed under a microscope. An emetic toxin assay has been developed based on its mitochondrial respiratory uncoupling activity on rat liver mitochondria.

A number of cell lines such the McCoy and Vero have been used to detect diarrheal enterotoxin. The enterotoxins cause cytopathic effects causing the destruction of cell monolayers. A CHO-based assay detects both the emetic and diarrheal toxins in 24–72 h by the MTT-based metabolic staining assay. *Bacillus* enterotoxin causes severe membrane damage, pore formation, and cellular detachment, which can be visualized by scanning electron microscopy (Fig. 11.5). A B-cell hybridoma, Ped-2E9-based cytotoxicity assay has been developed to detect *B. cereus* toxin-induced cell damage in 1–2 h either by Trypan blue staining or an alkaline phosphatase release assay.

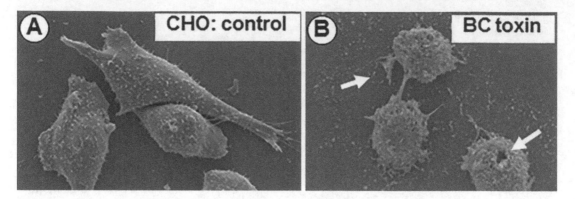

Fig. 11.5 Scanning electron microscopy photograph showing the *Bacillus cereus* toxin-induced CHO (Chinese hamster ovary) cell damage; pore formation, cell shortening, and detachment from plastic surface (Photo courtesy of P.P. Banada and Arun Bhunia)

Prevention and Control

The major contributing factors for *B. cereus* food-associated illness include improper food holding temperature, contaminated equipment, inadequate cooking, and poor sanitation. Furthermore, appropriate heat treatment used for cooking may not be adequate in destroying *B. cereus* spores. Thus, the storage temperature of food should be maintained well below the germination threshold of spores. This can be achieved by uniform quick chilling of the food to near 4 °C or holding the food above 60 °C. *B. cereus* is capable of producing emetic toxin at low temperature (8 °C); thus, prolonged storage at refrigerator should be avoided. Proper sanitary measures should be adopted to prevent cross-contamination while handling food. Live bacterial cells are necessary to show diarrheal symptoms; therefore, uniform reheating of a suspected food to above 75 °C before serving is recommended.

Detection of *Bacillus cereus*

Conventional Methods
Traditional plating and biochemical assays for detection of *Bacillus* species are time-consuming, and these methods do not indicate the toxin production capabilities of the isolates. Bacteria can be isolated by plating on selective or differential media containing egg yolk and mannitol, such as polymyxin–egg yolk–mannitol–bromothymol agar (PEMBA) and mannitol egg yolk polymyxin agar (MYP). Lecithinase activity is determined by observing white precipitate formation surrounding *Bacillus* colonies on MYP plates and BACARA™ plates. Emetic toxin-producing strains are both hemolytic and nonhemolytic. Spores can be counted indirectly on agar plates. To achieve that, food samples are subjected to heat (75–80 °C) for 10–15 min to inactivate the vegetative cells and to promote germination of spores. The samples are then plated to enumerate bacterial colonies and compare the counts with the sample that did not receive heat-shock.

PCR
Polymerase chain reaction (PCR) assay has been developed to examine the presence of toxin genes in *Bacillus* for strain characterization. Simplex or multiplex PCR-based detection is based on the presence of toxin genes listed in Table 11.2 and offers a high level of sensitivity and specificity, but the PCR assay does not reveal pathogenic potentials of strains since phenotypic expression of genes is not always warranted. PCR-based detection is often hindered by the presence of inhibitors in food matrices; therefore, bacterial cells may have to be separated prior to their analysis.

Antibody-Based Assay

Cereulide is nonantigenic; thus, there is no antibody-based test available for this toxin. For diarrheal toxin detection, two commercial kits, the *Bacillus* Diarrhoeal Enterotoxin Visual Immunoassay (BDE-VIA, Tecra) kit that detects only the 41 kDa subunit of NHE and the *Bacillus cereus* enterotoxin-reversed passive latex agglutination (BCET-RPLA, Oxoid) that detects the L_2 subunit of HBL, are available. It is known that *B. cereus* strains may produce both toxins, only one, and perhaps neither of these subunits; therefore, the tests could easily produce false positives in a complex medium such as food. Liquid chromatography and mass spectrometry have been developed for specific toxins.

Bacillus anthracis

Biology

Bacillus anthracis is a zoonotic pathogen associated with livestock and game animals. The disease is known as anthrax and is endemic in parts of Asia, Africa, and South America. *Bacillus anthracis* endospores are found in soil, dust, water, and pasture. Anthrax can be spread by biting flies, carnivore animals, and birds (vulture) who feed on the contaminated carcass; carnivores are generally resistant to anthrax. Animal handlers, farmers, butchers, wool processors, and veterinarians are in the greatest risk group and can contract the disease. The human form of anthrax is considered an occupational disease because of obvious predisposition. Anthrax also is referred to as wool sorter's disease, since wool can carry infective spores.

Bacillus anthracis is a large Gram-positive bacillus with centrally located ellipsoidal or cylindrical spore. Spores (1–2 μm in length) are highly resistant to heat, ultraviolet, and ionizing radiations, hydrostatic pressures, and chemical sanitizers. Vegetative cells typically form long chains. *B. anthracis* is a member of the *Bacillus cereus* group and is nonmotile, nonhemolytic, and sensitive to penicillin, and thus can be easily differentiated from the other members of the *Bacillus cereus* group (Table 11.1). Virulent strains produce capsules, while avirulent strains are capsule negative. Persistence of spores in soil is responsible for occasional outbreaks in animals, although vaccination can prevent infection. Medical intervention (antibiotic treatment) in the very early phase of infection can prevent fatalities.

Virulence Factors and Mechanism of Pathogenesis

Bacillus anthracis pathogenesis depends on the expression of key virulence factors which are encoded in two plasmids, pXO1 (181 kb) and pXO2 (96 kb) (Table 11.3). pXO1 carries genes for the bipartite lethal anthrax toxin. Anthrax toxin is an A–B-type toxin, and the B subunit is known as protective antigen (PA), while subunit A is composed of two alternative catalytic subunits: edema factor (EF) and lethal factor (LF). When PA binds to LF or EF, it is called lethal toxin (LT) or edema toxin (ET). PA is an 83 kDa protein that binds to a host cell receptor, called anthrax toxin receptor/tumor endothelial marker 8 (ATR/TEM8). PA spontaneously forms heptamer complex and facilitates entry of LF and EF into the cytosol through a clathrin-coated pit by endocytosis (Fig. 11.6). Protective antigen is a good vaccine candidate because of its obvious role in pathogenesis. In the endosome, the LF and EF are separated from PA and are released into the cytoplasm through the vesicle pores. LF is a 90 kDa zinc-dependent metalloprotease that cleaves the N-terminus of mitogen-activated protein kinase kinases (MAPKKs), thus disrupting signal transduction pathways and inducing cell death. LF is cytotoxic to endothelial cells, dendritic cells, and macrophages, promoting bacterial survival in the host. EF (89 kDa protein) is adenylate cyclase and converts intracellular ATP to cyclic adenosine monophosphate (cAMP) when introduced into the cytosol by PA and aids in fluid accumulation (edema) and swelling in the lungs. Edema factor also inhibits neutrophil-dependent phagocytosis and the oxidative burst of neutrophils.

Table 11.3 Virulence gene profiles in *Bacillus anthracis*

Virulence genes located in each plasmid	Genes	Molecular weight (kDa)	Function
pXO1 (181 kb)			
Protective antigen (PA)	*pagA*	83	Facilitates entry of LF and EF into cells
Edema factor (EF)	*cya*	89	Activates cAMP, fluid loss
Lethal factor (LF)	*lef*	90	Cytotoxic to macrophages, neutrophils, and dendritic cells
pXO2 (96 kb)			
Capsule	*capABC*		Inhibits phagocytosis

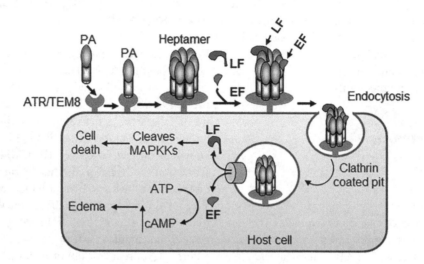

Fig. 11.6 Mechanism of anthrax toxin action on cells. Protective antigen (PA) first binds to anthrax toxin receptor/tumor endothelial marker 8 (ATR/TEM8) and then spontaneously forms heptamer and aids in the translocation of lethal factor (LF) and edema factor (EF) inside the cell by endocytosis. LF cleaves mitogen-activated protein kinase kinases (MAPKKs) to block signaling pathway to cause cell death, while the EF increases cAMP level (Adapted from Scobie and Young (2005). Curr. Opin. Microbiol. 8, 106–112)

pXO2 contains genes (*capABC*) for capsule biosynthesis. The capsule, composed of poly-γ-D-glutamic acid, inhibits phagocytosis. The noncapsulated strain has reduced virulence, and it has been used as a vaccine candidate. *B. anthracis* Sterne strain lacks pXO2 and thus has been used as a vaccine strain for veterinary application throughout the world. *B. anthracis* Pasteur strain is pXO1 negative but pXO2 positive. *B. anthracis* Ames strain is negative for both pXO1 and pXO2 and is considered avirulent.

Transcription of virulence genes is regulated by CO_2, carbonate, and temperature. Virulence regulatory genes are located on both plasmids and in a 44.5 kb pathogenicity island. Expression of EF, LF, and PA is regulated by regulatory proteins, AtxA and PagR, while capsule biosynthesis is regulated by *atxA* encoded on pXO1, and *acpA* and *acpB* present on pXO2. In a rare occasion, *B. cereus* strains may contain *B. anthracis* genes encoding for anthrax toxins and can exhibit illness similar to the inhalation anthrax.

Depending on the route of exposure, human anthrax is classified into (1) cutaneous, (2) inhalation, or (3) gastrointestinal. "Injectional anthrax" has been reported among the intravenous drug users during injection of *B. anthracis* spore-contaminated heroin. The mortality rate is very high among the users.

Cutaneous Anthrax

The cutaneous form, accounting for 90% of all cases, is generally associated with the occupational contact. Animal handlers, farmers, butchers, wool processors, and veterinarians are in the greatest risk group and can contract the disease. It has an average incubation period of 2–6 days and develops on the arms and hands eventually spreading to the face and neck. Initially, severe blister-like painless papules surrounded by edema develop. Within 5–6 days, ulceration of the papule causes the development of characteristic scabs of cutaneous anthrax. Clinical syndromes include high fever, toxemia, painful swelling of regional lymph nodes, extensive edema, shock, and death. The mortality rate of cutaneous form is about 1% with antibiotic treatment and 20% without.

Inhalation Anthrax

Inhalation anthrax is the most lethal form of anthrax, leading to death within 4–6 days after exposure. The mortality rate is 45% with antibiotics treatment and 97% without. The infective dose for inhalation anthrax is estimated to be 8000–10,000 spores. Two models have been proposed for systemic transfer of pathogens: Trojan horse model and Jailbreak model. In the Trojan horse model of infection, the ingested spores either invade epithelial barrier of the respiratory tract directly or are transferred by alveolar macrophages to the regional lymph nodes (mediastinal lymph nodes) where the spores germinate into vegetative cells. In the lymph node, bacteria multiply, produce toxins and destroy the structural integrity of the lymph node, and enter blood circulation leading to bacteremia and host death. In the Jailbreak model, spores germinate and become vegetative cells, produce toxins and cause damages to the epithelial barrier, cross the barrier, and trafficked to the regional lymph nodes without the help of phagocytes. In the lymph node, bacteria continue to replicate, produce toxins, cause tissue injury, and escape to the bloodstream leading to bacteremia and septicemia and host death.

The illness is biphasic. In the early phase, upper respiratory tract infection is evident, showing fever, chills, sweats, fatigue, or malaise with a nonproductive cough. Several days after infection, in the second phase, a sudden onset of high fever, rapidly progressive respiratory failure, and labored breathing, circulatory collapse, and shock are evident. In this phase, vegetative cells multiply and circulate in the blood, resulting in septicemia and toxemia. Within 24 h after the initiation of the second phase of infection, elevated levels of IL-1β and TNF-α are produced which are responsible for hypotension, edema, and fatal shock.

Gastrointestinal Anthrax

The incubation period of gastrointestinal anthrax is 3–7 days. In gastrointestinal anthrax, both Trojan horse and Jailbreak models have been proposed to aid in bacterial translocation across the epithelial barrier and trafficking to regional lymph nodes (mesenteric lymph nodes, MLN). In the Trojan horse model, dendritic cells help transfer the spores to the MLN. In the Jailbreak model, stomach acid may possibly facilitate germination of spores to form vegetative cells, and bacteria can translocate through the epithelial barrier by destroying the epithelial cells or through M-cells in the Peyer's patches.

The symptoms are fever, severe abdominal pain, nausea, vomiting, and bloody diarrhea. Septicemia, shock, and death may follow. The mortality rate of gastrointestinal anthrax is similar to that of the inhalation anthrax, 25–60%.

Animal Anthrax

Ruminant animals such as cattle and sheep are highly susceptible to anthrax, which appears within a few hours to a day. The symptom is per-acute or acute septicemia characterized by pyrexia, i.e., high body temperature 42 °C (108 °F) due to a massive release of inflammatory cytokines (TNF-α, IL-1), and sudden deaths in one or two animals in a herd. There is rapid bloating of the carcass, hemorrhage from natural orifices, and incomplete rigor mortis, and the spleen becomes very large (splenomegaly). Pigs and horses are moderately susceptible to anthrax. The disease is characterized as subacute with edematous swelling in the pharynx and regional lymphadenitis. Intestinal form of the disease is less common in pig but is fatal. The animal shows sudden death. Horses show a subacute form of anthrax with localized edema and septicemia with colic and enteritis. The carnivores are comparatively resistant and if they consume a massive dose of bacteria from consuming anthrax-infected carcass can lead to septicemia. Birds such as vultures and other carnivore birds, which feed on the anthrax-infected carcass, are resistant to the disease, possibly because of their natural high body temperature. Laboratory mice, guinea pigs, and rats are sensitive to anthrax toxins and are used to study the lethal effect of toxins.

Treatment and Prevention

Vaccination of animals with *B. anthracis* Sterne strain devoid of capsule biosynthesis gene is very effective in controlling animal anthrax. Early medical intervention with antibiotics (penicillin G, tetracycline, and doxycycline) is effective. Milk and meat from infected animals must be disposed of properly. Anthrax-infected carcasses should be disposed of properly, by incineration or deep burial, away from water streams or rivers. Scavengers should be controlled. Animals waste, feed, and bedding should be disinfected and disposed of. Contaminated building should be fumigated.

For controlling and treatment of human anthrax, early medical intervention with penicillin therapy is most effective. In addition, fluoroquinolones such as ciprofloxacin, levofloxacin, and moxifloxacin or doxycycline are also very effective. Antibiotics are also used as a preventive therapy. Supportive therapy should be initiated to restore fluid and electrolyte imbalance and to prevent septic shock.

Detection of *Bacillus anthracis*

Traditional Culture Methods
Culture-based methods are the gold standard for detection and identification of *B. anthracis*.

Specimens such as nasal swabs can be directly streaked on blood agar plates (Tryptic soy agar containing 5% sheep red blood cells), incubated overnight at 35 °C, and examined. Samples can be cultured in liquid media before plating onto blood agar plates to isolate colonies. Colonies appear large (4–5 mm in diameter), opaque, white to gray in color, with irregular boundaries. *Bacillus anthracis* colonies are nonhemolytic; therefore, no visible hemolysis is seen around the colonies. Gram-staining shows characteristic large bacilli (4–5 μm in length) often arranged in chains. Nonhemolytic or weakly hemolytic colonies are then transferred to semisolid motility medium to determine motility. *Bacillus anthracis* is nonmotile. Capsule presence is determined by staining cells with India ink followed by microscopic examination. Confirmatory tests of cultures are done by determining the hydrolysis pattern of casein, starch, and gelatin; acid production from salicin, inulin, and mannitol; production of arginine dihydrolase and indole; and nitrate reduction (Table 11.1). Lecithinase activity of the strains can be determined on egg yolk agar plates such as MYP and BACARA™ (bioMerieux). In cases of inhalation anthrax, blood cultures will be positive for anthrax bacilli. Antibiotic susceptibility testing helps to determine the antibiotics to be used for the treatment of patients.

Summary

Bacillus species are spore-forming aerobic rods; are natural inhabitants of soil, dust, water, and environment; and can contaminate milk, meat, rice, and pasta. Endospores are resistant to harsh environmental conditions or processing treatments. The majority of bacilli are nonpathogenic; however, several species produce multiple toxins and can cause various diseases in animals and humans. *Bacillus cereus* causes food poisoning characterized by a self-limiting gastrointestinal disorder: vomiting and diarrhea. *Bacillus thuringiensis* is pathogenic to insects and used as an insecticide, while *Bacillus anthracis* causes anthrax, a fatal disease in animals and humans. *B. cereus* produces two major toxins: emetic and diarrheagenic toxins. The emetic toxin (cereulide) is a low molecular weight circular peptide consisting of three modified amino acids and is resistant to heat, acids, and proteolytic enzymes. Emetic toxin is responsible for intoxication. Hemolysin BL (HBL), nonhemolytic enterotoxin (Nhe), and cytotoxin K are reported to be responsible for diarrheagenic action, and these are produced inside the gastrointestinal tract. Diarrheagenic toxin production is regulated by a regulatory protein, PlcR. *B. anthracis* causes three forms of the disease: cutaneous anthrax, inhalation anthrax, and gastrointestinal anthrax. *B. anthracis* produces a bipartite lethal anthrax toxin and a capsule, which are encoded in two virulent plasmids, pXO1 and pXO2. Anthrax toxin is an A–B-type toxin with A-component, known as protective antigen (PA), which binds to the host cell receptor, anthrax toxin receptor/tumor endothelial marker 8 (ATR/TEM8), and facilitates translocation of catalytic lethal factor (LF) and edema factor (EF) to the host cell cytosol. LF interferes with the signal transduction pathways by inactivating mitogen-activated protein kinases to induce cell death, while EF activates increased production of cAMP to allow fluid accumulation and edema. *B. cereus*-related gastrointestinal complications are self-limiting, whereas anthrax is fatal requiring immediate medical intervention with antibiotic therapy.

Further Readings

1. Bottone, E.J. (2010) *Bacillus cereus*, a volatile human pathogen. *Clin Microbiol Rev* **23**, 382–398.
2. Bravo, A., Likitvivatanavong, S., Gill, S.S. and Soberón, M. (2011) *Bacillus thuringiensis*: A story of a successful bioinsecticide. *Insect Biochem Mol Biol* **41**, 423–431.
3. Ceuppens, S., Rajkovic, A., Heyndrickx, M., Tsilia, V., van De Wiele, T., Boon, N. and Uyttendaele, M. (2011) Regulation of toxin production by *Bacillus cereus* and its food safety implications. *Crit Rev Microbiol* **37**, 188–213.
4. Ehling-Schulz, M. and Messelhausser, U. (2013) *Bacillus* "next generation" diagnostics: Moving from detection toward subtyping and risk-related strain profiling. *Front Microbiol* **4**, 32–32.
5. Friebe, S., van der Goot, F.G. and Bürgi, J. (2016) The ins and outs of anthrax toxin. *Toxins* **8**, 69.
6. From, C., Pukall, R., Schumann, P., Hormazabal, V. and Granum, P.E. (2005) Toxin-producing ability among *Bacillus* spp. outside the *Bacillus cereus* group. *Appl Environ Microbiol* **71**, 1178–1183.
7. Gray, K.M., Banada, P.P., O'Neal, E. and Bhunia, A.K. (2005) Rapid Ped-2E9 cell-based cytotoxicity analysis and genotyping of *Bacillus* species. *J Clin Microbiol* **43**, 5865–5872.
8. Irenge, L. and Gala, J.-L. (2012) Rapid detection methods for *Bacillus anthracis* in environmental samples: a review. *Appl Microbiol Biotechnol* **93**, 1411–1422.
9. Logan, N.A. (2012) *Bacillus* and relatives in foodborne illness. *J Appl Microbiol* **112**, 417–429.
10. McKillip, J.L. (2000) Prevalence and expression of enterotoxins in *Bacillus cereus* and other *Bacillus* spp., a literature review. *Antonie van Leeuwenhoek* **77**, 393–399.
11. Ngamwongsatit, P., Banada, P.P., Panbangred, W. and Bhunia, A.K. (2008a) WST-1-based cell cytotoxicity assay as a substitute for MTT-based assay for rapid detection of toxigenic *Bacillus* species using CHO cell line. *J Microbiol Methods* **73**, 211–215.
12. Ngamwongsatit, P., Buasri, W., Pianariyanon, P., Pulsrikam, C., Ohba, M., Assavanig, A. and Panbangred, W. (2008b) Broad distribution of enterotoxin genes (*hblCDA*, *nheABC*, *cytK*, and *entFM*) among *Bacillus thuringiensis* and *Bacillus cereus* as shown by novel primers. *Int J Food Microbiol* **121**, 352–356.
13. Pilo, P. and Frey, J. (2011) *Bacillus anthracis*: Molecular taxonomy, population genetics, phylogeny and patho-evolution. *Infection, Genetics and Evolution* **11**, 1218–1224.
14. Ramarao, N. and Sanchis, V. (2013) The pore-forming haemolysins of *Bacillus cereus*: A review. *Toxins* **5**, 1119–1139.
15. Rasko, D.A., Altherr, M.R., Han, C.S. and Ravel, J. (2005) Genomics of the *Bacillus cereus* group of organisms. *FEMS Microbiol Rev* **29**, 303–329.
16. Schoeni, J.L. and Wong, A.C.L. (2005) *Bacillus cereus* food poisoning and its toxins. *J Food Prot* **68**, 636–648.
17. Scobie, H.M. and Young, J.A. (2005) Interactions between anthrax toxin receptors and protective antigen. *Curr Opin Microbiol* **8**, 106–112.
18. Senesi, S. and Ghelardi, E. (2010) Production, secretion and biological activity of *Bacillus cereus* enterotoxins. *Toxins* **2**, 1690–1703.
19. Sun, J. and Jacquez, P. (2016) Roles of anthrax toxin receptor 2 in anthrax toxin membrane insertion and pore formation. *Toxins* **8**, 34.
20. Turnbull, P. (2008) *Anthrax in humans and animals*. 4th Edition, Geneva, Switzerland: World Organisation for Animal Health, WHO.
21. Weiner, Z.P. and Glomski, I.J. (2012) Updating perspectives on the initiation of *Bacillus anthracis* growth and dissemination through its host. *Infect Immun* **80**, 1626–1633.

Clostridium botulinum, Clostridium perfringens, Clostridium difficile

12

Introduction

The genus *Clostridium* has more than 80 species. *Clostridium* species are Gram-positive, anaerobic, and rod-shaped and form spores. These microorganisms are widely distributed in the environment particularly in soil, water, and decomposing plants and animals. *Clostridium* species are also found in human and animal intestine as part of their gut microbial community. *Clostridium* species cause a variety of diseases including food poisoning, neuroparalytic disease, gas gangrene, myonecrosis, necrotic enteritis, and antibiotic-associated diarrhea. In general, these organisms produce several exotoxins and enzymes, which contribute to the tissue damage and pathogenesis. Some *Clostridium* species are considered zoonotic pathogens (Table 12.1).

Clostridium botulinum is the most important pathogen of the genus, *Clostridium*, and the bacterium produces a potent heat-labile toxin, botulinum neurotoxin (BoNT). The toxin is responsible for a neuroparalytic disease, botulism, which occurs in humans and animals. BoNT has found a therapeutic use to treat dystonia and other muscular disorders. About 15 *C. botulinum* outbreaks are reported each year in the USA, affecting approximately 55 people and 9 deaths. Professor Emile Van Ermengem of Ghent University first isolated *Bacillus botulinus* in 1895, later renamed *Clostridium botulinum*, in the town of Ellezelles (Belgium) from a homemade raw salted ham that was consumed by 34 musicians who developed botulism. Three musicians died in a week. The culture supernatant, when injected into the laboratory animals, produced paralysis confirming the involvement of a deadly neurotoxin.

Clostridium perfringens is another significant microorganism of concern, responsible for foodborne toxicoinfection in humans and gas gangrene and enteritis in humans and animals. It is responsible for about 966,000 illnesses and 26 deaths every year in the USA. William H. Welch isolated a bacterium, *Bacillus aerogenes capsulatus*, in 1891 during a postmortem autopsy of a deceased man at John Hopkins Hospital (Baltimore, MD, USA). Later, Welch and George Nuttall in 1892 reported the organism as *Bacillus welchii*, which afterward was renamed as *Clostridium welchii*, also known as *Clostridium perfringens*.

Clostridium difficile has become a significant public health concern in recent years due to its nosocomial and hospital-acquired infections, *Clostridium difficile* antibiotic-associated diarrhea (CDAD), and life-threatening pseudomembranous colitis. The organism was first reported in 1934, but in 1978, John Bartlett at Tufts University School of Medicine (Boston, MA, USA) established it as the causative agent of CDAD. This microorganism is now routinely isolated from food animals and humans thus likely to have a foodborne implication.

© Springer Science+Business Media, LLC, part of Springer Nature 2018
A. K. Bhunia, *Foodborne Microbial Pathogens*, Food Science Text Series,
https://doi.org/10.1007/978-1-4939-7349-1_12

Table 12.1 Classification of *Clostridium* species

Organism	Disease
Clostridium botulinum	Botulism, infant botulism, wound botulism – affects peripheral nerves – flaccid paralysis
Clostridium baratii	Infant botulism, hidden botulism
Clostridium butyricum	Infant botulism
Clostridium chauvoei	Black leg – cattle and sheep
Clostridium difficile	Antibiotic-associated membranous colitis (diarrhea) in humans
Clostridium histolyticum	Gas gangrene (human)
Clostridium novyi A, B, C, D	Gas gangrene in humans, necrotic hepatitis in sheep
Clostridium perfringens A, B, C, D, E	Gas gangrene, food poisoning, clostridial myonecrosis, enteritis in animals and humans, enterotoxemia in sheep (struck), and pigbel in humans
Clostridium septicum	Gas gangrene
Clostridium sordellii	Gas gangrene and myonecrosis in humans and animals, liver disease in sheep
Clostridium tetani	Tetanus in humans and animals affects CNS causing spastic paralysis
Clostridium sporogenes	Nontoxigenic

Classification of *Clostridium* Species

Clostridium species are classified based on the shape of vegetative cells, cell wall structure, endospore formation, biochemical properties, 16S rRNA gene sequence homology, mol% G + C content of DNA, and PCR amplification of spacer regions of 16S and 23S rRNA genes. Genome sequence of *C. difficile*, *C. perfringens*, *C. tetani*, *C. botulinum*, and *C. acetobutylicum* are available which greatly increase our understanding of the molecular properties of these organisms. *Clostridium* spp. that are involved in human and animal diseases are summarized in Table 12.1.

Clostridium botulinum

Biology

Clostridium botulinum is an obligate anaerobe. It is motile, spore-forming Gram-positive rod, and the spores are located sub-terminally. It grows in the animal intestine, and the spores are found in feces, soil, and plants. *Clostridium botulinum* is a heterogeneous species and produces eight antigenically distinct botulinum toxin (BoNT) types A, B, C, D, E, F, G, and H. Type H is a new toxin discovered in 2014 by Jason Barash and Stephen Arnon. Each toxin is encoded in a toxin gene cluster that includes several nontoxin accessory genes. Two toxin gene clusters are known, hemagglutinin (*ha*) and *orfX*. The *ha* is found in *C. botulinum* types A, B, C, D, and G, while *orfX* is found in types A (different subtypes), E, F, and H (Fig. 12.1). The neurotoxin genes are located either on the chromosome, on the bacteriophage, or on a plasmid. The genome size varies from 3.76 to 4.26 Mb, and the G + C content is 27.4–28.5%. The genes for BoNT types A, B, E, F, and H are on the chromosome, type C and D on bacteriophages, and type G on an 81-MDa plasmid. The genes are regulated by a regulatory element, BotR. Sometimes a strain may produce two toxins. In such cases, the major toxin is designated by the uppercase letter and the minor toxin by the lower case letter, for example, *C. botulinum* types Bh, Ab, Ba, Bf, and so forth.

Toxin types A and B are common in the USA; C and D occur in farm animals; E, F, and G are also produced by non-botulinum species such as *C. argentinese* which produces type G botulinum toxin and *C. baratii* type F toxin. BoNT types A, B, E, H, and, in the rare cases, F cause human botulism, while BoNT types C and D cause animal botulism. Botulinum toxin is produced, when the clostridial cells are lysogenized by the bacteriophages or due to autolysis of the cells late in the growth cycle. Toxin production is influenced by nutrient composition, cultural conditions of the growth medium including the presence of metal ions, certain amino acids, peptides, pH, temperature, and the cell density.

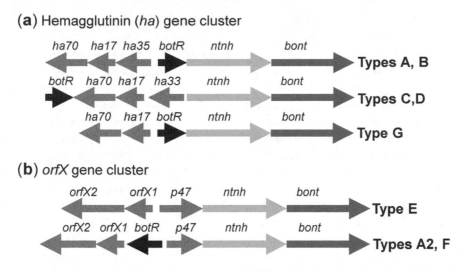

(a) Hemagglutinin (*ha*) gene cluster

(b) *orfX* gene cluster

Fig. 12.1 Botulinum neurotoxin gene clusters, **(a)** *hemagglutinin* (ha) and **(b)** *orfX* of *Clostridium Botulinum*. ha Hemagglutinin, *ntnh* nontoxic nonhemagglutinin, *bont* BoNT, and *botR* botulinum regulator (Adapted and modified from Johnson and Bradshaw 2001. Toxicon 39, 1703–1722)

Table 12.2 Proteolytic and non-proteolytic *Clostridium botulinum*

Properties	Proteolytic	Non-proteolytic
Neurotoxin type	A, B, F, H	B, E, F
Fermentation of glucose	+	+
Fermentation of fructose, maltose, mannose, sucrose	Variable to negative	+
Motility	+	+
Minimum growth temperature	10–12 °C	2.5–3.0 °C
Optimum growth temperature	37 °C	25 °C
Minimum pH for growth	4.6	5.0
Minimum a_w	0.94	0.97
NaCl conc. to prevent growth	10%	5%
Spore heat resistance	$D_{121 °C} = 0.21$ min	$D_{82.2 °C} = 2.4$ min

Adapted from Peck et al. (2011). Food Microbiol. 28, 183–191

Based on the enzymatic activity, *C. botulinum* strains are grouped into either proteolytic or non-proteolytic, and these strains are genetically distinct (Table 12.2). Proteolytic *C. botulinum* produces several extracellular enzymes to degrade proteins and carbohydrates in order to harvest nutrients which may result in food spoilage. The non-proteolytic strains do not have similar enzyme activity but can ferment different sugars and are called saccharolytic. Both proteolytic and non-proteolytic *C. botulinum* produce neurotoxins and are responsible for most foodborne botulisms. Besides, *C. baratii* and *C. butyricum* are also involved in foodborne botulism. Proteolytic strains may carry one or two neurotoxin genes, but a majority carry a single gene for neurotoxin types A, B, F, or H, while the majority of non-proteolytic strains produce types B, E, or F.

Proteolytic strains are mesophilic, and the optimum temperature for growth and toxin production is 35–37 °C; however, they can produce neurotoxins at temperatures as low as 12 °C. The non-proteolytic strains are psychrotrophic and can grow and produce neurotoxin at 3.0–3.3 °C, but the optimum temperature is 25 °C. Proteolytic strains do not grow below pH 4.6 and non-proteolytic below pH 5.

Spores are highly resistant to heat, and the decimal reduction time (D value) is 0.15–1.8 min at 110–121 °C. Spores of proteolytic strains are more resistant to heat than the spores of non-proteolytic strains. A standard heat treatment of 121 °C for 3 min ensures the safety of low-acid canned foods. Spores present in food subjected to heat treatment will germinate under favorable conditions to form vegetative cells. The amino acid, L-alanine alone or in combination with L-lactate, can trigger spore germination. Vegetative cells will grow under an anaerobic environment in food and produce botulinum toxin that would be a cause for concern. The botulinum toxin is sensitive to heat treatment (>85 °C for 15 min); therefore, thorough heating of the food prior to consumption will render food safe. *Clostridium botulinum* spores do not survive well in the healthy adult human gut environment.

Sources

Most *Clostridium* spp. survive in soil and grow in the animal intestine. *Clostridium* spores can be found in poorly or under-processed canned foods such as home-canned foods, spices, sewage, and plants. The spores are heat-resistant and can survive in the canned food when the temperature is used below 120 °C. Low acidity (pH above 4.6), low oxygen, and high water content favor spore germination and toxin production. Spices, herbs, and dehydrated mushrooms may serve as a potential source of spores. Home-canned vegetables – beans, peppers, carrots, corn, asparagus, potatoes, bamboo shoots – and fish are implicated in outbreaks. Foil-wrapped baked potatoes cooled at an inadequate rate and extent and served several hours later are responsible for a restaurant or at-home outbreaks. Yogurt, cream cheese, and jarred peanuts had also caused botulism outbreaks. Blood sausage caused frequent outbreaks in central Europe. Condiments – such as sautéed onions, garlic in oil, hot dog chili sauce, and commercial cheese sauce – were implicated in outbreaks. Bees can carry bacterial spores in addition to pollen and honey, which may contain clostridial spores, and the acceptable limit of clostridial spores in honey is less than 7 per 25 g.

Botulism

Clostridium botulinum produces botulinum toxin, which causes botulism in humans and animals. The name "botulism" came from the Latin word "botulus" meaning sausage. In central Europe, in the eighteenth century, the disease was frequently linked to the consumption of blood sausage. The foodborne botulism is an intoxication disease, not an infection since the preformed toxin causes the disease without the bacterium. The LD_{50} of active botulinum toxin is <0.01 ng in the mouse. In adult humans, intravenous or intramuscular administration of 0.9–0.15 μg or oral administration of 70 μg can cause death. There are five types of botulism: foodborne botulism, infant botulism, hidden botulism, inadvertent botulism, and wound botulism. Foodborne, wound, and infant botulism are caused primarily by *C. botulinum* toxin type A. Toxin B and E types also can cause such botulism but to a lesser frequency or extent.

Foodborne Botulism

Consumption of food contaminated with spores does not generally cause disease in healthy adults. Contaminated food after heating (at least 70 °C) and cooling at a slow rate promotes germination of spores. Spore germination may take place in the colon, but clostridia are unable to survive because of the resident microflora. Heating removes oxygen; thus the anaerobic environment created by the heated food allows germination and growth of the organism and subsequent botulinum toxin production. Ingestion of toxin is lethal, and the incubation period of botulism depends on the rate and the amount of toxin consumed, which can occur as early as 2 h or as long as 8 days, typically 12–72 h. Once consumed, the toxin is absorbed and reaches the blood circulation. The toxin blocks the release of neurotransmitter, acetylcholine, which prevents nerve impulse propagation in the peripheral neurons, affecting parasympathetic and sympathetic nervous system, and the neuromuscular junctions and causes flaccid paralysis (Fig. 12.2). In flaccid paralysis, the peripheral nerves, both sympathetic and parasympathetic nerves, are affected. Another

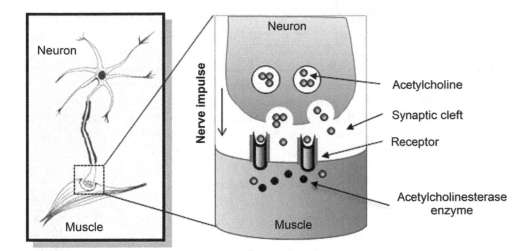

Fig. 12.2 Schematic drawing showing the neuromuscular junction and transmission of nerve impulse

neuroparalytic toxin, tetanus toxin, produced by *Clostridium tetani*, exhibits spastic paralysis, in which the central nervous system is affected. Once the botulinum toxin is bound to a neuron, external intervention has little or no effect.

The symptoms appear 4–36 h after ingestion of the toxin, which include nausea, vomiting, headache, double vision, slurred speech, muscle spasm (dystonia), and muscle weakness. It first affects the upper limbs then the lower limbs and exhibits breathing difficulty because of the pharyngeal and diaphragm muscle weakness. The heart function also weakens, and death follows. Autonomic symptoms include dry mouth, postural hypotension, urinary retention, and pupillary abnormalities. The fatality rate is 5–10%; however, the fatality is much higher for untreated patients.

Infant Botulism

Clostridium botulinum, *C. baratii*, and *C. butyricum* are responsible for infant botulism. Honey is the main source of the organism. Infants under 1 year of age are mostly susceptible. Since infants have not developed complete colonic microflora populations, there are no antagonistic effects from resident flora or from bile salts. The spores germinate in the intestinal tract of the infant, and the vegetative cells colonize the gut. The progression of the disease is very slow because of poor absorption of toxin through the colonic cell layers. Symptoms are very similar to the adult botulism; however, nausea and vomiting are absent. Early signs are weak cry, muscle weakness, difficulty in feeding, i.e., poor suckling ability, hypotonia (floppy baby syndrome), and a decrease in spontaneous movement. Constipation, tachycardia, and dry mouth are due to the blockade of parasympathetic nerve impulses. Death occurs in severe cases. Overall mortality rate is 5%. Most recover with adequate supportive therapy and interventions. Infant botulism cases are on the rise since 1990. Both bottle-fed and breast-fed infants are susceptible to infant botulism. Honey has been considered the significant risk factor, at least in 20% cases, for infant botulism; therefore, children should not be given honey in the first year of their lives.

Hidden Botulism

This adult form of botulism is hidden from the clinicians because of the lack of direct evidence for this type of botulism. Hidden botulism is neither food related, wound associated, nor drug use related. This adult form of infant botulism requires bacterial colonization in the gut. Antibiotic treatment due to other illnesses may disturb natural microflora balance in the gut thus allowing *Clostridium* species to colonize and produce toxins. Achlorhydria, prior surgery, or

Crohn's disease may also aid in the development of this form of botulism. *C. botulinum* or *C. baratii* produces toxin type F, which is responsible for hidden botulism in adults. Symptoms are similar to foodborne botulism, i.e., shortness of breath, dizziness, bradycardia, respiratory arrest, decreased voluntary movements, flaccid muscle tone, and so forth. Hidden botulism is diagnosed by isolation and identification of the *Clostridium* cells from feces.

Wound Botulism

Though the wound botulism is rare, it occurs in patients with traumatic and surgical wounds and intravenous drug users. It is also common for soldiers on the battlefield. In recent years, increased numbers of cases have been reported among the intravenous drug users. Wound botulism has also been reported in patients following intranasal cocaine abuse and the laboratory workers from inhalation of toxins. Spores lodge in the deep wound or in the injection sites of the drug users. The anaerobic environment created by tissue destruction and the growth of aerobic bacteria help germination and growth of *Clostridium botulinum*. The incubation period for wound botulism is 4–14 days. Botulinum toxin is produced and absorbed through the mucus membranes, broken skin, or wounds, leading to botulism. Note: *Clostridium tetani* can also enter through wound to cause tetanus in a similar manner.

Inadvertent Botulism

In recent years, botulinum toxin (such as Botox®, Dysport®, NeuroBloc®) is used for the treatment of dystonia and other movement-related disorders, strabismus (hyperactive extraocular muscles), and cosmetic enhancement. The patients treated with intramuscular injection of botulinum toxin have toxin circulating in the blood and can block the neurotransmitter release in adjacent muscle or in the autonomic nervous system. Cosmetic use of BoNT for removal of wrinkles, to improve skin or muscle tones, may serve as a possible risk factor for inadvertent botulism.

Virulence Factors and Mechanism of Pathogenesis

Clostridium botulinum produces three types of toxins: botulinum neurotoxin (BoNT), C2 toxin, and C3 toxin. Botulinum toxin affects neurons, while the C2 and C3 toxins induce epithelial cell damage possibly facilitating the spread of the BoNT to deeper tissues.

Botulinum Toxin

The botulinum toxins are classified as types A, B, C, D, E, F, G, and H and subtypes A1, A2, A3, A4, and A5, which are antigenically distinct. The amino acid sequence of each neurotoxin type differs by up to 70%, while each subtype differs by only 2.6–32%. BoNT types A, B, E, H, and, in rare cases, F cause human botulism, while BoNT types C and D cause animal botulism. Botulinum toxin is produced, when the clostridial cells are lysogenized by the bacteriophage or due to autolysis of cells late in the growth cycle. The BoNT toxin is produced in the culture as a large complex (about 900 kDa) consisting of BoNT, hemagglutinin (HA), and nontoxin nonhemagglutinin (NTNH). The proteolytic *C. botulinum* strains produce BoNT types A, B, F, and H, while the non-proteolytic strains produce types B, E, or F (Table 12.2).

The botulinum toxin is a 150 kDa polypeptide and is derived from a large progenitor inactive form of a toxin called derivative toxin or protoxin (Fig. 12.3). The derivative toxin produced by the proteolytic *C. botulinum* strain is activated by its own protease. In the case of non-proteolytic strains, the toxin is activated by host gastric proteases, since these strains do not produce protease. The active toxin is an A–B-type toxin, consisting of two subunits: B (heavy chain, 100 kDa) and A (light chain, 50 kDa). The heavy and light chains are joined by a disulfide bond (Fig. 12.3). The primary function of BoNT is to block neurotransmitter (acetylcholine) release at the neuromuscular junction, thus affecting nerve impulse propagation leading to flaccid paralysis.

Normally, acetylcholine release from the vesicles at the neuromuscular junction is aided by

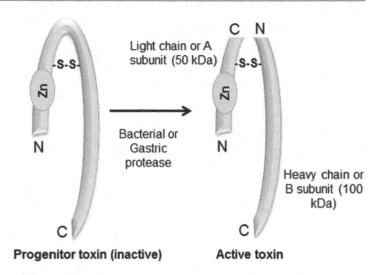

Fig. 12.3 Activation of botulinum toxin by the bacterial or gastric protease

Progenitor toxin (inactive)   Active toxin

SNARE (soluble N-ethylmaleimide-sensitive factor attachment protein receptor) proteins, which consist of synaptobrevin, SNAP-25 (synaptosomal-associated protein-25), and syntaxin. SNAP-25 and syntaxin are integral neuronal membrane bound, whereas the synaptobrevin is a vesicle-associated membrane protein (VAMP). During propagation of the nerve impulse, synaptobrevin associated with the synaptic vesicle fuses with the SNAP-25 and syntaxin to form a synaptic fusion complex. Synaptic vesicle with SNARE complex then fuses with the neuron membrane, and through the exocytosis process, the acetylcholine is released into the synaptic cleft. Acetylcholine binds to the acetylcholine receptor located on the muscle fiber, and acetylcholinesterase enzyme depolarizes the acetylcholine and aids in the propagation of nerve impulse (Figs. 12.2 and 12.4).

In the botulism patient, after absorption through the tissues, the toxin is transported by the blood to the peripheral cholinergic synapses, primarily in the neuromuscular junction. The heavy chain (B subunit) of the neurotoxin has two functional domains. The C-terminal domain facilitates binding of the neurotoxin to a specific receptor on synaptic membrane vesicle proteins at the nerve terminal. The receptor is a sialic acid containing glycoprotein or glycolipid, and it is found only on the neuron. While the N-terminal domain forms a channel in the neuron and delivers the light chain into the nerve cytosol. The light chain (A subunit) has a single zinc molecule and has zinc-dependent endopeptidase activity, which cleaves the SNARE proteins. The protein target may vary depending on the toxin type. The A subunit of BoNT types A and E cleaves SNAP-25; types B, D, F, and G cleave synaptobrevin or VAMP; and type C cleaves syntaxin. As a result, the SNARE complex does not fuse with the synaptic membrane vesicle, and exocytosis of the acetylcholine does not occur, interfering with the nerve impulse propagation (Fig. 12.4).

The impaired nerve impulse in both sympathetic and parasympathetic nerves results in flaccid paralysis, where the muscle is tensed and relaxed showing characteristic symptoms. In comparison, tetanus toxin prevents the release of γ-aminobutyric acid thus inhibits neuronal transmission. Muscle is tensed and not relaxed and results in spastic paralysis.

C2 Toxin

The C2 toxin is a binary toxin similar to A–B--type toxin produced by some *C. botulinum* strains, in particular, types C and D. C2 toxin is an ADP-ribosyltransferase. It causes depolymerization of the actin cytoskeleton in the host cell. C2 consists of two non-linked proteins, the enzyme component C2I (50 kDa) and the binding and translocation component C2II (90 kDa),

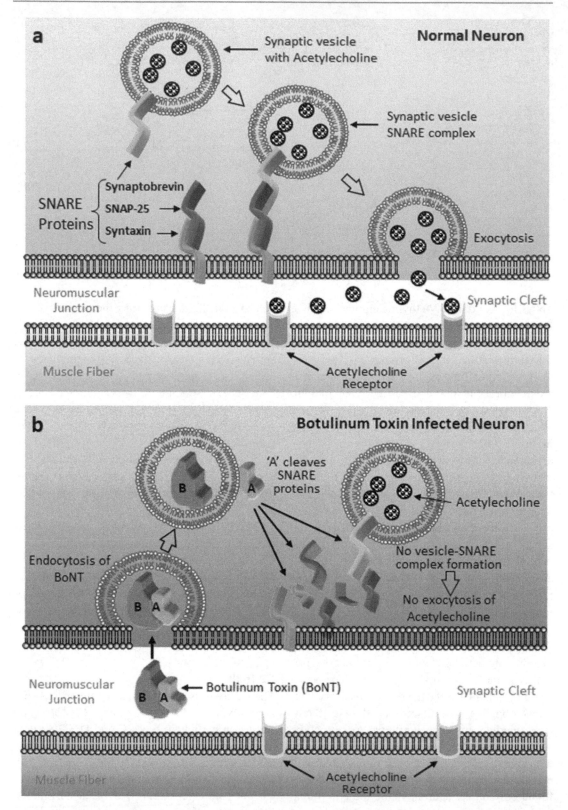

Fig. 12.4 Diagram showing the mechanism of action of botulinum neurotoxin (BoNT) (Arnon et al. 2001. J. Am. Med. Assoc. 285, 1059–1070)

which delivers C2I into the host cell. C2II binds to a variety of cells, and receptor-mediated endocytosis follows. The C2I with enzymatic activity translocates across the membrane and catalyzes the ADP-ribosylation of intracellular monomeric G-actin. C2I cleaves ADP-ribose from nicotinamide adenine dinucleotide (NAD^+) and attaches it to the actin protein, thus interferes with the microfilament formation. It is proposed that C2 toxin induces the formation of long microtubule-based protrusions forming a dense meshwork on the surface of intestinal epithelial cells providing a favorable environment for *C. botulinum* colonization and toxin production. C2 also disrupts the cytoskeleton structure leading to cell damage and increased vascular permeability and thus probably helps in dissemination of the botulinum toxin. C2 toxin also induces necrosis of the epithelial cells and edema in the lamina propria.

C3 Toxin

Clostridium botulinum C3 toxin is a 25 kDa protein. There are two isoforms, C3bot1 and C3bot2, with 60% sequence identity. The C3 proteins are also produced by *Clostridium limosum*, *Bacillus cereus*, and *Staphylococcus aureus*. C3 toxin ADP-ribosylates Rho, a GTP-binding signal transduction protein, which controls the actin polymerization. This toxin alters the cell morphology and leads to cell death. The pathological significance of C3 in the disease process is unclear; however, it has been preferentially internalized by monocytes and macrophages.

Symptoms

The clinical symptoms vary depending on the botulism type, i.e., foodborne, infant, wound, adult, or inadvertent form of botulism. Symptoms of foodborne botulism, which is most common, are visible 4–36 h after ingestion of the toxin, which include nausea, vomiting, headache, double vision, slurred speech, muscle spasm (dystonia), and muscle weakness. It first affects the upper limbs followed by the lower limbs and exhibits breathing difficulty because of pharyngeal and diaphragm muscle weakness. The heart

function also weakens, and death follows. Other symptoms include dry mouth, postural hypotension, urinary retention, and pupillary abnormalities. The fatality rate is 5–10%; however, fatality is much higher in the untreated patients.

Prevention and Treatment

Prevention strategies are most important to avoid botulism, which include proper canning of food products, boiling of homemade canned foods since homemade canned foods present serious dangers, restraining from feeding infants with honey, and early diagnosis. The botulinum toxin is heat-labile; therefore, heating of suspected products for at least 85 °C for 15 min before consumption can prevent botulism. Botulinum toxin formation in foods can be prevented by various methods. Store food at (1) <3.0 °C, (2) a pH ≤ 5.0 and a chilled storage, (3) a NaCl concentration ≥ 3.5% plus storage at chilled temperature, and (4) an a_w ≤ 0.97 plus chilled storage. (5) A 10-day rule should be followed, i.e., storage at ≤ 8 °C and a shelf life of <10 days, and (6) a heat treatment of 90 °C for 10 min or equivalent lethality (e.g., 80 °C for 129 min, 85 °C for 36 min for 6D process lethality) combined with the storage at chill temperature can be adopted.

Botulinum toxins are detected in patients sera, wound, or stool specimens. Sixty percent of botulism patients show a positive stool sample for *C. botulinum*. The suspected food should be tested by a mouse bioassay, which can detect as low as 0.03 ng of toxin, and results are obtained within 1–2 days. Antisera are used to type the toxin obtained from blood serum or other sources.

Treatment is unsuccessful if toxin has already entered the blood circulation and has bound to the receptors on the neuron. A successful intervention depends on the concentration of toxins being ingested and for how long. If the botulism is detected early, an antitoxin can be injected to neutralize the toxin that is still circulating in the blood. Toxins that are already bound to the receptor cannot be neutralized by the antitoxin. Mechanical respirators and life support are required to facilitate breathing, to counter neurological damage, and to regenerate the nerve endings. The recovering patients can suffer from a permanent neurological disorder (damage). The recovery period is generally very long. Dry mouth and general fatigue can last for weeks or months. An anticholinesterase drug such as edrophonium chloride may be beneficial to some patients. Guanidine and 4-aminopyridine (4-AP) have been reported to improve the ocular and limb muscle strength, but those compounds have a little or limited effect on the respiratory paralysis. In the case of the infant, wound, and hidden botulism, antibiotics are needed to clear the bacteria from the system.

Detection of *C. botulinum* or toxin

Culture Method

Blood, stool, or food samples can be tested for the presence of the toxin or the *C. botulinum* cells. Food samples are first enriched in the cooked meat medium or trypticase–peptone–glucose–yeast extract (TPGY), which is steamed for 10–15 min to remove dissolved oxygen and incubated at 35 or 28 °C for 5 days under a strict anaerobic condition. Gram staining or observations under a bright field microscopy should reveal the presence of typical cells with "tennis racket" appearance. For selective isolation of *C. botulinum*, enriched culture is either mixed with alcohol or heat-treated at 80 °C for 10–15 min, then streaked onto egg yolk agar or liver–veal–egg yolk agar, and incubated at 35 °C for 48 h under anaerobic conditions. The typical *C. botulinum* colonies appear as raised or flat, smooth or rough with some spreading, and irregular edges. On egg yolk agar, colonies show a luster zone (referred to also as a pearly zone), when observed under an oblique light. Often another subculture in TPGY or chopped liver broth under anaerobic conditions for 5 days followed by streaking onto egg yolk agar is needed to isolate the pure cultures of *C. botulinum*.

Immunoassays

Immunoassays are widely used for detection of the toxins. In a sandwich assay format,

polyclonal antibody to the toxin is first bound to the solid surface, and then the sample containing the toxin is added. A second antibody labeled with an enzyme (alkaline phosphatase or horseradish peroxidase) is added. The addition of an appropriate substrate will produce color. This assay is 10–100-fold less sensitive than that of the mouse bioassay, and negative results have to be reconfirmed by other assays. Lateral flow immunoassay has been developed which provides results in less than 30 min; however, the assay is less sensitive; thus the negative results have to be confirmed by the mouse bioassay.

Mouse Bioassay

Mouse bioassay is considered the gold standard but very expensive. The assay requires a large number of mice. Bacterial cell-free culture supernatants are treated with trypsin at 200 μg ml^{-1}, and 0.5 ml of each toxin preparation is injected intraperitoneally into mice. Typical symptoms include labored breathing, pinching of the waist, and paralysis, which develop in 1–4 days.

BoNT Enzyme Activity Assay

The BoNT has zinc endoprotease activity, and it degrades neuronal proteins that regulate the release of the acetylcholine. BoNT from type A strain cleaves SNAP-25, while type B cleaves VAMP or synaptobrevin. A synthetic VAMP peptide is used as a substrate to assay for the presence of BoNT enzyme activity colorimetrically. This assay is more sensitive and much faster than the mouse bioassay.

PCR and Oligonucleotide Microarray

A highly specific multiplex PCR assay has been developed for detection of *C. botulinum* types A, B, E, and F from food and fecal materials. The assay is able to detect types A, E, and F at 10^2 cells, while type B was detected at 10 cells per reaction mixture from naturally contaminated meat, vegetable, and fish. An oligonucleotide array has been developed to detect multiple foodborne pathogens including *C. botulinum*, *E. coli*, *Listeria monocytogenes*, *Shigella dysenteriae*, *Vibrio cholerae*, *Vibrio parahaemolyticus*, *Proteus vulgaris*, and *Bacillus cereus*.

Clostridium perfringens

Biology

Clostridium perfringens is a Gram-positive, rod-shaped, nonmotile, and spore-forming anaerobic but somewhat an aerotolerant bacterium. The genome size is 3.6 Mb, and all virulence genes are located either on the chromosome or on a plasmid. The bacterium can grow at a temperature between 15 °C and 50 °C with an optimal growth temperature of 43–45 °C. The growth rate is very fast, and the generation time is below 8 min at temperatures 33–49 °C. *C. perfringens* does not grow below 15 °C, but some rare strains can grow at 6 °C. The pH range for growth is 5–9.0. *C. perfringens* can survive and grow in the presence of curing salts consisted of 300 ppm of sodium nitrite and 4–6% NaCl. *C. perfringens* is a fastidious organism and requires more than 12 different amino acids and vitamins for their growth; thus it grows very well in the meat products. Meat and poultry are generally implicated in most outbreaks. Beef products are responsible for about 40% of *C. perfringens* outbreaks. Roast beef is a major vehicle of the outbreak because of improper handling, temperature abuse, and inadequate cooling after cooking.

C. perfringens is classified into five types, A, B, C, D, and E, based on the production of four types of extracellular toxins, alpha (α), beta (β), epsilon (ε), and iota (ι) (Table 12.3). Altogether, *C. perfringens* produces at least 20 different toxins, and the list is growing. In addition to the above four toxins, it produces several hydrolytic enzymes and toxins including lecithinase, hyaluronidase, collagenase (κ-toxin), DNAse, sialidases (affects sialic acid in the host cell membrane), amylase, *Clostridium perfringens* enterotoxin (CPE), hemolysin (perfringolysin O, PFO or theta toxin), necrotic enteritis toxin B (NetB), and so forth. The genes encoding the toxins are located either on the chromosome or on a plasmid. *C. perfringens* produces a double zone of hemolysis on blood agar plate. The clear inner zone is produced by PFO and the hazy outer zone by phospholipase C (α-toxin). *C. perfringens* type A is responsible for food poisoning (gastroenteritis), while type C and certain strains of type A cause necrotic enteritis. *C. perfringens* type A is also responsible for about 5–15% of all cases of antibiotic-associated diarrhea in humans and sudden infant death syndrome (SIDS).

Sources

Clostridium perfringens is found in soil, water, sludge, spices, dust, sewage, raw and processed foods, and contaminated equipment. It is present in low numbers (10^3–10^6 spores per gram) in human/animal intestine and feces. Ubiquitous distribution of the organism (spore), heat resistance of spore, and rapid growth and enterotoxin production made this organism a successful foodborne pathogen. It grows well in meat, especially in ground beef. During heating and cooling, if food is allowed to stand at room temperature – it will grow rapidly and produce toxin.

From 1992 to 1997, 248,520 human illnesses due to *C. perfringens* were documented by the Centers for Disease Control and Prevention (CDC). In the year 1993, at least 10,000 cases with 100 deaths were recorded. It is estimated that about 965,958 illnesses, 438 hospitalizations, and 26 deaths are associated with *C. perfringens* infection in the USA each year.

Virulence Factors and Mechanism of Pathogenesis

Clostridium perfringens causes food poisoning in humans and gas gangrene, myonecrosis, and enteritis in humans, animals, and birds. Food poisoning disease is characterized as toxicoinfection, not intoxication. Preformed *Clostridium perfringens* enterotoxin (CPE) in food is not responsible for the disease, since the toxin may be destroyed during passage through the stomach. CPE is heat-labile and destroyed at temperature 60 °C for 10 min. The infectious dose is about 10^7–10^9 *C. perfringens* cells. Raw foods or ingredients may be contaminated with spores. Spores are heat resistant and some will survive

Table 12.3 *Clostridium perfringens* classification and their toxin production profile

Type	Disease	Enterotoxin	α-toxin	β-toxin	ε-toxin	ι-toxin
A	Gas gangrene, food poisoning, myonecrosis	+	+	−	−	−
B	Dysentery in lambs; enteritis in calves, goats and foals; enterotoxemia in sheep	−	+	+	+	−
C	Necrotic enteritis in humans and animals (pigbel disease)	+	+	+	−	−
D	Pulpy kidney disease	+	+	−	+	−
E	Enteritis in animals	Variant	+	−	−	+
Mol. WT		35 kDa	43 kDa	35 kDa	33 kDa	48 and 74 kDa
Gene		*cpe*	*plc*	*cpb*	*etx*	*ia, ib*
Gene location		Plasmid/ chromosome	Chromosome	Plasmid	Plasmid	Plasmid

Adapted from Brynestad and Granum (2002). Int J Food Microbiol 74, 195–202

cooking. Slow cooling after cooking allows the spores to germinate since heating removes oxygen and creates an anaerobic environment. In addition, room temperature storage after cooking allows the vegetative cells to grow and multiply rapidly to reach an infective dose of $>10^7$ cells. Upon consumption, many vegetative cells die when exposed to stomach acid, but some survive and begin to form spores. After completion of the sporulation step, the mother cell undergoes lysis and releases the toxin and spores in the small intestinal lumen. The toxin causes epithelial cell damage, and pathology described below. *C. perfringens* spores with high heat resistance tend to produce increased levels of enterotoxin.

Clostridium perfringens Enterotoxin

Food Poisoning

Clostridium perfringens produces CPE, which is synthesized during sporulation in the intestine. CPE is a 319 amino acid containing the polypeptide of 35 kDa. The CPE toxin is sensitive to acidic pH but resistant to digestive enzymes, trypsin, and chymotrypsin, which remove about 25 and 37 amino acids, respectively, from the N-terminal end, thus making the toxin highly active. The *cpe* gene encoding the toxin is located either on the chromosome or on plasmids. About 5% of all *C. perfringens* isolates produce CPE. *C.*

perfringens type A strains are the primary producer of CPE; however, a majority of type C and D strain also produces CPE. Type E strain carries silent *cpe* or variant *cpe* gene. The CPE production by the type B strain has not been confirmed yet. CPE-positive type A strains are responsible for the great majority of human food poisoning cases.

CPE is a β-pore-forming toxin and has two important domains, (i) the C-terminal domain (amino acids 290–319) which interacts with the cellular receptor protein, claudin on enterocytes, and (ii) the N-terminal domain (amino acids 26–171) which oligomerizes in the membrane to form a pore. Claudin proteins are the member of a large family consisting of 27 proteins (20–27 kDa in size), which are integral to the tight junction (TJ) of epithelial and endothelial cells. Claudin-3, claudin-4, claudin-6, claudin-8, and claudin-14 have a high affinity for CPE and serve as receptors for CPE.

In the small intestine, the cytotoxic effect of CPE starts with binding of the toxin to the claudin proteins forming a complex of ~90 kDa referred to as "CPE small complex." This small complex then oligomerizes to form a hexamer of ~450 kDa pre-pore (CH-1). The CH-1 inserts into the membrane forming a complex with occludin designated CH-2 (~600 kDa), which forms an active pore in the membrane (Fig. 12.5) allowing an influx of calcium ions. The calcium ion influx

Fig. 12.5 Diagram showing the mechanism of action of *Clostridium perfringens* enterotoxin (CPE)

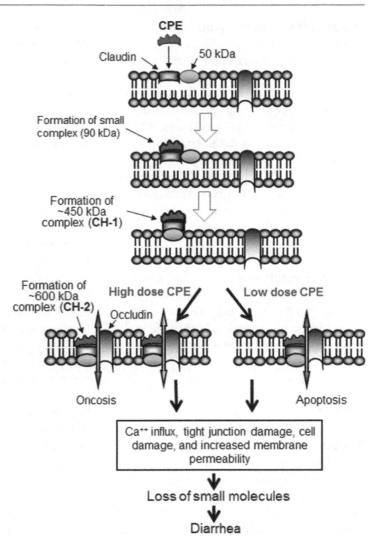

triggers apoptosis or oncosis, depending on the CPE concentration. CPE at low concentration (1 μg ml^{-1}) induces apoptosis through activation of caspase 3 and caspase 7; thus there is very little intestinal inflammation. The CPE at high concentration (10 μg ml^{-1}) induces necrosis (oncosis), which may contribute to the inflammatory process that is evident during infection. CPE modulates epithelial tight junction architecture, affects the membrane barrier function, and alters the paracellular membrane permeability, resulting in the loss of fluid and ions (Na$^+$, Cl$^-$) and the onset of diarrhea. Histopathological changes include desquamation of epithelial cells, damage to the villi tips, and shortening of villi. These changes affect the fluid and electrolyte transport, decreased intestinal absorptive capacity, and increased stool mass.

Antibiotic-Associated Diarrhea

The CPE-positive *C. perfringens* type A strain can cause non-foodborne diseases, such as antibiotic-associated diarrhea (AAD), which occurs as a consequence of the antibiotic treatment, such as penicillin, cephalosporins, trimethoprim, or co-trimoxazole. Antibiotic treatment inhibits resident microflora creating a favorable environment for *C. perfringens* bloom and CPE production in the intestine. The infectious dose is much lower than the food poisoning, and it is suggested that *C. perfringens* strain carrying the plasmid-borne *cpe* is involved in the development of AAD.

Sudden Infant Death Syndrome

The CPE-positive *C. perfringens* type A strains are also attributed to causing sudden infant death syndrome (SIDS). Infants under the age of 1 succumb to unexpected death possibly due to high numbers of type A strain in their intestine and consequent elevated CPE production. CPE can cause local damage in the intestine and can be absorbed into the blood circulation causing enterotoxemia. The toxin possibly exhibits superantigenic function (*see* Chap. 4) triggering the release of massive amounts of IL-2 by T lymphocytes. Toxemia can also affect the liver and kidney.

Enteritis Necroticans

The CPE-positive type C strain contributes to human enteritis necroticans (EN), though the β-toxin produced by these strains is essential for the disease, which is discussed under the section on β-toxin (see below).

α-Toxin

Gas Gangrene

Clostridium perfringens alpha-toxin (CPA), also known as phospholipase C, is produced by *C. perfringens* type A, B, C, D, and E strains. It is a 43 kDa toxin and is responsible for gas gangrene in humans and animals resulting in high fever, pain, edema, and myonecrosis. CPA and perfringolysin O (PFO) are essential for causing gas gangrene. The CPA polypeptide has two domains: the N-terminal zinc-binding domain has phospholipase C (PLC) and sphingomyelinase (SMasc) activities, while the C-terminal calcium-binding domain binds to the eukaryotic cell membrane to exert a cytotoxic effect. The substrates for PLC and SMase are phosphatidylcholine (PC) and sphingomyelin (SM), respectively. This toxin has significant homology with PLC from *Listeria monocytogenes* and *Bacillus cereus*. At high concentration, the α-toxin causes massive degradation of PC and SM in membranes leading to membrane disruption, and at low concentration, the damage is less severe.

PFO is a cholesterol-dependent cytolysin (CDC), and it binds to cholesterol on the membrane. The PFO monomers insert into the membrane, oligomerize, and form pore size of about 15 nm and cause cell lysis. The α-toxin induces production of IL-8, TNF-α, platelet activation factors, and the endothelial leukocyte adhesion molecules, which possibly contribute to the increased vascular permeability and swelling.

β-Toxin

Enteritis Necroticans or Pigbel

Clostridium perfringens β-toxin (CPB) is a 35 kDa toxin, which is produced by type C and B strains, and contributes to enteritis necroticans or pigbel in humans. The *cpb* gene is located on large virulence plasmids of ~ 65–110 kb, which also carry other toxin genes. CPB toxin is produced as a 336 amino acid-long protoxin, and cleavage of 27 amino acid leader sequence yields the active 35 kDa toxin. The toxin oligomerizes to form a cation-dependent channel in the host cell membrane. It forms channels of 12 Å in diameter in planar bilayers consisting of phosphatidylcholine and cholesterol. The channel is selective for transport of monovalent cations such as sodium and potassium.

The CPB-producing *C. perfringens* type C strain can cause fulminant disease in both humans and animals (piglet, chicken, calf, lamb, and goats). The necrotizing β-toxin acts on the autonomic nervous system causing arterial constriction leading to mucosal necrosis. The toxin also forms multimeric transmembrane pores and facilitates the release of cellular arachidonic acid and inositol. Additionally, the α-toxin with the phospholipase and sphingomyelinase activity is responsible for necrotizing effects. The enteritis necroticans or pigbel in humans is characterized by vomiting, abdominal pain, bloody stool, and, in severe cases, toxemia leading to rapid death. The disease is endemic in Southeast Asia. In animals, necrotic enteritis is characterized by hemorrhagic diarrhea and enterotoxemia similar to humans.

Epsilon Toxin

Dysentery, Enteritis, and Enterotoxemia

Epsilon toxin (ETX) is produced by *C. perfringens* type B and D strains and is responsible for dysentery, enteritis, and enterotoxemia in lambs, goats, calves, and foals. The *etx* gene encoding the toxin is located on a plasmid. The ETX toxin is produced as an inactive protoxin of 32.9 kDa and is activated by the host protease enzyme, trypsin, and chymotrypsin or *C. perfringens* λ-protease. Proteolytic cleavage removes 10–13 residues from the N-terminal end and 22–29 residues from the C-terminal end and makes the toxin highly active. The toxin affects brain and kidney cells and forms heptameric pores in the cell membrane. Edema in the brain increases the intracerebral pressure and occasionally causes the parenchymal brain necrosis. The toxin is similar to the aerolysin from *Aeromonas hydrophila* and parasporin-2 from *Bacillus thuringiensis*.

Iota Toxin

Enterotoxemia and Enteritis

C. perfringens type E strain produces iota toxin (ITX) and α-toxin, which are responsible for enterotoxemia and enteritis in cattle, sheep, and rabbits. The ITX is a binary toxin, Ia and Ib, similar to a classic A–B-type toxin; however, the genes encoding each toxin on the chromosome are separated by a noncoding sequence in between. ITX is similar to *C. botulinum* C2 toxin (C2I and C2II) and *C. difficile* ADP-ribosyltransferase (CdtA and CdtB). Ia is a 48 kDa enzyme, while the Ib is 74 kDa. Ib binds to the cellular receptor and helps in the translocation of Ia into the cytosol of the cell. Ia has enzymatic activity and catalyzes ADP-ribosylation of globular actin by inhibiting its synthesis resulting in cell rounding and cell death.

NetB Toxin

Avian Necrotic Enteritis

C. perfringens type A also produces necrotic enteritis toxin B (NetB), which is a β-pore-forming 33 kDa toxin that causes cell lysis. The NetB toxin production is controlled by a two-component VirS and VirR regulatory system. NetB is responsible for necrotic enteritis in chickens, primarily in broilers. The disease is of two types: acute and subclinical. Acute form kills broilers at the age of 5–6 weeks, the time of harvest, while the subclinical form lowers nutrient intake and thus decreases body weight. The intestinal lesions are characterized by a massive infiltration of granulocytes. The disease has a significant economic impact due to low productivity.

Genetic Regulation of Virulence

In *C. perfringens*, a majority of virulence genes are located on plasmids; however, the CPE production is both plasmid or chromosomal-linked. About 5% of *C. perfringens* isolates that are responsible for food poisoning carry *cpe* gene, which usually is located on the chromosome. CPE production is linked to sporulation. Inorganic phosphate or the bile activates master sporulation regulator Spo0A, which in turn activates SigF, a global transcription factor. SigF is essential for sporulation and CPA production. The two-component *virR* and *virS* locus also regulates many virulence genes in *C. perfringens* including phospholipase C (α-toxin), perfringolysin O (θ-toxin), and collagenase (κ-toxin) as well as many nontoxic proteins.

Animal and Mammalian Cell Culture Models

CPE action has been studied using rodents (mice, rats, rabbits), intestinal loop, and cell culture (Vero, Caco-2, and MDDK (Madin-Darby canine kidney)) models. Caco-2 is the most appropriate model since it is an enterocyte-like cell, representing cells of intestinal origin. Loss of membrane integrity due to the CPE action has been assayed by the release of [86]Rb. Toxic effects of β-toxin have been studied using a human umbilical vein endothelial cell line (HUVEC), in which the toxin affects the membrane permeability resulting in the loss of essential cellular materials.

Symptoms

The CPE-mediated gastroenteritis symptoms in humans are manifested by abdominal cramp and pronounced diarrhea, appearing within 8–12 h and in most cases resolve within 12–24 h. In healthy individuals, the disease is self-limiting; however, in malnourished persons and debilitated, elderly, and very young patients, the organism may colonize, invade, and cause severe ulceration and death. Death may occur due to severe dehydration. Sudden infant death or SID is associated with this pathogen due to high levels of toxins in the blood resulting in shock and sudden death.

Necrotic enteritis or enteritis necroticans is a rare but very serious disease in humans caused by *C. perfringens* type C and type A strains. The symptoms include diarrhea, abdominal cramps, vomiting, fever, and severe bowel necrosis, which can be fatal.

Prevention and Control

In food poisoning cases, treatment and prevention include bed rest and fluid supplement. In these patients, antibiotic therapy is discouraged, since the antibiotic will inhibit resident microflora and promote *Clostridium* bloom and toxin production in the colon. However, antibiotic is recommended for another form of the diseases including enteritis necroticans, gas gangrene, and myonecrosis in humans and animals.

A majority of *C. perfringens* outbreak is associated with meat from beef or poultry. Therefore, to prevent food poisoning, care should be given in food preparation, handling, and storage. Foods should not be allowed to stand at room temperature for an extended period. Adequate cooking, holding at hot temperatures (≥ 60 °C), or rapid cooling can avoid *C. perfringens*-related food poisoning. Organic acid salts, such as 1% sodium lactate, 1% sodium acetate, or 1% buffered sodium citrate (with or without sodium diacetate), can inhibit the germination and outgrowth of *C. perfringens* during the chilling process.

Detection of *C. perfringens* or Toxins

Culture and PCR Methods

Conventional culturing methods are used to isolate *C. perfringens* from stool or food samples. Samples are first suspended in fluid thioglycolate medium and then heat shocked for 70–75 °C for 15–20 min and then enriched at 37 °C for 18–24 h. The liquid culture is then streaked onto tryptose–sulfite–cycloserine agar containing egg yolk (10%) or brain heart infusion agar with 10% sheep blood and incubated under an anaerobic condition (such as an anaerobic jar) at 37 °C for 18–24 h. Isolates of pure colonies can be obtained from the plate.

Multiplex PCR methods have been developed to detect the presence of toxin genes, enterotoxin (*cpe*), alpha (*cpa*), beta (*cpb*), epsilon (*etx*), and iota (*itx*), in *C. perfringens* isolates. The toxin typing using PCR also allows the typing of isolates based on the presence of specific toxin-encoding genes.

Immunoassays

There are two antibody-based assay methods available commercially, reverse passive latex agglutination test from Oxoid and CPE receptor-based enzyme immunoassay from Tech Lab.

Clostridium difficile

Biology

Clostridium difficile is a Gram-positive, rod-shaped, spore-forming obligate anaerobe. *C. difficile* produces two major toxins called toxin A (TcdA, toxin clostridium difficile A) and toxin B (TcdB) that are encoded by *tcdA* and *tcdB* genes, respectively, located on a 19.6 kb pathogenicity island. *C. difficile* also produces a binary toxin, CDT (*C. difficile* toxin), and the toxin has high sequence similarity to the iota toxin produced by *C. perfringens*. *C. difficile* also produces adhesion and motility factors. The genome size of *C. difficile* is about 4.3 Mb.

Clostridium difficile causes toxicoinfection. It is traditionally considered a nosocomial (hospital-acquired) opportunistic human pathogen and is responsible for *Clostridium difficile* antibiotic-associated diarrhea (CDAD) and pseudomembranous colitis. *C. difficile* infection (CDI) is also associated with young, elderly, immunocompromised, and organ transplant patients and increasingly found in patients with inflammatory bowel diseases (IBD), ulcerative colitis (UC), and Crohn's disease (CD). Antibiotic therapy for treatment of other ailments is the predisposing risk factor for CDI in addition to immunosuppressed conditions. Broad-spectrum antibiotics, such as fluoroquinolones and cephalosporins, cause greater disruption of the gut microflora and are thus more likely to promote CDI.

The CDI cases have risen dramatically since 2003, and the CDI has caused numerous fatalities in both North America and Europe. In 2007, 3875 deaths due to CDI have been reported in the USA. The CDC reported that in a single year (2011), *C. difficile* infected almost half a million American patients and approximately 29,000 deaths. Of these 15,000 deaths were directly attributed to the *C. difficile* infection.

Source

Clostridium difficile colonizes human and animal intestines that are a major source of this pathogen. About 3% of adult humans and 50% of neonates are asymptomatic carriers of *C. difficile*. *C. difficile* has been isolated from many domestic and wild animals including pigs, calves, poultry, horses, donkeys, dogs, cats, seals, snakes, rabbits, and ostrich. The organism has been also isolated from ground meat (pork, beef, and poultry) at a prevalence rate of 3–12% or higher and less frequently from raw vegetables (cucumber, onion, radish, carrot, mushrooms, and ready-to-eat salads) and milk. *C. difficile* produces spores, which can be distributed widely due to fecal contamination of soil, food processing and packaging environment, and the food. *C. difficile* association with meat animals and routine isolation of this organism from meats provide a strong

evidence for its probable involvement as a foodborne pathogen.

Virulence Factors and Pathogenesis

Clostridium difficile spores are transmitted by fecal–oral route or through contaminated foods. In the small intestine, the bile acid (taurocholate) helps germination of spores and the vegetative cells colonize the intestine. *C. difficile* produces TcdA, TcdB, CDT, and adhesion and motility factors.

Flagella help in bacterial motility and also adhesion and colonization. Suppression of flagella synthesis also affects TcdA and TcdB production and bacterial virulence. In addition, fibronectin-binding protein A (FbpA) and cell wall proteins such as Cwp66, S-layer protein A, and Spo0A contribute to the bacterial adhesion and in the biofilm formation.

The TcdA and TcdB are the two major toxins responsible for pathology. TcdA is an enterotoxin, disrupts the intestinal epithelial lining, and allows the passage of TcdB to induce cytotoxicity. TcdB causes glycosylation of RhoGTPase, disrupts tight junctions and cellular architecture, and induces cell death. TcdB also activates a cascade of proinflammatory cytokines and leukotrienes production, such as TNF, IL-6, IL-8, IL-1β, leukotrienes B4, and IFN-γ, resulting in inflammation and enhanced epithelial permeability, epithelial apoptosis, ulceration, and diarrhea.

Some strains, especially the hypervirulent ones, may express the binary toxin, CDT. CDT gene is not encoded in the pathogenicity locus. CDT is composed of two proteins, CDTa and CDTb. CDTa (53 kDa) has enzymatic activity, while the CDTb (98.8 kDa) has the adhesion function and binds to a receptor. CdtB binds to LSR (lipolysis-stimulated lipoprotein receptor), forms pores in the endosome, and facilitates the transfer of CDTa to the cytosol. CDTa is an ADP-ribosyltransferase that ribosylates actin protein in the eukaryotic cells, affecting the cell structure. Actin depolymerization helps the formation of microtubule-like structures that facilitate further bacterial adherence.

Symptoms

The clinical symptoms and severity of the disease vary among the patients, which may include (1) *C. difficile* diarrhea, (2) *C. difficile* colitis, (3) pseudomembranous colitis, and (4) fulminant colitis.

(1) *C. difficile* diarrhea is mild self-limiting and non-bloody. Symptoms begin to appear shortly after initiation of the antibiotic therapy and cease when the antibiotic therapy is stopped. (2) *C. difficile* colitis presents a severe form of CDI without pseudomembrane formation. The symptoms include nausea, anorexia, severe abdominal cramp, fever, malaise, and high-volume watery diarrhea with traces of blood in the stool. (3) The pseudomembranous colitis presents a severe form of CDI, associated with bloody diarrhea, fever, and abdominal pain. About 50% of the CDI patients show pseudo-membranous colitis, due to the disruption of cytoskeletal structure, ulceration, serum proteins leakage, inflammatory cell accumulation, and mucus forming plaques on the mucosa. (4) Fulminant colitis is associated with life-threatening toxic megacolon and intestinal perforation. About 3% of the patients account for the most serious form of CDI, i.e., the fulminant colitis characterized by systemic inflammatory syndromes including diffuse abdominal pain, with or without diarrhea, high fever, chills, hypotension, tachypnea, and marked leukocytosis.

Diagnosis and Detection of *C. difficile*

The endoscopy (colonoscopy) is used for examination of pseudomembrane formation, but it is difficult to distinguish from IBD-induced colitis. Stool culture for isolation of *C. difficile* cells followed by analysis of toxin production is the most reliable method for diagnosis. Enzyme immunoassay specific for toxins A (TcdA) and B (TcdB) and PCR methods are used as diagnosis and detection tools. For food sample testing, culture methods employing a suitable enrichment broth and the solid agar media supplemented with antibiotics, cefoxitin and cycloserine or moxalactam and norfloxacin are used for pathogen isolation.

Treatment and Prevention

Metronidazole (500 mg three times daily orally for 10–14 days) is recommended for treatment of mild-to-moderate cases of CDI. Fidaxomicin at 200 mg twice daily in adults is also approved for CDI treatment. Vancomycin (125 mg, four times a day orally) is recommended for more severe cases. Vancomycin followed by rifaximin (400 mg, twice daily for 10–14 days) is successful in treating recurrent CDI. Oral administration of probiotics such as *Saccharomyces boulardii* and *Lactobacillus* has been found to be effective, but results are inconsistent. Fecal bacteriotherapy to repopulate the patient gut with healthy microbiota through nasogastric or rectal route is highly efficacious for recurrent CDI. The surgical removal of a section of infected intestine is also used to treat severe and recurrent CDI patients.

To prevent CDI, prescribing practices of antimicrobial drugs such as the type of antibiotics, frequency, and duration of therapy should be revised. The patients should be secluded in a separate room, and healthcare professionals should take precautions from the unintentional spread of the pathogens throughout the hospital. Disinfection of hospital facilities using sporicidal agents such as the chlorine-based disinfectants (sodium hypochlorite) and high concentration of hydrogen peroxides should be practiced. Ammonium- and alcohol-based sanitizers are effective against the vegetative cells, but this treatment may facilitate sporulation thus is discouraged.

Summary

Members of the genus *Clostridium* cause a variety of diseases in humans and animals, sometimes with fatal consequences. These organisms are anaerobic spore-forming rod-shaped bacteria and mostly associated with soil and sediments.

Three species, *Clostridium botulinum*, *C. perfringens* and *C. difficile*, have a significant importance because these pathogens are responsible for neuroparalytic botulism (intoxication), food poisoning (toxicoinfection), and antibiotic-associated diarrhea and pseudomembranous colitis (infection), diseases, respectively. *Clostridium botulinum* strains are grouped into proteolytic and non-proteolytic due to their ability to produce proteases. *C. botulinum* produces eight antigenically distinct botulinum toxins (A, B, C, D, E, F, G, and H). In foodborne botulism, the botulinum toxin is produced in the food during anaerobic growth. Botulinum toxin is an A–B-type toxin with a zinc-dependent endopeptidase activity. It cleaves SNARE protein complex, which is responsible for the release of neurotransmitter, acetylcholine, from the synaptic vesicles into the neuromuscular junction for transmission of nerve impulse. Lack of acetylcholine release impedes nerve impulse propagation resulting in the onset of flaccid paralysis. The symptoms appear as early as 2 h after ingestion of toxin, and the severity and progression of the disease depend on the amount of toxins being ingested. Early medical intervention involves administration of antibotulinal antisera. *C. perfringens* causes food poisoning, necrotic enteritis, gas gangrene, myonecrosis, and toxemia. It produces at least 20 different toxins and causes toxicoinfection. There are five types of *C. perfringens* (A, B, C, D, and E), classified based on the production of four types of extracellular toxins: alpha (α), beta (β), epsilon (ε), and iota (ι). *C. perfringens* type A strain is generally associated with the foodborne disease. After consumption of vegetative cells, the bacterium begins to sporulate as it encounters acidic pH of the stomach. The enterotoxin (CPE) is produced during sporulation. The enterotoxin binds to the claudin receptor in the tight junction (TJ) and forms a large protein complex with other membrane proteins to form a pore in the membrane that alters the membrane permeability to cause Ca^{2+} influx and fluid and ion (Na^+, Cl^-) losses. CPE alters the paracellular membrane permeability and promotes diarrhea. Food poisoning is generally self-limiting requiring bed rest and fluid therapy, but in rare cases, myonecrosis and necrotic enteritis

diseases could be life-threatening thus require hospitalization and antibiotic therapy. *Clostridium difficile* is a nosocomial (hospital-acquired) human pathogen and causes *Clostridium difficile* antibiotic-associated diarrhea (CDAD) and pseudomembranous colitis. It produces toxin A (TcdA), toxin B (TcdB), and CDT which cause diarrhea and mucus membrane damage, inflammation leading to diarrhea, and sometimes life-threatening pseudomembranous colitis and megacolon and intestinal perforation. *C. difficile* association with meat animals and routine isolation from meats support its possible involvement as a foodborne pathogen. Prevention of *C. difficile* infection is possible by revising the antibiotic prescription practices such as the type of antibiotics, frequency, and duration of use. Probiotics supplement and fecal bacteriotherapy to repopulate the patient's gut with healthy microbiota, and the surgical removal of infected section of the intestine are used to control recurrent infection.

Further Readings

1. Abt, M.C., McKenney, P.T. and Pamer, E.G. (2016) *Clostridium difficile* colitis: Pathogenesis and host defence. *Nat Rev Microbiol* **14**, 609–620.
2. Arnon, S.S., Schechter, R., Inglesby, T.V., Henderson, D.A., Bartlett, J.G., Ascher, M.S., Eitzen, E., Fine, A.D., Hauer, J., Layton, M., Lillibridge, S., Osterholm, M.T., O'Toole, T., Parker, G., Perl, T.M., Russell, P.K., Swerdlow, D.L. and Tonat, K. (2001) Botulinum toxin as a biological weapon: medical and public health management. *JAMA* **285**, 1059–1070.
3. Barash, J.R. and Arnon, S.S. (2014) A novel strain of *Clostridium botulinum* that produces Type B and Type H botulinum toxins. *J Infect Dis* **209**, 183–191.
4. Barth, H. and Aktories, K. (2011) New insights into the mode of action of the actin ADP-ribosylating virulence factors *Salmonella enterica* SpvB and *Clostridium botulinum* C2 toxin. *Eur J Cell Biol* **90**, 944–950.
5. Barth, H., Fischer, S., Möglich, A. and Förtsch, C. (2015) Clostridial C3 toxins target monocytes/macrophages and modulate their functions. *Front Immunol* **6**, 339.
6. Bartlett, J. G., Chang, T. W., Gurwith, M., Gorbach, S. L. and Onderdonk, A. B. (1978) Antibiotic-associated pseudomembranous colitis due to toxin-producing *Clostridia*. *N Engl J Med* **298**, 531–534.
7. Böhnel, H. and Gessler, F. (2005) Botulinum toxins – cause of botulism and systemic diseases? *Vet Res Commun* **29**, 313.

8. Bokori-Brown, M., Savva, C.G., da Costa, S.P.F., Naylor, C.E., Basak, A.K. and Titball, R.W. (2011) Molecular basis of toxicity of *Clostridium perfringens* epsilon toxin. *FEBS J* **278**, 4589–4601.

9. Bruggemann, H. (2005) Genomics of clostridial pathogens: implication of extrachromosomal elements in pathogenicity. *Curr Opin Microbiol* **8**, 601.

10. Brynestad, S. and Granum, P.E. (2002) *Clostridium perfringens* and foodborne infections. *Int J Food Microbiol* **74**, 195–202.

11. Freedman, J., Shrestha, A. and McClane, B. (2016) *Clostridium perfringens* enterotoxin: Action, genetics, and translational applications. *Toxins* **8**, 73.

12. García, S. and Heredia, N. (2011) *Clostridium perfringens*: A dynamic foodborne pathogen. *Food Bioprocess Technol* **4**, 624–630.

13. Gerding, D.N., Johnson, S., Rupnik, M. and Aktories, K. (2014) *Clostridium difficile* binary toxin CDT: mechanism, epidemiology, and potential clinical importance. *Gut Microbes* **5**, 15–27.

14. Gorbach, S.L. (2014) John G. Bartlett: Contributions to the Discovery of *Clostridium difficile* Antibiotic-Associated Diarrhea. *Clin Infect Dis* **59**, S66–S70.

15. Johnson, E.A. and Bradshaw, M. (2001) *Clostridium botulinum* and its neurotoxins: a metabolic and cellular perspective. *Toxicon* **39**, 1703–1722.

16. Li, J., Uzal, F. and McClane, B. (2016) *Clostridium perfringens* sialidases: Potential contributors to intestinal pathogenesis and therapeutic targets. *Toxins* **8**, 341.

17. Lindström, M., Heikinheimo, A., Lahti, P. and Korkeala, H. (2011) Novel insights into the epidemiology of *Clostridium perfringens* type A food poisoning. *Food Microbiol* **28**, 192–198.

18. Lindstrom, M. and Korkeala, H. (2006) Laboratory diagnostics of botulism. *Clin Microbiol Rev* **19**, 298–314.

19. Lucey, B.P. and Hutchins, G.M. (2004) William H. Welch, MD, and the discovery of *Bacillus welchii*. *Arch Pathol Lab Med* **128**, 1193–1195.

20. McClane, B.A. and Chakrabarti, G. (2004) New insights into the cytotoxic mechanisms of *Clostridium perfringens* enterotoxin. *Anaerobe* **10**, 107.

21. Mitchell, L.A. and Koval, M. (2010) Specificity of interaction between *Clostridium perfringens* enterotoxin and claudin-family tight junction proteins. *Toxins* **2**, 1595–1611.

22. Novak, J., Peck, M., Juneja, V. and Johnson, E. (2005) *Clostridium botulinum* and *Clostridium perfringens*. In *Foodborne Pathogens: Microbiology and Molecular Biology* eds. Fratamico, P., Bhunia, A. and Smith, J. pp.383–407. Norfolk, UK: Caister Academic Press.

23. O'Horo, J.C., Jindai, K., Kunzer, B. and Safdar, N. (2014) Treatment of recurrent *Clostridium difficile* infection: a systematic review. *Infection* **42**, 43–59.

24. Peck, M.W., Stringer, S.C. and Carter, A.T. (2011) *Clostridium botulinum* in the post-genomic era. *Food Microbiol* **28**, 183–191.

25. Rossetto, O., Pirazzini, M. and Montecucco, C. (2014) Botulinum neurotoxins: genetic, structural and mechanistic insights. *Nat Rev Microbiol* **12**, 535–549.

26. Rupnik, M. and Songer, J.G. (2010) *Clostridium difficile*: Its potential as a source of foodborne disease. *Adv Food Nutr Res* **60**, 53–66.

27. Sakurai, J., Nagahama, M., Oda, M., Tsuge, H. and Kobayashi, K. (2009) *Clostridium perfringens* Iota-toxin: Structure and function. *Toxins* **1**, 208–228.

28. Sinh, P., Barrett, T.A. and Yun, L. (2011) *Clostridium difficile* infection and inflammatory bowel disease: A review. *Gastroenterol Res Practice* **2011**, 11.

29. Sobel, J. (2005) Botulism. *Clin Infect Dis* **41**, 1167–1173.

30. Songer, J.G. (2010) Clostridia as agents of zoonotic disease. *Vet Microbiol* **140**, 399–404.

31. Ting, P.T. and Freiman, A. (2004) The story of *Clostridium botulinum*: from food poisoning to Botox. *Clin Med* **4**, 258–261.

32. Uzal, F.A., Freedman, J.C., Shrestha, A., Theoret, J.R., Garcia, J., Awad, M.M., Adams, V., Moore, R.J., Rood, J.I. and McClane, B.A. (2014) Towards an understanding of the role of *Clostridium perfringens* toxins in human and animal disease. *Future Microbiol* **9**, 361–377.

33. Wise, M.G. and Siragusa, G.R. (2005) Quantitative detection of *Clostridium perfringens* in the broiler fowl gastrointestinal tract by real-time PCR. *Appl Environ Microbiol* **71**, 3911–3916.

Introduction

In 1926, E.G.D. Murray reported the isolation of *Bacterium monocytogenes* from a rabbit, and in 1940, Harvey Pirie changed the name of the bacterium to *Listeria monocytogenes* in honor of Sir Joseph Lister. Historically, this pathogen has been recognized as an animal pathogen, which predominantly infects ruminants (cattle, sheep, and goats), pigs, dogs, and cats. The disease in animals is often referred to as "circling disease," since the affected animal walks in a circle, exhibits uncoordinated posture, and is unable to stand without support. In ruminants, *Listeria* also causes keratoconjunctivitis and uveitis (silage eye), and these strains are generally resistant to lysozyme present in tears. In the 1950s, human infections involving neonates with sepsis and meningitis were reported. In the late 1970s and in the early 1980s, *Listeria monocytogenes* emerged as a foodborne pathogen, infecting humans through contaminated foods in North America. Now, it is well recognized globally as a causative agent for a rare but invasive fatal systemic disease, called listeriosis, affecting primarily the immunocompromised populations: pregnant women, neonates, human immunodeficiency virus (HIV)-infected patients, and organ transplant recipients. In rare occasions, *L. monocytogenes* causes gastroenteritis in immunocompetent individuals. Being a facultative intracellular pathogen, *L. monocytogenes* is used as a good model to study the intracellular parasitism and the host immune response. Furthermore, the ability of *Listeria* to invade and maintain intracellular lifestyle became a subject of much investigation for its use in delivering foreign genes as a vaccine for other diseases including cancers.

Classification

The genus *Listeria* has 17 species and many of them are added to the list in recent years. Six species show high genetic relatedness; hence they are clustered in *Listeria* sensu stricto (i.e., in the narrow or strict sense) group: *L. monocytogenes*, *L. ivanovii*, *L. seeligeri*, *L. welshimeri*, *L. innocua*, and *L. marthii*. The remaining *Listeria* spp. are termed *Listeria* sensu lato, which are phylogenetically divergent from the species of *Listeria* sensu stricto: *L. floridensis*, *L. aquatic*, *L. cornellensis*, *L. riparia*, *L. grandensis*, *L. booriae*, *L. rocourtiae*, *L. newyorkensis*, *L. weihenstephanensis*, *L. fleischmannii*, and *L. grayi*. Among these, *L. monocytogenes* is pathogenic to humans and ruminants, while *L. ivanovii* is pathogenic to ruminants; however, in rare occasions, it may infect humans causing foodborne outbreaks. All other *Listeria* species are considered nonpathogenic. *L. ivanovii* has two subspecies: *L. ivanovii* subspecie *ivanovii* and *L. ivanovii* subspecies *londoniensis*. *L. seeligeri* is a nonpathogenic species, but it possesses a part of the virulence gene cluster or the pathogenicity island, which is present in both *L. monocytogenes* and *L. ivanovii*.

Listeria monocytogenes is grouped into 13 serotypes based on the O-antigenic patterns: 1/2a, 1/2b, 1/2c, 3a, 3b, 3c, 4a, 4b, 4c, 4d, 4e, 4ab, and 7. The serotypes 1/2a, 1/2b, and 4b are responsible for 98% of the foodborne outbreaks, and serotype 4b is considered the most virulent. A subpopulation of serotype 4b also exists, and one or more of those subtypes are considered epidemic clones. *L. monocytogenes* are grouped into three lineages based on their ribopatterns and their association with outbreaks. Lineage I has the highest pathogenic potential and the serotypes belonging to this group are involved in most epidemic outbreaks. Lineage II has intermediate pathogenic potential and possibly responsible for sporadic outbreaks while the lineage III has a low pathogenic risk and rarely causes human infection (Table 13.1). Based on the

Table 13.1 Classification of *Listeria monocytogenes* based on genomic fingerprint patterns and association with epidemic outbreaks

Groups	Outbreaks	Pathogenic potential	Predominant serotypes
Lineage I	Epidemic clones and responsible for most outbreaks	High	1/2b, 3b, 4b, 4d, 4e
Lineage II	Sporadic listeriosis cases	Medium	1/2a, 1/2c, 3c, 3a
Lineage III	Rarely cause human disease	Low	4a, 4c
IIIA (Rham +ve)			4a (avirulent) and 4c (virulent)
IIIB (Rham −ve)			Virulent nonserotype 4a and nonserotype 4c
IIIC (Rham −ve)			Virulent 4c

Rham Rhamnose fermentation property

flagellar (H) antigen types, *Listeria* serovars are classified into A, B, C, D, and E, and this typing scheme has rarely been used during the outbreak investigation.

Biology

Listeria species are Gram-positive, rod-shaped, and nonspore-forming bacteria. *Listeria* cells are 1–2 μm long and exist as single or double cells. Occasionally, *Listeria* may display long chains depending on the growth conditions, temperatures, and exposure to stress factors. *Listeria* is ubiquitous in nature, forms biofilm, and survives in extreme environments, including a broad pH range (4.1–9.6), high salt (10% or higher), and in the presence of antimicrobial agents. They are psychrophilic and grow at a wide temperature range (1–45 °C), including the refrigeration temperature. *Listeria* is catalase positive, oxidase negative, and esculin hydrolase positive.

Listeria expresses peritrichous flagella and is motile, exhibiting typical tumbling motion when examined under a microscope. Flagellum expression is temperature-dependent with a maximum observed at 20–30 °C, least at 37 °C or higher. The flagellum consists of a 29 kDa subunit flagellin protein, encoded by *flaA* gene, and its transcription is regulated by *flaR*. Flagellin is a dimer consisting of two identical proteins with the same molecular mass and assembles on the bacterial surface. The globular subunits are packed around the long axis of the flagellum to provide a stable structure. It remains unclear whether flagella are important for pathogenesis since their expression is low at normal body temperature (37 °C). Flagellin is highly immunogenic and the anti-flagellar antibodies have been used for serotyping and for bacterial detection.

Some *Listeria* species secrete hemolysins. *L. monocytogenes*, *L. ivanovii*, and *L. seeligeri* are hemolytic and produce a zone of β-hemolysis on blood agar. *L. ivanovii* produces a strong hemolysis containing two distinct zones. A clear zone nearest to the colony is due to the action of hemolysin called ivanolysin O (ILO), and the partial zone of lysis in the periphery is caused by sphingomyelinase and lecithinase, while *L. monocytogenes* or *L. seeligeri* produces a less intense zone of hemolysis by listeriolysin O (LLO), which may not be readily visible. These three hemolytic bacteria, however, can be differentiated by using CAMP (Christie, Atkinson, Munch–Peterson) test. In this test, *L. monocytogenes* (or *L. seeligeri*) culture is streaked on a blood agar plate, perpendicular to the line of a hemolysin-producing *Staphylococcus aureus* strain. A clear zone of hemolysis resembling an arrowhead is produced at the junction of two cultures after an incubation period of 8–18 h. Similarly, *L. ivanovii* produces a zone of hemolysis when grown in the presence of *Rhodococcus equi* in a similar fashion to *S. aureus* on the blood agar plate.

LLO is produced in low concentrations at about 20–200 hemolysin units (HU) ml^{-1} in conventional broth media (one HU equals 1 ng of purified toxin). However, a high level (1500 HU ml^{-1}) can be achieved after growth in a resin (Chelex)-treated broth. LLO production is also regulated by growth temperature, yielding higher amounts in the presence of a constant glucose level of 0.2% at 37 °C than at 26 °C.

Fermentation properties of carbohydrates such as rhamnose, xylose, and mannitol are unique among the members of the genus *Listeria* and are used for identification of the species (Table 13.2). Pathogenic properties of *Listeria* are examined on in vitro cell culture model for invasion and cell-to-cell spread and mouse for pathogenicity (see Chap. 5).

Source

Soil, decaying vegetation, silage, sewage, and animal intestine are natural habitats of *Listeria*. Therefore, ready-to-eat (RTE) foods of animal or plant origin are the primary source, such as meat, dairy products, and fruits and vegetables. Post-processing contamination of products is a major concern since those products are consumed without further cooking. Foods that are involved in outbreaks include lunchmeats, hotdogs, pâté, salad, smoked fish, milk, soft cheese made from unpasteurized milk, ice cream, cantaloupe,

frozen vegetables, and caramel apple. *Listeria*-associated gastroenteritis has been linked to rice salad, cold corn and tuna salad, corned beef and ham, cold smoked trout, and cheese. There is a "zero tolerance" policy for *L. monocytogenes* in ready-to-eat foods in the USA. Other countries like Canada and some European countries allow a limit of 100 cfu in 25 g food, provided the food environment does not support bacterial multiplication during the period of products' shelf life. The US FDA has categorized RTE foods based on the risk. The very-high-risk foods include the deli meats and frankfurters (unheated); and the high-risk foods are high-fat-containing dairy products such as soft unripened cheese, unpasteurized fluid milk, pâté and meat spreads, and smoked seafood. The medium-risk products are pasteurized fluid milk, fresh soft cheese, semisoft cheese, soft ripened cheese, cooked RTE crustaceans, deli salads, dry/semidry fermented sausages, frankfurters (reheated), fruits, and vegetables; and the very low-risk products are cultured milk products, hard cheese, ice cream and frozen dairy products, and processed cheese.

About 1500 Americans, especially the immunocompromised ones, suffer from invasive listeriosis each year. The mortality rate is highest of all foodborne pathogens and is usually 20–30%, and as high as 50%. Annually, about 255 people succumb to death from listeriosis in the USA. Strict *Listeria* regulations in food often result in food recalls that cost the food industry millions of dollars and overall economic cost due to outbreaks; human lives and product recalls are about 2.8 billion dollars annually. Examples of some of the outbreaks in the USA include deli turkey meat in the year 2000, which sickened 30, killed 4, and miscarried 3; in 2000–2001, Mexican-style soft cheese (Queso Fresco) made from unpasteurized milk caused 12 illnesses and 5 miscarriages; in 2002, ready-to-eat turkey deli meat caused 54 illnesses, 8 deaths, and 3 stillbirths. In 2009, in Canada, tainted meat products infected 57 and killed 22. In Europe, in 2010, Quargel sour milk curd cheese infected 27 and killed 8; in the USA, in 2011, cantaloupe infected 147 and killed 33; and in 2015, caramel apple infected 35 and killed 7, and ice cream infected 10 and killed 3.

Table 13.2 Biochemical properties of *Listeria* species

Characteristics	L. monocytogenes	L. innocua	L. ivanovii	L. welshimeri	L. seeligeri	L. marthii	L. rocourtiae	L. grayi
β-hemolysin	+	–	+	–	+	–	–	–
CAMP (*S. aureus*)	+	–	–	–	+	–	–	–
CAMP (*R. equi*)	–	–	+	–	–	–	–	–
Rhamnose	+	v	–	–	–	v	v	–
Xylose	–	–	+	+	+	–	+	
Mannitol (α-methyl-D-mannoside)	+	+	–	–	–	–	+	
Cytotoxicity	+	–	+	–	–	–	–	–
Invasion assay	+	–	+	–	–	–	–	–
Mouse virulence	+	–	+	–	–	–	–	–

CAMP Christie, Atkinson, Munch–Peterson hemolysis assay; + positive; – negative; *v* variable

Disease

Listeria monocytogenes is an intracellular pathogen and infects immunocompetent as well as immunosuppressed populations. In immunocompetent individuals, this organism can cause rare gastroenteritis and fever. However, in immunocompromised individuals, the disease is highly invasive. The infective dose is not known; however, it is estimated to be between 100 cfu and 10^6 cfu depending on the immunological status of the host. The incubation period varies from about 3 days to 3 months depending on the immune status of the host and the number of bacterial cells ingested.

In ruminant animals (bovine and ovine species), *Listeria* causes keratoconjunctivitis, referred to as silage eye (uveitis), and abortion in sheep. *L. monocytogenes* is found in healthy and infected cattle eyes, and the persistence in eye depends on the bacterial resistance to lysozyme. The bovine tear may carry lysozyme up to 580 μg/ml, sheep up to 600 μg/ml, and human 2 mg/ml. Resistance to lysozyme is largely due to the synthesis of peptidoglycan-modifying enzymes by *L. monocytogenes* encoded in *pgdA*, *pbpX*, and *oatA* genes. These enzymes modify the sugar backbone of peptidoglycan making the enzyme ineffective.

In humans, *Listeria* causes three forms of the disease: (1) gastrointestinal form, (2) systemic listeriosis, and (3) abortion and neonatal listeriosis.

Gastrointestinal Form

Since 1993, *L. monocytogenes* was responsible for seven outbreaks of gastroenteritis worldwide. The serotype 1/2a, 1/2b, and 4b have been implicated in those outbreaks. The foods involved were rice salad, cold tuna salad, cheese, trout, and chocolate milk. The infectious dose associated with those outbreaks was very high at 10^6–10^8 cfu; however, dosage as low as 100 cfu g^{-1} and as high as 10^{11} cfu g^{-1} has been associated with the gastrointestinal form of the disease. Adults were more susceptible than the children were, and the individuals who were on gastric acid suppressive medications were at the greatest risk. The incubation period was less than 24 h but varied from 6 h to 10 days. The exact mechanism of gastroenteritis is not known; however, it appears that the organism causes damage to the absorptive villi affecting absorption of nutrients and promoting fluid secretion. The gastroenteritis symptoms are characterized by fever, headache, nausea, vomiting, abdominal pain, and diarrhea. Diarrhea is non-bloody but watery.

Systemic Listeriosis

Listeria monocytogenes causes invasive listeriosis, a rare but fatal disease primarily affecting immunocompromised individuals such as young, old (above 60 years of age), pregnant, and immunologically challenged hosts (YOPI). High-risk individuals include AIDS (acquired immunodeficiency syndrome) patients, cancer patients receiving chemotherapy, organ transplant patients, diabetic, alcoholics, and patients with cardiovascular diseases. The organism may colonize for a short duration in the intestine and passes through the intestinal barrier reaching the blood circulation and the lymphatic system. A majority of the organism is distributed in the liver (90%), spleen (10%), and the mesenteric lymph nodes (MLN) within 24 h. After a brief bacteremic phase, *Listeria* crosses the blood–brain barrier and infects the brain causing meningitis and encephalitis. In pregnant women, *Listeria* crosses the placental barrier and infects the fetus (see below for detail) (Fig. 13.1). Systemic listeriosis is characterized by fever, headache, malaise, septicemia, meningitis, brainstem encephalitis, ataxia, bacteremia, and liver abscess. Sometimes *L. monocytogenes* may be associated with Crohn's disease.

Abortion and Neonatal Listeriosis

In pregnant women, *L. monocytogenes* can cause complications late in the third trimester of pregnancy resulting in stillbirth, or premature birth of

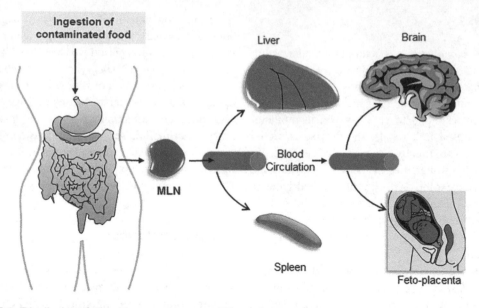

Fig. 13.1 Distribution of *Listeria monocytogenes* to different tissues and organs

an infected fetus. Pregnancy is associated with suppression of cell-mediated immunity to protect the fetus from rejection, which renders pregnant women highly susceptible to listeriosis. During pregnancy, systemic listeriosis often results in the intrauterine infection. *L. monocytogenes* infects trophoblastic cells in the placenta then rapidly disseminates to trophoblastic structures (syncytiotrophoblast) in the villous core in the labyrinthine zone and crosses the placental barrier. Inflammation at the site is characterized by infiltration of polymorphonuclear cells. Although the intracellular lifestyle of *Listeria* has been well understood, relatively little is known about how or why this organism infects the fetoplacental unit. It has been shown that *L. monocytogenes* moves from maternal organs to the placenta and then returns to the maternal organs. The placenta is relatively protected from *L. monocytogenes* colonization; however, if a single bacterium enters the placenta and finds a niche, it will colonize the placenta and act as a source of infection for the mother. It is possible that preterm labor or spontaneous abortion is a protective mechanism for the mother. This allows expulsion of the infected fetoplacental unit, removing the source of infection and providing the mother with a greater opportunity for survival. The symptoms

of infection begin as flu-like and gradually progress to a severe headache. Decreased fetal movement and early labor and premature birth, spontaneous abortion, stillbirth, or neonatal infections are common.

Two forms of neonatal listeriosis are reported: early- and late-onset forms. In the early-onset form (1.5 days), the organism is acquired in the uterus, and the infection is widespread exhibiting very high mortality rate. In the late forms (14 days), the infection is possibly acquired during birth from the vaginal tract or from the environment. In this form, meningitis is the predominant clinical feature. The symptoms of neonatal listeriosis include respiratory distress, pneumonia, shortness of breath, hyperexcitability, conjunctivitis, rash, vomiting, cramps, and hypo- or hyperthermia. The mortality rate in neonatal listeriosis is about 36%.

Mechanism of Pathogenesis

The pathogenic mechanism of *L. monocytogenes* is a complex process, which is subdivided into two phases: intestinal and systemic. The intestinal phase of infection involves initial bacterial colonization in the intestine and consequent translocation through the mucosal barrier to blood circulation or to the lymphatic system for systemic dissemination. During systemic spread, dendritic cells or macrophages transport the organism to the liver, spleen, lymph nodes, brain, and to the placenta (in pregnant women). Many surface-associated and secreted proteins (Table 13.3; Fig. 13.2) are critical for bacterial persistence in the intestinal tract, to enter the host cells for intracellular movement and cell-to-cell spread and to evade the host immune system.

A 9 kb pathogenicity island (PAI) carries the majority of the virulence genes in a cluster comprised of *prfA–plcA–hlyA–mpl–actA–plcB*. These genes, together with internalin genes (*inlA* and *inlB*), *hpt*, and *bilE*, are regulated by the first gene *prfA* (positive regulatory factor A). In addition, expressions of many of these genes that are essential during bacterial transit through the gastrointestinal tract such as *gadA*, *hpt*, and *bilE* are coregulated by an alternate sigma factor, sigma B (σ^B).

Intestinal Phase of Infection and Systemic Spread

Most listeriosis cases are foodborne; thus the bacterial passage through the stomach and the small intestine is a prerequisite for infection. The glutamate decarboxylase (GAD) system of *L. monocytogenes* protects the bacterium from

Table 13.3 Major virulence proteins in *Listeria monocytogenes*

Virulence factors	Size (kDa)	Host cell receptor	Function
Protein regulatory factor (PrfA)	27	–	Regulation of many virulence protein expression
Adhesion proteins			
Listeria adhesion protein (LAP)	104	Hsp60 (chaperone protein)	Adhesion to intestinal epithelial cells; disruption of intestinal epithelial barrier
Autolysin amidase (Ami)	99	Unknown	Adhesion to host cells
Fibronectin-binding protein (FbpA)	55.3, 48.6, 46.7, 42.4, and 26.8 kDa	Fibronectin	Adhesion to cells and also serve as chaperone to stabilize and secretion of LLO, InlB
Internalin J (InlJ)	92	Mucus	Adhesion to epithelial cells and binds to MUC2 (mucus)
LapB	180	–	Adhesion and invasion into host cells
Invasion			
Internalin (InlA)	88	E-cadherin (tight junction protein)	Responsible for invasion into intestinal epithelial cells and placenta during pregnancy
Internalin B (InlB)	65	Met (tyrosine kinase), gC1q-R/p32	Entry into hepatocytes and hepatic phase of infection
Virulence invasion protein (Vip)	43	Gp96 (chaperone protein)	Invasion of epithelial cells
Lysis of vacuole			
Listeriolysin (LLO) *hlyA*	58–60	Cholesterol	A hemolysin aids in bacterial escape from phagosome inside the cell
Phospholipase (*plcA*–PI-PLC; *plcB*–PC-PLC)	29–33	–	Lyses of vacuole membrane
Cell-to-cell spread			
Actin polymerization protein (ActA)	90	–	Nucleation of actin tail for bacterial movement inside the cytoplasm
PC-PLC			Lyses of vacuole membrane
Metalloprotease (Mpl)	29	–	Helps synthesis of PLC
Miscellaneous			
p60 (cell wall hydrolase)	60	–	Adhesion/invasion
Bile salt hydrolase (BSH)	36	–	Survival in the gut

Fig. 13.2 (**a**) Approximate location of *L. monocytogenes* virulence genes in the chromosome (3 Mb) and the location of gene products in the extracellular milieu, cell wall (*dark shade*), and cytoplasmic membrane (*light shade*) (Schematic is based on Dussurget, O. et al. 2004. Annu. Rev. Microbiol. 58:587–610). Panel (**b**) is showing virulence gene clusters in 9 kb pathogenicity island (PAI) and the internalin gene family. *Open circles* represent promoters

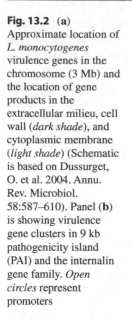
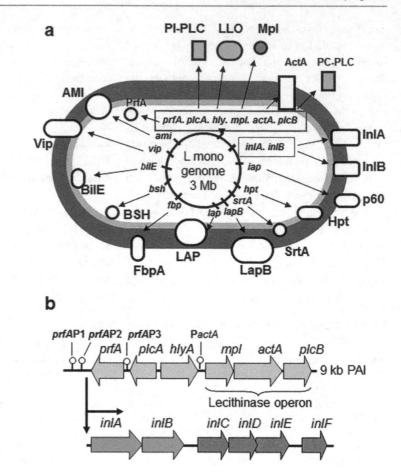

gastric acid, and food particles help neutralize pH to ensure safe passage to the small intestine. Bile salt hydrolase (BSH) and the bile exclusion system (BilE) protect bacteria against bile salts, and OpuC provides osmotolerance. The organism crosses the intestinal epithelial barrier by four possible pathways: (1) through M cells overlying Peyer's patches, (2) dendritic cells sampling the lumen of the intestine, (3) transepithelial translocation by disrupting epithelial barrier by LAP-dependent mechanism, and (4) active invasion of epithelial cells by InlA−/InlB-dependent mechanism (Fig. 13.3). Dendritic cells (DC) and/or macrophages located beneath the epithelial cell lining, i.e., in the lamina propria, can process and present antigen to helper T cells (CD_4^+) or cytotoxic T cells (CD_8^+) for an adaptive immune response. DC can transport bacteria to the extraintestinal sites such as lymph nodes, the liver, the spleen, the brain, and the fetoplacental junction in pregnant women (Fig. 13.1).

Bacterial translocation through naturally phagocytic M cells and DC-assisted pathways is a passive system and common to many enteric pathogens (see Chap. 4), while the intracellular (transcellular) and intercellular (paracellular) pathways through enterocytes are active processes where bacterial interaction with the host cell receptors initiates a cascade of signaling events that modify the cellular architecture to promote bacterial passage into the lamina propria (Fig. 13.4).

The cellular mechanism of *Listeria* adhesion and invasion involves six major steps: (1) attachment, (2) invasion, (3) lysis of vacuole (phagosome), (4) intracellular growth, (5) cell-to-cell spread, and (6) lysis of double-membrane vacuole (Fig. 13.5).

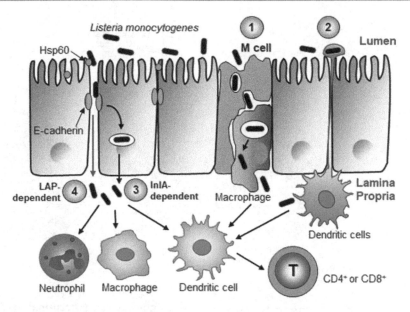

Fig. 13.3 *Listeria monocytogenes* translocation through intestinal cell lining (barrier). Four possible pathways are proposed: (1) translocation through M cells, (2) entrapment by dendritic cells, (3) active invasion through the epithelial cell by internalin A (InlA)–E-cadherin pathway, and (4) LAP-mediated transepithelial pathway. Immune cells (macrophage, neutrophil, DC) interact with bacteria for innate and adaptive responses

Fig. 13.4 Schematics showing *Listeria monocytogenes* adhesion and invasion factors and their corresponding host cell receptors. Interaction of bacterial proteins with the host cell receptor triggers signaling cascade to facilitate bacterial crossing of epithelial barrier by intracellular or paracellular route

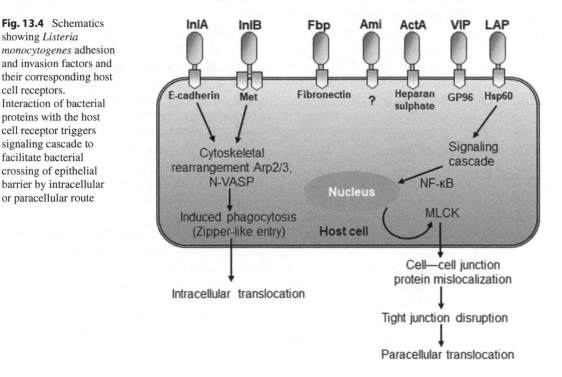

Fig. 13.5 Mechanisms of *Listeria monocytogenes*: (1) adhesion, (2) invasion, (3) lysis of vacuole, (4) intracellular multiplication, (5) cell-to-cell spread, and (6) lysis of double-membrane vacuole

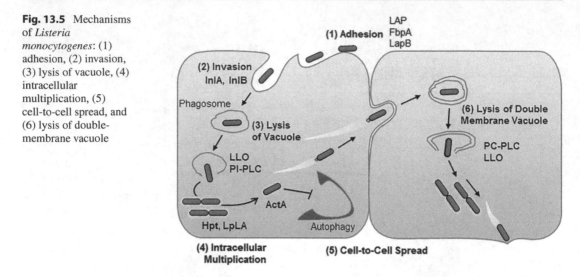

Attachment

Adhesion to the mucosal membrane in the gastro-intestinal tract is critical in the initial phase of *L. monocytogenes* infection. Several adhesion factors have been identified including *Listeria* adhesion protein (LAP), LapB, autolysin amidase (Ami), fibronectin-binding protein (FbpA), Cysteine transport protein (Ctap), P60, and lipoteichoic acid (LTA) (Table 13.3; Fig. 13.4); however, only a few are studied in detail.

Listeria Adhesion Protein

Listeria adhesion protein (LAP) is an 866-amino acid-long polypeptide (~104 kDa), encoded by *lap* gene (*lmo1634*), which is the leading gene in an operon containing an unknown protein (*lmo1635*), an ATP-binding protein (*lmo1636*), and a membrane protein (*lmo1637*). LAP contains two major domains: the N-terminal domain containing an acetaldehyde dehydrogenase and the C-terminal domain consisting of an iron-binding alcohol dehydrogenase. LAP is a bifunctional membrane-bound enzyme and is essential for harvesting energy during growth under anaerobic condition.

It does not have an N-terminal leader sequence thus its secretion is dependent on SecA2, a protein that helps protein translocation across the cytoplasmic membrane (see Chap. 4). LAP interacts with the eukaryotic receptor, heat-shock protein 60 (Hsp60), a molecular chaperone. Normally, Hsp60 regulates folding of mitochondrial proteins and proteolytic degradation of misfolded and damaged proteins in cells. LAP displays enhanced affinity for intestinal epithelial cells and plays a major role during the intestinal phase of infection. Nutrient-limiting and anaerobic environment enhance surface expression of LAP, which promotes the LAP-mediated adhesion and bacterial translocation across the epithelial barrier during the intestinal phase of infection. LAP interaction with the host cell Hsp60 initiates a signaling cascade that involves NF-κB activation, and proinflammatory cytokines (TNF-α and IL-6) release promoting intestinal epithelial barrier dysfunction. LAP also activates cellular myosin light-chain kinase (MLCK), which in turn pulls the major epithelial tight junction proteins such as claudin-1, occludin, and adherence junction protein E-cadherin away from the cell membrane making the epithelial barrier vulnerable and allowing *Listeria* passage across the membrane to access lamina propria (Fig. 13.6).

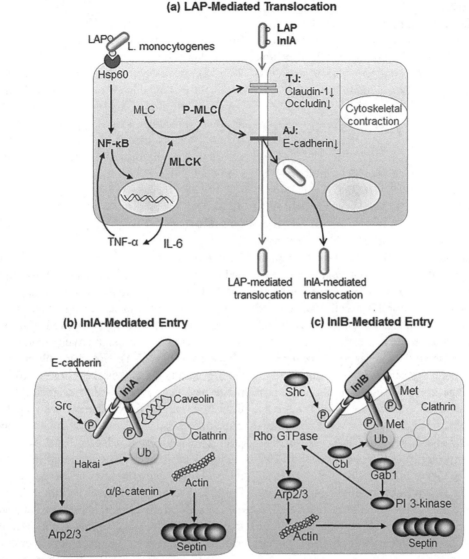

Fig. 13.6 Schematic showing (**a**) LAP-mediated epithelial tight junction compromise and bacterial translocation, (**b**) InlA and (**c**) InlB-mediated *L. monocytogenes* invasion of cells by zipper mechanism. *MLC*, Myosin light chain, *MLCK*, myosin light-chain kinase, *TJ*, tight junction, *AJ*, adherence junction (Schematics of InlA- and InlB-mediated cell entry concepts are based on Stavru et al. (2011). Immunol Rev. 240, 160–184)

Autolysin Amidase

Autolysin amidase (Ami) is an autolytic enzyme and serves as an adhesion protein. The size of Ami varies among the serotypes. In serotype 1/2a, Ami is a 917-residue-long polypeptide, whereas Ami from serotype 4b is 770 residue long. It has three domains: a 30-residue-long signal sequence; N-terminal alanine amidase, homologous to Atl autolysin of *Staphylococcus aureus*; and a C-terminal cell wall-anchoring (CWA) domain containing GW (Glycine-Tryptophan) modules. The N-terminal amidase domain is highly conserved between the two serotypes, but the C-terminal sequence is variable. Serotype 1/2a contains eight GW modules while 4b contains six GW modules. Ami uses its CWA domain to anchor to the cell wall where it exerts its autolytic action. It also promotes binding of *L. monocytogenes* to mammalian cells using the CWA domain.

P60

P60 is a major extracellular protein containing 484 amino acid residues (60 kDa) and is known as cell wall hydrolase (CWH), which assists in bacterial cell division. In the absence of P60, bacteria form a long chain and a rough colony phenotype. It was originally reported to be an invasion-associated protein (iap) and thought to be required for invasion into the mouse fibroblast cell. The gene coding for P60 is called *iap*, and its expression is controlled at the posttranscriptional level. P60 synthesis and expression are not regulated by *prfA*, and it does not appear to be directly involved in adhesion and invasion during *Listeria* infection. The *iap* gene shows strong homology with a protein with bacteriolytic or autolytic activity from *Streptococcus faecalis*. P60 is present in all *Listeria* spp. but is heterogeneous among the species. The 120 amino acid residues in the N- and C-terminal ends are conserved in all *Listeria* spp., whereas the middle part, comprising of 240 amino acids, are highly variable. The P60 from *L. monocytogenes* carries two unique sequence regions, which are absent in other *Listeria* spp. P60 production is significantly higher in cells when grown at 37 vs. 26 °C.

LapB

LapB is a 180 kDa polypeptide and helps *L. monocytogenes* adhesion to and entry into mammalian epithelial cell lines and contributes to pathogenesis during the intravenous or oral challenge of mice. The N-terminal domain possesses adhesion function that probably interacts with an unknown host cell receptor. LapB expression is regulated by PrfA.

FbpA

Five fibronectin-binding proteins (Fbp) with molecular masses of 55.3, 48.6, 46.7, 42.4, and 26.8 kDa have been identified in *L. monocytogenes*. All Fbp bind to fibronectin, a 450 kDa glycoprotein present on the epithelial cell surface. The high molecular weight FbpA is thought to be required for colonization of the mouse liver and contributes to the bacterial adherence to human epithelial cells.

Invasion

Several proteins have been implicated in adhesion and invasion including internalin A (InlA), internalin B (InlB), LpeA (lipoprotein promoting entry), Auto (autolytic protein containing a signal peptide and an N-terminal N-acetylglucosaminidase activity), Vip (virulence invasion protein), and so forth (Table 13.3; Fig. 13.4). However, only a few are studied in detail, and their roles in in vivo animal model have been established. Only a select few are discussed below. InlA interacts with epithelial cadherin (E-cadherin), a transmembrane glycoprotein located in the adherence junction of the epithelial cell, and promotes bacterial entry during intestinal and uteroplacental infection. Internalin B (InlB) interacts with tyrosine kinase Met and aids in *Listeria* invasion in hepatocytes and endothelial cells. InlA–E-cadherin and InlB–Met interactions cause receptor ubiquitination, recruitment of clathrin, rearrangement of the cellular cytoskeleton, and the pathogen uptake. *L. monocytogenes* interaction with the host cell receptors triggers signaling cascade leading to activation of phosphatidylinositol 3-kinase (PI3K), type II PI4-kinases, and mitogen-activated protein kinase (MAPK) pathways to recruit proteins for actin cytoskeletal rearrangement to promote bacterial entry by a mechanism termed "zipper like." It is proposed that the synergistic action of both InlA and InlB is needed to achieve cell invasion to various target sites of bacterial tropism. A virulence infection protein (Vip) also plays an important role during intestinal phase of infection.

Internalin A

Internalin A is a member of the internalin multigene family containing internalin A–J. The *inlA* and *inlB* are located on the same locus, while the remainders are on different loci. InlA is 800-amino acid-long (88 kDa) cell surface protein. It has a signal sequence and leucine-rich repeats (LRRs), which are critical for bacterial entry into the eukaryotic cells. The C-terminal end contains a cell wall-anchoring motif, L (Leucine) P (Proline) X (any) T (Threonine) G (Glycine), which forms cross-link with the peptidoglycan. The LPXTG sequence helps InlA to anchor to the cell wall.

InlA synthesis is regulated at the transcriptional level by both PrfA-dependent and PrfA-independent mechanisms. InlA helps *L. monocytogenes* to invade epithelial cells following interaction with the cellular receptor, E-cadherin, a transmembrane glycoprotein located in the adherence junction of epithelial cells at the basolateral side, which is inaccessible when bacteria are located in the intestinal lumen (Figs. 13.5, 13.6, and 13.7). It is proposed that InlA access to E-cadherin occurs during the villous epithelial cell extrusion (an event when the dead epithelial cells are replaced by new cells at the tip of the villi) and following exocytosis of mucus from the goblet cells. The junctional E-cadherin is exposed during each event facilitating InlA interaction and consequent *Listeria* invasion. An alternative proposed mechanism for E-cadherin accessibility involves LAP-dependent disruption of epithelial barrier to promote bacterial crossing through InlA-dependent and InlA-independent mechanisms (Fig. 13.6). In general, the InlA– E-cadherin interaction is a species-specific event, which is seen in humans, gerbils, and guinea pigs but not in mice or rats, due to an amino acid sequence variation at the position 16 in E-cadherin, where proline is substituted by glutamic acid in mice and rats.

InlA binding to E-cadherin induces its clustering into the lipid rafts in a caveolin-dependent fashion. E-cadherin undergoes Src-mediated tyrosine phosphorylation, followed by ubiquitylation via the Hakai E3 ubiquitin ligase resulting in the recruitment of clathrin and other components for actin polymerization for bacterial uptake by zipper mechanism. To promote actin polymerization, E-cadherin in the adherence junction interacts first with β-catenin, followed by α-catenin, which in turn interacts with actin and actin effectors such as Src. Src activates the Arp2/3 complex and recruits actin followed by septin (Fig. 13.6).

Internalin B

Internalin B is a 630 amino acid-containing polypeptide (~65 kDa) and is located on the surface of cells. It possesses N-terminal signal sequence and LRR repeats, and the C-terminal domain carries 3 repeats of 80-amino acid-residue-long segment starting with GW (Glycine-Tryptophan) repeats (called GW module). The GW module acts as an anchor and remains attached to the membrane lipoteichoic acid of the cell wall. InlB facilitates entry of *Listeria* to a variety of host cells. InlB binds to the cellular receptor Met, a family of tyrosine kinase. Upon binding to InlB, Met is phosphorylated and recruits adapter proteins, Cbl, Shc, and Gab1, which activate PI3K, and a small GTPase, Rac, which in turn activates actin nucleator protein Arp2/3 for actin polymerization followed by septin recruitment, which alters cytoskeletal configuration to promote

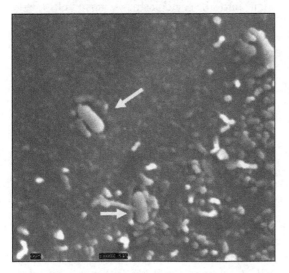

Fig. 13.7 Scanning electron microscopic photograph showing the early stage of *L. monocytogenes* invasion to Caco-2 cell (Courtesy Jennifer Wampler and Arun Bhunia)

bacterial entry by zipper mechanism (Fig. 13.6). Phosphorylated Met is also ubiquitinated by the ubiquitin ligase Cbl and recruits clathrin and other components for actin polymerization and bacterial uptake. InlB has also been shown to interact with the globular part of 33 kDa complement protein C1q (gC1q-R/p32), and this serves as a co-receptor. The GW module interacts with the gC1q-R. InlB interaction is also found to be species-specific. It does not interact with the Met receptor from guinea pigs or rabbits.

Vip

Virulence infection protein (Vip) is a surface protein carrying LPXTG motif responsible for anchoring to the cell wall peptidoglycan, an action that is mediated by the transpeptidase, sortase. Vip contains 399 amino acids (~43 kDa) with an N-terminal signal sequence and a C-terminal sorting signal. It has a proline-rich region (amino acids 268–318), and its expression in bacteria is regulated by *prfA*. Vip binds to the cellular receptor GP96, a chaperone protein found in the endoplasmic reticulum (ER). Vip is an invasion protein which is functional only in select cell lines, including enterocyte-like Caco-2 and mouse fibroblast L2071, but not in Vero cells. It has been suggested that Vip is responsible for invasion during intestinal phase of infection as well as in the later stages of infection.

LpeA

Lipoprotein promoting entry (LpeA; 35 kDa) is a listerial cell surface-exposed lipoprotein, which shares significant sequence similarity (~60%) to pneumococcal surface adhesin A (PsaA), a known adhesin from *Streptococcus pneumoniae*. LpeA is required for invasion of nonprofessional phagocytic cells but not macrophages.

Lysis of Vacuole (Phagosome)

After entry into the host cell, *L. monocytogenes* is trapped inside a vacuole (phagosome), also called *Listeria*-containing vacuole (LCV). Before the lysosomal fusion with the vacuole can take place, the bacterium escapes by rupturing the vacuolar membrane with the help of listeriolysin O (LLO) and a phosphatidylinositol-specific phospholipase (PI-PLC) enzyme (Fig. 13.5).

Listeriolysin

Listeriolysin O (LLO) is a sulfhydryl (SH)-activated, pore-forming hemolysin with a molecular mass of 58–60 kDa. LLO is encoded by *hly* gene, which is under the direct control of *prfA*. Mutation in the *prfA* gene suppresses *hly* and other genes located downstream, such as *plcA* (PI-PLC) and *mpl* (metalloprotease), which are essential for maintenance of the virulence status of *L. monocytogenes*. LLO is a member of the cholesterol-dependent pore-forming cytotoxins (CDC) and is related genetically and antigenically to streptolysin O from *Streptococcus pyogenes*, ivanolysin O from *L. ivanovii*, pneumolysin from *Streptococcus pneumoniae*, and so forth.

LLO molecules oligomerize in the membrane forming a pore resulting in the lysis of the cell. About 30–40 molecules of LLO are required to cause lysis of a single erythrocyte. The maximum cytolytic activity of LLO is at pH 5.5, which coincides with the phagosomal pH and aids in vacuole disruption for the release of the bacterium into the cytoplasm. A host cell factor, γ-interferon-inducible lysosomal thiol reductase (GILT) enzyme, activates LLO and promotes phagosomal escape. In addition, *L. monocytogenes* delays vacuole maturation by inhibiting GTPase Rab5 activity by using glyceraldehyde-3-phosphate dehydrogenase (GAPDH), providing another opportunity to escape the vacuole.

LLO has a PEST (P, Proline; E, Glutamic acid; S, Serine; T, Threonine)-like sequence, which is the target for degradation by the host cell and aids bacterial persistence in the cytoplasm. LLO lacking the PEST sequence induces apoptosis and kills the host cell, thus prevent listerial cell-to-cell movement.

LLO also induces strong signaling events in cells leading to apoptosis in hepatocytes, dendritic cells, and lymphocytes, activates NF-κB, upregulates cytokine secretion, and activates MAP kinase pathways and protein kinase C (PKC). LLO also induces intracellular calcium accumulation and stimulates lipid raft aggregation. LLO can also inhibit ubiquitination of bacterial proteins by the host cell. Ubiquitination is a process, in which the ubiquitin protein binds to target proteins for degradation by the host proteasome. LLO can decrease ubiquitination or SUMOylation, where the posttranslational modification of ubiquitin allows the covalent addition of protein SUMO (small ubiquitin-like modifier) to the enzyme Ubc9 and consequent suppression of host immune response.

Phosphatidylinositol-Specific PLC

Two types of phospholipase C are produced by *L. monocytogenes*, phosphatidylinositol-specific phospholipase C (PI-PLC) and phosphatidylcholine-specific phospholipase C (PC-PLC), which are responsible for membrane disruption. PI-PLC works synergistically with LLO to destroy the lipid bilayer membrane of phagosome (LCV) allowing *L. monocytogenes* escape. PI-PLC is a 33–36 kDa enzyme, encoded by the *plcA* gene, which is regulated by *prfA* and is present only in *L. monocytogenes* and *L. ivanovii*. The substrate for PI-PLC is phosphatidylinositol (PI), and it has no activity on phosphatidylethanolamine (PE), phosphatidylcholine (PC), or phosphatidylserine (PS). The PC-PLC aids in the destruction of the double-membrane vacuole during the cell-to-cell spread of *Listeria*. The substrate for PC-PLC is phosphatidylcholine.

Intracellular Growth

After the escape from the vacuole, *L. monocytogenes* multiplies, accumulates F-actin surrounding the cell before initiating cell-to-cell movement (Fig. 13.5). Glucose is the preferred carbon source, and *Listeria* expresses hexose phosphate translocase (Hpt) to scavenge glucose-1-phosphate, glucose-6-phosphate, fructose-6-phosphate, and mannose-6-phosphate from the host cell cytoplasm. *Listeria* also expresses lipoate protein ligase (LpLA1) to utilize the host-derived lipoic acid, a cofactor required for the function of pyruvate dehydrogenase. LpLA1 ligates exogenous lipoic acid to the E2 subunit of the pyruvate dehydrogenase to form E2-lipoamide, which is important during aerobic metabolism. *Listeria* also secretes superoxide dismutase to defend against reactive oxygen species (ROS).

Cell-to-Cell Spread

Inside the cytoplasm, *Listeria* expresses actin polymerization protein (ActA) at one pole, which helps recruitment of actin protein to create a comet-tail-like structure to help bacterial movement. ActA is structurally similar to the host Wiskott–Aldrich syndrome protein (WASP); thus it is able to recruit actin-related protein 2/3 (Arp2/3) and actin polymerization machinery (Fig. 13.5).

ActA

ActA is a 639 amino acid-containing cell surface protein (~90 kDa) and is encoded by *actA* gene, which is under the direct control of *prfA*. ActA contains three domains: the N-terminal domain interacts with the Arp2/3 complex to initiate actin accumulation. The centrally located proline-rich domain interacts with the members of Enabled (Ena)/vasodilator-stimulated phosphoprotein (VASP) family proteins and helps directional actin assembly, and the C-terminal domain anchors to the bacterial cell wall. The actin tail acts as a fixed platform for propulsive movement inside the host cell cytoplasm. Microfilament cross-linking proteins, such as α-actinin, fimbrin, and villin are found around the bacteria as soon as actin filaments are detected on the bacterial surface. Other proteins associated with the actin tail include tropomyosin, plastin (fimbrin), vinculin, talin, and profilin. The *actA* mutant strain is unable to accumulate actin and fails to infect adjacent cells.

ActA, and to a lesser extent, PLC help *Listeria* to avoid killing by autophagy, through ActA-induced recruitment of actin component, Arp2/3, and Ena/VASP. In autophagy, cells degrade and recycle many cellular materials including whole organelles and plays an important role in defense against intracellular pathogens.

Microheterogeneity in the molecular mass of the ActA polypeptide exists among the strains of *L. monocytogenes* serotypes. *L. ivanovii* also polymerizes actin filaments when present in the cytosol and the actin synthesis is encoded by the *iactA* gene, which is homologous to the *L. monocytogenes actA* gene. The IactA is a 1044-amino acid-long polypeptide containing a proline-rich repeat sequence, similar to the ActA protein.

Lysis of Double-Membrane Vacuole

The moving bacterium forms a pseudopod-like protruding structure, which extends into the neighboring cell. It is suggested that the protruding structure is engulfed by the neighboring cell facilitating bacterial cell-to-cell spread. Bacteria trapped in the double membrane (the outer membrane is from the newly infected cell and the inner membrane from the previously infected cell), is escaped with the aid of phosphatidylcholine-specific PLC (PC-PLC) and LLO (Fig. 13.5).

Phosphatidylcholine-Specific PLC

PC-PLC is a 29 kDa enzyme that exhibits both lecithinase and sphingomyelinase activities and assists in the lysis of vacuole formed in the neighboring cell with the help of LLO (Fig. 13.5). PC-PLC requires zinc as a cofactor and is active at a pH range of 6–7. PC-PLC has a weak hemolytic activity but does not lyse sheep red blood cells because these cells are devoid of phosphatidylcholine. The *plcB* gene encodes a 33 kDa proenzyme precursor protein, which is processed to form a mature PC-PLC (29 kDa) with the help of a zinc metalloprotease. Zinc metalloprotease is encoded by *mpl*, a proximal gene in the lecithinase operon. Mutation of *mpl* causes reduced virulence of *L. monocytogenes* in mice. The *plcB* gene is under the direct control of *prfA*.

Regulation of Virulence Genes

The 9 kb pathogenicity island contains a virulence gene cluster, *prfA–plcA–hly–mpl–actA–plcB*, which is regulated by the first gene, *prfA* (Fig. 13.2). The PrfA protein binds to a palindromic *prfA* recognition sequence (*prfA*-box; −TTAACANNTGTTAA-) located at position −40 from the transcription start site to initiate the transcription of virulence genes. Three promoters contribute to the *prfA* expression. The two promoters *prfAp1* and *prfAp2* are located immediately upstream of the *prfA* coding region. The third promoter is located upstream of *plcA* gene. PrfA positively regulates its own expression through the activation of *plcA* transcription. The PrfA box sequence is present in all *prfA*-regulated virulence genes. PrfA expression is dependent on the growth phase and temperature, and it is controlled at the transcription, translation, or post-translational level. During transcription, bicistronic transcripts of *plcA* and *prfA* form loops, which increase *prfA* transcription. One *prfA* promoter, *prfA2*, is controlled by the alternative sigma factor B (SigB).

Readily metabolizable sugars, like glucose, cellobiose, fructose, mannose, trehalose β-glucosides, and maltose negatively regulate the *prfA*-dependent genes. Such negative regulation is possibly due to the presence of a common catabolite repression (CR) or spontaneous mutation or another molecule responding specifically to β-glucosides. Surprisingly, the expression of PrfA is induced in the presence of cellobiose, suggesting a posttranscriptional modification of PrfA takes place. The addition of charcoal in the growth medium could overcome repression by sugars since charcoal can sequester a diffusible autorepressor substance that is released by *L. monocytogenes*.

Stress response genes such as those necessary for growth during osmotic stress, low pH stress, growth at low temperature, and carbon starvation are regulated by sigma B. Sigma B has been shown to regulate about 55 genes including several virulence genes that are necessary during the intestinal phase of infection such as *bsh*, *gadA*, *opuCA*, *inlA*, *inlB*, and *prfA*.

Immunity to *Listeria monocytogenes*

Innate and cell-mediated adaptive immunity are most effective against *L. monocytogenes* infection (Fig. 13.8). During infection, 90% of *L. monocytogenes* cells arrive in the liver and 10% in the spleen within 24–48 h. Intestinal epithelial cells (IEC) play important role in innate immunity early in the intestinal phase of infection. Initial recognition of *Listeria* is dependent on the presence of pattern recognition receptors (PRRs), i.e., toll-like receptors (TLRs), NOD-like receptors (NLRs), and retinoic acid-inducible gene 1 (RIG-I)-like receptors, which are expressed on the cell surface membrane, endosomal membranes, or in the cytosol of the cell.

Fig. 13.8 Innate and adaptive immune response against *Listeria monocytogenes* infection. TCR T-cell receptor, MHC major histocompatibility complex

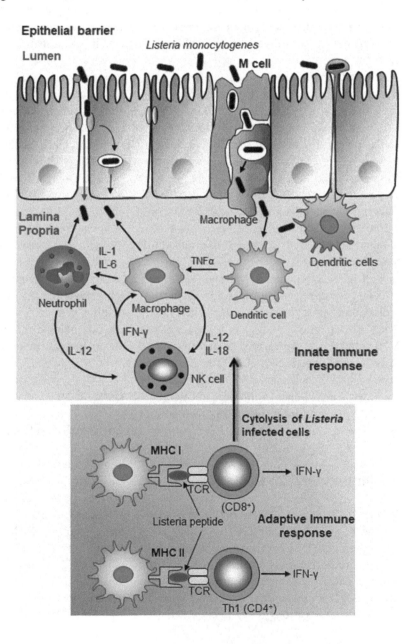

Binding of pathogen-associated molecular patterns (PAMP) to their respective PRR activates a signaling cascade, which induces transcription of immune response genes through activation of NF-κB. The epithelial cytokines, TNF-α, IL-1β, and IL-6, are produced which in turn recruit and activate neutrophils, macrophages, and dendritic cells. *Listeria* flagellum is recognized by TLR5, which activates NF-κB triggering production of TNF-α for immunity. Similarly, LLO, LAP, and InlB may activate NF-κB and produce cytokines, TNF-α, IL-1, and IFN-β, to activate macrophages and dendritic cells in the lamina propria for bacterial clearance. In the liver, the neutrophil influx is high and recognizes infected hepatocytes. Neutrophils and resident macrophages, i.e., Kupffer cells, are responsible for initial control of infection through production of ROS. TNF-α, IL-1, and IFN-β are produced by Kupffer cells and also activated macrophages for clearance of bacteria. Complement activation in the alternative pathway produces C3b to facilitate *L. monocytogenes* phagocytosis. IFN-γ produced by NK cells and a subpopulation of neutrophil also plays a role in innate immunity. *L. monocytogenes*-infected macrophages secrete TNF-α and IL-12 to activate NK cells.

Cell-mediated immunity involving CD_8^+ T cells and activated macrophages provide the best protection against listeriosis. CD_8^+ T cell recognizes *Listeria* antigens such as LLO and P60 via MHC class I when presented by infected target cells. In addition, IFN-γ, IL-6, and IL-10 produced by CD_4^+ cells activate macrophages for increased phagocytosis. Antibodies to LLO, P60, and LAP are present in blood, but their role in protective immunity is not clear.

Prevention and Control

In 2004, an *L. monocytogenes* risk analysis study conducted jointly by the US Food and Drug Administration (US FDA) and the US Department of Agriculture (USDA) grouped various foods into risk categories to provide a warning system and to educate consumers about the risk levels associated with various ready-to-eat foods (see the section on Source). The immunocompromised populations including the elderly and pregnant women should avoid foods that belong to very high to medium risk categories to prevent the contraction of listeriosis.

Implementation of Hazard Analysis and Critical Control Point (HACCP) strategies are mandatory for all processing plants that produce ready-to-eat products to reduce or eliminate *L. monocytogenes* from food processing facilities. If a product has been implicated in an outbreak, it is recalled to prevent further incidences. As a precaution, producers are also asked to recall products if any of their products are found to carry *L. monocytogenes* after retail distribution.

Pasteurization or heating of processed foods before consumption has been suggested for certain products. Heating of foods at 71 °C for 1 min kills the bacterium. Control of *Listeria* in the farm, animal feed, dairy cows, silage, and in the processing plant is needed to prevent *Listeria* in raw and processed products. Routine surveillance of production facilities, implementation of an effective sanitization scheme, and improved product formulation to include antimicrobial inhibitors are some of the best practices to reduce *Listeria* in products.

Antibiotic therapy with gentamicin, ampicillin, amoxicillin, and penicillin G are effective in treating systemic listeriosis. Supplemental therapy consisting of fluid and electrolytes is needed during the gastrointestinal form of listeriosis.

Detection

Conventional Culturing Method

Food or clinical specimen (cerebral spinal fluid, tissue from medulla and pons, aborted fetus, abomasal content, uterine discharge) can be used for isolation of *Listeria* after a primary enrichment step in UVM (University of Vermont Medium) broth followed by the secondary enrichment in FB (Fraser Broth). Enrichment is also done on *Listeria* repair broth (LRB) containing antimicrobial supplement or buffered *Listeria* enrichment broth (BLEB) to recover stressed or injured cells. Enriched cultures are then streaked on MOX (modified Oxford medium), PALCAM (polymyxin acriflavine lithium chloride ceftazidime aesculin mannitol agar), or Rapid'L.Mono agar plates for isolation of presumptive colonies. In immunomagnetic separation (IMS) system, paramagnetic particles coated with *Listeria*-specific antibodies have been used for capture and concentration of *Listeria* from enriched samples. Captured *Listeria* cells are plated on selective agar plates for isolation and used for identification by PCR. Presumptive *Listeria* colonies can be examined for hemolytic activity on blood agar plate, CAMP test, and sugar and amino acid utilization by using biochemical tests such as API *Listeria* test that can give a result within 24 h.

Immunoassays and Lateral Flow Assay

Antisera to O and H antigens are used for serotyping of *Listeria* isolates. Lateral flow-based dipstick assays are used for bacterial detection directly from the *Listeria* enrichment broths.

PCR and Whole Genome Sequencing

Several commercial PCR assay kits are available for detection and identification of *L. monocytogenes*. Pulsed-field gel electrophoresis (PFGE), ribotyping, and whole genome sequencing are used for identification and source tracking.

Summary

Listeria monocytogenes is an opportunistic intracellular pathogen. Historically, it has been recognized as a causative agent for listeriosis in animals; however, in the past few decades, it has been responsible for fatal foodborne diseases in immunocompromised individuals. It is one of the leading foodborne pathogens and has been implicated in numerous outbreaks. *L. monocytogenes* belonging to lineage I and II are generally associated with outbreaks. It causes three forms of the disease: gastroenteritis in healthy adults, systemic listeriosis in immunocompromised populations, and abortion and neonatal listeriosis in pregnant mothers and their fetuses. The intestinal phase of infection is a complex process, and the mechanism is not fully understood. Similarly, transmission to the brain or to the fetus is less clear. *L. monocytogenes* crossing of the intestinal epithelial barrier, invasion into mammalian cells, survival inside the phagosome, and escape into the cytoplasm, growth, and cell-to-cell spread is well understood. These events are orchestrated by numerous virulence factors such as *Listeria* adhesion protein (LAP), internalin A (InlA), InlB, listeriolysin O (LLO), actin polymerization protein (ActA), phospholipase C (PLC), metalloprotease (Mpl), hexose phosphate transport permease (Hpt), and lipoprotein ligase (LpL). Immune response to *L. monocytogenes* is largely dependent on the innate immunity involving neutrophils, dendritic cells, NK cells, and macrophages and cell-mediated immunity involving CD_8^+ T-cell subsets. Humoral immunity possibly has limited or no role in the immunity. Risk analysis study has identified several ready-to-eat foods to be of the high-risk category: hotdogs, sliced deli meats, soft cheeses especially those made with unpasteurized milk and pâté and meat spread, and smoked seafood. Immunocompromised or high-risk individuals, especially those who are pregnant, should avoid high-risk foods.

Further Readings

1. Bakardjiev, A., Theriot, J. and Portnoy, D. (2006) *Listeria monocytogenes* traffics from maternal organs to the placenta and back. *PLoS Pathog* **2**, e66.
2. Bhunia, A.K. (1997) Antibodies to *Listeria monocytogenes*. *Crit Rev Microbiol* **23**, 77–107.
3. Burkholder, K.M. and Bhunia, A.K. (2013) *Listeria monocytogenes* and Host Hsp60 – An invasive pairing. In *Moonlighting Cell Stress Proteins in Microbial Infections, Heat Shock Proteins* ed. Henderson, B. pp.267–282. Dordrecht, Germany: Springer Science+Business Media.
4. Camejo, A., Carvalho, F., Reis, O., Leitao, E., Sousa, S. and Cabanes, D. (2011) The arsenal of virulence factors deployed by *Listeria monocytogenes* to promote its cell infection cycle. *Virulence* **2**, 379–394.

5. Disson, O., Grayo, S., Huillet, E., Nikitas, G., Langa-Vives, F., Dussurget, O., Ragon, M., Le Monnier, A., Babinet, C., Cossart, P. and Lecuit, M. (2008) Conjugated action of two species-specific invasion proteins for fetoplacental listeriosis. *Nature* **455**, 1114–1118.

6. Disson, O. and Lecuit, M. (2013) *In vitro* and *in vivo* models to study human listeriosis: mind the gap. *Microbes Infect* **15**, 971–980.

7. Dussurget, O., Pizarro-Cerda, J. and Cossart, P. (2004) Molecular determinants of *Listeria monocytogenes* virulence. *Annu Rev Microbiol* **58**, 587–610.

8. Farber, J.M. and Peterkin, P.I. (1991) *Listeria monocytogenes*, a food-borne pathogen. *Microbiol Rev* **55**, 476–511.

9. Freitag, N.E., Port, G.C. and Miner, M.D. (2009) *Listeria monocytogenes* from saprophyte to intracellular pathogen. *Nat Rev Microbiol* **7**, 623–628.

10. Gahan, C.G.M. and Hill, C. (2005) Gastrointestinal phase of *Listeria monocytogenes* infection. *J Appl Microbiol* **98**, 1345–1353.

11. Hamon, M., Bierne, H. and Cossart, P. (2006) *Listeria monocytogenes*: a multifaceted model. *Nat Rev Microbiol* **4**, 423–434.

12. Jagadeesan, B., Koo, O.K., Kim, K.P., Burkholder, K.M., Mishra, K.K., Aroonnual, A. and Bhunia, A.K. (2010) LAP, an alcohol acetaldehyde dehydrogenase enzyme in *Listeria* promotes bacterial adhesion to enterocyte-like Caco-2 cells only in pathogenic species. *Microbiology* **156**, 2782–2795.

13. Kazmierczak, M.J., Wiedmann, M. and Boor, K.J. (2006) Contributions of *Listeria monocytogenes* {sigma}B and PrfA to expression of virulence and stress response genes during extra- and intracellular growth. *Microbiology* **152**, 1827–1838.

14. Lomonaco, S., Nucera, D. and Filipello, V. (2015) The evolution and epidemiology of *Listeria monocytogenes* in Europe and the United States. *Infect Gen Evol* **35**, 172–183.

15. Ogawa, M., Yoshikawa, Y., Mimuro, H., Hain, T., Chakraborty, T. and Sasakawa, C. (2011) Autophagy targeting of *Listeria monocytogenes* and the bacterial countermeasure. *Autophagy* **7**, 310–314.

16. Ooi, S.T. and Lorber, B. (2005) Gastroenteritis due to *Listeria monocytogenes*. *Clin Infect Dis* **40**, 1327–1332.

17. Regan, T., MacSharry, J. and Brint, E. (2014) Tracing innate immune defences along the path of *Listeria monocytogenes* infection. *Immunol Cell Biol* **92**, 563–569.

18. Stavru, F., Archambaud, C. and Cossart, P. (2011) Cell biology and immunology of *Listeria monocytogenes* infections: novel insights. *Immunol Rev* **240**, 160–184.

19. Swaminathan, B. and Gerner-Smidt, P. (2007) The epidemiology of human listeriosis. *Microbes Infect* **9**, 1236–1243.

20. Vazquez-Boland, J.A., Kuhn, M., Berche, P., Chakraborty, T., Dominguez-Bernal, G., Goebel, W., Gonzalez-Zorn, B., Wehland, J. and Kreft, J. (2001) *Listeria* pathogenesis and molecular virulence determinants. *Clin Microbiol Rev* **14**, 584–640.

21. Xayarath, B. and Freitag, N.E. (2012) Optimizing the balance between host and environmental survival skills: lessons learned from *Listeria monocytogenes*. *Future Microbiol* **7**, 839–852.

22. Drolia, R., Tenguria, S., Durkes, A.C., Turner, J.R., and Bhunia, A.K. (2018) *Listeria* adhesion protein induces intestinal epithelial barrier dysfunction for bacterial translocation. *Cell Host & Microbe* **23**, 470–484.

23. Radoshevich, L., and Cossart, P. (2018) *Listeria monocytogenes*: towards a complete picture of its physiology and pathogenesis. *Nat Rev Microbiol* **16**, 32–46.

Escherichia coli

14

Introduction

Theodor Escherich, a German–Austrian pediatrician and a professor at universities in Graz and Vienna, Austria, reported, in 1885, the isolation of a bacterium called *Bacterium coli* from a fecal sample. Later, in 1919, the bacterium was renamed *Escherichia coli*. *E. coli* is a Gram-negative, short motile rod that inhabits the intestinal tract of animals and humans since birth. *E. coli* has been used extensively as a model organism to study bacterial physiology, metabolism, genetic regulation, signal transduction, and the cell wall structure and function. The bacterium is one of the natural microflora of human and animal gut microbial community. Hence, fecal shedding and contamination of water and food with *E. coli* or coliforms are common and are often used as indicators for hygiene monitoring. A majority of *E. coli* strains are nonpathogenic; however, only a small subset is pathogenic and causes a variety of diseases in humans and animals. The diseases include gastroenteritis, dysentery, hemorrhagic colitis (HC), hemolytic uremic syndrome (HUS), urinary tract infection (UTI), septicemia, pneumonia, and meningitis. In recent years, however, the major concern has been the increasing numbers of foodborne outbreaks, caused by pathogenic *E. coli* in the industrialized countries due to consumption of contaminated meat, fruits, and vegetables.

Biology

Escherichia coli is a member of the genus *Escherichia* and family *Enterobacteriaceae*. Other lesser-known species in the genus *Escherichia* are *E. albertii*, *E. blattae*, *E. hermannii*, *E. vulneris*, and *E. fergusonii*. The *Enterobacteriaceae* family consists of 40 genera and about 180 species, and they are either primary pathogens, opportunistic pathogens, or commensals (Fig. 14.1). The primary pathogens are also recognized as enteric pathogens, and they include *Escherichia coli*, *Salmonella enterica*, *Shigella* species, and *Yersinia* species. The opportunistic pathogens are *Proteus*, *Klebsiella*, *Enterobacter*, *Edwardsiella*, *Morganella*, *Hafnia*, *Serratia*, and so forth. The members of *Enterobacteriaceae* family ferment glucose and lactose, are oxidase negative and catalase positive, and reduce nitrate to nitrite. The members are resistant to bile salts; therefore, they can be readily isolated using violet red bile glucose agar (VRBGA), brilliant green bile agar (BGBA), and MacConkey agar (MAC) media.

Escherichia coli is a Gram-negative, short-rod (1–2 µm in length), facultative anaerobe and is motile. It expresses peritrichous flagella, fimbriae or pili, and curli. Some strains express capsules. The optimum growth temperature is 35–37 °C. They are lactose fermenter and produce pink colonies when grown on MacConkey agar plate.

© Springer Science+Business Media, LLC, part of Springer Nature 2018
A. K. Bhunia, *Foodborne Microbial Pathogens*, Food Science Text Series,
https://doi.org/10.1007/978-1-4939-7349-1_14

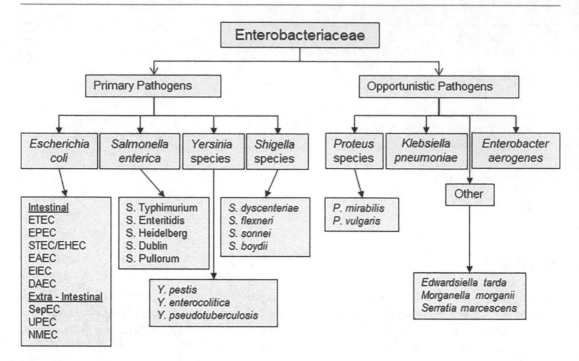

Fig. 14.1 *Enterobacteriaceae* family and its major pathogenic members (genus and species)

Some pathogenic strains are also acid tolerant. The genome size of a pathogenic strain could be 1 Mb larger than the nonpathogenic strains because of the presence of virulence genes. Note: the genome size of nonpathogenic *E. coli* K12 strain is 4.6 Mb. In pathogenic strains, the virulence genes are clustered in pathogenicity islands (PAIs), plasmids, or prophages. The PAIs are flanked by mobile genetic elements – bacteriophages, insertion sequences, or transposons. Horizontal gene transfer facilitates host adaptation and emergence of new pathogenic strains such as enterohemorrhagic *E. coli* (EHEC), Shiga-toxin producing *E. coli* (STEC), and enteroaggregative *E. coli* (EAEC) pathovars (Fig. 14.2).

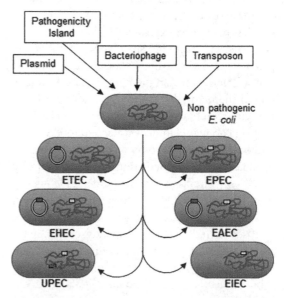

Fig. 14.2 Evolution of pathogenic *Escherichia coli* through horizontal gene transfer

Sources

Escherichia coli is a member of the intestinal microflora of humans, animals, and birds (Fig. 14.3). The bacterium is routinely shed into the environment through feces, and it can contaminate drinking water, irrigation water, and soil, consequently fresh fruits and vegetables, especially if untreated manures are used as fertilizers. Contaminated fresh produce may even

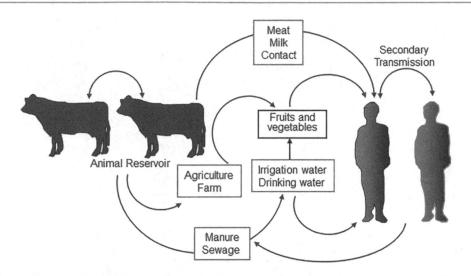

Fig. 14.3 Mode of transmission of *Escherichia coli* to humans. The primary vehicle of transmission is meat, but the bacteria can be transported via animal-to-person contact, milk, contaminated water, and vegetables contaminated with cow manure. Bacteria can move from cow-to-cow and also from wild animals such as deer, caribou, and domestic sheep. Person-to-person transmission occurs directly or via contaminated water such as in the swimming pool

harbor some bacteria inside the plant tissues. Some *E. coli* pathotypes such as enterohemorrhagic *E. coli* (EHEC) can be transferred through meat, which may be acquired during slaughter through fecal and hide contact. EHEC outbreaks have been associated with meat (especially ground beef), dairy products, mayonnaise, apple cider, sprouts, lettuce, and spinach. Outbreaks are also associated with various establishments including swimming pools, nursing homes, restaurants, petting zoo, and day-care facility. Travelers in the endemic region may experience *E. coli*-mediated diarrhea and dysentery. Human-to-human transmission can also happen.

Classification

Serotypes

Somatic "O" antigens are one class of *E. coli* serogroup determinants, consisting of lipopolysaccharide (LPS), and there are 174 O antigens, numbered 1–181, with numbers 31, 47, 67, 72, 93, 94, and 122 omitted. In addition, there are 53 serotypes of "H" or flagellar antigens (H1–H53). Strains that lack flagella are nonmotile (NM).

E. coli isolates can have a variety of antigen combinations. Thirty serovars are reported to be responsible for diarrheal diseases, and the first serogroup identified was O111 and was isolated from a child. The "O" antigen identifies serogroup, while the "H" identifies serotype. For example, two strains with same O antigens but different H antigens such as O111:H4 and O111:H12 have same serogroup but different serotype. There are also 80 capsular antigens or "K" antigens.

Virotypes

Virotype or pathotype classification is based on the presence of certain virulence factors and their interaction with the mammalian cells and cell signaling events for effective adhesion to and/or invasion of mammalian cells and toxin production (Fig. 14.4). Pathogenic *E. coli* are classified broadly into two groups: gastrointestinal tract-infecting *E. coli* that cause diarrhea and extraintestinal tract-infecting *E. coli* that affect the kidney, urinary tract, brain, and circulatory system leading to septicemia (Table 14.1). These *E. coli* strains are designated septicemic *E. coli*

Fig. 14.4 Virotype classifications of *Escherichia coli* based on their interaction with intestinal villous epithelial cells: enterotoxigenic *E. coli* (ETEC), enteropathogenic *E. coli* (EPEC), enterohemorrhagic *E. coli* (EHEC), enteroaggregative *E. coli* (EAEC), enteroinvasive *E. coli* (EIEC), and diffusely adhering *E. coli* (DAEC). *LT* Heat-labile toxin, *ST* heat-stable toxin, *bfp* bundle-forming pili, *A/E lesion* attachment/effacement lesion showing actin accumulation, *Stx* Shiga toxin

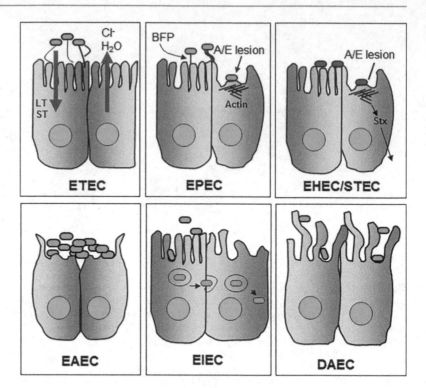

Table 14.1 Colonization sites of pathogenic *E. coli* in the human body

Organs and tissues affected	Pathotypes
Small intestine	ETEC, EPEC, DAEC, EAEC
Large intestine	EHEC (STEC), EIEC, EAEC
Kidney	UPEC
Bladder	UPEC
Brain	NMEC
Bloodstream	UPEC, NMEC, SepEC

(SepEC), uropathogenic *E. coli* (UPEC), and neonatal meningitis-causing *E. coli* (NMEC). Diarrheagenic *E. coli* are again divided into six pathotypes (Fig. 14.4): (1) enterotoxigenic *E. coli* (ETEC), (2) enteropathogenic *E. coli* (EPEC), (3) Shiga toxin-producing *E. coli* (STEC) and their most pathogenic subset enterohemorrhagic *E. coli* (EHEC), (4) enteroaggregative *E. coli* (EAEC), (5) enteroinvasive *E. coli* (EIEC), and (6) diffusely adhering *E. coli* (DAEC).

1. Enterotoxigenic *E. coli* (ETEC) adhere to epithelial cells and produce several toxins including heat-labile (LT) and/or heat-stable toxins

(ST), but they do not invade epithelial cells. The predominant serogroups are O6, O8, O11, O15, O20, O25, O27, O78, O128, O148, O149, O159, and O173.

2. Enteropathogenic *E. coli* (EPEC) adhere to epithelial cells intimately, produce attachment/effacement (A/E) lesion, and are noninvasive. They do not produce any heat-labile (LT), heat-stable (ST), or Shiga toxins (Stx). The notable serogroups are O26, O55, O86, O111, O114, O119, O125, O126, O127, O128, O142, and O158.

3. Shiga toxin-producing *E. coli* (STEC) and their most pathogenic subset enterohemorrhagic *E. coli* (EHEC): EHEC bind strongly to the epithelial cells, produce attachment/effacement lesions, and produce Stx. The serogroups are O4, O5, O16, O26, O45, O55, O91, O103, O111ab, O113, O121, O117, O145, O157, O172, O176, O177, O178, O180, and O181.

4. Enteroaggregative *E. coli* (EAEC) adhere to epithelial cells; form aggregates, appearing like "stacked brick;" and produce toxin but do

not invade. The serogroups in this virotype are O3, O15, O44, O86, O77, O104, O111, and O127.

5. Enteroinvasive *E. coli* (EIEC) cells also adhere, invade cells, and move from cell to cell but do not produce toxin. The pathogenicity of EIEC resembles infection caused by *Shigella* spp., and the predominant symptom is dysentery. The EIEC serogroups are O28, O29, O112, O124, O136, O143, O144, O152, O159, O164, and O167.

6. The diffusely adhering *E. coli* (DAEC) adhere to epithelial cells, but they neither invade nor produce toxin. Details of each virotype are described below.

Enterotoxigenic *E. coli*

Characteristics

Enterotoxigenic *E. coli* (ETEC) cause toxicoinfection. ETEC were first reported to cause cholera-like illness in adults and children in Calcutta (India) in 1956. In general, ETEC commonly cause infectious diarrhea in people living in the tropical climate, and children under the age of 2 are highly susceptible. Infants in many developing countries suffer from ETEC due to poor sanitary conditions. ETEC also affect travelers since the water and food consumed by them may be contaminated with ETEC, and native people may be resistant because of their frequent exposure to the organism. Thus, ETEC-mediated diarrhea is also known as "traveler's diarrhea" because of the bacterial association with international travel. The same disease has gained regional names such as "Montezuma's revenge," in travelers who travel to South American countries, especially Mexico, and "Delhi belly" for those who acquire the disease while traveling to India. Globally, about 200 million cases and approximately 380,000 deaths are associated with ETEC.

ETEC cells adhere to the mucosal epithelial cells and produce heat-labile toxin (LT) and/or heat-stable diarrheal toxins (ST). LT resembles cholera toxin (CT) and produces symptoms similar to the *Vibrio cholerae* infection.

Virulence Factors and Pathogenesis

Adhesion Factors

ETEC express at least 25 colonization factors (CFs) that target the intestinal mucosa, and the genes (*cfa*) for CFs are located mostly on the plasmid. CFs are proteinaceous fimbrial and afimbrial (fibrillar) structures (lengths, 1–20 μm) that allow bacteria to attach to the intestinal mucosa. Fimbrial colonization factors are also known as colonization factor antigens or colonization fimbrial antigen (CFA), and three types are reported: CFA/I, CFA/II, and CFA/III. CFA/I is rigid rod shaped; CFA/II is flexible fimbriae present alone or in association with other rod-shaped fimbriae; and CFA/III is a bundle-forming flexible pilus (BFP), called longus, because of its unusual length (about 40 μm long). One afimbrial colonization factor, CFA/IV, is present in ETEC. The tip of CFA is hydrophobic and promotes ETEC adhesion to epithelial cells.

Other non-fimbrial adhesion factors include TibA and Tia. TibA is a 104 kDa afimbrial adhesin, encoded on the chromosome. TibA aids in bacterial aggregation on epithelial cells and promotes biofilm formation. Tia, a 25 kDa outer membrane protein encoded on a 46 kb pathogenicity island, is involved in adhesion. A Tia homolog has been found in other *E. coli* (EPEC and EAEC), suggesting it has a broader role in pathogenesis.

ETEC express peritrichous flagella and each of which is comprised of 20,000 individual flagellin proteins. Flagella contribute significantly to intestinal adhesion and colonization.

Toxins

ETEC produce two types of toxins: (i) heat-labile toxin (LT; LT-I and LT-II) and (ii) heat-stable toxin (ST; STa and STb).

Heat-Labile Toxins

Heat-labile toxin (LT) is of two types, LT-I and LTII, and they are encoded by the *eltAB* operon. The genetic organization of LT-I is similar to the cholera toxin. LT-I is expressed in *E. coli* that induce disease in both humans and animals, while LT-II is found mainly in animal isolates. LT-II has same basic structure and mechanism of action as LT-I. LT-I is an A–B-type toxin with a molecular mass of 87.5 kDa. The A subunit is 27 kDa, while the B subunit consists of five identical subunits of 11.7 kDa (pentameric form), and the subunits are arranged in a ring. The A subunit consists of two domains: A1 and A2. A1 domain is the active toxin molecule, and the A2 anchors to B pentamer subunit via disulfide bridge. This A–B$_5$ complex is also called holotoxin, because of covalent association of A and B subunits, and the amino acid sequence is similar to that of the cholera toxin.

After synthesis, the toxin is transported across the outer membrane by a two-step process using SecA2 and the type II secretion system (T2SS). SecA2 translocates LT across the cytoplasmic membrane, and the T2SS helps translocate across the outer membrane (OM). The secreted toxin may remain associated with the LPS in the OM. During infection of the host cell, the B subunit binds a GTP-binding protein and ganglioside (GM1) receptor on the epithelial cells and triggers endocytosis of the holotoxin. [Note: cholera toxin interacts with the same receptor and acts in the same manner; see Chap. 18.] The A subunit then causes ADP-ribosylation of ganglioside (Gs) protein and activates adenylate cyclase resulting in an increase in cyclic adenosine monophosphate (cAMP) level in the cytoplasm. cAMP then activates cAMP-dependent protein kinase A (PKA), which in turn causes phosphorylation of cystic fibrosis transmembrane regulator (CFTR), a chloride ion transporter protein, increases Cl$^-$ secretion in the crypt cells, and decreases absorption of Na$^+$ and Cl$^-$ by absorptive cells (Fig. 14.5).

In addition, the A subunit is involved in arachidonic acid metabolism leading to the formation of prostaglandin E$_2$ (PGE$_2$) and 5-hydroxytryptamine (5-HT), which promote electrolyte and water release from intestinal cells. The LT-II action is similar to LT-I with an exception where it binds to ganglioside GD1 receptor instead of GM1.

Heat-Stable Toxin

Heat-stable toxin (ST) consists of a family of small peptide toxin of 2 kDa, which is stable at 100 °C for 30 min. Two major types of ST, (i) STa

(STh), a methanol-soluble toxin isolated from human (h), and (ii) STb (STp), a methanol insoluble toxin isolated from pig (p), are also found in human isolates. ST is produced as a 72-amino acid-long precursor protein, which is stored in the periplasm. After N-terminal 19 amino acids are removed, a 54-amino acid-long peptide is secreted to the periplasm by SecA2 and through outer membrane by TolC protein exporter. The peptide is further cleaved to a 17–19-amino acid-long active peptide with a molecular mass of approximately 2 kDa. STa (STh) binds to the membrane guanylyl cyclase C (GC-C) receptor on the brush border of epithelial cells of the small intestine and colon. STa binding to GC-C activates protein kinase G (PKG) and protein kinase C (PKC), increasing inositol 1,4,5-trisphosphate (IP$_3$)-mediated Ca^{2+} to increase intracellular cyclic guanidine monophosphate (cGMP). Increased cGMP activates calcium ion channel, CFTR, and increases the concentration of Cl$^-$ ions in the extracellular space. As a result, electrolyte balance in the bowel is disrupted, which leads to fluid accumulation within the lumen of the intestine resulting in diarrhea (Fig. 14.5).

STb is a 48 amino acid containing a peptide, found primarily in swine isolates but also in human ETEC isolates. STb has no sequence homology with STa, and the receptor for STb is sulfatide, a widely distributed glycosphingolipid. After endocytosis of STb, it activates GTP-binding regulatory protein, which increases the efflux of Ca^{2+}, opens ion channels, and activates protein kinase C. Increased Ca^{2+} levels regulate phospholipases (A2 and C) that release arachidonic acid from membrane phospholipids, leading to the formation of prostaglandin E$_2$ (PGE$_2$) and 5-hydroxytryptamine (5-HT), which mediate electrolyte and water release from the intestinal cells. Unlike STa, STb also stimulates the secretion of bicarbonate from epithelial cells. STb also reported to forming a multimeric structure on epithelial membrane forming pores and increased membrane permeability.

Other Toxins

ETEC also secretes 38 amino acid containing enteroaggregative heat-stable toxin 1 (EAST1), which has been isolated from ETEC strains of

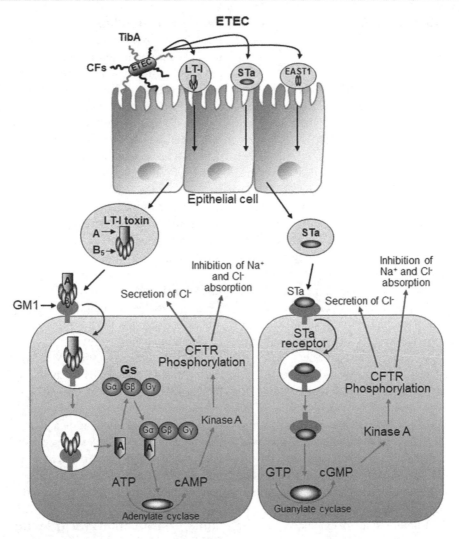

Fig. 14.5 Mechanism of action for enterotoxigenic *Escherichia coli* (ETEC)-mediated diarrhea. After arrival in the intestine, ETEC binds to the epithelial cells using colonization factors (CFs) and/or TibA and produces several toxins including LT-I, LT-II, STa, STb, EAST1, and ClyA. Mechanism of action of LT-I and STa is presented.

LT increases cAMP level, while ST increases cGMP, and both mediate phosphorylation of CFTR (cystic fibrosis transmembrane conductance regulator), a chloride ion transporter protein, which increases Cl^- secretion in crypt cells and decreases absorption of Na^+ and Cl^- by absorptive cells

human and animal origin. The toxin activates cGMP, induces fluid accumulation, and possibly plays a role in the onset of diarrhea. In addition, ETEC also produces EatA, a serine protease autotransporter, which plays a role in pathogenesis by damaging the epithelial cell surface. Some strains of ETEC may secrete ClyA, a cholesterol-dependent cytolysin (CDC), and form pore in the cell membrane. It binds to cholesterol on the membrane and shows lytic activity against erythrocytes, macrophages, and HeLa cells.

Symptoms

ETEC infection does not show any apparent histological changes in the mucosal layer, and there is little or no inflammation in the intestine. The symptoms may include watery diarrhea, vomiting, sunken eyes, massive dehydration, and a collapse of the circulatory system. Diarrhea lasts for 3–4 days and is self-limited. Diarrhea may be lethal in young children and infants, with a mortality rate of less than 1%.

Enteropathogenic *E. coli*

Characteristics

Enteropathogenic *E. coli* (EPEC) was the first virotype of *E. coli* to be described, and the bacterium primarily causes fatal diarrhea in children under the age of 5. EPEC is also pathogenic to calves, pigs, rabbits, and dogs. EPEC expresses bundle-forming pilus (BFP) or EPEC adherence factor (EAF) encoded by *bfpA* gene located on a plasmid, for adhesion and colonization. EPEC also expresses intimin encoded by the *eae* gene located on 35 kb pathogenicity island designated locus of enterocyte effacement (LEE) to produce attachment and effacement (A/E) lesion. EPEC does not express Shiga toxin (Stx). Based on the virulence gene distribution, EPEC is classified into two groups: typical EPEC (tEPEC) and atypical EPEC (aEPEC). tEPEC is *eae+bfpA+stx−* and produces localized adherence (LA) phenotype, while aEPEC is *eae+bfpA−stx−* and produces localized-like (LAL) diffuse adherence phenotype. tEPEC adheres intimately to epithelial cells and exhibits a "patchy" pattern of adherence. Adherence promotes a dramatic change in the ultrastructure of epithelial cells, resulting in the formation of an attaching and effacing lesion, which is characterized by the formation of a "cuplike" or "pedestal" structure due to extensive cellular actin rearrangements in the architecture. Microvilli structures gradually disappear and the epithelium lose the ability to absorb nutrients. EPEC strains are highly invasive and cause an inflammatory response and potentially fatal diarrhea in children and infants.

Virulence Factors and Pathogenesis of EPEC

EPEC interaction with the host cells occurs in four stages: (1) expression of adhesion factors, (2) initial localized adherence, (3) signal transduction and intimate contact, and (4) cytoskeletal rearrangement and pedestal formation (Fig. 14.6).

Expression of Adhesion Factors

Initially, the bacterium binds to epithelial cells but the adhesion is non-intimate. The adhesion is mediated by the type IV adhesion fimbriae, BFP or EAF, which is similar to Tcp pilus of *V. cholerae*. BFP is a ropelike structure, which interacts with other bacterial cells to form microcolonies for localized adherence and with N-acetyllactosamine-containing receptors on epithelial surface. In addition, DsbA, a periplasmic enzyme, encoded by the *dsbA* gene, facilitates the disulfide bond formation between proteins that are involved in localized adherence. EPEC also produces intimin and a short, surface-associated filament, EspA.

Localized Adherence

EPEC adheres to epithelial cells using BFP, intimin, and EspA and the type III secretion system (T3SS, a molecular syringe). T3SS injects translocated intimin receptor (Tir, 78 kDa) and several other effector molecules (EspB, EspD, EspG, EspF, EspH) directly into the host cell. The effector molecules activate cell signaling pathways, allowing actin polymerization and depolymerization to alter the cytoskeletal structure. Tir is then phosphorylated by protein kinases and inserted into the host cell membrane to facilitate binding with bacterial intimin (Figs. 14.6 and 14.7).

Signal Transduction and Intimate Contact

The attachment of EPEC to mammalian cells triggers host cell signal transduction pathways and activates host cell tyrosine kinase, which causes the release of two signaling molecules: inositol triphosphate (IP_3) and intracellular Ca^{2+}. These activate calcium-dependent actin depolymerization enzyme and trigger phosphorylation of host cell proteins–myosin light chain (MLC) and the 90 kDa epithelial Hp90 protein. Intracellular Ca^{2+} can inhibit Na^+ and Cl^- absorption and stimulate Cl^- secretion.

Intimate contact is mediated by adhesion of intimin (Eae), a 95 kDa outer membrane protein, to the Tir receptor on the host cell membrane. Intimin protein has 27 variants (eae alleles) based on the heterogeneity in the C-terminal sequence

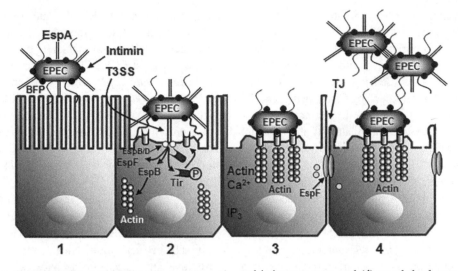

Fig. 14.6 Schematic diagram showing the sequence of events for enteropathogenic *Escherichia coli* (EPEC) on enterocytes during infection. The pathogenic event can be grouped into four stages (**1**) expression of adhesion factors, (**2**) initial localized adherence, (**3**) signal transduction and intimate contact, and (**4**) cytoskeletal rearrangement and pedestal formation. *Esp E. coli* secreted protein, *BFP* bundle-forming pilus, *T3SS* type III secretion system, *Tir* translocated intimin receptor, *TJ* tight junction

Fig. 14.7 Delivery of virulence effectors of enterohemorrhagic *Escherichia coli* (EHEC) and enteropathogenic *E. coli* (EPEC) to the host cell by the type III secretion system (T3SS) (Redrawn from Hayward et al. 2006. Nat. Rev. Microbiol. 4:358–370)

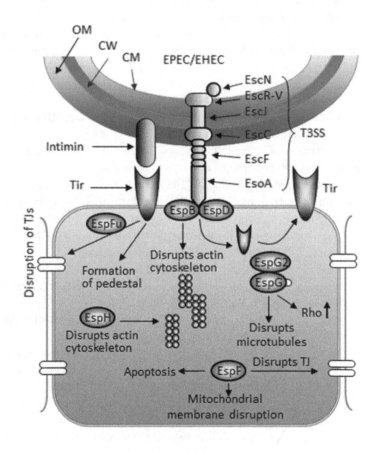

of the protein. The intimin subtypes α-, β-, and γ-type are most relevant clinically. EHEC also expresses intimin and has 83% amino acid sequence homology with intimin from EPEC. EPEC uses T3SS to inject Tir into the host cytosol, which is initiated by Ca^{2+} sensing, and Tir is expressed on the surface to interact with intimin. Tir is then phosphorylated by the host cell kinases, which recruit Nck to the site of attachment and activate neural Wiskott–Aldrich syndrome protein (N-WASP) and the actin-related protein 2/3 (Arp2/Arp3) complex to ochestrate actin rearrangements and pedestal formation.

Extensive rearrangement of actin causes abnormalities in cytoskeletal structure, resulting in the formation of the characteristic attaching and effacing (A/E) lesions. Effacement refers to the loss of microvilli. The epithelial cell membrane beneath the bacterium forms a pedestal, a pseudopod-like structure, due to massive cellular cytoskeletal protein rearrangements, which include actin filament and its cross-linker, talin, ezrin, α-actinin, and MLC. In this phase, bacteria lose EspA filaments from the surface (Figs. 14.6 and 14.7).

Cytoskeletal Rearrangement and Pedestal Formation

As mentioned above, intimate contact results in the formation of a "pedestal-like" structure and attachment and effacement lesion due to massive accumulation of actin filaments. The lesion is further characterized by deformation and loss of microvilli due to depolymerization of the actin filament in microvilli. The effector protein EspF translocated through the T3SS also affects tight junction (TJ) proteins and mitochondrial function and increases epithelial membrane permeability. As a result, there is malabsorption of nutrients and ions, cell death, and the onset of osmotic diarrhea. EspF and EspB also inhibit phagocytosis.

LEE and Regulation of Virulence Genes

Virulence factors for EPEC pathogenesis are located primarily on the 35 kb LEE pathogenicity island and on non-LEE (Nle)-encoded genes. LEE is integrated into the chromosome near tRNA gene, *selC*. LEE has five polycistronic operons (LEE1–LEE5), which contain genes for A/E lesion (*map*, *espF*, *espG*, *espZ*, *espH*, and *espB*), intimin (*eae*), Tir (*espE*), the T3SS (*espA*, *espB*, and *espD*), and the exported proteins. LEE is also present in EHEC and the organization of genes is similar to EPEC, but the size and the number of genes may vary between EHEC and EPEC.

Nle effector proteins are involved in dampening the host immune response. NleB, NleC, NleD, NleE, and NleH have all been shown to inhibit NF-κB activation. Nle effectors such as EspJ have antiphagocytic activity, while NleA alters host protein secretion and tight junction integrity.

EPEC virulence genes are also located on a large plasmid, EAF (*E. coli* adherence factor). Two important operons are present: *bfp* and *per* (plasmid-encoded regulator). *bfp* encodes for BFP, while Per is the transcriptional activator, which regulates manifestation of A/E activity and BFP.

Symptoms

EPEC causes diarrhea in children under the age of 5 and is associated with high mortality. Diarrhea may be acute or persistent, and the latter type is the most common form of clinical presentation. In severe cases, diarrhea may be bloody and the infection may persist for several days.

Enterohemorrhagic *E. coli*

Characteristics

Enterohemorrhagic *E. coli* (EHEC) is a subset of a broadly classified Shiga toxin-producing *E. coli* (STEC) that inflict bloody diarrhea or hemorrhagic colitis (HC) and hemolytic uremic syndrome (HUS) and is prevalent in industrialized countries. Broadly, any *E. coli* strain that produces Stx is called STEC. Not all STEC strains are pathogenic to humans; however, the EHEC strains that express Stx and Eae are the most virulent pathotypes to humans. EHEC, a subset of STEC, carries the LEE pathogenicity island and displays attaching and effacing lesion. The STEC strains that lack the LEE are less virulent. EHEC does not express bundle-forming pili (BFP); instead, the EHEC plasmid carries a homolog of the *lifA* gene encoding lymphostatin, as well as the genes encoding T3SS, catalase, peroxidase, a serine protease, and hemolysin.

In 1977, Konowalchuk and his colleagues demonstrated that Shiga toxin infects Vero cells derived from the kidney; thus, this toxin is also referred to as verotoxin (VT), and the verotoxin-producing *E. coli* is called VTEC. Both VTEC and STEC terminologies have been used interchangeably. Karmali and his colleagues in 2003 divided STEC strains into five seropathotypes (A–E), based on the type of outbreaks and the severity of infection. Pathotype A (O157:H7, O157:NM) is the common outbreak group responsible for severe HUS and HC, and pathotype B (O26:H11, O103:H2, O111:NM, O121:H19, O145:NM) is responsible for occasional outbreak and can cause HUS and HC. Pathotypes C and D rarely cause outbreaks and may cause HUS and HC, while pathotype E is not implicated in outbreaks.

The principal serotype associated with EHEC group is *E. coli* O157:H7 and nonmotile serovar, O157:NM, and the first outbreak of O157:H7 was reported in 1982–1983. Other important non-O157 EHEC serovars include O26, O45, O103, O111, O121, and O145 all of which express Stx and Eae. The USDA-FSIS (Food Safety Inspection Service) has imposed a zero tolerance for these six serovars plus O157:H7 in ground beef and meat trimmings and considered these serovars to be adulterant if present in meat. Likewise, the European Food Safety Authority (EFSA) has also placed greater emphasis on a slightly modified list of serovars that include O157:H7, O26, O103, O145, O111, and O91.

As opposed to other commensal strains, *E. coli* O157:H7 generally does not ferment sorbitol and does not have β-glucuronidase activity (GUD). It grows rapidly at 30–42 °C, grows poorly at 44–45 °C, and does not grow at 10 °C or below. Strains resistant to pH 4.5 or below (pH 3.6–3.9) have been identified, and acid resistance is mediated by RpoS, a sigma factor. The organism is destroyed by pasteurization (at 64.3 °C in 9.6 sec), but the cells survive well in food at −20 °C.

Several selective and chromogenic media are available for isolation of EHEC especially the serovar, O157:H7: sorbitol MacConkey agar (SMAC) supplemented with cefixime–tellurite, Rainbow® Agar, CHROMagar™, and R&F® *E. coli* (Fig. 14.8). The US Food and Drug Administration (FDA)'s *Bacteriological Analytical Manual* (BAM) provides details of diarrheagenic *E. coli* detection (https://www.fda. gov/Food/FoodScienceResearch/ LaboratoryMethods/ucm2006949.htm).

Food Association and Outbreaks

Cattle are the natural reservoir of STEC. STEC is generally present in the intestine of animals without causing disease. STEC also have been isolated from the feces of chickens, goats, sheep, pigs, dogs, cats, and sea gulls. Foods of animal origin, especially ground beef, have been implicated in many outbreaks in the USA, Europe, and Canada; however, in late 2006, a major outbreak of O157:H7 involving 26 states was associated with spinach. The affected people (199 with 3 deaths) were found to have consumed spinach. In the 1993 outbreak, affecting over 500 people and causing 4 deaths, consumption of undercooked hamburgers served by a fast-food chain in Washington, Nevada, Oregon, and California was implicated. In addition to ground beef, other foods, such as uncooked sausages, fermented hard salami, raw milk, yogurt, mayonnaise, raw milk cheese, apple cider, fruits, sprouts, and salad, have been implicated in epidemic and sporadic outbreaks. STEC has been routinely isolated from many different types of foods of animal origin, such as ground beef, pork, poultry, lamb, and raw milk (Fig. 14.3). The organism was also isolated in low frequencies from dairy cows, calves, and chickens. However, in some

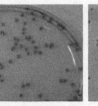
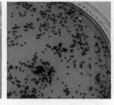
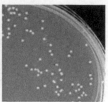

| Sorbitol MacConkey agar (SMAC) | Rainbow® Agar O157 | CHROMagar™ O157 | R&F® *E. coli* O157:H7 | BHI |

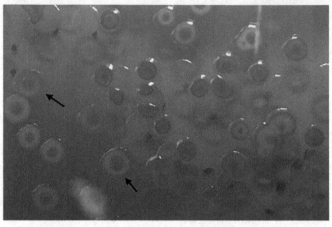

E. coli O157:H7 colonies on Sorbitol MacConkey agar+CT

Fig. 14.8 Colonies of *E. coli* O157:H7 on various selective agar media. *BHI* brain heart infusion, *CT* cefixime–tellurite

cases, a high percent of feedlot cattle shed *E. coli* O157:H7, some of which are persistent shedders. Interestingly, calves after weaning shed in higher frequency than before weaning.

EHEC Pathogenesis

EHEC reaches the intestine from contaminated food or water and colonizes the intestine. The prototype EHEC strain, *E. coli* O157:H7, is acid resistant; it can pass through the stomach unharmed and reaches the small intestine. A small infectious dose of 50–100 cells is sufficient to cause infection. In addition, preexposure of cells to mild acid, as with acidic foods, such as apple cider or fermented hard salami, the bacterium becomes more resistant to low pH and ensures better survival during transit through the stomach.

Attachment and Effacement

EHEC causes characteristic attaching and effacing lesion similar to EPEC and occurs in three stages: localized adherence, signaling event, and intimate contact. In the first stage, adhesion of bacteria to the microvilli of the intestinal epithelial cells is mediated by fimbriae (encoded on the 60 MDa plasmid), type IV pilus called the hemorrhagic coli pilus, and flagella. Unlike EPEC, EHEC does not express BFP. In the second stage, a signal is transmitted to the host cell via T3SS (Fig. 14.7), and phosphorylation of eukaryotic protein occurs, leading to actin polymerization, cytoskeletal rearrangement, and effacement of microvilli. In the third stage, intimate contact is mediated by intimin protein encoded by the *eae* gene located on the LEE pathogenicity island, similar to EPEC. There are 27 intimin variants, based on the heterogeneity in the

C-terminal sequence of the protein, and γ-subtype intimin is associated with EHEC *E. coli* O157:H7 and EPEC O55:H- and O55:H7 strains. Intimin binds to the Tir receptor, and subsequent signaling events amplify the cytoskeletal rearrangement of proteins beneath the adherent bacteria. Increased actin filament accumulations are mediated by Arp2/Arp3 complex, which is regulated by N-WASP, and form pedestal with the loss of microvilli structure. The A/E pathology causes enterocyte sloughing, inflammation, and possibly diarrhea, which may result from the inhibition of sodium and chloride absorption, activation of the chloride channel, loosening of tight junction, increased paracellular permeability, inflammatory response, and cytokine production.

T3SS and Delivery of Effector Proteins

T3SS plays a crucial role during EHEC pathogenesis, and it delivers virulence effector proteins directly inside the host cell cytoplasm that are responsible for A/E lesion (Fig. 14.7). The T3SS needle complex is composed of several Esc proteins (EscN, EscR-V, Esc J, EscC, EscF, and EspA), which spans from the bacterial cytoplasmic membrane (CM) to outer membrane (OM). The T3SS injects effector proteins known as Esp (*E. coli* secreted proteins), which perform various functions: EspB and EspD form a plasma membrane translocon for effective delivery of effector proteins. EspB also affects cytoskeletal structure by disrupting the actin cytoskeleton. Another effector protein, EspH, also promotes disruption of the actin cytoskeleton. EspG and EspG2 disrupt microtubule and activate a small GTPase protein, Rho. EspF causes membrane disruption in the mitochondria, disrupts tight junction proteins, and causes increased membrane permeability. Bacteria also inject Tir (also known as EspE in EHEC) which binds to the intimin. After translocation into the host cell, unlike EPEC, Tir is not tyrosine phosphorylated, and pedestal formation is independent of Nck protein.

Shiga Toxin

STEC produces two types of Stx, Stx1 and Stx2, and the genes are located on bacteriophage (prophage). The Stx1 sequence is highly conserved and exhibits a high sequence similarity to Stx produced by *Shigella dysenteriae* type 1 (see Chap. 19). The antibody developed against the Stx from *S. dysenteriae* type 1 can neutralize Stx1 from STEC but not the Stx2. Stx1 has three subtypes: Stx1a, Stx1c, and Stx1d. Stx2 has seven subtypes: Stx2a, Stx2b, Stx2c, Stx2d, Stx2e, Stx2f, and Stx2g (Table 14.2). STEC strains do not have a secretory system for Stx; thus, secretion is dependent on the cell lysis mediated by the lytic bacteriophages. Stx2 is highly toxic, and if the isolate also has *eae*, there are greater risk of causing hemorrhagic colitis (HC) and hemolytic uremic syndrome (HUS) than Stx1. Strains of STEC with *eae* and producing Stx2 cause more severe disease than the strains producing only Stx1 or both Stx1 and Stx2.

As mentioned above, the gene for Stx production is encoded in a temperate bacteriophage (*stx* phage) related to the classic λ phage, which integrates into the chromosome and maintains in the lysogenic state. The genome size of *stx* phage ranges from 29.7 to 68.7 kb, mostly above 60 kb. The *stx* phages can be induced to enter from lysogenic to lytic cycle, and the resulting free phages can transfer *stx* genes horizontally to *E. coli* or other members of the *Enterobacteriaceae* family. Most lysogens stably maintain *stx* phage; however, only a small subpopulation is induced spontaneously. Induction of *stx* prophages is controlled by RecA, a regulator of the SOS bacterial response during DNA damage. RecA-dependent UV irradiation and mitomycin C can induce *stx* phases, while RecA-independent ethylenediamine tetra acetic acid (EDTA) can induce stx_2 phages. Other factors that regulate the lysogeny switch and lysis include H_2O_2, sodium citrate, high temperature plus UV, amino acid starvation, phenethyl isothiocyanate, colicins, sodium chloride, nitric oxide, gamma irradiation, and antibiotics (azithromycin,

Table 14.2 Shiga toxin types associated with STEC

Shiga toxin types	Receptor types	Description
Stx1	GB3	STEC and identical to Stx from *Shigella dysenteriae*
Stx1a	GB3	Linked to serious human disease
Stx1c	GB3	Found in *eae*-negative STEC, common in sheep STEC isolates, mild diarrhea in humans
Stx1d	GB3	Mild diarrhea or asymptomatic in humans
Stx2a	GB3	High virulence and HUS in humans
Stx2b	GB3	Not involved in a serious disease
Stx2c	GB3	Diarrhea and HUS in humans, common in ovine STEC, less toxic to Vero cells
Stx2d	GB3	It can be activated by mucus and cause severe diarrhea and HUS in humans even in the absence of Eae. It is less toxic to Vero cells
Stx2e	GB3 and GB4	Edema disease in pigs, rare in humans, low pathogenicity
Stx2f	GB4	Pigeon isolates, rare in humans, pathogenicity – uncertain
Stx2g	GB3	Low pathogenicity in humans

ciprofloxacin, fosfomycin, imipenem, gentamicin, norfloxacin, and rifampicin).

The Stx molecules are A–B_5 heterohexamer toxins of 70 kDa, in which the A subunit (StxA) is about 32 kDa and the B subunit (StxB) is 7.7 kDa each. StxA is an enzyme and StxB interacts with the receptor, and both are secreted into the bacterial periplasm and assemble via a noncovalent bond, hence called holotoxin. A single enzymatic A subunit remains associated with a pentamer of B subunits. StxB interacts with the host cell receptor, globotriaosylceramide (Gb3), a glycolipid consisting of galactose α (1–4), galactose β (1–4), glucose ceramide, or globotetraosylceramide (Gb4), and is abundant in the endothelium and the kidney tubule.

Following the binding of the StxB subunit to the receptor (GB_3 or GB_4), the StxA subunit is internalized by receptor-mediated endocytosis and transported to the Golgi and then to the endoplasmic reticulum (ER) (Fig. 14.9). The StxA1 subunit is activated after cleavage of the 4 kDa C-terminal A2 peptide, which remains associated with the StxB subunit. The resulting active A1 portion has *N*-glycosidase activity and cleaves a purine residue (adenine base) in VI domain of 28S ribosomal RNA of host ribosome, altering the function of ribosome such that it no longer can interact with the elongation factors EF-1 and EF-2 required for chain elongation thus inhibiting protein synthesis. A lack of protein synthesis leads to cell death. The severity of infection and damage to the tissues depend on the number of receptors present.

Both Stx1 and Stx2 toxins bind to Gb3. Stx1 causes localized damage to the colonic epithelium because of high binding affinity to Gb3, whereas Stx2 has low affinity for Gb3. Stx1 and Stx2 can reach the circulatory system and the kidneys. Human kidney tubules have high Gb3 and are the major target for toxin-induced damage resulting in hemolytic uremic syndrome. Stx subtypes, Stx2e and Stx2f, use Gb4 as the preferred receptor (Table 14.2). Stx2 has significantly higher toxicity than Stx1. Stx2 together with intimin (Eae) pose the highest risk of developing HUS. The crystal structure of toxin reveals a greater accessibility of the active site of Stx2 than Stx1, and this possibly contributes to the enhanced cytotoxicity.

Stx has been shown to possess nephrotoxic, cytotoxic, enterotoxic, and neurotoxic effects. Stx causes nephrotoxicity following enteric infection, resulting in massive damage to the kidney tubules, bloody urine, and the hemorrhagic uremic syndrome. Since Stx causes chronic kidney damage, there is a need for dialysis and kidney transplant. HUS is also characterized by thrombocytopenia and hemolytic anemia. Though calves are susceptible to EHEC infection, they do not develop HUS because they lack the receptors in the endothelial cells of blood vessels. Stx also acts as a neurotoxin, causing a neurological disorder called thrombotic thrombocytopenic purpura (TTP), which is characterized by hemolysis, thrombocytopenia, renal failure, and fluctuating fever. Stx is also reported to have enterotoxin activity resulting in fluid accumulation and diarrhea. Stx is cytotoxic, inhibits protein synthesis, and induces pro-

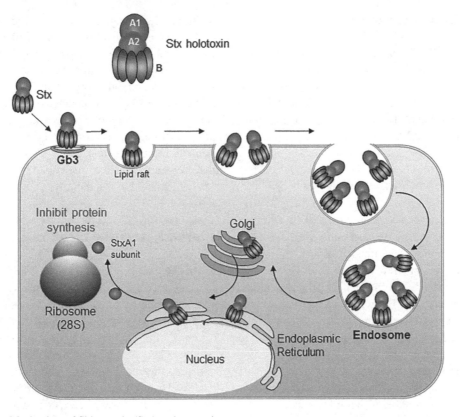

Fig. 14.9 Mechanism of Shiga toxin (Stx) pathogenesis

grammed cell death. Stx subtypes (Stx2e, Stx2f) are also found in pigs, humans, or pigeons and can cause edema disease, bloody diarrhea, HC, and HUS.

Inflammation

Inflammation is very prominent in the intestine during infection with *E. coli* O157:H7. Flagellin (H7) is thought to be responsible for the inflammatory response. It binds to the toll-like receptor 5 (TLR-5) on epithelial cells; activates p38, ERK (extracellular signal-regulated kinase)–MAP (mitogen-activated protein kinase) kinase, and NF-κB; and increases expression of proinflammatory cytokine, IL-8. The inflammation likely disrupts epithelial barrier function and facilitates Stx passage from the lumen to the submucosal layer. LPS (O157 antigen) activates platelets and

together with Stx may cause endothelial cell injury and may contribute to the thrombocytopenia observed in HUS. An LPS-mediated release of cytokines IL-1 and TNF-α from activated macrophages can cause vascular damage during the renal failure in the HUS patients.

Enterohemolysin

Enterohemolysin (Ehly or Ehx) has been isolated from EHEC, and it belongs to the family of RTX (repeats in toxin). It is a monomeric pore-forming toxin and its role in pathogenesis is unclear. It is encoded by four genes (*ehxC*, *ehxA*, *ehxB*, and *ehxD*) and is located on the 60 MDa plasmid. Ehly is secreted by type I secretion system (T1SS). RTX may cause localized lesions or affect cells of renal tubules, but its involvement in pathogenesis is undetermined.

Other Virulence Factors

EHEC O157:H7 genome sequence also revealed the presence of several putative virulence factors: fimbrial adhesins such as Lpf and SfpA, non-fimbrial adhesins (EfaI, Iha, OmpA, and ToxB), toxins (cytolethal distending toxin; CDT), proteases (EpeA, EspP/PssA), and urease.

Regulation of Virulence Genes

Regulation of genes located on LEE is complex and involves non-LEE-encoded (Nle) and LEE-encoded genes. The transcriptional regulators, Ler (LEE-encoded regulator) and Grl (global regulator of LEE activator), positively regulate the genes on LEE. EHEC uses a quorum sensing regulatory system to recognize the intestinal environment and activate genes that are required for colonization in the gut. Autoinducers, like epinephrine and norepinephrine, also regulate genes for flagella and motility, which allow bacteria to find a suitable niche in the gut.

The *stx* genes are located in the lysogenic lambdoid phage and are highly expressed when the lytic cascade of the phage is activated. *Stx* gene expression is regulated by iron concentration where higher concentration suppresses expression.

Symptoms

Symptoms of EHEC/STEC infection appear 3–9 days after ingestion of contaminated food and generally last for 4–10 days. The colitis symptoms include a sudden onset of abdominal cramps, watery diarrhea (which in 35–75% of cases turns bloody), and vomiting. Damage to the blood vessels in the colon is responsible for bloody diarrhea and hemorrhagic colitis. Stx may damage endothelial cells in the kidney and hemolytic uremic syndrome that develops in 5–10% of STEC-infected patients. The HUS is characterized by acute renal failure, hypertension, microangiopathic hemolytic anemia, and thrombocytopenia. EHEC infection can be fatal, particularly in

children under the age of 5 and the elderly. About 1–2% of patients die during the acute phase of the disease, and about 30% of patients exhibit long-term renal damage. Though the kidney is the primary target organ, other organs such as the lungs, central nervous system, pancreas, and heart are also affected. Thrombotic thrombocytopenic purpura (TTP) may result from the blood clot in the brain, eliciting seizures, coma, and death. The most severely affected patients require blood transfusion and dialysis therapy.

Enteroaggregative *E. coli*

Characteristics

Enteroaggregative *E. coli* (EAEC) causes persistent diarrhea, lasting more than 14 days, in children and adults, and is prevalent in developing countries. It is also increasingly responsible for persistent diarrhea in HIV-infected persons in developed countries. The bacterium causes mostly sporadic cases, but recent data show it also causes outbreaks and traveler's diarrhea. EAEC is a highly heterogeneous group, and 40 different O types have been identified. Persistent diarrhea in children is similar to ETEC, characterized by mild but significant mucosal damage. EAEC also possesses pathogenicity islands that carry genes for enterotoxin and mucinase activity.

A Stx-producing EAEC serovar O104:H4 was involved in a large outbreak in Germany in 2011. Fenugreek sprout was the vehicle, and the seeds were imported from Egypt. About 3800 people were infected, of which 2987 suffered from gastroenteritis and bloody diarrhea, and 855 people exhibited hemolytic uremic syndrome with severe kidney damage and 54 deaths.

Virulence Factors and Pathogenesis

EAEC pathogenesis involves three major steps: (1) adhesion to the mucosal surface, (2) biofilm formation, and (3) signal transduction and toxin production.

Adhesion

In the first stage of pathogenesis, EAEC strains express aggregative adherence fimbriae (AAF), and there are four variants with the distinct structure of pilin subunits: AAF/I, AAF/II, AAF/III, and AAF/IV. Fimbriae bind to intestinal epithelial cell matrix proteins such as laminin, collagen, cytokeratin, and fibronectin. In addition, EAEC strains express the surface protein dispersin, which is encoded by *aap* gene. The bacterium also expresses 18 and 30 kDa outer membrane adhesin proteins. The 18 kDa adhesin protein is a thin filamentous (fibrillar) structure and is called GVVPQ fimbria (G, glycine; V, valine; P, proline; and Q, glutamine). This sequence is located near the N-terminal end and may be responsible for "clumping" of cells or adherence to each other promoting autoaggregation, rather than facilitating bacterial attachment to the host cell. Aggregated EAEC cell adhesion appears as a characteristic "stacked brick." The genes for adhesion are encoded on the 60 MDa plasmid. The virulence genes located on the plasmid or the chromosome are regulated by *aggR* located on the plasmid.

Biofilm Formation

In the second stage of pathogenesis, EAEC forms a thick aggregating biofilm on the mucosal layer. This helps persistent colonization with prolonged diarrhea. Biofilm production is regulated by AggR, and it requires other gene products that are involved in biofilm formation. Fis, a DNA-binding protein, is involved in bacterial growth regulation, and YafK (28 kDa) and Shf (32.8 kDa) proteins are both involved in biofilm formation.

Toxins

The third stage of pathogenesis involves the production of toxins by EAEC, which elicit an inflammatory response, mucosal toxicity, and intestinal secretion. EAEC produces several toxins:

1. Heat-stable, ST-like toxin, also called EAST (enteroaggregative ST), encoded by *astA* gene, is responsible for fluid loss similar to STa of ETEC.

2. The plasmid-encoded toxin (Pet), a serine protease autotransporter that cleaves spectrin protein within the cytoskeleton of the epithelium, resulting in cell elongation and exfoliation.

3. Sat (secreted autotransporter toxin) affects cellular tight junctions in the kidney cells and vacuolation in both the kidney cells and the bladder cells.

4. Pic (protein involved in intestinal colonization), a mucinase that interferes with the integrity of the mucus membrane.

5. *Shigella* enterotoxin I (ShET1) similar to *Shigella* enterotoxin that induces intestinal cAMP- and cGMP-mediated secretion, hemorrhagic necrosis and shortening of villi, enlarged crypt openings, and formation of crypt abscesses.

Mechanism of Pathogenesis

EAEC adhere to the enterocytes forming aggregates, and adherence is characterized by a "stacked brick." EAEC also enhance mucus secretion from the goblet cells and trap themselves in mucus-forming biofilms. EAEC do not invade epithelial cells, but the toxins are responsible for the histopathological damage. The toxins induce shortening of villi and hemorrhagic necrosis of villous tips, increased epithelial cell extrusion, and inflammation that is characterized by the infiltration of mononuclear cells to the submucosa. Virulence factors induce levels of fecal cytokines and inflammatory markers, such as TNF-α, IL-1, IL-6, IL-8, IFN-γ, lactoferrin, fecal leukocytes, and occult blood. IL-8 also recruits neutrophils to the epithelial mucosa without mucosal injury and facilitates intestinal fluid secretion. Flagellin (*fliC*) binding to TLR-5 on monocytic cells initiates a signaling cascade through p38, MAPK, and NF-κB to induce the production of IL-8. Infection results in mucoid stool and persistent diarrhea. EAEC also infect immunocompromised hosts, and bacteria are isolated frequently from HIV–AIDS patients stool.

The Shiga toxin-producing EAEC serovar O104:H4 produces Stx2a, which may have acquired the *stx* gene through a bacteriophage and considered a hybrid strain of EAEC and STEC strain. This strain also expresses aggregative adherence fimbriae (AAF) for attachment and carries several virulence genes (*aggA*, *aggR*, *set1*, *pic*, *aap*, and *stx2*) that are encoded on the phage and on a plasmid. The colonization factor and virulence genes are common in both EHEC and EAEC thus explains its increased virulence properties. The pathogenesis of O104:H4 is attributed to increased adhesion, colonization, and damage of the intestinal epithelial cells followed by increased dissemination of Stx to blood circulation and the kidney resulting in high numbers of HUS patients that was associated with the outbreak. This strain is sensitive to antibiotics carbapenems but resistant to penicillin and cephalosporin.

Symptoms

The EAEC-mediated disease could be acute or chronic (>14 days) in nature. Symptoms of EAEC infection include watery, mucoid secretory diarrhea, abdominal pain, nausea, vomiting, and low-grade fever. Some patients show grossly bloody stools. HUS, kidney failure, and death can occur due to infection with the newly emerged EAEC O104:H4 strain.

Enteroinvasive *E. coli*

Characteristics

Enteroinvasive *E. coli* (EIEC) is a facultative intracellular pathogen and causes bacillary dysentery similar to *Shigella* species (see Chap. 19). *Shigella* was discovered by a Japanese physician and bacteriologist, Kiyoshi Shiga, in 1897 from an epidemic in Japan, where the bacterium infected more than 91,000 people and greater than 20% mortality. EIEC was discovered 50 years later, and it is genetically, biochemically, and pathogenetically related to *Shigella* spp. and produces watery diarrhea and dysentery. EIEC strains that are generally lysine decarboxylase negative, nonmotile, and lactose negative can be misidentified as *Shigella*. Sporadic outbreaks are common; however, occasional foodborne outbreaks may occur. An outbreak in the USA, as early as 1971, was recognized from the consumption of imported camembert cheese contaminated with serotype O124:H17. An outbreak was also reported in a restaurant in Texas involving 370 people.

Disease and Symptoms

Ingestion of as many as 10^6 EIEC cells may be necessary for an individual to develop the disease. Mechanism of infection is similar to *Shigella*, but the infective dose of shigellosis is 10–100 cells (see Chap. 19), and EIEC disease is less severe. In colonic mucosa, EIEC first binds and invades epithelial cells, lyses the endocytic vesicle, multiplies in the cytoplasm, moves inside the cytoplasm directionally, and projects toward adjacent cells to spread from cell-to-cell. The genes responsible for invasion are encoded in a 140 MDa plasmid, pInv. A toxin of 63 kDa, encoded by *sen* gene located on the plasmid, has been linked to causing watery diarrhea. Extensive cell damage due to invasion and cell-to-cell spread elicits a strong inflammatory response and bloody mucoid diarrhea, similar to bacillary dysentery caused by *Shigella*. Human carriers, directly or indirectly, also spread the disease.

The symptoms appear as abdominal cramp, profuse diarrhea, headache, chills, and fever. Some patients may develop dysentery. A large number of pathogens are excreted in the feces. The symptoms can last for 7–12 days, but a person may be a carrier and shed the pathogens in feces for a prolonged period.

Diffusely Adhering *E. coli*

Diffusely adhering *E. coli* (DAEC) causes infantile diarrhea and produces a diffuse adherence (DA) to cultured HEp-2 cell lines, which is mediated by a fimbrial adhesin, designated F1845. The genes encoding the fimbriae are located on the chromosome or on a plasmid. DAEC also expresses afimbrial adhesins (Afa) belonging to the Afa/Dr. family of adhesins, which include AfaE-I, AfaE-II, AfaE-III, AfaE-V, Dr., Dr.-II, F1845, and NFA-I adhesins. Adhesion leads to cytoskeleton rearrangement, destroying or partially rearranging microvilli structure. Adhesion also affects paracellular permeability by rearranging occludin and ZO-1 proteins in the TJ. During adhesion, flagellin interaction with TLR-5 stimulates mitogen-activated protein kinases, p38, ERK1/2 (extracellular signal-regulated kinases), and Jun-C kinase and activates NF-κB to produce proinflammatory cytokine, IL-8. Locus of enterocyte effacement (LEE) has been isolated from DAEC, and it possibly carries genes required for the attachment/effacement lesions and signaling events similar to EPEC.

DAEC causes watery diarrhea in children without blood or fecal leukocytes. The DAEC-induced diarrhea is age related and increases with age from 1 year to 4–5 years. The older adults become the asymptomatic carrier of DAEC.

Animal and Cell Culture Model Used for Diagnosis of E. coli

A ligated rabbit ileal loop (RIL) assay has been used for detection of diarrheal toxins produced by different virotypes/serotypes. For diagnosis of ETEC, calves and piglets are used since no small animal models are available. ETEC causes diarrhea in gnotobiotic pig (e.g., specific pathogen-free pig), while EPEC causes attaching and effacing lesions in the piglet intestine.

Tissue culture models have been used extensively to study specific traits. For example, attachment and effacement phenomenon of EPEC has been studied using HEp-2 (laryngeal cells) and HeLa (cervical cancer) cell lines. Caco-2 cells and HT-29 (colon cancer cells) are used to study ETEC attachment. Interestingly, ETEC does not adhere to HEp-2 cells. HEp-2 cells are also used to study diffuse adherence phenotype of DAEC. Vero cells (African green monkey kidney) have been used to study cytotox-

icity (Fig. 5.1) of EHEC/STEC and HEp-2 for attaching/effacing (A/E) assay. The HEp-2 adherence assay is the gold standard for identification of EAEC, although a PCR assay has been developed to detect pathogenic E. coli isolates.

Prevention, Control, and Treatment

Fatalities from diarrheal diseases are due to the extensive dehydration and electrolyte imbalance (loss). Oral hydration and the electrolyte replenishment are the most important therapy. Antibiotics can shorten the duration of infection. Antibiotic therapy is less effective and is not recommended for STEC; however, for ETEC and EAEC infection, fluoroquinolones (e.g., ciprofloxacin, norfloxacin, and ofloxacin) and rifaximin are recommended for treatment. As a preventive measure, travelers can use doxycycline, rifaximin, and trimethoprim–sulfamethoxazole before a scheduled trip to the endemic region. Concerns of antibiotic resistance discourage the use of antibiotics as a prophylactic measure; therefore, travelers are advised to avoid potential hazardous food and water. Water should be boiled and food should be properly cooked to prevent infection. The antidiarrheal drug, Imodium, is effective against diarrhea.

Proper sanitation, cooking or heating at appropriate temperatures, proper refrigeration, and prevention of cross-contamination should be practiced in order to control the presence of EHEC E. coli O157:H7 in a ready-to-eat food. EHEC is a heat-sensitive organism and is inactivated at 62.8 °C for 0.3 min in ground beef. The USDA-FSIS has provided several guidelines to control EHEC−/STEC-related foodborne illnesses: Use only pasteurized milk; quickly refrigerate or freeze perishable foods; never thaw a food at room temperature or keep a refrigerated food at room temperature over 2 h; wash hands, utensils, and work areas with hot soapy water after contact with raw meat and meat patties; cook meat or patties until the center is gray or brown or internal temperature reaches to 68.3 °C (155 °F); and prevent fecal–oral contamination through proper personal hygiene. Routine surveilance of cattle for the presence of EHEC should be carried out, and cattle should be tested for pathogen presence before slaughter. HACCP (hazard analysis critical control points) principles should be incorporated into the slaughtering and processing operations. Consumers should be educated for safe handling of raw meats and should avoid cross-contamination of cooked products.

Summary

Most *Escherichia coli* are a harmless inhabitant of the intestinal tract, and only a small percentage of strains are considered pathogenic. However, a recent surge in the enterohemorrhagic *E. coli* (EHEC), a highly virulent subset of Shiga toxin-producing *E. coli* (STEC) outbreaks, suggests a possible increased horizontal or vertical transfer of pathogenic genes among bacterial species. There are six virotypes of *E. coli* (EHEC, EPEC, ETEC, EIEC, DAEC, and EAEC), of which EHEC, EPEC, and ETEC are known to cause severe disease worldwide. Increased insight into their genetic and phenotypic properties of virulence factors and their pathogenic mechanisms should help in formulating appropriate preventive or therapeutic measures. The common themes shared by all *E. coli* virotypes include the following: they adhere to the epithelial cells and cause damage to the cells by initiating signaling events that lead to blockage of protein synthesis, alter the cytoskeletal structure leading to attachment and effacement lesion, affect ion pumps, increase fluid loss, or cause cell death. In recent years, however, the research focus is geared more toward EHEC group because of their continued association with serious foodborne outbreaks from a wide variety of foods, including fruits, vegetables, meats, and dairy products. Analysis of recent outbreak strains indicates association of Stx2 and Eae to be the most important virulence factor of EHEC/STEC, causing hemorrhagic colitis (HC), severe hemolytic uremic syndrome (HUS), and kidney damage. Association of this pathogen with fresh vegetables presents a serious problem because these products are minimally processed and, apparently, the processing conditions are inadequate for complete removal or inactivation. Furthermore, these organisms probably have developed strategies to utilize nutrients from plants for prolonged survival inside the plant tissues, and they are resistant to washing and disinfections. Diarrheal diseases are preventable by adopting proper sanitary condition during the preparation of food, by thorough cooking, and by avoiding foods that might be the potential source of the organism. Dehydration and electrolyte loss result from diarrhea, which can be fatal; thus, hydration is the most important therapy against diarrheal diseases. The most severely affected patients suffering from EHEC/STEC require blood transfusion and dialysis therapy.

Further Readings

1. Bergan, J., Dyve Lingelem, A.B., Simm, R., Skotland, T. and Sandvig, K. (2012) Shiga toxins. Toxicon 60, 1085–1107.
2. Bettelheim, K.A. and Goldwater, P.N. (2013) Shigatoxigenic *Escherichia coli* in Australia: a review. *Rev Med Microbiol* **24**, 22–30.
3. Beutin, L. (2006) Emerging enterohaemorrhagic *Escherichia coli*, causes and effects of the rise of a human pathogen. *J Vet Med Series B* **53**, 299–305.
4. Beutin, L. and Martin, A. (2012) Outbreak of Shiga toxin-producing *Escherichia coli* (STEC) O104:H4 infection in Germany causes a paradigm shift with regard to human pathogenicity of STEC strains. *J Food Prot* **75**, 408–418.
5. Clarke, S.C., Haigh, R.D., Freestone, P.P.E. and Williams, P.H. (2003) Virulence of enteropathogenic *Escherichia coli*, a global pathogen. *Clin Microbiol Rev* **16**, 365–378.
6. Croxen, M.A., Law, R.J., Scholz, R., Keeney, K.M., Wlodarska, M. and Finlay, B.B. (2013) Recent advances in understanding enteric pathogenic *Escherichia coli*. *Clin Microbiol Rev* **26**, 822–880.
7. Estrada-Garcia, T. and Navarro-Garcia, F. (2012) Enteroaggregative *Escherichia coli* pathotype: a genetically heterogeneous emerging foodborne enteropathogen. *FEMS Immunol Med Microbiol* **66**, 281–298.
8. Feng, P., Weagant, S.D. and Jinneman, K. (2014) BAM: diarrheagenic *Escherichia coli*. US Food and Drug Administration.
9. Feng, P., Weagant, S.D. and Monday, S.R. (2001) Genetic analysis for virulence factors in *Escherichia coli* O104: H21 that was implicated in an outbreak of hemorrhagic colitis. *J Clin Microbiol* **39**, 24–28.

10. Fleckenstein, J.M., Hardwidge, P.R., Munson, G.P., Rasko, D.A., Sommerfelt, H. and Steinsland, H. (2010) Molecular mechanisms of enterotoxigenic *Escherichia coli* infection. *Microbes Infect* **12**, 89–98.

11. Gyles, C.L. (2007) Shiga toxin-producing *Escherichia coli*: An overview. *J Anim Sci* **85**, E45–62.

12. Hayward, R.D., Leong, J.M., Koronakis, V. and Campellone, K.G. (2006) Exploiting pathogenic *Escherichia coli* to model transmembrane receptor signalling. *Nat Rev Microbiol* **4**, 358–370.

13. Hebbelstrup Jensen, B., Olsen, K.E.P., Struve, C., Krogfelt, K.A. and Petersen, A.M. (2014) Epidemiology and clinical manifestations of entero-aggregative *Escherichia coli*. *Clin Microbiol Rev* **27**, 614–630.

14. Isidean, S.D., Riddle, M.S., Savarino, S.J. and Porter, C.K. (2011) A systematic review of ETEC epidemiology focusing on colonization factor and toxin expression. *Vaccine* **29**, 6167–6178.

15. Johannes, L. and Romer, W. (2010) Shiga toxins - from cell biology to biomedical applications. *Nat Rev Microbiol* **8**, 105–116.

16. Kaper, J.B., Nataro, J.P. and Mobley, H.L.T. (2004) Pathogenic *Escherichia coli*. *Nat Rev Microbiol* **2**, 123–140.

17. Karmali, M.A., Mascarenhas, M., Shen, S., Ziebell, K., Johnson, S., Reid-Smith, R., Isaac-Renton, J., Clark, C., Rahn, K. and Kaper, J.B. (2003) Association of genomic O island 122 of *Escherichia coli* EDL 933 with verocytotoxin-producing *Escherichia coli* sero-pathotypes that are linked to epidemic and/or serious disease. *J Clin Microbiol* **41**, 4930–4940.

18. Konowalchuk, J., Speirs, J. and Stavric, S. (1977) Vero response to a cytotoxin of *Escherichia coli*. *Infect Immun* **18**, 775–779.

19. Krüger, A. and Lucchesi, P.M.A. (2015) Shiga toxins and stx phages: highly diverse entities. *Microbiology* **161**, 451–462.

20. McWilliams, B.D. and Torres, A.G. (2014) EHEC adhesins. *Microbiol Spectrum* **2**, EHEC-0003-2013.

21. Nataro, J.P. and Kaper, J.B. (1998) Diarrheagenic *Escherichia coli*. *Clin Microbiol Rev* **11**, 142–201.

22. Ochoa, T.J. and Contreras, C.A. (2011) Enteropathogenic *Escherichia coli* infection in children. *Curr Opin Infect Dis* **24**, 478–483.

23. Qadri, F., Svennerholm, A.-M., Faruque, A.S.G. and Sack, R.B. (2005) Enterotoxigenic *Escherichia coli* in developing countries: epidemiology, microbiology, clinical features, treatment, and prevention. *Clin Microbiol Rev* **18**, 465–483.

24. Servin, A.L. (2005) Pathogenesis of Afa/Dr diffusely adhering *Escherichia coli*. *Clin Microbiol Rev* **18**, 264–292.

25. Shulman, S.T., Friedmann, H.C. and Sims, R.H. (2007) Theodor Escherich: The first pediatric infectious diseases physician? *Clin Infect Dis* **45**, 1025–1029.

26. Sperandio, V. and Pacheco, A.R. (2012) Shiga toxin in enterohemorrhagic *E. coli*: Regulation and novel antivirulence strategies. *Front Cell Infect Microbiol* **2**.

27. van den Beld, M.J.C. and Reubsaet, F.A.G. (2012) Differentiation between *Shigella*, enteroinvasive *Escherichia coli* (EIEC) and noninvasive *Escherichia coli*. *Eur J Clin Microbiol Infect Dis* **31**, 899–904.

Salmonella enterica

15

Introduction

Daniel E. Salmon first reported the isolation of *Salmonella* from a pig in 1885 and named the organism *Bacterium choleraesuis*. The bacterium is currently known as *Salmonella enterica* serovar Choleraesuis. *Salmonella* causes typhoid fever and gastroenteritis, and it is one of the major foodborne pathogens of significant public health concern in both developed and developing countries. Meat, poultry, eggs, nuts, fruits and vegetables, and humans are the major source of infection. The primary mode of infection is the fecal–oral route. *Salmonella enterica* serovar Typhi and serovar Paratyphi are highly invasive and cause typhoid fever, and the infection is restricted to the human host. On the other hand, non-typhoidal *Salmonella* (NTS) disease such as febrile gastroenteritis is seen in broad vertebrate host range and is caused by *Salmonella enterica* serovars Typhimurium, Enteritidis, Newport, Heidelberg, and so forth. Many serovars have narrower host range such as *Salmonella enterica* serovar Dublin which infects cattle, while *S. enterica* serovar Choleraesuis infects swine; however, these two serovars can also infect humans resulting in a more invasive form of the disease. In the majority of immunocompetent hosts, NTS causes self-limiting colitis, while in the immunocompromised hosts, such as in malnourished children, and HIV-infected and malaria-infected persons, the disease is highly invasive leading to a systemic infection refer to as invasive non-typhoidal salmonellosis (iNTS). Sub-Saharan Africa has the highest cases of iNTS, while South and Southeast Asia show the median burden of the disease. Worldwide there are over 21 million annual cases of typhoid fever, 1.3 billion cases of gastroenteritis, and 3 million deaths attributed to the *Salmonella* infection. In the United States, annually there are about 1 million cases and 378 deaths with an economic loss of about 3 billion dollars.

Biology

The genus *Salmonella* is a member of the *Enterobacteriaceae* family and is a Gram-negative, nonspore-forming bacillus. Salmonellae are motile (except *Salmonella enterica* serovar Pullorum and serovar Gallinarum) and express peritrichous flagella. They are facultative anaerobes that can grow in a temperature range of 5–45 °C with an optimum temperature of 35–37 °C. The desirable pH range for *Salmonella* growth is between 6 and 7, but it can grow at pH 4.1. In some cases, *Salmonella* is able to grow at low pH (3.5–3.8) in commercial salad dressings and mayonnaise. *Salmonella* is generally sensitive to increasing concentrations of salt (0.5–5%). *Salmonella* forms a long filamentous chain when grown at temperature extremes of 4–8 °C or at

© Springer Science+Business Media, LLC, part of Springer Nature 2018
A. K. Bhunia, *Foodborne Microbial Pathogens*, Food Science Text Series,
https://doi.org/10.1007/978-1-4939-7349-1_15

44 °C and when grown at pH 4.4 or 9.4. All sal-monellae are a facultative intracellular pathogen and considered pathogenic and can invade macro-phages and dendritic and epithelial cells. The vir-ulence genes responsible for invasion, survival, and extraintestinal spread are located on *Salmonella* pathogenicity islands (SPIs).

Source and Transmission

Salmonellae are present in the intestinal tract of birds, reptiles, turtles, insects, farm animals, and humans. Poultry is the major source for human foodborne salmonellosis, in part due to high-density farming operations, which allow colo-nized birds to spread salmonellae to other birds within a flock. Intestinal colonization by salmo-nellae increases the risk for contamination during slaughter and eviscerations. Eggs are also the res-ervoirs for *Salmonella*, particularly the serovar Enteritidis, as this organism can invade and colo-nize the ovary of the laying hen. Likewise, serovars Typhimurium and Heidelberg have also shown transovarian transmission. In such case, the bacteria are present in the egg before the egg-shell is formed in the oviduct. Eggs stored at room temperature or temperature-abusive conditions, *Salmonella* can attain as high as 10^{11} cells per egg.

Human salmonellosis is generally foodborne and is contracted through consumption of con-taminated food of animal origins such as meat, milk, poultry, fish (tuna), and eggs. Dairy prod-ucts including cheese and ice cream are also implicated in *Salmonella* outbreaks. However, fruits and vegetables and nuts, such as lettuce, tomatoes, cilantro, alfalfa sprouts, peanuts, almonds, cantaloupe, and mango, have also been implicated in recent outbreaks. In addition, flour, cookie dough, and spices have been implicated in outbreaks. Animal-to-human or human-to-human transmission can also occur.

Classification

Historically, the Latin binomial form of bacterial naming such as *Salmonella typhimurium* was based on one species-one serovar concept. Many of the species were also named based on the places of origin, such as *S. london* (originally iso-lated at London, England), *S. miami*, *S. rich-mond*, *S. dublin*, *S. indiana*, *S. kentucky*, *S. tennessee*, and so forth. This has been discontin-ued due to the close relatedness among the *Salmonella* serovars. *Salmonella* has been also classified based on their susceptibility to differ-ent bacteriophages (called phage typing). More than 200 definitive phage types (DT) have been reported. Those include phase type (PT) 1, 4, 8, 13, 13a, 23, DT104, DT108, DT204, and so forth. Resistance to different antibiotics has also been used as a means of classification. For example, DT104 is resistant to multiple antibiotics includ-ing ampicillin, chloramphenicol, streptomycin, spectinomycin, sulfonamides, florfenicol, and tetracycline. It is also reported to be resistant to nalidixic acid and ciprofloxacin. The emerging strain DT204 is resistant to eight to nine antibiot-ics and is a major human health concern. Salmonellae have been also grouped based on their somatic (O), flagellar (H), and capsular (Vi) antigenic patterns and there are over 2500 serovars.

To simplify *Salmonella* classification, the genus *Salmonella* has been divided into two major species: *Salmonella enterica* and *Salmonella bongori*. A third species, *Salmonella subterra-nean*, has been added to the list. *Salmonella enterica* contains 2443 serotypes and *S. bongori* contains 20 serotypes. *S. enterica* now has six subspecies, which are designated by roman numerals: I (*enterica*), II (*salamae*), IIIa (*arizo-nae*), IIIb (*diarizonae*), IV (*houtenae*), and VI (*indica*). For example, a *Salmonella* isolate is des-ignated as *Salmonella enterica* subspecies I (*enterica*) serovar Enteritidis. Under the modern nomenclature system, often, the subspecies infor-mation is omitted, and the organism is called *Salmonella enterica* serovar Enteritidis, and in its subsequent appearance, it is written as *S.* Enteritidis. Scientists are thus encouraged to follow this system of *Salmonella* classification and nomenclature to bring uniformity in reporting and to avoid further confusions. The *Salmonella* pathogens that infect birds and mammals primar-ily belong to *S. enterica* subspecies *enterica* (subspecies I). According to the Centers for

Disease Control and Prevention (CDC), the top-20 *Salmonella enterica* serovars that are responsible for the majority of outbreaks in recent years in the United States include (high to low) Enteritidis, Typhimurium, Newport, Javiana, I 4,[5],12:i-, Heidelberg, Montevideo, Oranienburg, Saintpaul, Muenchen, Braenderup, Infantis, Thompson, Mississippi, Paratyphi B, Typhi, Agona, Schwarzengrund, Bareilly, and Hadar.

Salmonellosis and Animal Specificity

Salmonella enterica (Table 15.1), especially the subspecies *enterica*, infects humans and warm-blooded animals. It causes three forms of the disease: typhoid fever, gastroenteritis, and septicemia (bacteremia). *Salmonella enterica* serovar Typhi is the most invasive serovar and it causes typhoid fever, a systemic disease in humans. This serovar is unable to infect other mammals. *S. enterica* serovar Paratyphi causes typhoid-like infection in humans. While the non-typhoidal *Salmonella* (NTS), such as *S. enterica* serovar Typhimurium and serovar Enteritidis, cause self-limiting gastroenteritis or enterocolitis, which is mostly localized to the gastrointestinal tract, these two are the most common serovars responsible for 60% of the total *Salmonella*-related outbreaks globally. The NTS serovars can also cause serious systemic infection in immunocompromised hosts. *S. enterica* serovar Typhimurium causes typhoid-like infection in mice. *S. enterica* serovar Choleraesuis, a swine-adapted pathogen,

causes septicemia (paratyphoid) in pigs. A bovine-adapted *S. enterica* serovar Dublin causes bacteremia, typhoid-like infection, inflammation in the digestive tract, and abortion in cows, and serotype Arizonae infects reptiles. Importantly, the serovars Choleraesuis, Dublin, and Arizonae occasionally cause invasive infections in humans. On the other hand, *S. enterica* serovars Pullorum and Gallinarum cause infection only in poultry.

Salmonella bongori has been primarily associated with the cold-blooded animals such as reptiles; however, it can also infect humans but very rarely.

Septicemic Salmonellosis

Systemic salmonellosis in animals may occur in all age groups, but calves, piglets, and foals are most susceptible, and the diseases are characterized by sudden onset of high fever, depression, and recumbence (lying down). Animals may die within 48 h if no therapeutic intervention is applied; however, the survivors may develop persistent diarrhea, arthritis, meningitis, or pneumonia.

Swine Salmonellosis

Salmonella enterica serovar Choleraesuis causes bluish discoloration of ears and snout in pigs. Pigs may simultaneously suffer from swine fever virus (intercurrent infections) and thus may complicate the disease.

Table 15.1 The major *Salmonella* serovars and their target hosts

Serovar	Pathogen specific to	Disease
Typhi	Humans	Typhoid fever
Paratyphi	Humans	Typhoid fever-like
Typhimurium	Many animals and humans	Gastroenteritis
Enteritidis	Humans	Gastroenteritis
Choleraesuis	Swine	Enterocolitis and septicemia
Dublin	Cattle	Enterocolitis and septicemia, typhoid fever
Pullorum	Chicken	Bacillary white diarrhea
Gallinarum	Chicken	Fowl typhoid
Arizonae	Turkeys	Paracolon infection
Brandenburg	Sheep	Abortion

Pullorum Disease

The disease in poultry is called "pullorum disease" or "bacillary white diarrhea," which affects chicks and turkey poults of 2–3 weeks of age, and the mortality rate is very high, often 100%. The infected birds may huddle under a heat source and become anorexic and depressed and show whitish fecal paste around their vents. Characteristic pathologic lesions in lungs and focal necrosis of the liver and spleen are evident. The survivors may be asymptomatic and can transfer the pathogen via eggs. Other avian species such as guinea fowl, quail, pheasants, sparrows, parrots, canaries, and bullfinches may suffer from this disease.

Fowl Typhoid

Salmonella enterica serovar Gallinarum is the causative pathogen for fowl typhoid, which may be either acute or chronic. The pathologic lesions in chicks and poults are similar to the pullorum disease. In endemic region, septicemic disease can occur in adult birds leading to sudden death. The infection is characterized by enlarged liver and spleen.

Typhoid Fever

Salmonella enterica serovar Typhi causes a systemic febrile illness called typhoid fever, and the pathogen is transmitted through food and water that have been contaminated by human feces. Human-to-human transmission is the most common vehicle. Individuals recovering from the typhoid fever act as chronic carriers and shed bacteria for months. For this reason, it is important that food handlers abstain from working during and immediately following *S.* Typhi infections. The incubation period for *S.* Typhi infection is 1 week to 1 month. Following consumption of contaminated food, *S.* Typhi translocates via mucosal microfold (M) cells or dendritic cells and multiplies in the submucosal layer. The migratory dendritic cells (DCs) transfer *S.* Typhi

to the liver and spleen, where the bacteria are localized in hepatocytes, splenocytes, reticuloendothelial cell, macrophages, neutrophils, and DCs. Bacteria leave the liver and spleen and enter blood circulation in large numbers and reach the gall bladder. From the gall bladder, bacteria are shed into the intestine for the second round of infection through M cells (Fig. 15.1). In some cases, ulceration of the intestine is seen.

S. Typhi produces Vi antigen, which consists of capsular polysaccharide composed of *N*-acetylglucosamine uronic acid. Vi antigen is anti-phagocytic and allows bacterial survival inside phagocytes. Vi antigen also helps scavenge the reactive oxygen species (ROS). Vi antigen has been used as a vaccine candidate to provide protection against typhoid fever, and antibodies to Vi antigen are detected in typhoid fever patients.

Symptoms of typhoid fever appear within 1–2 weeks and last for 2–3 days. High fever is seen due to high levels of LPS-mediated cytokine release. Other symptoms include malaise, headache, nausea, myalgia, anorexia, constipation, chills, convulsions, and delirium (delusion, restlessness, and temporary disturbance of consciousness). The mortality rate of typhoid fever is as high as 10%. In chronic carriers, the organism is shed from the gall bladder for months to years. Antibiotic treatment with fluoroquinolones is the most effective therapy; however, nalidixic acid, ampicillin, and trimethoprim/sulfamethoxazole are also found to be effective. Two types of vaccines are currently employed: the injectable form of live-attenuated bacteria and Vi antigens and oral form of an attenuated strain of *S.* Typhi Ty21a.

Gastroenteritis

The non-typhoidal *Salmonella* (NTS) causes gastroenteritis which is often self-limiting in healthy individuals and rarely causes systemic infection. However, the disease could be invasive, and the bacterium can spread systemically to extraintestinal sites in immunocompromised individuals such as children, malnourished persons, or a

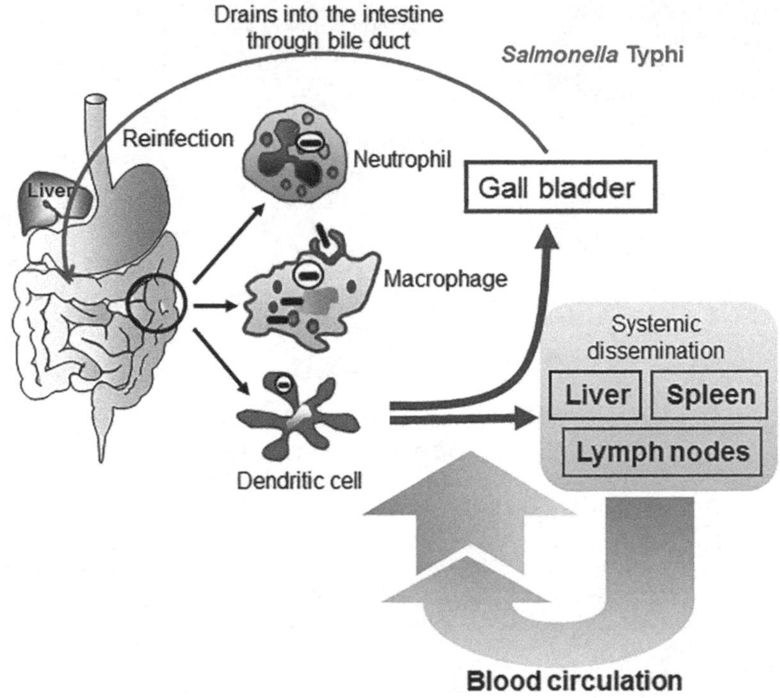

Fig. 15.1 Diagram showing *Salmonella enterica* serovar Typhi spread and interaction with the phagocytic cells during systemic infection

person suffering from HIV or other immunosuppressive diseases. In healthy individuals, symptoms include fever, diarrhea, abdominal pain, and sometimes vomiting. These symptoms are self-resolving in healthy individuals and usually subside within 3–4 days.

The infectious dose of salmonellae is rather broad and may vary from 1 to 10^9 cfu g^{-1}. Studies conducted with the human adult volunteers have indicated that a dose range of 10^5–10^{10} organisms is required to cause disease. The infectious dose decreases if consumed with the liquid food that traverses stomach rapidly or food such as milk and cheese that neutralize the stomach acid. Individuals with underlying conditions (e.g., immunocompromised individuals) are susceptible to a low dosage of bacteria.

In the intestinal tract, salmonellae exhibit tropism for intestinal lymphoid tissue and pass through the M cells present in the follicle-associated epithelium (FAE) overlying Peyer's patches (PP) (Fig. 15.2). M cells and DCs located in the lamina propria ingest luminal microbiota to maintain gut homeostasis, and this process facilitates salmonellae to cross the epithelial barrier. M cells located in the solitary intestinal lymphoid tissue (SILT) distributed throughout the intestinal lining have also been proposed to be the invasion site for salmonellae. SILTs are composed of isolated lymphoid follicles (ILF) that contain B cells and M cells. SILTs are ordered structures that can develop into cryptopatches (lymphoid aggregations mostly filled with stem cell-like cells) or larger ILFs that resemble PP and vice versa. In the lamina propria, salmonellae are engulfed by the resident dendritic cells or macrophages and replicate inside the host cells and induce apoptosis.

Independent of M-cell-mediated entry, salmonellae invade the apical epithelium in the ileum, cecum, and proximal colon and elicit a significant inflammation at the site, which is characterized by neutrophil infiltration, necrosis, edema, and fluid secretion. Massive neutrophil infiltration occurs within 1–3 h of infection. In some cases, salmonellae also disseminate to extraintestinal sites: mesenteric lymph nodes, the liver, and the spleen. Lipopolysaccharide (LPS)-induced "inflammation" is seen during the invasion. LPS also induces abdominal pain, fever, and gastroenteritis. Inflammation and damage to the mucosal cells cause diarrhea and fluid loss (Fig. 15.2).

Fig. 15.2 Sequence of events leading to *Salmonella enterica* Typhimurium-induced pathogenesis and diarrhea

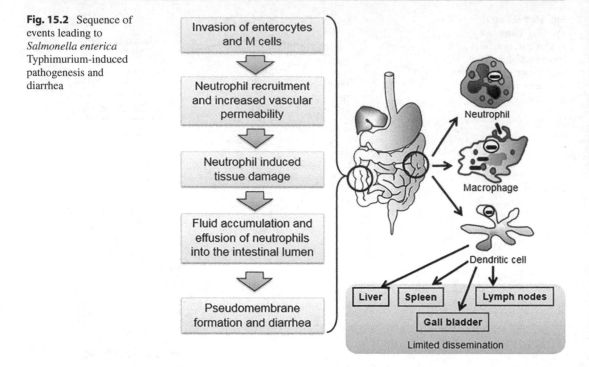

Symptoms appear within 6–24 h as nausea, vomiting, abdominal pain (resembling appendicitis), headache, chills, and bloody or non-bloody diarrhea followed by muscular weakness, muscle pain, faintness, and moderate fever. Symptoms can persist for 2–3 days. The mortality rate for salmonellosis is about 4%. One to 5% of the recovering patients may serve as a chronic carrier and shed bacteria from 3 months to 1 year. Systemic forms of the disease may be seen in children or immunocompromised adults, including cancer, malaria, and acquired immunodeficiency syndrome (AIDS) patients.

Animal Models for Salmonellosis

Inbred mice are the most common model to study *Salmonella* infection (**see** Chap. 5). Healthy mice do not show diarrheal symptoms; therefore, the calf model especially the ligated loop is used to study intestinal inflammation and fluid secretion. In mice, *S.* Typhimurium causes systemic infection, resembling human typhoid fever, thus considered a convenient model to study human typhoid fever. In mice, non-typhoidal *Salmonella*

can cross the epithelial lining over the Peyer's patches and disseminate to the liver, spleen, and bone marrow and reside and replicate inside the resident macrophages and dendritic cells. *S.* Typhi does not infect mice, but humanized mice can be used to study this pathogen. To study non-typhoidal serovar-mediated gastroenteritis in mice, the antibiotic streptomycin is administered to reduce microbiota burden in mice gut before the oral administration of *Salmonella*. This strategy has been widely accepted to study *Salmonella*-induced pathology including inflammatory colitis. The infectious dose for animal models may vary from 10^5 to 10^{10} cfu per animal.

Pathogenicity Islands and Virulence Factors

Salmonella enterica virulence gene clusters are located in 12 pathogenicity islands (SPIs), and some of them are located near the tRNA genes (Table 15.2). Some SPIs are acquired by horizontal gene transfer, while others are conserved throughout the genus. Some SPIs may be specific for certain serovars. Virulence genes that are

Table 15.2 *Salmonella* pathogenicity islands (SPIs) at a glance

Islands	*Salmonella*	Length (kb)	GC (%)	Function
SPI-1	*Salmonella enterica* and *S. bongori*	43	47	T3SS, invasion, iron uptake
SPI-2	*S. enterica*	40	44.6	T3SS, invasion, systemic infection
SPI-3	*S. enterica* and *S. bongori*	17	39.8–49.3	Mg^{2+} uptake, macrophage survival
SPI-4	*S. enterica* and *S. bongori*	27	37–54	Macrophage survival
SPI-5	*S. enterica* and *S. bongori*	7.6	43.6	Enteropathogenicity
SPI-6	*S. enterica* subspecies enterica serovars	59	51.5	Fimbriae
SPI-7	Serovars Typhi, Dublin, Paratyphi	133	49.7	Vi antigen
SPI-8	Serovar Typhi	6.8	38.1	Unknown
SPI-9	*S. enterica* and *S. bongori*	16.3	56.7	Type I secretion system and RTX-like toxin
SPI-10	Serovars Typhi and Enteritidis	32.8	46.6	Sef fimbriae
SGI-1	Serovars Typhimurium (DT104), Paratyphi, and Agona	43	48.4	Antibiotic-resistant genes
HPI	*S. enterica* subspecies IIIa, IIIb, IV	?	?	High-affinity iron uptake, septicemia

Adapted from Hensel, M. (2004) . Int. J. Med. Microbiol. **294**, 95–102

involved in the intestinal phase of infection are located on SPI-1 and SPI-2 (Table 15.3), and the remaining SPIs are required for causing systemic infection, intracellular survival, fimbrial expression, antibiotic resistance, and Mg^{2+} and iron uptake.

SPI-1

SPI-1 is a 43-kb segment that is presumably acquired from other pathogenic bacteria by horizontal gene transfer during evolution. It contains 31 genes that are responsible for the invasion of nonphagocytic cells and components of the type III secretion apparatus (T3SS, more specifically T3SS-1) designated as the Inv/Spa-T3SS apparatus. The major genes located on the SPI-1 are *invA*, *invB*, *invC*, *invF*, *invG*, *hilA*, *sipA*, *sipC*, *sipD*, *spar*, *orgA*, *sopB*, and *sopE* (Table 15.3). InvG is an outer membrane protein secreted by T3SS, and it plays a critical role in bacterial uptake and protein secretion. InvA is an inner membrane protein and is involved in the formation of a channel through which the polypeptides are exported. InvH and HilD (hyperinvasive locus) are accessory proteins involved in the adhesion of *Salmonella*. There are two types of effector proteins secreted by T3SS: InvJ/SpaO and SipB/SipC. SipB and SipC are the major proteins, which interact with the host cytoskeletal proteins, modulate host cytoskeleton, and

promote *Salmonella* uptake. Inv/Spa is also responsible for macrophage apoptosis. SipA is an actin-binding protein. SopB is an inositol phosphate phosphatase and SopE activates GTP-binding proteins. HilA is the central transcriptional regulator of genes located in SPI-1. SPI-1 is absent in *S. bongori*.

SPI-2

SPI-2 is a 40-kb segment that encodes 32 genes. The majority of the genes are expressed during bacterial growth inside the host. It carries genes for Spi/Ssa and T3SS apparatus (more specifically T3SS-2), i.e., SpiC, which inhibits the fusion of the *Salmonella*-containing vacuole (SCV) and lysosome. The gene products are essential for causing systemic infection and mediate bacterial replication rather than survival within the host macrophages (Table 15.3). SPI-2 is also absent in *S. bongori*.

SPI-3

SPI-3 is a 17-kb locus and has ten genes, which are conserved in *S. enterica* serovar Typhi and Typhimurium. SPI-3 is present in *S. bongori*. The SPI-3 encodes gene for MgtCB, which is required for Mg^{2+}-dependent growth, and is essential for survival inside the macrophage.

Table 15.3 Summary of *Salmonella* virulence proteins encoded in SPI-1 and SPI-2

SPI	Proteins encoded	Functions
SPI-1	SpaO, InvJ, InvG, PrgI, PrgJ, PrgK, SipB, SipC	Proteins involved in T3SS needle assembly and secretion of virulence proteins
	SopE, SopE2, SopB (or SigD)	Affect actin cytoskeletal rearrangement via Rho GTPases and disrupts tight junction
	SipA	Interferes with actin polymerization
	SipC (or SspC)	Modulation of actin cytoskeleton by actin nucleation
	SptP	Acts on SopE, SopE2, and SopB to reverse cytoskeletal deformation
	SopA, SopD, SopB	Fluid accumulation in intestine
	SopB, SopE	Modulate chloride ion channel and induce diarrhea
	AvrA, SspH1	Inhibit NF-κB activity and IL-8 secretion
	SipA, SopA	Aids in transmigration of polymorphonuclear leukocytes (PMNL)
	SipB (or SspB)	Activate caspase 1 and autophagy in macrophages
SPI-2	SpiB, SpiC, SpiD	T3SS needle assembly
	SseB, SseC, SseD	Virulence factors and effector protein delivery to host cell cytosol
	SpiC	Interferes with endosome trafficking
	SifA	Salmonella-containing vacuole (SCV) integrity and stability maintenance
	SifA, SseF, Sseg, SopD2, PipB2	*Salmonella*-induced filament (Sif) formation and microtubule bundling
	SspH2	Inhibits actin polymerization and cellular architecture
	SpvB	Downregulates Sif formation
	SseI	Host cell dissemination
	TtrABC, TtrRS	Anaerobic respiration by reducing tetrathionate

Adapted from Garai et al. (2012). Virulence 3, 377–388

SPI-4

SPI-4 is a 27-kb locus, which is located next to a putative tRNA gene, and contains 18 genes. It is thought to encode genes for type I secretion system (T1SS), and the gene products are required for survival in the macrophage.

SPI-5

SPI-5 is a 7.6-kb region and encodes six genes, which encode effector proteins for T3SS. SopB, an inositol phosphatase, is translocated by T3SS and involved in triggering fluid secretion to cause diarrhea. Thus, it is thought that SPI-5 is possibly responsible for the enteric infection.

SPI-6

SPI-6 is a 59-kb locus and is present in both serovars Typhi and Typhimurium. It contains *saf*

gene cluster for fimbriae, *pagN* for invasion, and several genes with unknown function.

SPI-7 or Major Pathogenicity Island (MPI)

SPI-7 is a 133-kb locus and is specific for serovars Typhi, Dublin, and Paratyphi. It encodes a gene for Vi antigen, a capsular polysaccharide, which elicits high fever during typhoid fever. SPI-7 also carries *pil* gene cluster for type IV pilus synthesis and encodes a gene for SopE effector protein of T3SS.

SPI-8

SPI-8 is a 6.8-kb locus and is found only in serovar Typhi. It carries genes for putative bacteriocin biosynthesis, but functional attributes have not been fully investigated.

SPI-9

SPI-9 is about 16-kb locus and it carries genes for T1SS and a large putative RTX (repeat in toxin)-like toxin.

SPI-10

SPI-10 is a 32.8-kb locus, found in serovars Typhi and Enteritidis, and encodes genes for Sef (*Salmonella* enteritidis fimbriae).

Salmonella Genomic Island-1 (SGI-1)

SGI-1 is a 43-kDa locus and encodes genes for antibiotic resistance. It was identified in *S.* Typhimurium DT104, Paratyphi, and Agona, which are resistant to multiple antibiotics. The DT104 strain has been implicated in outbreaks worldwide. The insertion site is flanked by direct repeats (DR) and is not associated with tRNA gene. The genes for five antibiotic-resistant phenotypes (ampicillin, chloramphenicol, streptomycin, sulfonamides, and tetracycline) are clustered in a multidrug resistance region and are composed of two integrons.

High-Pathogenicity Island (HPI)

HPI encodes genes for siderophore biosynthesis, which are required for iron uptake. HPI is present in *S. enterica* and other pathogens such as *Yersinia enterocolitica* and *Y. pseudotuberculosis*.

Type III Secretion System

The T3SS is also called a molecular syringe and is responsible for contact-dependent secretion or delivery of virulence proteins into the host cells for eliciting intestinal inflammation. This molecular apparatus is present in *Salmonella*, *Shigella*, *Escherichia coli*, and other pathogens. In *Salmonella*, there are two type III secretion systems, T3SS-1 and T3SS-2, encoded in SPI-1 and SPI-2, respectively. The genes encoding the synthesis of T3SS have distinct chromosomal locations, and the gene products include several proteins that form a needle-like organelle on the bacterial envelope. The molecular syringe (T3SS) has four parts: a needle, outer rings, neck, and inner rings (Fig. 15.3). The needle is made of PrgI and a putative inner rod protein, PrgJ. InvG is part of the outer rings structure, neck consists of PrgK, and the base is made of PrgH that makes up the inner rings. The inner membrane components are made of InvC, InvA, SpaP, SpaQ, SpaR, and SpaS proteins (Fig. 15.3). The effector proteins delivered through the T3SS are responsible for salmonellae entry, survival, and replication inside the host cells. T3SS-1 transports effector proteins required for bacterial invasion and inflammation, while T3SS-2 transports effector proteins that are responsible for bacterial intracellular survival and vacuolar movement within SCV.

Pathogenic Mechanism

Adhesion and Colonization

Salmonella enterica has a large number of adhesive proteins (Table 15.4), and the function of many adhesins is not fully understood. Some adhesins may be restricted to specific host tissues or function outside a mammalian host organism. Adhesins are subdivided into fimbrial and nonfimbrial. The fimbrial adhesins include Fim (fimbriae), Lpf (long polar fimbriae), Tafi (thin aggregative fimbriae), Pef (plasmid-encoded fimbriae), and curli (thin aggregative fimbriae). The fimbriae promote adhesion to M cells or colonization of intestinal epithelial cells. Fim binds to α-D-mannose receptor on the host cell; Lpf binds to cells in the Peyer's patch; and Tafi and Curli interact with extracellular matrices such as fibronectin and laminin and aid in adhesion to intestinal epithelial cells. Tafi can also interact with both biotic and abiotic surfaces to form biofilms and can facilitate animal-to-animal transmission. Curli also helps bacteria in autoaggregation,

Fig. 15.3 Type III secretion system (T3SS) apparatus (molecular syringe) in *Salmonella* (Schematic is based on Galan and Wolf-Watz (2006). Nature 444, 567–573; Galan and Collmer (1999). Science 284, 1322–1328)

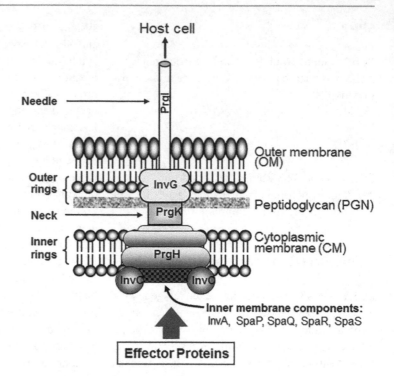

Table 15.4 Adhesins of *Salmonella*

Adhesins	Receptor, tissue tropism
Fimbrial adhesins	
Fim (fimbriae)	Mannose residues
Lpf (long polar fimbriae)	Cells of Peyer's patches
Tafi (thin aggregative fimbriae)	Fibronectin or laminin
Pef (plasmid-encoded fimbriae)	Blood group antigen, LewisX
Curli	Fibronectin or laminin
Non-fimbrial adhesin (T1SS-associated secreted proteins)	
ShdA (>200 kDa)	Fibronectin
MisL	Fibronectin
SadA	Unknown specificity
BapA (biofilm-associated protein)	Biofilm formation and host colonization
SiiE (giant adhesin; 595 kDa)	N-acetyl-glucosamine (GlcNAc) and/or α 2,3-linked sialic acid
T3SS-associated effector proteins	
SipB, SipC, SipD	Intimate attachment to epithelial cells

which enhances survival in the presence of stomach acid or biocides. In addition, lectin-like adhesion molecules bind to the glycoconjugate receptor Gal β (1–3) Gal NAc found on enterocytes.

The non-fimbrial adhesins, also known as autotransporter adhesins, include ShdA, MisL, and SadA and type I secretion system (T1SS) substrates such as BapA and SiiE. Both ShdA and MisL bind to extracellular matrix protein, fibronectin, on host cells for colonization. The *shd*A locus is present in *S. enterica* subspecies I, i.e., all avian and mammalian serovars of *Salmonella*. ShdA localizes to the bacterial outer membrane and interacts with host extracellular matrix, fibronectin. Likewise, MisL promotes bacterial colonization and persistence and shedding in mice. The trimeric SadA interacts with host cells with unknown specificity. Two large repetitive proteins, BapA and SiiE, are substrates of T1SS. BapA is required for biofilm formation and host colonization, while SiiE enables *Salmonella* to breach the intestinal epithelial barrier. SiiE is a giant adhesin protein with a size of 595 kDa. It is the largest protein of the *Salmonella* proteome consisting of 53 repetitive bacterial immunoglobulin (BIg) domains, each containing several conserved residues. SiiE is encoded by genes located on SPI-4. SiiE-mediated host cell binding takes place at the apical side of the epithelial cell and is specific for glycostructures with terminal N-acetyl-glucosamine (GlcNAc) and/or α 2,3-linked sialic acid.

Invasion and Intracellular Growth

Invasion of *Salmonella* is mediated by three mechanisms: phagocytosis by M cells, phagocytosis by dendritic cells, and induced phagocytosis by epithelial cells (Fig. 15.4).

Phagocytosis by M Cells

The preferential route of *Salmonella* translocation is through the M cells overlying Peyer's patches. *Salmonella* expresses several invasin genes, *invABCD* located in the salmonella pathogenicity island 1 (SPI-1), that promote bacterial attachment and invasion of M cells to cross the epithelial barrier.

Phagocytosis by Dendritic Cells

Dendritic cells located in the lamina propria project their dendrites through the epithelial lining to the lumen to sample the intestinal environment and internalize luminal bacteria to transport them to the basolateral side of the epithelial lining. Phagocytosis of *Salmonella* by the dendritic cells

Fig. 15.4 Diagram showing *Salmonella* invasion through the host mucosal membrane. *Salmonella* translocates by three possible pathways: (1) translocation through M cells, (2) dendritic cells, and (3) induced phagocytosis via membrane ruffling. *Salmonella* multiply inside the phagocytic cells such as macrophages, neutrophils, and dendritic cells and induce apoptosis and promote inflammation. Dendritic cells are responsible for systemic dissemination of bacteria to lymph nodes, liver, and spleen

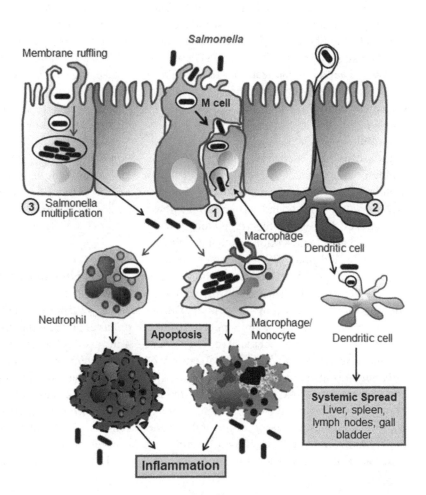

thus does not affect the cellular integrity of DC. When enclosed in the DC vacuole, *Salmonella* does not proliferate but still secretes effector proteins and is transported to extraintestinal sites. DC, rather than macrophages, is thought to be the primary phagocytes in the subepithelial region, which are responsible for dissemination of bacteria to extraintestinal sites.

Induced Phagocytosis by Epithelial Cells

Binding of *Salmonella* to epithelial cells triggers membrane ruffling, a mechanism that allows bacteria to be internalized by the nonprofessional phagocytic cells such as epithelial cells; hence, it is called the "trigger mechanism." Membrane ruffling is a well-orchestrated event that allows the formation of a "ruffle-like" lamellipodial appearance on the surface of the host cell membrane (Fig. 15.5). Two major events take place during invasion: GTPase activation and actin polymerization. The gene products required for invasion of epithelial cells are delivered by both T3SS-1 and T3SS-2. During contact with the host cell, the T3SS-1 transported effector protein SopB initiates a signaling cascade that activates the Rho GTPase regulator that normally maintains the cellular architecture. Rho GTPase converts inactive GDP to active GTP-bound conformation, which is regulated by yet another regulatory protein called guanine nucleotide exchange factors (GEFs). In addition, T3SS-1 effector proteins, SopE, SopE2, and SopB, activate Rho GTPase, Cdc42, and Rac, which activate Arp2/3 complex for initiation of actin polymerization. During this process, globular G-actin is polymerized to highly ordered F-actin necessary for engulfment of bacteria (Fig. 15.5). Internalization of *Salmonella* is also promoted by T3SS-1 translocon, SipC and SipA proteins, which directly bind to actin for cytoskeletal rearrangement to promote membrane ruffling. SipA inhibits actin depolymerization and bundling of actin, and SipC nucleates actin and bundles to promote

invasion. T3SS-2 delivers gene products (SseG, SifA) that promote bacterial survival and replication inside the SCV.

Salmonella invasion also stimulates tyrosine phosphorylation of epidermal growth factor receptor (EGFR), which triggers the cascade of phosphorylation and dephosphorylation reactions, eventually activating the phospholipase A2 (PLA2). Phosphorylated PLA2 helps produce arachidonic acid. The enzyme 5-lipoxygenase converts arachidonic acid into leukotrienes (inflammatory mediators), which increase membrane permeability, causing increased fluid accumulation resulting in diarrhea. Leukotrienes also control calcium channels and allow intracellular accumulation of Ca^{2+} influx, which activates actin polymerization and cytoskeletal rearrangement allowing bacterial entry through the ruffled membrane. After entry, *Salmonella* reverses the actin cytoskeletal rearrangements in the cell. The internalized bacteria are trapped in a membrane-bound vesicle, SCV (Fig. 15.5).

The sequential events leading to *Salmonella* uptake and cell lysis include (1) formation of the membrane ruffle appearing as a splash, (2) actin rearrangement to allow cytoskeletal rearrangement, (3) formation of pseudopod to entrap *Salmonella* inside SCV, (4) bacterial multiplication inside SCV, (5) coalescence of multiple SCV-containing bacteria to form a large vesicle, and (6) lysis of the vesicles to allow *Salmonella* release. *Salmonella* then enters the circulation for systemic infection (bacteremia or septicemia).

Survival in Phagocytes

Salmonella present in the subcellular lamina propria is engulfed either by macrophages or by dendritic cells for extraintestinal disseminations. *Salmonella* expresses approximately 40 proteins that aid their survival inside a macrophage. The macrophage attempts to control intracellular *Salmonella* growth via the inducible nitric oxide synthase (iNOS) and NADPH oxidase-dependent respiratory burst. However, salmonellae produce catalase, which inactivates lysosomal H_2O_2 and

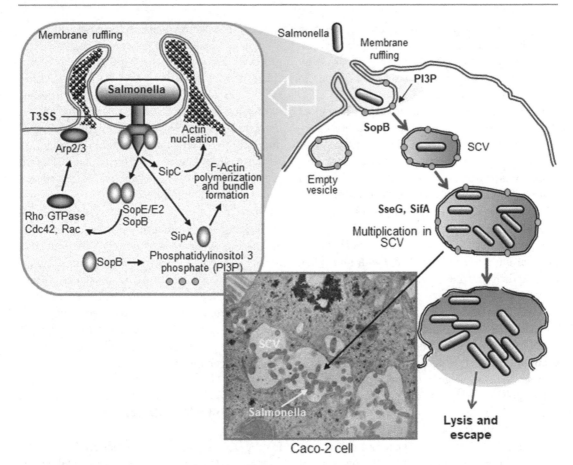

Fig. 15.5 Cellular mechanism of *Salmonella* invasion and multiplication. The sequence of events leading to *Salmonella* uptake and cell lysis include (1) membrane ruffling, (2) cytoskeletal rearrangement, (3) formation of pseudopod to entrap *Salmonella* inside SCV (salmonella-containing vacuole), (4) bacterial multiplication inside SCV, (5) coalescence of multiple SCV-containing bacteria to form a large vesicle, and (6) lysis of the vesicles to allow *Salmonella* release and systemic spread

superoxide dismutase (SOD) that inactivates reactive oxygen.

Salmonella also expresses a two-component signal transduction system, the PhoP–PhoQ system, to promote bacterial survival inside the macrophage. PhoQ is a sensor and PhoP is a transcriptional activator that expresses different genes, required for bacterial survival inside the macrophage, as well as various stresses including carbon and nitrogen starvation, low pH, low O_2 levels, and the action of defensins. In addition, PhoP regulates genes such as *spiC* and *tassC* that prevent lysosome fusion with the SCV. PhoQ regulon activates *pags* genes that are essential for adaptation during the intracellular lifecycle.

Inflammation

Salmonella-induced intestinal inflammation provides a growth advantage for the bacteria during luminal existence. *Salmonella* SPI-1-encoded effector proteins, SopA, SopB, SopE, SopE2, SipA, and SipC, promote the production of inflammatory cytokine, IL-8, through activation of MAPK (mitogen-activated protein kinase) and NF-κB pathways. Inflammatory cytokine also destabilizes epithelial tight junction and allows neutrophil migration into the lumen. Macrophages are activated due to LPS through toll-like receptor (TLR) recognition and enhance *Salmonella* killing. Inflammation also activates inflammatory caspases (caspase 1, caspase 4, and caspase 5 in humans). Caspase 1 activation triggers proinflammatory cytokine IL-18 and IL-1β release by macrophages, which in turn activates T cells to produce IL-17 and IL-22 to amplify inflammation. Once *Salmonella* is internalized, it reduces the inflammatory response by using several effector proteins SptP (tyrosine phosphatase), AvrA, SpvC, and GogB and many more effector proteins to downregulate the MAPK and NF-κB pathways, thus reducing production of IL-8, TNF, and other proinflammatory cytokines.

During the invasion of epithelial cells, reactive oxygen species (ROS) is produced and released into the lumen. The ROS reacts with thiosulfate produced by the gut microbiota and converts it into tetrathionate, which is used by *Salmonella*, as a terminal electron acceptor for growth under an anaerobic condition in the gut.

During intestinal inflammation, lipocalin-2 is accumulated in the lumen. Lipocalin is a host protein that sequesters enterochelin (bacterial siderophore) used by the gut microbiota to acquire iron for growth. Salmochelin, a *Salmonella* siderophore, is not sequestered by lipocalin; therefore, it provides a growth advantage for *Salmonella* in the gut in the presence of natural gut microbiota.

Immunity to *Salmonella*

The innate immune response to *Salmonella* involves epithelial cells and local phagocytes including macrophages and dendritic cells. They recognize *Salmonella* antigens such as LPS and flagella using pattern recognition molecules, TLR4 and TLR5, respectively. NOD-like receptors are also important and induce inflammation. To initiate an adaptive immune response, DC also serves as antigen-presenting cells and activates CD4+ or CD8+ T cells by presenting antigens using MHC class II or class I molecules, respectively. Recognition of *Salmonella* by DC is mediated by TLR, and DC carrying *Salmonella* will activate naïve T cells to release cytokines and chemokines, which in turn activate NK cells and other T cells for IFN-γ production for a specific immune response. NK cells are thought to play important role in controlling bacterial growth early in the infection process. Inflammatory cytokines, IL-6, IL-12, IL-18, and IL-23, activate CD4+ T cells and favor differentiation toward CD4+ Th1 and Th17 lineages in the intestine. Cytokines IL-17 and IL-22 produced by these cells help maintain tight junction barrier and induce production of antimicrobial protein, lipocalin-2, respectively (discussed above). B cells producing *Salmonella*-specific antibody can also provide protective immunity during secondary infection by acting as an opsonin or complement activator. B cells can also induce protective effect by serving as antigen-presenting cells or cytokine producer during antibody-independent immunity. In general, protective immunity against attenuated *Salmonella* vaccine requires CD4+ Th1, CD8+ T cells, and B cells in antibody-dependent and antibody-independent mechanisms.

Regulation of Virulence Genes

In order to cause successful infection, *Salmonella* must survive under diverse environmental conditions that include the acid in the stomach, bile salts, oxygen limitations, nutrient starvation, antimicrobial peptides, mucus, and natural microbiota in the intestine. During invasion and growth in macrophages, salmonellae also encounter lysosomal enzymes, hydrogen peroxide, reactive oxygen radicals, iNOS, and defensins. Thus, a large array of gene expression is required to establish an infection in a host. Invasion-associated genes are maximally expressed at 37 °C, at neutral pH, at high osmolarity, and during the late phase of growth and the gene expression is regulated by HilA (hyperinvasive locus), encoded in SPI-1. HilA is the central transcriptional regulator of SPI-1. Some strains may be hyperinfectious (hypervirulent) displaying altered transcription of genes within the PhoP–PhoQ, PhoR–PhoB, and ArgR regulons. Thus, this may result in changes in the expression of classical virulence functions such as SPI-1 and SPI-2 effector proteins and those involved in cellular physiology, metabolism, and acid stress. Hfq is a global regulatory RNA-binding protein, and it activates SPI-1 gene transcription including HilA and InvF.

RpoS Regulator

At the level of gene transcription, stress responses are controlled by the association of sigma factors such as RpoS/σ^S with the RNA polymerase. The sigma factors interact with the core RNA polymerase to reprogram the promoter recognition specificities to express different sets of genes suitable for survival during environmental changes. Sigma factors are thus considered to be global regulators of stress response that connect many signaling networks with downstream regulatory cascades that ultimately control the expression of genes required for the survival and virulence of the bacteria. RpoS is a well-characterized sigma factor of *Salmonella*, which is produced during changes in growth conditions such as starvation, pH, and temperature changes. RpoS controls the expression of more than 60 proteins such as Nuv, KatE-encoded catalase, acidic phosphatase, and so forth. It has been reported that the production of sigma factors during stress increases the survivability of the cells and promotes cross-protection to additional stresses yet to be encountered. The RpoS may also regulate the virulence genes, which play a role during the intestinal phase of infection. There is an increased likelihood that prior to ingestion, the bacterium might be in the stationary phase where RpoS is upregulated.

RpoS also regulates the virulence genes located on the plasmid, *spv* (*Salmonella* plasmid virulence), which is essential for systemic infection in mice. RpoS is also believed to regulate unknown chromosomally encoded genes, which play a significant role in bacterial virulence. It has been demonstrated that the expression of other sigma factors such as σ^E and σ^H respond to extracytoplasmic shock and heat stress, respectively, which may also be dependent on the expression of RpoS.

ATR Response

Acid tolerance response is critical for bacterial transit through the stomach during the gastrointestinal phase of infection or for survival inside the acidic phagosomal environment. Bacteria exposed to milder acidic pH (pH 5) induce a set of genes (as many as 50 genes) that are essential for adaptation of bacteria to more acidic pH (~pH 3). Global regulators such as RpoS, PhoP–PhoQ, OmpR, and Fur play an important role in this process. Two types of ATR responses are identified: log-phase ATR involved during exposure to organic and inorganic acids and stationary-phase ATR-induced during the late phase of growth.

Treatment and Prevention of Gastroenteritis

Non-typhoidal *Salmonella*-induced gastroenteritis is self-limiting in healthy patients. Antibiotic therapy is not recommended for "uncomplicated gastroenteritis." However, treatment with chloramphenicol may be needed to clear the *Salmonella* during systemic infection. Prevention of salmonellae includes proper food handling, avoiding cross-contamination, implementing personal hygiene, and educating the public about the source and safe handling of foods and proper sanitation. Eggs should be kept refrigerated until eaten to prevent the multiplication of bacteria in the yolk. Proper cooking with a minimum pasteurization temperature of 71.7 °C for 15 s followed by prompt cooling to 3–4 °C or freezing within 2 h would eliminate *Salmonella* from food.

Detection

Culture Methods

The traditional *Salmonella* culture method involves pre-enrichment, selective enrichment, isolation of pure culture, biochemical screening, and serological confirmation, which requires 5–7 days to complete. The USDA- and

FDA-recommended methods involve a 6–24-h pre-enrichment step in a nonselective broth such as lactose broth, tryptic soy broth, nutrient broth, skim milk, or buffered peptone water. The selective enrichment step requires an additional 24 h in Rappaport–Vassiliadis semisolid medium, selenite cystine broth, or Muller–Kauffmann tetrathionate broth. Bacterial cells are isolated from selective agar plates such as Hektoen enteric agar (HEA), xylose lysine deoxycholate (XLD), and/or brilliant green agar (BGA). If necessary, biochemical testing is done using triple sugar iron agar and lysine iron agar, which require an additional 4–24 h.

Immunological Methods

Immunological methods including enzyme-linked immunosorbent assay (ELISA), surface adhesion immunofluorescent technique, dot-blot immunoassay, surface plasmon resonance (SPR) biosensor, piezoelectric biosensor, time-resolved immunofluorescence assay (TRF), and fiber-optic sensor have been used for detection of *Salmonella*, and the detection limit for these assays is in the range of 10^5–10^7 cells. These assays require sample enrichment and a concentration step that may include immunomagnetic separation or centrifugation/filtration.

Nucleic Acid-Based Assays

Real-time quantitative polymerase chain reaction using PCR (Q-PCR), reverse transcriptase PCR (RT-PCR), and nucleic acid sequence-based amplification (NASBA) have been used for detection of *Salmonella* from various food matrices. *Salmonella enterica* was detected at 1 cfu ml^{-1} after a culture enrichment of 8–12 h in the TaqMan-based Q-PCR using *invA* gene as target. NASBA method has been used for detection of viable *Salmonella* cells, and it has been demonstrated to be more sensitive than RT-PCR, and moreover, it requires fewer amplification cycles than the conventional PCR methods.

Summary

Salmonella enterica causes gastroenteritis, typhoid fever, and bacteremia. Worldwide there are 16 million annual cases of typhoid fever, 1.3 billion cases of gastroenteritis, and 3 million deaths. Poultry, egg, meat, dairy products, fish, nuts, and fruits and vegetables serve as vehicles of transmission. Non-typhoidal *Salmonella* (NTS) causes gastroenteritis, and the inflammation is primarily localized in the gastrointestinal tract; however, it can cause invasive disease in immuno-compromised host with underlying conditions. To induce gastroenteritis, *Salmonella* passes through the M cells overlying Peyer's patches, through dendritic cells, or through the epithelial lining in the lower part of the small intestine or the proximal colon to arrive in the subepithelial location. *Salmonella* induces apoptosis of macrophages and dendritic cells (DCs), and localized infection is characterized by neutrophil infiltration, tissue injury, fluid accumulation, and diarrhea. During invasive disease, *Salmonella* is transported to extraintestinal sites such as the liver, spleen, and mesenteric lymph nodes (MLN). Typhoidal *Salmonella* is an invasive pathogen and disseminates to extraintestinal sites with the aid of DC and infects the liver, spleen, and MLN and resides in the gall bladder for reinfection and to promote fecal–oral transmission.

Salmonella can invade intestinal epithelial lining by induced phagocytosis and survive and multiply inside the *Salmonella*-containing vacuole (SCV). Invasion of epithelial cells is a complex process involving multiple virulence factors, which orchestrate events that lead to membrane ruffling, actin polymerization, bacterial localization and replication inside SCV, and cell lysis. *Salmonella* injects virulence proteins directly inside the epithelial cell cytoplasm using the type III secretion system (T3SS), a syringe-like apparatus, to induce its own internalization. Genes encoding the T3SS are located on *Salmonella* pathogenicity islands SPI-1 and SPI-2. Genes located on other pathogenicity islands are responsible for survival inside the macrophages, enteropathogenicity, iron uptake, and antibiotic resistance. During the intestinal phase of infection, *Salmonella* has to survive under harsh environmental conditions including stomach acid, bile salts, oxygen limitations, nutrient starvation, antimicrobial peptides, mucus, and the presence of natural microbiota. A global regulator, sigma factor like RpoS is thought to regulate the expression of more than 60 proteins, which possibly promote bacterial survival under these conditions. In addition, the transcriptional regulator, HilA, controls genes required for invasion, and PhoP–PhoQ is required for bacterial survival inside macrophages. Antibiotic therapy is not recommended for self-limiting gastroenteritis, but needed for invasive salmonellosis caused by both NTS and *Salmonella* Typhi. Proper food handling, avoiding cross-contamination, implementing personal hygiene, and educating the public about safe handling of foods and proper sanitation can help reduce salmonellosis cases.

Further Readings

1. Ao, T.T., Feasey, N.A., Gordon, M.A., Keddy, K.H., Angulo, F.J. and Crump, J.A. (2015) Global burden of invasive nontyphoidal *Salmonella* disease, 2010. *Emerg Infect Dis* **21**, 941–949.
2. Chiu, C.-H., Su, L.-H. and Chu, C. (2004) *Salmonella enterica* serotype Choleraesuis: epidemiology, pathogenesis, clinical disease, and treatment. *Clin Microbiol Rev* **17**, 311–322.
3. Clements, M., Eriksson, S., Tezcan-Merdol, D., Hinton, J.C.D. and Rhen, M. (2001) Virulence gene regulation in *Salmonella* enterica. *Ann Med* **33**, 178–185.
4. Coburn, B., Grassl, G.A. and Finlay, B.B. (2007) *Salmonella*, the host and disease: A brief review. *Immunol Cell Biol* **85**, 112–118.
5. Darwin, K.H. and Miller, V.L. (1999) Molecular basis of the interaction of *Salmonella* with the intestinal mucosa. *Clin Microbiol Rev* **12**, 405–428.
6. Galan, J.E. and Collmer, A. (1999) Type III secretion machines: Bacterial devices for protein delivery into host cells. *Science* **284**, 1322–1328.
7. Galan, J.E. and Wolf-Watz, H. (2006) Protein delivery into eukaryotic cells by type III secretion machines. *Nature* **444**, 567–573.
8. Galan, J.E. and Zhou, D. (2000) Striking a balance: Modulation of the actin cytoskeleton by *Salmonella*. *Proc Nat Acad Sci USA* **97**, 8754–8761.
9. Garai, P., Gnanadhas, D.P. and Chakravortty, D. (2012) *Salmonella enterica* serovars Typhimurium and Typhi as model organisms. *Virulence* **3**, 377–388.
10. Gilchrist, J.J., MacLennan, C.A. and Hill, A.V.S. (2015) Genetic susceptibility to invasive *Salmonella* disease. *Nat Rev Immunol* **15**, 452–463.
11. Groisman, E.A. and Ochman, H. (2000) The path to *Salmonella*. *ASM News* **66**, 21–27.
12. Hensel, M. (2004) Evolution of pathogenicity islands of *Salmonella enterica*. *Int J Med Microbiol* **294**, 95–102.
13. Humphrey, T. (2004) Science and society - *Salmonella*, stress responses and food safety. *Nat Rev Microbiol* **2**, 504–509.
14. LaRock, D.L., Chaudhary, A. and Miller, S.I. (2015) Salmonellae interactions with host processes. *Nat Rev Microbiol* **13**, 191–205.
15. Lopez-Garrido, J., Puerta-Fernandez, E., Cota, I. and Casadesus, J. (2015) Virulence gene regulation by L-arabinose in *Salmonella enterica*. *Genetics* **200**, 807–819.
16. Maciorowski, K.G., Herrera, P., Jones, F.T., Pillai, S.D. and Ricke, S.C. (2006) Cultural and immunological detection methods for *Salmonella* spp. in animal feeds - A review. *Vet Res Com* **30**, 127–137.
17. Marzel, A., Desai, P.T., Goren, A., Schorr, Y.I., Nissan, I., Porwollik, S., Valinsky, L., McClelland, M., Rahav, G. and Gal-Mor, O. (2016) Persistent infections by nontyphoidal *Salmonella* in humans: Epidemiology and genetics. *Clin Infect Dis* **62**, 879–886.
18. Michael, G.B. and Schwarz, S. (2016) Antimicrobial resistance in zoonotic nontyphoidal *Salmonella*: An alarming trend? *Clin Microbiol Infect* **22**, 968–974.
19. Nieto, P.A., Pardo-Roa, C., Salazar-Echegarai, F.J., Tobar, H.E., Coronado-Arrázola, I., Riedel, C.A., Kalergis, A.M. and Bueno, S.M. (2016) New insights about excisable pathogenicity islands in *Salmonella* and their contribution to virulence. *Microbes Infect* **18**, 302–309.
20. Patel, J.C. and Galan, J.E. (2005) Manipulation of the host actin cytoskeleton by *Salmonella* - all in the name of entry. *Curr Opin Microbiol* **8**, 10–15.
21. Santos, R.L. and Baumler, A.J. (2004) Cell tropism of *Salmonella enterica*. *Int J Med Microbiol* **294**, 225–233.
22. Winter, S.E., Thiennimitr, P., Winter, M.G., Butler, B.P., Huseby, D.L., Crawford, R.W., Russell, J.M., Bevins, C.L., Adams, L.G., Tsolis, R.M. and Roth, J.R. (2010) Gut inflammation provides a respiratory electron acceptor for *Salmonella*. *Nature* **467**, 426–429.
23. Srikanth, C.V., Mercado-Lubo, R., Hallstrom, K. and McCormick, B.A. (2011) *Salmonella* effector proteins and host-cell responses. *Cell Mol Life Sci* **68**, 3687–3697.
24. Tindall, B.J., Grimont, P.A.D., Garrity, G.M. and Euzeby, J.P. (2005) Nomenclature and taxonomy of the genus *Salmonella*. *Int J Syst Evol Microbiol* **55**, 521.
25. Troxell, B., Petri, N., Daron, C., Pereira, R., Mendoza, M., Hassan, H.M. and Koci, M.D. (2015) Poultry body temperature contributes to invasion control through reduced expression of *Salmonella* pathogenicity island 1 genes in *Salmonella enterica* serovars Typhimurium and Enteritidis. *Appl Environ Microbiol* **81**, 8192–8201.
26. Wagner, C. and Hensel, M. (2011) Adhesive Mechanisms of *Salmonella enterica*. In *Bacterial Adhesion: Chemistry, Biology and Physics* eds. Linke, D. and Goldman, A. pp.17–34.

Campylobacter and Arcobacter

16

The *Campylobacteraceae* family consists of three genera, *Campylobacter*, *Arcobacter*, and *Helicobacter*. The members of the genus *Campylobacter* and *Arcobacter* were initially recognized as commensals in birds and domestic mammals and had veterinary importance causing abortion in animals. However, in recent years, these pathogens are increasingly implicated in foodborne outbreaks and are considered significant zoonotic pathogens. *Campylobacter* causes enteritis, bacteremia, endocarditis, and periodontal diseases in humans and animals, and the infection often leads to chronic sequelae, such as Miller Fisher syndrome, reactive arthritis, and Guillain–Barré syndrome in humans. *Arcobacter* has been identified relatively recently to cause diarrhea in humans and abortion in animals. *Helicobacter pylori* is a human pathogen, which can be carried asymptomatically in humans for decades. In 1982, two Australian physicians, Barry J. Marshall and J. Robin Warren, discovered *Helicobacter* in the human stomach to be related to the genus *Campylobacter*. Later, they reclassified the organism as *Helicobacter pylori* and demonstrated it to be responsible for gastritis and peptic ulcer. Both scientists won the Nobel Prize in Physiology and Medicine in 2005. Characteristics of three genera are summarized in Table 16.1.

Campylobacter

Introduction

Campylobacter means "curved rod" in Greek, and the bacterium was discovered in the late nineteenth century (1886) by Theodor Escherich, a German–Austrian pediatrician. He isolated the bacterium from an infant who died of cholera, and he called the disease "cholera infantum." Later in 1913, McFayden and Stockman identified an organism from an aborted sheep and called it *Vibrio fetus*, which is renamed as *Campylobacter fetus*. Since then *Campylobacter* is considered a significant animal pathogen. In 1972, Dekyser and Butzler in Belgium isolated a *Campylobacter* strain from the blood and feces of a woman who suffered from hemorrhagic enteritis. The development of selective growth media in the 1970s permitted more laboratories to test stool specimens for the presence of *Campylobacter*. Soon *Campylobacter* spp. were established as common human pathogens. In the last 40 years, *Campylobacter* has been recognized as a leading pathogen causing diseases in both animals and humans and considered a zoonotic pathogen. Human campylobacteriosis disease may vary from mild, noninflammatory,

© Springer Science+Business Media, LLC, part of Springer Nature 2018
A. K. Bhunia, *Foodborne Microbial Pathogens*, Food Science Text Series,
https://doi.org/10.1007/978-1-4939-7349-1_16

Table 16.1 Classification of the *Campylobacteraceae* family based on biochemical properties

Characteristics	*Arcobacter*	*Campylobacter*	*Helicobacter*
Aerobic growth at 25 °C	+	–	–
Catalase	+	+ (*C. concisus* and *C. upsaliensis* are negative)	+
Oxidase	+	+	+
Urease	–	– (*C. lari* is positive)	– (*H. pylori* is positive)

Adapted from Lehner et al. (2005). Int. J. Food Microbiol. 102, 127–135

self-limiting diarrhea to severe, inflammatory, bloody diarrhea lasting for several weeks. In some cases, the disease may progress to the development of immunoreactive arthritis and peripheral neuropathies, such as Guillain–Barrè syndrome and Miller Fisher syndrome.

Campylobacter is involved in 400–500 million illnesses annually worldwide. In developing countries, *Campylobacter* infection is often limited to the children and culminates into watery diarrhea. In industrialized countries, *Campylobacter* is the most reported bacterium associated with an acute inflammatory enteric infection. *Campylobacter jejuni* infection is now the leading cause of bacterial gastroenteritis reported in the USA. The Centers for Disease Control and Prevention (CDC) reports that between 1999 and 2008, *Campylobacter* was responsible for 4936 outbreaks in the USA. It is now estimated that the *Campylobacter*-related infection is the highest among all the foodborne bacterial infections in the USA with an estimated 1.9 million cases per year, and many of these cases are associated with consumption of chicken products. Approximately 95% of these human infections are caused by *C. jejuni* or *C. coli*. Immunocompromised individuals are at the highest risk of infection, and the infection is 40–100% more common in the AIDS (acquired immunodeficiency syndrome) patients than in the immunocompetent individuals.

Biology

Campylobacter species are Gram-negative, non-spore-forming, curved, S-shaped, or spiral–helical rods with approximately 0.5–5.0 μm in length (Fig. 16.1). Bacteria also express capsules, which primarily consist of 6-methyl-D-glycero-α-L-

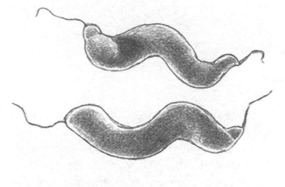

Campylobacter jejuni

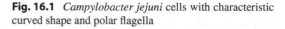

Fig. 16.1 *Campylobacter jejuni* cells with characteristic curved shape and polar flagella

glucoheptose, β-D-glucuronic acid modified with 2-amino-2-deoxyglycerol, β-D-Gal*f*NAc, and β-D-ribose. The capsule and the outer membrane also contain lipooligosaccharides (LOS) which are composed of a core oligosaccharide and lipid A. Most *Campylobacter* species display a single polar flagellum at one end or both ends and exhibit corkscrew-like motility, except *Campylobacter gracilis*, which is nonmotile. Flagella are also important virulence factor. In older cultures, the bacterium may actually appear as spherical or coccoid bodies, which correspond to a dormant, viable but nonculturable (VBNC) state. These highly successful foodborne pathogens are actually quite fastidious and have a stringent set of growth requirements. *Campylobacter* is microaerophilic that requires oxygen concentrations of 3–5% and carbon dioxide of 3–10% and have a respiratory type of metabolism. However, several species, including *C. concisus*, *C. curvus*, *C. rectus*, *C. mucosalis*, *C. showae*, and *C. gracilis*, require hydrogen or formate as an electron donor for microaerobic growth. Some *Campylobacter* spp. such as *C. coli*, *C. jejuni*,

C. upsaliensis, and *C. lari* are thermophiles and grow at 37–42 °C and optimally at 42 °C and will not grow below 30 °C. The non-thermophilic *Campylobacter* spp. include *C. concisus*, *C. curvus*, and *C. fetus*. Growth is further limited by the osmotic stress (2% NaCl concentration), desiccation, and pH less than 4.9. *Campylobacter* utilizes amino acids instead of carbohydrates for energy and is resistant to bile. The genome of *C. jejuni* is relatively small, 1.6–1.9 Mbp, indicating the presence of fewer genes compared to other bacterial pathogens, thus reflecting its requirement for complex growth media. Some strains of *C. jejuni* carry a plasmid, pVir, which encodes genes for type IV secretion system (T4SS), and the effector proteins may have a role in the host cell invasion and pathogenicity.

Classification

The *Campylobacteraceae* family belongs to the order *Campylobacterales*, the class *Epsilonproteobacteria*, and the phylum *Proteobacteria*. The *Campylobacteraceae* family consists of three genera: *Campylobacter*, *Arcobacter*, and *Helicobacter*. The genus *Campylobacter* contains 26 species and 9 subspecies. Of these 12 species are considered pathogenic including *C. jejuni*, *C. coli*, *C. fetus*, *C. upsaliensis*, *C. concisus*, *C. curvus*, and *C. lari*.

Infection by these pathogens may lead to gastroenteritis; however, *C. jejuni* has also been implicated in systemic infection. *C. jejuni* is the most recognized *Campylobacter* that is involved in the 95% of the outbreaks and sporadic illnesses. A typing system based on the heat-labile antigenic factors has identified 100 serotypes of *C. jejuni*, *C. coli*, and *C. lari*, whereas the heat-stable LPS O-antigen typing system classifies these pathogens into 60 serotypes. Biochemical properties are used routinely in the laboratory for classification of *Campylobacter* species (Table 16.2).

Sources

Mammals and birds are the main reservoirs, but *Campylobacter* can be found in rabbits, birds, sheep, horses, cows, pigs, poultry, and even domestic pets. This organism is also found in vegetables, shellfish, and water. Poultry is the natural host, and poultry products serve as the major source of *Campylobacter* contributing about 80% of the human campylobacteriosis cases. However, outbreak investigations have also implicated unpasteurized milk, food handler, and contaminated surface water as infection sources. *Campylobacter* spp. colonize in the caeca of broiler chicken with an average of 10^6–10^7 cfu g^{-1} of cecal content, and *C. jejuni* and *C. coli* are reported to be the predominant species.

Table 16.2 Classification of *Campylobacter* species based on their biochemical properties

Characteristic	C. jejuni	C. jejuni subsp. doylei	C. coli	C. lari	C. fetus subsp. fetus	C. upsaliensis
Growth at 25 °C	−	±	−	−	+	−
Growth at 35–37 °C	+	+	+	+	+	+
Growth at 42 °C	+	±	+	+	+	+
Nitrate reduction	+	−	+	+	+	+
H₂S, lead acetate strip	+	+	+	+	+	+
Catalase	+	+	+	+	+	−
Oxidase	+	+	+	+	+	+
Motility (wet mount)	+	+	+	+	+	+
Hippurate hydrolysis	+	+	−	−	−	−
Nalidixic acid	S	S	S	R	R	S
Cephalothin	R	R	R	R	S	S

Adapted from FDA/CFSAN – BAM – *Campylobacter* (http://www.cfsan.fda.gov/~ebam/bam-7.html)
+ positive, − negative, S sensitive, R resistance

Antibiotic Resistance

The emergence of antibiotic resistance in thermophilic campylobacters especially in *C. jejuni* and *C. coli* is becoming a major concern worldwide, because of their resistance to fluoroquinolones and macrolide. Antibiotic resistance is also prevalent among the poultry isolates of campylobacters; thus, the US Food and Drug Administration (FDA) has banned the use of fluoroquinolones as growth-promoting supplement in poultry production. *Campylobacter* species generally cause self-limiting diarrhea; thus, antibiotic treatment is usually not necessary. However, antibiotic is needed for severe cases with prolonged or systemic infection. Campylobacters are zoonotic pathogens; thus, their transmission to humans will raise a serious concern since the most popular antibiotics would be ineffective against Campylobacteriosis.

Virulence Factors and Mechanism of Pathogenesis

The infective dose is thought to be 500–10,000 *Campylobacter* cells, and the dosage often correlates with the intensity of the attack. Immunocompromised individuals are at the higher risk of infection. The incubation period for *Campylobacter* spp. is 1–7 days, but 24–48 h is most common. Following ingestion, *Campylobacter* reaches the lower gastrointestinal tract and invades the epithelial cells in the distal ileum and colon, resulting in cell damage and severe inflammation (Fig. 16.2). For bacterial colonization and invasion, chemotaxis and motility and quorum sensing are critical. For survival and growth, iron acquisition, oxidative stress defense, and resistance to bile salts are required. Tissue damage and inflammatory response are mediated by bacterial toxins. *Campylobacter* adhesion, invasion and toxin-induced cell damage and inflammation involving macrophage, neutrophil and dendritic cell recruitment, NF-κB activation, and IL-8 secretion are depicted in Fig. 16.2.

Intestinal Colonization

Motility

Bacterial motility requires flagella and a chemosensing system to respond to the gastrointestinal environment. Hence, the flagellum is an important virulence factor, which is involved in the bacterium motility toward the epithelial cell surface, adhesion and colonization, and invasion of host epithelial cells. The helical shape of one or two polar flagella contributes to the corkscrew propulsive movement in the viscous mucus. Flagellum also possesses the type III protein secretion (T3SS) responsible for transport of proteins necessary for bacterial host interaction. Flagellum consists of a woven structure of the flagellin proteins consisting of a hook–basal body and the extracellular filament. The hook–basal body is composed of proteins that include FliF, the T3SS proteins (FlhA, FlhB, FliO, FliP, FliQ, FliR), motor switch proteins (FliG, FliM, FliN, FliY), motor components (MotA and MotB), and minor hook components (FlgI, FlgH, FlgE, FliK, FliM, FliM). The extracellular filament is composed of FlaA and FlaB proteins. The mutation in key genes such as *flaA*, *flaB*, and *flhA* and *flhB* prevents the production of FlaA or FlaB, ultimately halting motility, invasion, and pathogenesis.

Chemotaxis

Chemotaxis is a physiological response of the motile bacterium to move toward chemoattractants using chemosensors. The chemosensors are two components: histidine kinase-dependent signal transduction system composed of chemotaxis proteins, CheA, CheB, CheR, CheW, CheY, and CheZ, and methyl-accepting chemotaxis proteins (MCPs). The chemosensing proteins regulate flagellar proteins thus control directional movement of the bacterium. *Campylobacter* movement toward mucins and glycoproteins on the mucosal surface help bacterial colonization in the gut. Other chemoattractants include α-ketoglutarate, aspartate, asparagine, cysteine, glutamate, pyruvate, serine, formate, malate, lactate, succinate, and so forth.

Fig. 16.2 Mechanism of *Campylobacter jejuni*-induced epithelial cell damage in the intestine. Steps in pathogenesis include (1) chemotaxis and motility; (2) adhesion, invasion, and growth inside the vacuole; and (3) production of cytolethal distending toxin (CDT). Cell damage and inflammation lead to fluid loss and diarrhea

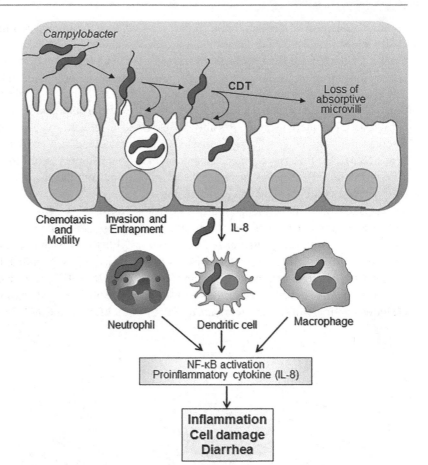

Adhesion

Several *Campylobacter* adhesion proteins have been identified that contribute to bacterial adhesion. *Campylobacter* uses CadF protein (37 kDa) that binds to fibronectin, a 220 kDa glycoprotein commonly found in locations of cell-to-cell contact on epithelial cells in the gastrointestinal tract. This interaction triggers a signaling cascade that leads to the activation of the GTPases Rac1 and Cdc42, which promote *Campylobacter* cell internalization through actin-mediated induced phagocytosis.

Campylobacter also uses periplasmic binding protein (Peb1; 21 kDa protein), *Campylobacter* adhesion protein A (CapA), fibronectin-like protein A (FlipA), and a lipoprotein, JlpA (42.3 kDa) for adhesion. JlpA binds to the eukaryotic Hsp90 (90 kDa) and induces signal transduction in the host cell. *Campylobacter* also adheres to the host

cell H-2 antigen for colonization. The lipopolysaccharide (LPS) and lipooligosaccharide (LOS) also contribute to bacterial adhesion and serum resistance (Fig. 16.2).

Campylobacter heat-shock proteins, GroESL, DnaJ, DnaK, and ClpB, aid in bacterial thermotolerance and survival in the bird intestine since the gut temperature is about 42 °C. Among the heat-shock proteins, only DnaJ was shown to directly contribute to bacterial colonization.

Invasion

The flagellum is believed to play an important role during the host cell invasion by facilitating secretion of nonflagellar proteins through its T3SS channel. The FlaC and CiaB (*Campylobacter* invasion antigen B; 73 kDa) proteins are delivered through T3SS to the host cell cytoplasm and are required for adhesion and invasion. Other invasion and intracellular survival factors are CiaC, CiaI, invasion-associated protein A (IamA), and HtrA, a chaperone protein. Bacterial binding to the host cells triggers host cell cytoskeletal rearrangements through activation of microfilaments and microtubules that allow bacterial internalization possibly by two mechanisms: zipper and trigger mechanisms (Fig. 16.3). In the zipper mechanism, high-affinity binding of bacterial surface adhesins to their receptors initiates signaling event leading to the cytoskeleton-mediated zippering of the host cell plasma membrane around the bacterium. The bacterium is subsequently internalized into a vacuole. In the trigger mechanism, the bacterium injects effector proteins through T3SS and initiates a signaling event to activate small Rho GTPases and cytoskeletal reorganization to induce membrane ruffling. The bacterium is subsequently internalized into the vacuole. *Campylobacter* is unable to escape the membrane-bound vacuole and replicates at least one cycle inside. Survival inside the vacuole is facilitated by the production of superoxide dismutase (SOD) to inactivate oxygen radical and the catalase to protect against oxidative stress from the host. Some studies have reported escape of the bacterium from the vacuole.

Toxin

Campylobacter produces several toxins, but the cytolethal distending toxin (CDT) is the main toxin and is encoded by a three-gene operon (*cdtABC*). CDT consists of three similar-sized molecular weight toxins, CdtA (30 kDa), CdtB (29 kDa), and CdtC (21 kDa). Hence, it is called

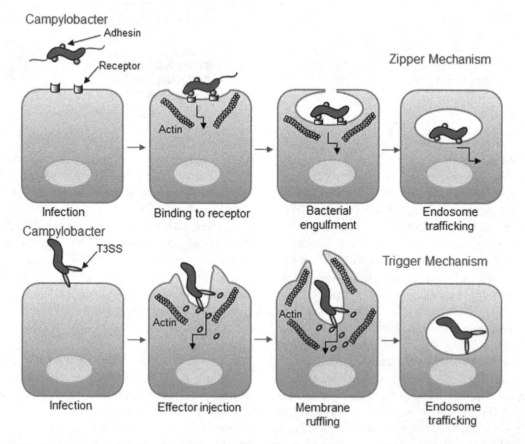

Fig. 16.3 *Campylobacter jejuni* invasion strategy: zipper mechanism and trigger mechanism (Adapted from Tegtmeyer et al. (2012). Microbial Food Safety. pp.13–31: Springer)

a tripartite "AB$_2$" toxin, in which CdtB is the active toxic unit with enzymatic activity, while the CdtA and CdtC constitute the "B$_2$" subunit, which is responsible for binding to the cell receptor and internalization of the active CdtB. CdtB has a nuclease activity that triggers DNA damage resulting in cell cycle arrest, especially in the G2/M phase of mitosis affecting cell division. CDT also causes cell distention over 72 h and detachment in target cells such as HeLa, Vero, Caco-2, and CHO cell lines. CDT is thought to disturb the maturation of crypt cells into functional villous epithelial cells; thus, it temporarily ceases the absorptive function and induces diarrhea. CDT is heat-labile (70 °C for 30 min) and trypsin-sensitive. Other toxins produced by *Campylobacter* are cholera-like enterotoxin that activates cAMP, a Shiga-like toxin that inhibits protein synthesis, pore-forming hemolysin, and hepatotoxin.

Iron Acquisition

The ability of *Campylobacter* to acquire iron from the host transferrin in serum and lactoferrin from mucus is important for their survival and pathogenesis. The bacterium does not produce any siderophores of its own but uses siderophores, ferrichrome, and enterochelin produced by other bacteria to acquire iron. However, *Campylobacter* expresses the iron-uptake receptor, enterobactin FeEnt, encoded by *cfrA* and *cfrB* genes. Fur (ferric uptake regulator) and PerR (peroxide stress regulator) are two proteins that regulate gene products necessary for iron acquisition.

Regulation of Virulence Genes

Thermotolerance and colonization in the gut are regulated by a two-component regulatory system consisting of a histidine kinase (HPK) sensor and a response regulator (RR). The RR is phosphorylated by HPK and regulates the expression of CheY, RacR, and other proteins that are responsible for colonization and thermotolerance at 37–42 °C. Furthermore, the FlgS/FlgR two-component signal transduction system regulates the *flaA* regulon, responsible for flagella synthesis in *C. jejuni*. As mentioned before, the iron acquisition is regulated by Fur and PerR proteins.

Symptoms

Gastrointestinal Disease
Campylobacter infects children and young adults, and the symptoms appear within 24–72 h following ingestion. The symptoms are acute enterocolitis characterized by severe abdominal cramp mimicking appendicitis, nausea, general malaise, fever, muscle ache, headache, and acute watery to bloody diarrhea lasting for 3–4 days. The symptoms usually begin within 24–72 h following ingestion, but may take longer to develop when infected with a low dosage of the bacterium. Majority cases are self-limiting; however, *C. jejuni* infection can be severe and may even lead to death in patients with immunocompromised conditions: AIDS, cancer, and liver diseases.

Campylobacter infection is also associated with other gastrointestinal conditions, such as inflammatory bowel disease (IBD), Barrett's esophagus (BE), colorectal cancer, and cholecystitis. In the esophageal disease, BE, flow-back of stomach acid or the bile can cause mucosal damage thus increases the risk of BE onset. *Campylobacter* (often *C. concisus*) can colonize the damaged mucosa and exaggerates esophageal reflux. Cholecystitis refers to inflammation of the gallbladder, which can be induced by *Campylobacter*. When the cystic duct is blocked by the gallstones, bile accumulates within the gallbladder resulting in cholecystitis.

Extraintestinal Diseases
Campylobacter infection is also involved in extra-gastrointestinal diseases, such as Guillain–Barré syndrome (GBS), Miller Fisher syndrome (MFS), and reactive arthritis. *Campylobacter* infection can cause other inflammatory conditions such as appendicitis, endocarditis, peritonitis (inflammation of peritoneum), brain abscess and meningitis, and bacteremia.

Guillain–Barré Syndrome

Guillain–Barré syndrome (GBS) is one of the common consequences of *Campylobacter* infection, especially by *C. jejuni*. GBS is a neurologic condition characterized by a progressive weakness and flaccid paralysis in the limbs and can also affect respiratory muscles. GBS was first described in 1916 by the French neurologists Guillain, Barré, and Strohl. *Campylobacter* possesses lipooligosaccharide (LOS) in its capsule. The LOS molecule is composed of a core oligosaccharide and lipid A and helps bacteria to evade the immune system and promotes host cell adhesion and invasion. During the infection, sialylation of the LOS occurs, and the sialylated LOS mimics the ganglioside structure of the myelin sheath on nerve cells. During campylobacter infection, the immune system develops an antibody against sialylated LOS, and the LOS-reactive antibodies cause demyelination of nerve, block nerve impulses, and culminate into progressive weakness in limbs and the respiratory muscles, resulting in paralysis.

Miller Fisher Syndrome

Charles Miller Fisher discovered a clinical variant of GBS, which is characterized by an acute onset of ophthalmoparesis (ophthalmoplegia), areflexia (hyporeflexia), and ataxia. In this case, the antibody to ganglioside GQ1b is developed after exposure to the LOS from certain bacterial pathogens, including *C. jejuni*. Such molecular mimicry leads to the development of Miller Fisher syndrome (MFS) characterized by oculomotor weakness and the vision problem.

Reactive Arthritis

Reactive arthritis develops in about 4 weeks after gastrointestinal or genitourinary infections, in which bacterial antigens disseminate into the synovial fluids in the joints, such as knees and ankles, as well as the eyes and the genital, urinary, and gastrointestinal systems. Enteric pathogens, *Campylobacter*, *Salmonella*, *Shigella*, *Vibrio*, and *Yersinia* infections, and urinary tract infective agent, *Chlamydia trachomatis*, often cause reactive arthritis. Antibody–antigen reaction due to molecular mimicry may be involved

in the process. Symptoms include inflammation of joints which may vary from mild mono- or oligo-arthralgia to a severely disabling polyarthritis. Symptoms are seen a month after the infection and resolve within a year but may persist for up to 5 years. Reactive arthritis is generally associated with patients who are positive for HLA-B27 (human leukocyte antigen B27) and are susceptible to the autoimmune diseases.

Arcobacter

Biology

The genus *Arcobacter* is closely related to the genus *Campylobacter*; however, the members of the genus *Arcobacter* are aerotolerant and are able to grow at temperature below 30 °C. *Arcobacter* (means "arc-shaped" in Latin) is a Gram-negative helical rod (1–3 μm × 0.2–0.4 μm) and sometimes may produce unusually long cells (>20 μm). These bacteria express single polar flagellum and display typical corkscrew motility. *Arcobacter* is microaerophilic, grows at a pH range of 6.8–8.0, and has a temperature range of 15–37 °C but not at 42 °C. *Arcobacter* spp. grow well in brain heart infusion (BHI) agar containing 0.6% yeast extract and 10% blood.

There are 18 recognized species in the genus *Arcobacter*, of which *Arcobacter butzleri*, *A. cryaerophilus*, and *A. skirrowii* are recognized human and animal pathogens. *Arcobacter butzleri* and *A. cryaerophilus* are responsible for gastroenteritis in humans and occasional bacteremia and peritonitis. *A. skirrowii* has also been routinely isolated from the patients with diarrhea, despite a lack of evidence for its direct involvement in enteritis. An *Arcobacter*-related outbreak was first reported in 1983 in Italy affecting children in a primary school caused by *A. butzleri*. Ten children showed the symptoms of abdominal cramp without causing any diarrhea.

Arcobacter has been isolated from food, drinking water, animals, humans, and aborted animal fetuses. The bacterium has been isolated from poultry carcasses; however, it is not considered a natural reservoir.

Pathogenesis

The pathogenic mechanism of *Arcobacter* is not fully elucidated, but the genome sequence revealed the presence of several key virulence genes, which contribute to the pathogenesis: *cadF*, *ciaB*, *pldA*, *tlyA*, *hecA*, and *irgA*. Bacterial flagella and chemotaxis are important in cell motility and chemotaxis and involved in adhesion to the host epithelial cells. Bacterial adhesion to the intestinal cells is attributed to the presence of CadF encoded by *cadF*, which is involved in adhesion to the host fibronectin protein. CadF interaction also triggers bacterial internalization by the host epithelial cells. CiaB (*ciaB*) homologous to *Campylobacter* species is involved in the invasion. The *pldA* gene encodes phospholipase A, and the *tlyA* encoding a hemolysin have hemolytic activities. The *hecA* gene encodes an adhesin protein, hemagglutinin (20 kDa), a glycoprotein, which possibly interacts with a glycan receptor containing D-galactose for bacterial adhesion. The *irgA* gene encodes an iron-regulated outer membrane protein. Some species produce cytotoxin, enterotoxin, or vacuolizing toxins responsible for cell rounding and vacuole formation. Evidence for the production of cytolethal distending toxin (CDT) by *Arcobacter* species is inclusive. *Arcobacter* induces a strong inflammatory response similar to campylobacter and induces secretion of IL-8, which is a strong chemoattractant for inflammatory cells at the site of infection (Fig. 16.2).

Arcobacter causes abortion and stillbirth in cows, sheep, and pigs. The organism has been isolated from the uterus, oviduct, and placental tissues. The organism is also isolated from the stomach of pigs showing gastric ulcer. In humans, especially in children, they cause gastroenteritis and has been routinely isolated from diarrheal patients. *Arcobacter* is also thought to cause chronic enteritis in adults. The symptoms include nausea, abdominal pain, vomiting, fever, chills, and diarrhea. Arcobacter-associated gastrointestinal illness is self-limiting; thus, antibiotic treatment may not be necessary.

Prevention, Control, and Treatment of *Campylobacter* and *Arcobacter*

Campylobacter spp. are considered normal inhabitant of livestock, poultry, and wild animals. Therefore, it is rather difficult to control the access of *C. jejuni* to raw foods, especially foods of animal origin. However, proper sanitation can be used to reduce its load in raw foods during production, processing, and future handling. Preventing consumption of raw foods of animal origin; heat treatment of a food, when possible; and preventing post-heat contamination are important to control *Campylobacter* in foods. Contamination of vegetables can be controlled by applying treated animal manures as fertilizer and washing produce in chlorinated or ozonated water. Good personal hygiene and sanitary practices must be maintained by food handlers to avoid campylobacter-related diarrhea. *Campylobacter* spp. are heat-sensitive with a decimal reduction time at 55 °C of 1 min.

Avoiding fecal contamination during slaughter can reduce the pathogen load in food. *Campylobacter* and *Arcobacter* are temperature-sensitive, and thus cold storage of meat at or near 4 °C can reduce bacterial counts. Storage in the presence of sodium lactate, sodium citrate, and sodium triphosphate can be effective in controlling *Arcobacter*. Heating of food to an internal temperature of 70 °C and irradiation with 0.27–0.3 kGy for 10 s can inactivate *Arcobacter*.

Experimental approach with fucosylated human milk oligosaccharides has demonstrated the inhibition of *C. jejuni* binding to the intestinal H-2 antigen. This and similar strategies could be used to control *Campylobacter* infection in humans.

Since the campylobacteriosis involves self-limiting diarrhea, antibiotic therapy is not required, but maintenance of hydration and electrolyte balance is advised. Antibiotic therapy is, however, needed for immunocompromised patients to control bacteremia and sepsis. Erythromycin and newer macrolides, azithromycin, and clarithromycin are effective against *C. jejuni* infection. Increased resistance of *Campylobacter* to fluoroquinolone discourages its therapeutic application.

Detection of *Campylobacter* and *Arcobacter*

Several selective isolation media have been formulated to isolate *Campylobacter* spp. from environmental, fecal, and food samples. Campylosel media used cefoperazone, vancomycin, and amphotericin B as selective agents. The CCDA (charcoal cefoperazone, deoxycholate agar) and CAT (cefoperazone, amphotericin B, teicoplanin) media have been used for isolation of *Campylobacter* at 37 °C under microaerophilic conditions, i.e., under oxygen concentrations of 5–10% and a CO_2 concentration of 1–15%. Sometimes a gas mixture of 15% carbon dioxide, 80% nitrogen, and 5% oxygen is used to create the microaerophilic environment.

PCR-based assays have been developed for detection of *Campylobacter* species, and the target gene included flagellin (*flaA*), 16S rRNA, and 16S/23S intergenic spacer region. *Campylobacter* species identification and typing have been done by using ribotyping, restriction fragment length polymorphism (RFLP), amplified fragment length polymorphism (AFLP), pulsed-field gel electrophoresis (PFGE), and randomly amplified polymorphic DNA (RAPD)-PCR methods. Whole genome sequencing is now used for identification and characterization of *Campylobacter* and *Arcobacter* species.

Arcobacter spp. have been isolated from food samples, poultry carcasses, drinking water, animals, humans, and aborted animal fetuses. An enrichment broth containing cefoperazone, bile salts, thioglycolate, and sodium pyruvate is used. Enrichment at 25 °C under aerobic environment is commonly practiced, which requires about 4–5 days. In addition, a commercial medium called modified charcoal cefoperazone deoxycholate agar (mCCDA) is used for isolation of *Arcobacter* species.

Identification of *Arcobacter* has been done by using various molecular tools that use 16S or 23S rRNA as probes including ribotyping, RFLP, AFLP, PFGE, and RAPD. Multiplex PCR method targeting the 16S and 23S rRNA genes has been developed for detection and identification of *Arcobacter* species.

Summary

Historically, *Campylobacter* and *Arcobacter* species are considered animal pathogens; however, in the last 40 years, both were implicated in outbreaks causing gastroenteritis in humans. These two pathogens are fastidious curved rods and have stringent growth requirements. *Campylobacter* is microaerophilic, and several of the species are thermophilic and are unable to grow below 30 °C. *Arcobacter* is aerotolerant and can grow below 30 °C. Both *Campylobacter* and *Arcobacter* are routinely isolated from livestock, poultry, and water. The outbreak of *Campylobacter* is associated with meat, poultry, and milk. Of 26 species of *Campylobacter*, *C. jejuni* is responsible for 95% of the outbreaks and is considered the most dominant pathogen. *Campylobacter* pathogenesis depends on the expression of several virulence factors that control their motility, chemotaxis, quorum sensing, bile resistance, adhesion, invasion, toxin production, growth inside the host cells, and iron acquisition. Bacteria possibly induce their own internalization through signaling events and rearrangement of the host cytoskeletal structure and survive inside the epithelial cells by expressing superoxide dismutase and catalase to deactivate host oxidative stress defense. Cytolethal distending toxin (CDT) arrests cell cycle division, disrupts the absorptive function of villous epithelial cells, and promotes diarrhea. The *Campylobacter*-induced diarrhea is mostly self-limiting; however, *Campylobacter* may cause fatal infection in immunocompromised patients. The patients suffering from *C. jejuni* infection may also develop Guillain–Barré syndrome characterized by generalized paralysis and muscle pain, and the reactive arthritis is characterized by arthritis in knee joints or lower back. The pathogenic mechanism of *Arcobacter* (*A. butzleri*) is not fully elucidated, but the mechanism is similar to *Campylobacter* infection as to the tissue tropism, adhesion, invasion, tissue damage, and inflammation. *Arcobacter* causes diarrhea in humans (in children) and abortion and stillbirth in cows, sheep, and pigs. Preventing consumption of raw foods of animal origin and heat treatment of a food and preventing post-heat contamination are important to control *Campylobacter* in foods. In most cases, the campylobacteriosis is a self-limiting disease; thus, antibiotic therapy is not required; however, antibiotic is needed for immunocompromised patients to control bacteremia and sepsis.

Further Readings

1. Bolton, D.J. (2015) *Campylobacter* virulence and survival factors. *Food Microbiol* **48**, 99–108.
2. Collado, L. and Figueras, M.J. (2011) Taxonomy, epidemiology, and clinical relevance of the genus *Arcobacter*. *Clin Microbiol Rev* **24**, 174–192.
3. Cróinín, T.Ó. and Backert, S. (2012) Host epithelial cell invasion by *Campylobacter jejuni*: trigger or zipper mechanism? *Front Cell Infect Microbiol* **2**.
4. Dasti, J.I., Tareen, A.M., Lugert, R., Zautner, A.E. and Groß, U. (2010) *Campylobacter jejuni*: A brief overview on pathogenicity-associated factors and disease-mediating mechanisms. *Int J Med Microbiol* **300**, 205–211.
5. Ferreira, S., Queiroz, J.A., Oleastro, M. and Domingues, F.C. (2016) Insights in the pathogenesis and resistance of *Arcobacter*: A review. *Crit Rev Microbiol* **42**, 364–383.
6. Josefsen, M.H., Bhunia, A.K., Engvall, E.O., Fachmann, M.S.R. and Hoorfar, J. (2015) Monitoring *Campylobacter* in the poultry production chain - From culture to genes and beyond. *J Microbiol Methods* **112**, 118–125.
7. Kaakoush, N.O., Castaño-Rodríguez, N., Mitchell, H.M. and Man, S.M. (2015) Global epidemiology of *Campylobacter* infection. *Clin Microbiol Rev* **28**, 687–720.
8. Lehner, A., Tasara, T. and Stephan, R. (2005) Relevant aspects of *Arcobacter* spp. as potential foodborne pathogen. *Int J Food Microbiol* **102**, 127–135.
9. Lertsethtakarn, P., Ottemann, K.M. and Hendrixson, D.R. (2011) Motility and chemotaxis in *Campylobacter* and *Helicobacter*. *Annu Rev Microbiol* **65**, 389–410.
10. Paulin, S.M. and On, S.L.W. (2010) *Campylobacter* fact sheet: Taxonomy, pathogenesis, isolation, detection and future perspectives. *Quality Assurance Safety Crops Foods* **2**, 127–132.
11. Sahin, O., Kassem, I.I., Shen, Z., Lin, J., Rajashekara, G. and Zhang, Q. (2015) *Campylobacter* in poultry: Ecology and potential interventions. *Avian Dis* **59**, 185–200.
12. Silva, J., Leite, D., Fernandes, M., Mena, C., Gibbs, P.A. and Teixeira, P. (2011) *Campylobacter* spp. as a foodborne pathogen: A review. *Front Microbiol* **2**.
13. Snelling, W.J., Matsuda, M., Moore, J.E. and Dooley, J.S.G. (2005) *Campylobacter jejuni*. *Lett Appl Microbiol* **41**, 297–302.
14. Snelling, W.J., Matsuda, M., Moore, J.E. and Dooley, J.S.G. (2006) Under the microscope: *Arcobacter*. *Lett Appl Microbiol* **42**, 7–14.
15. Tegtmeyer, N., Rohde, M. and Backert, S. (2012) Clinical presentations and pathogenicity mechanisms of bacterial foodborne infections. In *Microbial Food Safety*. pp.13–31: Springer.
16. Umaraw, P., Prajapati, A., Verma, A.K., Pathak, V. and Singh, V.P. (2017) Control of *Campylobacter* in poultry industry from farm to poultry processing unit: A review. *Crit Rev Food Sci Nutr* **57**, 659–665.
17. van Vliet, A.H.M. and Ketley, J.M. (2001) Pathogenesis of enteric *Campylobacter* infection. *J Appl Microbiol* **90**, 45S–56S.
18. Wassenaar, T.M. (1997) Toxin production by *Campylobacter* spp. *Clin Microbiol Rev* **10**, 466–476.
19. Young, K.T., Davis, L.M. and DiRita, V.J. (2007) *Campylobacter jejuni*: molecular biology and pathogenesis. *Nat Rev Microbiol* **5**, 665–679.

Yersinia enterocolitica and Yersinia pestis

17

Introduction

A French bacteriologist, Alexandre Yersin in 1894 first described a bacterium, called *Pasteurella*, which was later renamed *Yersinia pestis*. The genus *Yersinia* belongs to the phylum *Proteobacteria*, class *Gammaproteobacteria*, order *Enterobacteriales*, and family *Enterobacteriaceae*. The genus has 17 species: *Yersinia enterocolitica, Y. pseudotuberculosis, Y. pestis, Y. frederiksenii, Y. intermedia, Y. kristensenii, Y. mollaretii, Y. bercovieri, Y. aldovae, Y. aleksiciae, Y. entomophaga, Y. massiliensis, Y. mollaretti, Y. nurmii, Y. pekkanenii, Y. rhodei, Y. similis*, and *Y. ruckeri*. The three most important species that cause infections in humans are *Yersinia enterocolitica, Y. pseudotuberculosis*, and *Y. pestis*. These are zoonotic pathogens, and among them, *Y. enterocolitica* is responsible for the greatest number of cases of the zoonotic disease. All the three species are facultative intracellular pathogens, harbor a 70 kb virulence plasmid (pYV), and exhibit tropism for lymphoid tissues.

Yersinia enterocolitica is associated with foodborne infections resulting in gastroenteritis, terminal ileitis, mesenteric lymphadenitis, and septicemia. The bacterium emerged as a human pathogen during the 1930s. The Centers for Disease Control and Prevention (CDC) estimates about 97,000 cases of human diseases occur due to *Y. enterocolitica* infection annually in the USA with 533 hospitalizations and 29 deaths. *Yersinia pseudotuberculosis* also causes gastrointestinal disorders, septicemia, and mesenteric adenitis. *Yersinia pestis* causes bubonic or pneumonic plague, and the organism can be transmitted through contact with wild rodents and their fleas. Plague is an old-world disease and often referred to as "Black Death" and occurs in the bubonic or pulmonary forms.

Yersinia species are Gram-negative coccobacilli (short rods) nonspore former and can be differentiated based on their biochemical properties (Table 17.1). *Yersinia* spp. are catalase-positive, oxidase-negative, glucose fermentative organisms, and these may be isolated on MacConkey agar and cefsulodin–irgasan–novobiocin (CIN) agar media. Most *Yersinia* species are noncapsulated except *Y. pestis*, which develops an envelope at 37 °C. In addition, all three species also share *Yersinia* outer membrane proteins (YOPs), V (immunogenic protein), and W (nonprotective lipoprotein) antigens. The fraction 1 envelope antigen (F1) is produced at 37 °C and has two major components: fraction 1A (polysaccharide) and 1B (protein). *Y. pestis* has been identified as a subspecies of *Y. pseudotuberculosis* based on the 16S rDNA sequence. These two species also share 11 common antigens.

© Springer Science+Business Media, LLC, part of Springer Nature 2018
A. K. Bhunia, *Foodborne Microbial Pathogens*, Food Science Text Series,
https://doi.org/10.1007/978-1-4939-7349-1_17

Table 17.1 Classification of *Yersinia* species based on biochemical properties

Characteristics	*Y. pestis*	*Y. pseudotuberculosis*	*Y. enterocolitica*
Motility at 22 °C	−	+	+
Lipase at 22 °C	−	−	v
Ornithine decarboxylase	−	−	v
Urease	−	+	+
Citrate at 25 °C	−	−	−
Voges–Proskauer	−	−	v
Indole	−	−	v
Xylose	+	+	v
Trehalose	+	+	+
Sucrose	−	−	v
Rhamnose	+	+	−
Raffinose	−	−	v

Adapted from Smego et al. (1999). Eur. J. Clin. Microbiol. Infect. Dis. 18, 1–15
+ positive, − negative, v variable

Yersinia enterocolitica

Biology

Yersinia enterocolitica is a Gram-negative short rod and facultative anaerobe. *Y. enterocolitica* grows between 0 °C and 44 °C with an optimum growth at 25–28 °C. Growth occurs in milk and raw meat at 1 °C, but at a slower rate. Bacteria can grow in 5% NaCl and at a pH above 4.6 (range pH 4–10). *Yersinia* expresses peritrichous flagella at a lower temperature (25 °C), but it is nonflagellated (nonmotile) at 37 °C. Bacterial swimming and swarming motility are thought to be regulated by bacterial quorum-sensing ability. *Yersinia* is equipped to maintain biphasic lifestyle, one in the aquatic environment/food system and the other in human host. *Yersinia* grows slowly on sheep blood agar, MacConkey agar, and Hektoen enteric agar producing pinpoint colonies after 24 h of incubation. It ferments sucrose but not xylose or lactose. For selective isolation of the bacterium, cefsulodin–irgasan–novobiocin (CIN) and virulent *Yersinia enterocolitica* (VYE) agar can be used.

Classification

Yersinia enterocolitica has been classified into six major biotypes based on their pathogenicity, and ecologic and geographic distributions: 1A, 1B, 2, 3, 4, and 5. 1A is considered the nonpathogenic biotype; however, a minority of biotype 1A strains are found to cause gastroenteritis, while all remaining biotypes are pathogenic, of which biotypes 1B is highly pathogenic and is commonly associated with human infections. The pathogenic biotypes carry virulence plasmid, pYV. *Y. enterocolitica* has about 70 serotypes. Select serogroups for each biogroup include 1A (O:5; O:6, 30; O:7, 8; O:18; O:46), 1B (O:8; O:4; O:13a, 13b; O:18; O:20; O:21), 2 (O:9; O:5, 27), 3 (O:1,2,3; O:5,27), 4 (O:3), and 5 (O:2,3). The predominant serogroups that cause most human infection worldwide are O:3, O:8, O:9, and O:5,27.

Y. enterocolitica is again grouped into two subspecies, *Y. enterocolitica* subspecies *enterocolitica* and *Y. enterocolitica* subspecies *palearctica*. *Y. enterocolitica* subspecies *enterocolitica* contains biotype 1B, and the strains are highly pathogenic and are commonly termed the North American strains. *Y. enterocolitica* subspecies *palearctica* include strains of 1A, 2, 3, 4, and 5 and distributed throughout the world.

Sources

Yersinia enterocolitica is widely distributed in nature, including foods, water, sewage, and animals (cattle, sheep, goats, dogs, cats, rodents); however, the pig is the primary reservoir (bacteria present as a commensal) of pathogenic strains such as serotype O:3. This serotype has been frequently isolated from pigs and responsible for infections in humans (Fig. 17.1). Thirty-five to 70% of swineherds and 4.5–100% of individual pigs harbor pathogenic *Y. enterocolitica*. Environmental isolates are generally nonpathogenic and belong to the biogroup 1A. The first reported foodborne outbreak of *Y. enterocolitica* occurred in New York state in 1976 affecting 222 children due to consumption of chocolate milk and was caused by serotype O:8. *Y. enterocolitica* outbreaks have also been associated with pasteurized and unpasteurized fluid milk. Chitterlings, a product made of swine intestines, also are implicated in outbreaks in infants in the USA and other countries. Chitterlings are prepared by boiling the intestine of swine and are a traditional winter holiday food for many African American families in the USA. Though the fats and fecal contents are removed before boiling the final product, the chitterlings preparation requires substantial handling, and children in the household may be exposed during its preparation.

Virulence Factors

Yersinia enterocolitica is an invasive intracellular enteric pathogen and interestingly not all strains are virulent. Environmental isolates are generally nonpathogenic, and pathogenic strains are predominant in pigs. The pathogenic strains vary in serological characteristics. In the USA, the most common serovar implicated in yersiniosis is serovar O:8. Pathogenic strains carry several virulence factors encoded on the chromosome, and in a 70 kb virulence plasmid (pYV), which are required for adhesion, invasion, and colonization of intestinal epithelial cells and lymph nodes, growth and survival inside the macrophages, killing of neutrophils and macrophages, and serum resistance (Table 17.2). The nonpathogenic strain (biotype 1A) lacks the virulence plasmid, pYV.

The chromosomal-linked virulence proteins include invasin, attachment invasion locus (Ail), *Yersinia* stable toxin (Yst) (enterotoxin), and siderophore. The environmental isolates are shown to be negative for *inv*, and *ail* genes and are not associated with human disease. The plasmid-linked virulence genes encode *Yersinia* outer membrane proteins (YOPs) that are responsible for bacterial adhesion, a type III secretion system (T3SS) to deliver virulence proteins to the host cell cytosol to allow bacterial growth inside macrophages, serum resistance, and septicemia. YOPs are present on the cell surface as well as secreted into the medium. YOPs are secreted in the presence of very low levels of Ca^{2+}; YOP expression is temperature dependent and occurs

Fig. 17.1 *Yersinia enterocolitica* transmission to humans. Swine is the primary reservoir

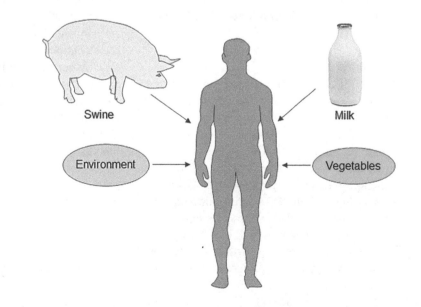

Table 17.2 *Yersinia enterocolitica* key virulence factors and their genetic origin

Origin	Protein size	Function
Chromosome		
Invasin (Inv)	92 kDa	Inv binds to β1-integrin and promotes adhesion and invasion
Attachment invasion locus (Ail)	17 kDa	Attachment and invasion; serum resistance
Yersinia stable toxin (Yst)	7.5 kDa	Yst increases cGMP level and fluid secretion
Yersiniabactin (catechol-type)	482 Da	Siderophore, an iron-binding protein
Virulence plasmid (pYV) 70 kb	–	
Ysc (Yop secretion)	–	A type III secretion system (T3SS) made of 28 proteins
YopH	51 kDa	Dephosphorylates host proteins, modulates signaling pathway, and prevents phagocytosis
YopM	41.6 kDa	Kinase activity; inhibits caspase-1
YopD	33.3 kDa	Responsible for translocation of YopE and other effector proteins (YopH, YopM, YopO, etc.) across the membrane
YopE	219-amino acid protein (~22 kDa)	Inactivates Rho family of GTPase, disrupts actin cytoskeleton, and prevents phagocytosis
YopP	33 kDa	Macrophage apoptosis; alters the expression of cytokines
YopT	35.5 kDa	Interferes with actin cytoskeleton formation by inactivating Rho GTPase
YadA (adhesion protein)	160–240 kDa	Adhesion to epithelial cells by interacting with β1-integrin; blocks complement-mediated killing; serum resistance
YopB	41.8 kDa	Inhibits cytokine release from macrophages
LcrV	37.2 kDa	Low calcium response

mostly at 37 °C, which is critical for pathogenesis in the host. Description of important virulence factors is presented below.

Chromosome-Linked Virulence Gene Products

Invasin

Invasin (Inv) is a 103 kDa *Yersinia* outer membrane protein and binds to the β1-integrin receptor located at the apical side of the M cells or the basolateral side of epithelial cells and is an important virulence factor for the early phase of intestinal infection. Invasin is encoded by *inv* gene located on the chromosome, and its expression is high when *Yersinia* is grown in media with a neutral pH at 25 °C but low at 37 °C. Invasin is expressed at high levels during the stationary phase of growth at low temperatures. However, when *Y. enterocolitica* is grown

at low pH (equivalent to intestinal pH of 5.6–6) at 37 °C, the invasin expression is enhanced. The *inv* expression is regulated by *rovA* (regulator of virulence A), which is located on the chromosome. RovA is a 143-amino acid protein and is present in all three pathogenic *Yersinia* species. Invasin facilitates bacterial colonization and translocation through M cells located in the follicle-associated epithelium overlying Peyer's patches. Invasin activates multiple signaling cascades including cSrc kinase, Rac1, MAP (mitogen-activated protein kinase) kinases, and transcription factor NF-κB and promotes proinflammatory immune response (IL-8, monocyte-chemoattractant protein-1 (MCP-1), and granulocyte-macrophage colony stimulating factor (GM-CSF)). As a result, chemotactic cytokines are made which recruit neutrophils and macrophages at the site of infection. Macrophages ingest *Yersinia* and disseminate to regional lymph nodes, liver, and spleen.

Attachment and Invasion Locus

Attachment and invasion locus (Ail) is a 17 kDa membrane protein and is involved in adhesion, in invasion, and in the serum resistance. It possibly acts by inactivating complement by-products C3-convertase (C4b2a) thus preventing the formation of C3b and the membrane attack complex (MAC), which are required for lysis of bacterial cells. Ail is the primary serum resistance factor in *Y. pestis* and *Y. pseudotuberculosis* and blocks alternative and lectin pathways thus promoting bacterial growth to high densities during infection.

Iron Acquisition

Iron acquisition is achieved by siderophore such as yersiniabactin (catechol-type), which is encoded by a chromosomal high-pathogenicity island (HPI). Under the iron-starvation condition, *Yersinia* produces large amounts of iron receptors, FoxA and FcuA on the outer membrane, which bind the siderophores to sequester iron. Yersiniabactin also reduces production of reactive oxygen species (ROS) by macrophages and neutrophils thus decreasing bacterial killing during innate immune response.

Yersinia Stable Toxin (Yst)

Y. enterocolitica produces an enterotoxin, Yst, which is a heat-stable (100 °C for 15 min) 7.5 kDa protein and remains active at pH range of 1–11 at 37 °C for 4 h and is methanol-soluble. Yst is structurally and functionally homologous to the heat-stable enterotoxin (ST) of enterotoxigenic *Escherichia coli* (*see* Chap. 14) and is encoded by the chromosomal *yst* gene. Yst is involved in diarrhea. Three subtypes of Yst exist: Yst-a, Yst-b, and Yst-c. Each Yst-a and Yst-b subtype is made of 30 amino acids, while the Yst-c consists of 53 amino acids.

LPS and Flagella

LPS located in the bacterial outer membrane consists of antigenic O specific oligosaccharide, core oligosaccharide, and lipid A. Lipid A of LPS is recognized by host TLR-4 (macrophage) and initiates a signaling cascade resulting in the production of proinflammatory cytokine (TNF-α) and recruitment of macrophages and neutrophils. High levels of LPS in the blood can induce septic shock. LPS produced by bacteria at 21 °C predominantly contains hexa-acylated lipid A, which is recognized by TLR-4 and stimulates monocytes to secrete TNF-α; when bacteria is grown at 37 °C, LPS contains tetra-acylated lipid A, which is a weak TLR-4 agonist and is unable to elicit an immune response by the macrophages. This stealth strategy helps *Y. enterocolitica* to avoid recognition by TLR-4 and consequent immune activation.

Flagella consist of flagellin proteins and are involved in bacterial adhesion. Flagellin proteins bind to TLR-5 on monocytes and induce an innate immune response. Aflagellated strains fail to adhere and stimulate the immune system to produce proinflammatory cytokines (TNF-α and IL-17) and thus unable to recruit macrophages. *Y. enterocolitica* synthesizes flagella only at 22–30 °C and loses when grown at 37 °C. This strategy possibly helps the bacterium to avoid host innate immune defense by avoiding recognition by TLR-5.

Plasmid (pYV)-Linked Virulence Gene Products

pYV (70 kb) encodes for 12 major proteins termed YOPs (*Yersinia* outer membrane proteins) and the two outer membrane proteins called YadA (*Yersinia* autotransporter adhesin) and YlpA (*Yersinia* lipoprotein).

Yersinia Adhesion Protein

Yersinia adhesion protein (YadA) is an important virulence factor and facilitates the bacterial attachment to host cells and protects *Yersinia* from nonspecific immune systems such as phagocytosis and complement-mediated cell lysis. YadA is an outer membrane protein encoded by pYV. It is present in both *Y. enterocolitica* and *Y. pseudotuberculosis* but is nonfunctional in *Y. pestis*. YadA is a 160–240 kDa protein composed of three monomers, each 44–47 kDa, and appears as fibrillar (or lollipop-like) structure covering the entire bacterial cell surface. Each fibrilla is of

50–70 nm in length and 1.5–2.0 Å in diameter. YadA has three parts: N-terminal head, intermediate stalk, and the C-terminal anchor domain. The N-terminal domain contains 25-amino acid-long signal sequence. The stalk binds to host cell receptor and resists host immune system, and the C-terminal domain anchors to the bacterial outer membrane.

YadA serves as a major adhesin and binds to several extracellular matrices (ECM) including cell surface-associated collagen and laminin. YadA also promotes bacterial internalization by interacting with the epithelial β1-integrin proteins by zipper mechanism. Interaction with β1-integrin initiates signaling events that orchestrate actin recruitment to alter the cytoskeletal structure to promote bacterial entry.

YadA is expressed at 37 °C but not at 25 °C. Expression of YadA is regulated by two different gene products: VirF (virulence) and LcrV (low calcium response). VirF senses the optimal temperature, i.e., 37 °C required for protein synthesis, and LcrV regulates the *yadA* expression depending on the availability of extracellular calcium concentration. Furthermore, *yadA* expression is not affected by pH, salt, or sugar concentration.

YadA also disrupts the host cell signaling pathways to block the release of a proinflammatory cytokine such as IL-8, which is a chemoattractant for neutrophils. YadA inhibits oxidative burst in neutrophils and activates YopH, a phosphotyrosine phosphatase that blocks the phagocytic mechanism. YadA also plays a major role in serum resistance. It inhibits the formation of C3b and membrane attack complex (MAC) by activating the proteolytic enzyme, factor H, which degrades the C3b.

YopB and YopD

YopB, a 41 kDa protein, suppresses the secretion of macrophage-derived cytokines, IL-1 and TNF-α, which are important in inflammatory response against infection. YopB also regulates YopD, a 33 kDa protein that inhibits respiratory burst in macrophages. YopD is also responsible for translocation of YopE and other effector proteins (YopH, YopM, YopO) across the membrane.

Type III Secretion System

The type III secretion system (T3SS) apparatus, called Ysc, is responsible for the formation of a supramolecular apparatus called injectisome/injectosome and is essential for secretion of YOPs to extracellular milieu, across the outer membrane, and to the host cell cytosol. It is made of 28 proteins and the genes for which are encoded by the pYV plasmid, and their expression is temperature dependent (37 °C). The T3SS is responsible for delivery of YopH, YopO, YopT, YopP/J, YopE, and YopM effector proteins into the host cell cytosol. These effector proteins affect signaling events and alter actin cytoskeletal structure to induce bacterial entry by zipper mechanism, phagocytosis, apoptosis, and inflammatory response.

YopH is a 51 kDa protein and interferes with the signal transduction pathway to block phagocyte-mediated killing. Phagocytes (macrophages and neutrophils) are important components of the innate immunity and provide the first line of defense against *Yersinia* infection. The process of phagocytosis involves actin rearrangement to form pseudopods for bacterial internalization. YopH dephosphorylates host phosphotyrosine-containing proteins that are involved in actin polymerization to form pseudopods. YopH also interferes with the calcium signaling in neutrophils and downregulates respiratory bursts in the macrophages and neutrophils. In adaptive immune response, YopH suppresses B-cell activation and production of cytokines by the T cells.

YopO/YpkA is an 82 kDa secreted protein and possesses kinase activity and displays cytotoxic action. YpkA induces phosphorylation of VASP (vasodilator-stimulated phosphoprotein), a regulator of actin dynamics, thus disrupts actin polymerization and impairs phagocytosis. YopT possibly has Rho GTPase activity. Rho families of GTPases (Rac1, RhoA, and Cdc42) regulate actin cytoskeleton formation and interference of GTPase activity prevents phagocytosis of bacteria. YopP/J is a 33 kDa protein that blocks inflammation and induces apoptosis in macrophages by inhibiting MAPK signaling pathways and NF-κB pathway. As a result, TNF-α and IL-8 production are downregulated. YopE has cytotoxic action causing rounding and detachment of host cells from the extracellular matrix, and it regulates the inflammatory response. Controlling activation of serine protease caspase-1 is critical for the progression of the disease. Caspase-1 aids in IL-1β and IL-18 synthesis and induces pyroptosis, a lytic form of cell death. YopM is a 41 kDa protein that inhibits caspase-1 and is required for initiation of pyroptosis.

Pathogenic Mechanism

Foods such as chocolate milk or water that are incriminated in yersiniosis are generally cycled through refrigeration or below 25 °C. Generally, a high dose (10^7–10^9 cells) is required for the disease. Once ingested, bacteria travel through the stomach to the small intestine, and the primary site of infection is the terminal ileum and proximal colon. Survival in the stomach acidity has

been proposed to be neutralized by urease produced by the pathogen. It is speculated that, initially, bacteria use chromosomally encoded virulence gene products to colonize the intestine until the temperature shift to 37 °C and then initiate the expression of pYV-encoded gene products. Virulence factors help bacteria to (i) colonize and invade, (ii) prevent activation of cell death pathways, (iii) perturb inflammatory processes, and (iv) evade both innate and adaptive immune response to promote disease.

Bacteria bind to mucus membrane using Invasin, Ail, and YadA and enter through M cells overlying the Peyer's patches (Fig. 17.2). Invasin interacts with the β1-integrin receptor located abundantly on the M cells on the luminal side (apical side). YadA also aids in the invasion by interacting with the β1-integrin as well as the collagen and laminin. Engulfed bacteria are then released from the M cells into the basal layer in the lamina propria, multiply within the lymphoid follicle, and cause necrosis and abscess in Peyer's patches. Bacteria are able to reinvade epithelial

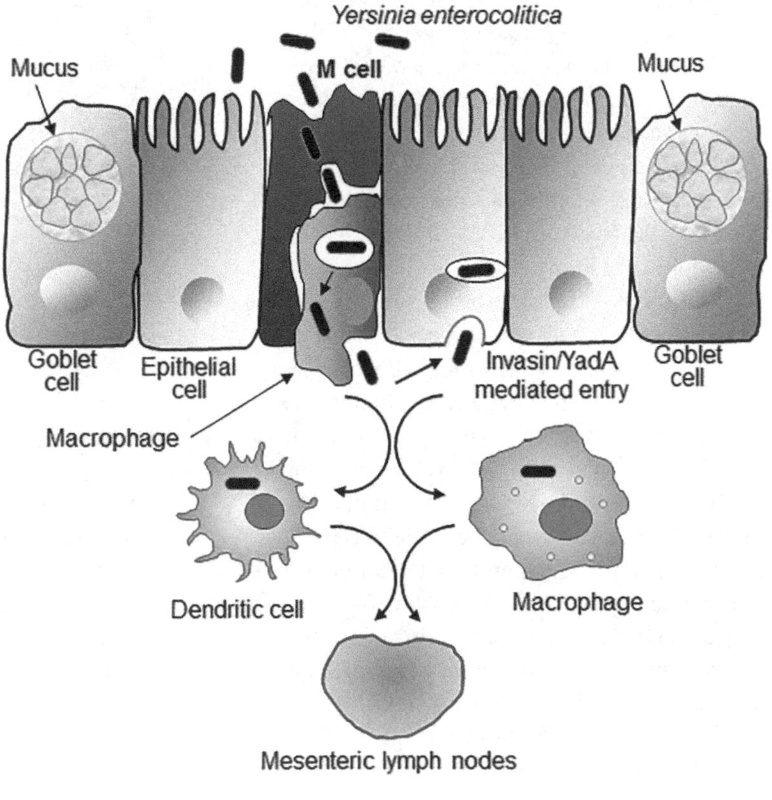

Fig. 17.2 *Yersinia enterocolitica* translocation through intestinal epithelial barrier. After entry into the basal layer via M cells, bacteria invade epithelial cells through interaction with host cell β1-integrin. Macrophage/dendritic cells transport *Yersinia* to mesenteric lymph nodes and to the liver. *Yersinia* prevents phagocytosis, also induces macrophage apoptosis, and prevents cytokine production

cells using the β1-integrin receptor located at the basolateral side. From the Peyer's patches, dendritic cells help bacteria to spread to regional mesenteric lymph nodes to induce characteristic lymphadenitis. Bacteria also disseminate to liver, spleen, and lungs and survive by resisting phagocytosis by macrophages and polymorphonuclear leukocytes. Survival inside macrophage is facilitated by the delivery of YOPs to the macrophage by T3SS that interferes with the cellular signaling events by blocking the phagocytosis process, phagocytic oxidative burst, and apoptosis or pyroptosis (Fig. 17.3). The superoxide dismutase (SOD) enzyme also helps bacterial survival inside the macrophage. Survival inside macrophage is orchestrated by two regulators, OmpR (outer membrane protein R) and GsrA (global stress response A protein, 49.5 kDa).

The lipid A component of *Yersinia* LPS undergoes temperature-dependent modification to avoid detection by TLR-4 and subsequent signaling event to release proinflammatory cytokines (TNF-α, IFN-γ, IL-8) from macrophages and other immune cells thus inhibiting inflammation. IL-8 is a chemoattractant for neutrophils. Bacteria also block complement activation thus C3b- and MAC-mediated inactivation are avoided. Overall, the T3SS system blocks phagocytosis and suppresses immune system thereby ensuring bacterial survival in the lymphoid tissue. In the intestine, bacteria promote the formation of an abscess in the Peyer's patches and induce damage to the epithelial lining.

The enterotoxin, Yst, promotes fluid secretion from cells. All three Yst components Yst-a, Yst-b, and Yst-c stimulate the membrane-bound guanylate cyclase leading to increased accumulation and activation of intracellular cyclic guanosine monophosphate (cGMP), followed by an activation of cGMP-dependent protein kinase, culminating in the final biological event, i.e., inhibition of Na^+ absorption and stimulation of Cl^- secretion. The biological activity of Yst enterotoxin is determined by the suckling mouse bioassay (*see* Chap. 5).

The predominant protective immunity against yersiniosis is the development of CD8+T cell response against YopE antigen in *Y. pseudotuberculosis* and *Y. pestis*.

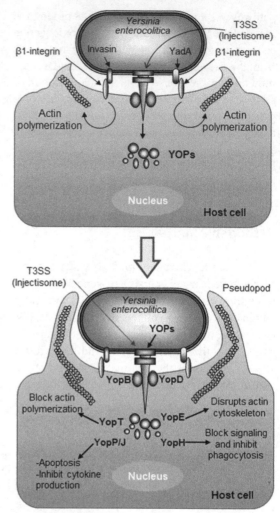

Fig. 17.3 Cellular mechanism of interaction of *Yersinia enterocolitica* with macrophage. After binding to macrophage, bacteria deliver YOPs through type III secretion system (T3SS) or injectisome into the host cell cytosol

Symptoms

Children are more susceptible than adults to foodborne yersiniosis showing acute enteritis. The highest incidence of infection is reported in children under the age of four who typically present with self-limiting diarrhea. Symptoms include severe abdominal pain at the lower quadrant of the abdomen mimicking appendicitis, diarrhea, nausea, vomiting, and fever. It also causes enterocolitis, mesenteric lymphadenitis, and terminal ileitis. Symptoms generally appear within 24–30 h after consumption of the contaminated food and last for 3–28 days for infants and 1–2 weeks for adults. The disease can be fatal in rare cases. The severity of infection is pronounced in immunocompromised host or individuals with underlying diseases resulting in septicemia, pneumonia, meningitis, and endocarditis and can be fatal. The severity of the disease is also dependent on the serotype of the organism involved. For example, the disease caused by serotype O:8 (biotype 1B) is more severe than that of other serotypes. *Yersinia* is also known to cause nosocomial infection in hospital patients exhibiting symptoms of diarrhea. *Yersinia* infection can result in sequelae in some patients leading to reactive arthritis or Reiter's syndrome.

Prevention, Control, and Treatment

Yersinia enterocolitica is a psychrotroph, and therefore refrigeration is ineffective in controlling growth. Good sanitation at all stages of handling and processing of food and proper heat treatment are important to control the occurrence of foodborne yersiniosis. Chitterling-related outbreaks can be avoided by strict hygienic practices, preventing cross-contamination, and proper cooking. Consumption of raw milk or meat cooked at low temperatures should be avoided. *Yersinia enterocolitica* is susceptible to heat and pasteurization and can be easily destroyed by ionizing radiation, UV radiation, and other food preservation procedures.

Yersinia produces two types of β-lactamases (enzymes that hydrolyze the β-lactam ring of the β-lactam antibiotics) and is thus resistant to the penicillin group of antibiotics; however, the newer β-lactam antibiotics (ceftriaxone, ceftazidime, moxalactam) are found to be effective. *Yersinia enterocolitica* is also sensitive to imipenem and aztreonam antibiotics. Broad-spectrum cephalosporins are also effective against extraintestinal infections.

Detection of *Yersinia enterocolitica*

Culture Methods
Cold and selective enrichment has been used for isolation of *Yersinia*. Cold enrichment (at 4 °C for 3 weeks) is generally used for *Y. enterocolitica* and *Y. pseudotuberculosis*, and the media used are buffered peptone water (BPW), phosphate buffered saline (PBS), PBS containing 1% mannitol and 0.15% bile salts (PMB), and PBS containing 0.25% peptone and 0.25% mannitol (PMP). For selective enrichment, irgasan–ticarcillin–potassium chlorate (ITC) broth at 25–30 °C for 48 h has been used to increase bacterial numbers. *Yersinia* is isolated on several commonly used enteric media: MacConkey agar, Hektoen enteric (HE) agar, *Salmonella–Shigella* with sodium deoxycholate and calcium chloride (SSDC), and xylose–lysine deoxycholate (XLD) agars. Other selective media include cefsulodin–irgasan–novobiocin (CIN) and modified virulent *Yersinia enterocolitica* (mVYE) medium. mVYE agar contains CIN agar, 0.1% esculin and 0.05% ferric citrate. Of all the selective media, CIN agar is the most effective. The mVYE agar can differentiate virulent *Y. enterocolitica* (esculin nonhydrolyzing and produced red colonies) from avirulent environmental *Y. enterocolitica* or other *Yersinia* species that produce dark colonies.

Serodiagnosis
Agglutination-based serodiagnosis of *Yersinia* using host serum has been used. However, it is found to be an inconsistent diagnostic tool because of the cross-reaction of antiserum with several other pathogens. Though ELISA has improved sensitivity, cross-reactions and the false-positive rates are very high. Indirect immunofluorescence assay is used with biopsy specimens. It appears that culture confirmation, in conjunction with serodiagnosis, may be used to correctly diagnose a patient suffering from yersiniosis.

Molecular Detection Method

Standard PCR method that targets virulence genes, namely, *yadA* and *virF* located on pYV, and the 16S rRNA genes is used. Real-time quantitative PCR employing *ail* gene as the target has been used for detection of *Y. enterocolitica* in pig feces and meat.

Molecular Typing Methods

Serotyping based on O (somatic) and H (flagellar) antigens, multilocus enzyme electrophoresis (MLEE), phage typing, and DNA-based schemes have been used to study typing, taxonomy, and epidemiology of *Yersinia enterocolitica*. Among the DNA-based schemes, pulsed-field gel electrophoresis (PFGE), ribotyping, amplified fragment length polymorphism (AFLP), randomly amplified polymorphic DNA (RAPD) and multilocus sequence typing (MLST), and whole genome sequencing (WGS) are used that produce reproducible typing information. MLST has also been used to understand the global epidemiology of yersiniosis.

Yersinia pestis

Introduction

Yersinia pestis causes plague, which is either bubonic or pneumonic (pulmonary). The plague was described as early as the 430 BC in Athens, Greece, and is called an old-world disease. Plague is often referred to as "Black Death." *Y. pestis* has been responsible for three pandemics and over 200 million deaths within the last 1500 years. Plague can be transmitted through contact with wild rodents and their fleas, which act as the vector (Fig. 17.4). *Yersinia pestis* has a high affinity for lymphoid tissues and causes acute inflammation, abscess, and swelling of lymph nodes. These inflamed and pus-emitting lymph nodes are called "buboes" (hence bubonic). In recent years, *Y. pestis* has gained significant interest as a possible agent of biological warfare. Pulmonary (pneumonic) plague is of major concern because the infection can be acquired directly by inhalation of infectious

aerosols generated by the infected person. Camels and goats are susceptible to *Y. pestis* infection and can transmit to human. *Yersinia pestis* is not a major foodborne concern; however, consumption of camel meat has been implicated in pharyngeal plague and in a rare occasion, gastroenteritis.

Biology

Yersinia pestis is a nonmotile, nonspore-forming, coccobacillary organism measuring 0.5–0.8 μm by 1–3 μm. The optimum growth temperature is 28 °C with a range of 4–40 °C. *Yersinia pestis* is a facultative intracellular organism, and it has three biotypes: antiqua, medievalis, and orientalis. Biotypes can be distinguished by their ability to utilize certain substrates. Antiqua is glycerol-, arabinose-, and nitrate-positive. Medievalis is glycerol- and arabinose-positive but nitrate-negative. Orientalis is glycerol-negative and arabinose- and nitrate-positive. The existence of another biotype called microtus has been proposed. Microtus may have lost genes essential for virulence determinants and host adaptation. Microtus is glycerol-positive and arabinose- and nitrate-negative.

Virulence Factors and Pathogenesis

Yersinia pestis infection starts from fleabite or contact with rats, or through inhalation. Bacteria multiply rapidly inside the midgut (stomach) of the rat flea (*Xenopsylla cheopis*), aided by phospholipase D and form a mass at around 26 °C. During feeding of blood (37 °C) on the host (rat or humans), the fleas regurgitate the bacilli to the wound. Wild animals, particularly rats, are the major reservoir and can be infected asymptomatically. In rural areas, humans coming in contact with the rodents or fleas (vector) carrying *Y. pestis* become infected. Rat-to-human transmission is possible and human-to-human transmission is mostly by the airborne route or by human-adapted flea (*Pulex irritans*) (Fig. 17.4). The bubonic form is most common and starts with the bite of

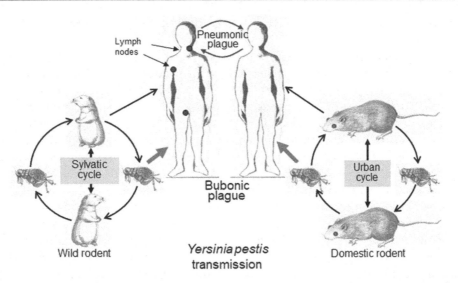

Fig. 17.4 Pathways showing transmission of *Yersinia pestis* to humans. A flea can transmit *Y. pestis* to wild rodents (sylvatic cycle), domestic rats (urban cycle), or humans. Infected flea also helps maintain both sylvatic and urban cycles. Fleabite or direct contact of rodents with a human may result in the development of bubonic plague. Human-to-human transmission via aerosol/droplets may cause pneumonic plague

an infected flea. The incubation period is 2–8 days.

Yersinia pestis virulence genes are located on the chromosome and on three plasmids (pCD1, pMT1, pPCP1). The virulence factors are F1 (fraction 1 antigen) capsule, plasminogen activator (Pla) protein, low calcium response (Lcr) stimulon, pesticin, and a *Yersinia* murine toxin (Ymt). These proteins are used as markers for biotyping of *Y. pestis*. The gene for Lcr is present on pCD1, F1 and the Ymt (phospholipase D) are present on pMT1, and Pla and pesticin are located on pPCP1. Chromosome-encoded invasin and pCD1-encoded YadA are inactive in *Y. pestis* due to mutation. Pla serves as an adhesin, invasin, and proteolytic factor and helps in systemic dissemination of bacteria. F1 blocks phagocytosis. Psa is an adhesion pilus (15 kDa) and forms a capsule-like structure on the surface. It is expressed at high levels at 37 °C and a pH range of 5.8–6.0. It is positively regulated by RovA and negatively by Fur. The virulence plasmid, pCD1, encodes genes for T3SS for secretion of YOPs (YopE, YopH, YopM, YopO/YpkA, YopP/J, and YopT) directly into the host cells to inhibit phagocytosis and blockade of proinflammatory signals.

During the early stage of infection, *Y. pestis* invasion and survival inside macrophages and nonprofessional phagocytes (host epithelial cells) are critical, which are mediated by Ail and Pla, respectively. In nonprofessional phagocytic cells, bacteria enter by zipper mechanism (Fig. 17.3). In the later stage of infection, the bacterium exerts anti-phagocytic activity which is attributed to Psa, F1 antigen (capsule), and YOPs (YopH, YpkA, YopE, and YopT). The key to the successful infection is a switch of bacterial initial invasive status to the anti-phagocytic status at the later stage of infection. It is speculated that initial intracellular lifestyle inside the macrophage may trigger the genes that are responsible for upregulation of surface components to promote bacterial resistance to subsequent phagocytosis during the second stage of the infection process. Temperature shift from 28 °C to 37 °C in vivo also induces F1, Psa, and YOPs expression, which collectively block phagocytosis. Low pH in the phagosome also upregulates Psa expression on the surface of the bacterium. In addition, Psa-mediated binding enhances contact with host cells to facilitate YOP translocation via T3SS to block further phagocytosis.

Symptoms

Clinical manifestation starts with fever, chills, and headache. There are a gradual swelling and enlargement of inguinal (groin) and submaxillary lymph nodes called "buboes." Inflammation or cellulitis may develop around the buboes. Large carbuncles may also develop. Gastroenteritis characterized by abdominal pain, nausea, vomiting, and diarrhea may develop in some cases. Septicemia, disseminated intravascular coagulation, and shock can develop in some patients. Secondary pneumonic plague may result in bronchopneumonia with the production of bloody purulent sputum. Pneumonic plague is highly contagious and can be transmitted easily by the airborne route. In the primary septicemic plague, buboes are absent but bacteremia develops and the mortality rate is 30–50%. In the secondary pneumonic phase, bacteria disseminate to the respiratory tract and develop severe bronchopneumonia, cavitation, and formation of bloody purulent (containing pus) sputum. This form of plague is highly contagious and fatal.

Treatment and Prevention

Early intervention with antibiotics within 1–3 days of exposure may be effective. Streptomycin, gentamicin, chloramphenicol, and tetracycline are effective. Antibiotic-resistant strains have been reported from Madagascar where plague is endemic. These resistant strains are also a serious public health threat. Use of ciprofloxacin as a prophylactic antibiotic has been found to be effective but is not good for treatment. Treatment with other fluoroquinolones has been reported to be promising.

A killed "whole cell" vaccine is available but requires multiple boosters. Efficacy of the vaccine against pneumonic plague is questionable. Other preventive measures include public education about the transmission of the disease by rodents and fleas. Access of rodents to food for human consumption should be prevented. Fleabite should be avoided by wearing protective clothing or by using insect repellant. Insecticides can be used to eliminate fleas from home or pet populations.

Detection of *Yersinia pestis*

Body fluids such as blood, sputum, bubo aspirates, or cerebrospinal fluid can be stained for bipolar staining (Giemsa or Wayson's) to directly visualize the organism under a microscope. The test samples can be streaked onto blood agar plates, brain heart infusion (BHI) agar plates, or MacConkey agar plates and incubated at 28 °C for 48 h (note: at 37 °C, the organism develops an envelope and becomes highly virulent). The colonies are characteristically opaque, smooth, and round-shaped with irregular edges. The bacteria can be extremely slow growing; thus, culture plates should be examined for 1 week before discarding. Biochemical characterization of *Yersinia* is reliable and can be done with diagnostic kits (Table 17.1). Serological tests such as passive hemagglutination (PHA) test could be used for diagnosis. Other immunoassays such as ELISA, direct immunofluorescence assay, and dipstick assays that target F1 antigen or Pla protein are very useful rapid tools. A real-time 5′ nuclease PCR that targets the *pla* gene of *Y. pestis* has been developed for analysis of human sputum or respiratory swabs samples. Whole genome sequencing is a powerful tool in *Y. pestis* identification.

Summary

The genus *Yersinia* consists of 17 species, of which *Y. enterocolitica*, *Y. pseudotuberculosis*, and *Y. pestis* are pathogenic to humans. The former two are enteropathogenic and responsible for gastroenteritis, and the latter one is responsible for the plague. *Y. enterocolitica* and *Y. pseudotuberculosis* are transmitted through food, and pigs serve as the primary reservoir while *Y. pestis* infection is transmitted by fleabite, and rodents act as the intermediate host. All three pathogens carry chromosomal- and plasmid-encoded virulence factors, which are required for adhesion, invasion, and colonization of intestinal epithelial cells and lymph nodes, growth inside macrophages, macrophage apoptosis, and serum resistance. In *Y. enterocolitica*, chromosome-encoded virulence gene products include invasin, attachment invasion locus (Ail), siderophore (yersiniabactin), and an enterotoxin (Yst), which are important during initial colonization and invasion of intestinal M cells and enterocytes in the intestine. The pYV plasmid-encoded virulence factors include *Yersinia* outer membrane proteins (YOPs) that are responsible for bacterial virulence protein translocation by T3SS to the host cell cytosol and resist macrophage and neutrophil-mediated inhibition and serum resistance. Expression of YOPs is temperature-dependent and occurs mostly at 37 °C, which is critical for bacterial pathogenesis while inside a host. During passage through the digestive tract, bacteria invade M cells overlying Peyer's patches, multiply in the lymphoid follicle, and are engulfed by macrophages. *Y. enterocolitica* is resistant to phagocytic killing by neutrophil, DC, and macrophages, and these cells help disseminate the organism to mesenteric lymph nodes, liver, and spleen.

Y. pestis causes the bubonic and pneumonic form of plague by colonizing the lymphoid tissues of the gastrointestinal and the respiratory tracts. The organisms acquired by either fleabite or aerosol are transported to the lymph nodes by macrophages where the bacteria survive and resist the killing by macrophages and neutrophils by producing several virulence factors: F1 (fraction 1 antigen) capsule, plasminogen activator (Pla), and low calcium response (Lcr) stimulon. In the bubonic form, the submaxillary lymph nodes are enlarged and appear as buboes. In this form, fever, chill, headache, and septicemia develop, and the mortality rate in untreated cases is 30–50%. In the secondary pneumonic phase, the bacteria disseminate to the respiratory tract and develop severe bronchopneumonia with bloody purulent sputum. This form of plague is highly contagious and invariably fatal, if not treated.

Further Readings

1. Atkinson, S. and Williams, P. (2016) *Yersinia* virulence factors - a sophisticated arsenal for combating host defences. *F1000Research* **5**, F1000 Faculty Rev-1370.

2. Bhaduri, S., Wesley, I.V. and Bush, E.J. (2005) Prevalence of pathogenic *Yersinia enterocolitica* strains in pigs in the United States. *Appl Environ Microbiol* **71**, 7117–7121.

3. Bhagat, N. and Virdi, J.S. (2011) The enigma of *Yersinia enterocolitica* biovar 1A. *Crit Rev Microbiol* **37**, 25–39.

4. Bottone, E.J. (1999) *Yersinia enterocolitica*: overview and epidemiologic correlates. *Micres Infect* **1**, 323–333.

5. Chung, L.K. and Bliska, J.B. (2016) *Yersinia* versus host immunity: how a pathogen evades or triggers a protective response. *Curr Opin Microbiol* **29**, 56–62.

6. Dhar, M.S. and Virdi, J.S. (2014) Strategies used by *Yersinia enterocolitica* to evade killing by the host: thinking beyond Yops. *Microbes Infect* **16**, 87–95.

7. Drummond, N., Murphy, B.P., Ringwood, T., Prentice, M.B., Buckley, J.F. and Fanning, S. (2012) *Yersinia enterocolitica*: a brief review of the issues relating to the zoonotic pathogen, public health challenges, and the pork production chain. *Foodborne Pathog Dis* **9**, 179–189.

8. Fredriksson-Ahomaa, M., Stolle, A. and Korkeala, H. (2006) Molecular epidemiology of *Yersinia enterocolitica* infections. *FEMS Immunol Med Microbiol* **47**, 315–329.

9. Fukushima, H., Shimizu, S. and Inatsu, Y. (2011) *Yersinia enterocolitica* and *Yersinia pseudotuberculosis* detection in foods. *J Pathog* **2011**, Article ID 735308.

10. Gupta, V., Gulati, P., Bhagat, N., Dhar, M. and Virdi, J. (2015) Detection of *Yersinia enterocolitica* in food: an overview. *Eur J Clin Microbiol Infect Dis* **34**, 641–650.

11. Ke, Y., Chen, Z. and Yang, R. (2013) *Yersinia pestis*: mechanisms of entry into and resistance to the host cell. *Front Cell Infect Microbiol* **3**.

12. Navarro, L., Alto, N.M. and Dixon, J.E. (2005) Functions of the *Yersinia* effector proteins in inhibiting host immune responses. *Curr Opin Microbiol* **8**, 21–27.

13. Pujol, C. and Bliska, J.B. (2005) Turning *Yersinia* pathogenesis outside in: subversion of macrophage function by intracellular yersiniae. *Clin Immunol* **114**, 216–226.

14. Smego, R.A., Frean, J. and Koornhof, H.J. (1999) Yersiniosis I: Microbiological and clinicoepidemiological aspects of plague and non-plague *Yersinia* infections. *Eur J Clin Microbiol Infect Dis* **18**, 1–15.

15. Virdi, J.S. and Sachdeva, P. (2005) Molecular heterogeneity in *Yersinia enterocolitica* and 'Y. enterocolitica-like' species - Implications for epidemiology, typing and taxonomy. *FEMS Immunol Med Microbiol* **45**, 1–10.

Vibrio cholerae, Vibrio parahaemolyticus, and Vibrio vulnificus

18

Introduction

Filippo Pacini, an Italian physician, discovered *Vibrio cholera* in 1854. In the same year, John Snow, a British physician (obstetrician), unraveled that the cholera outbreak in London that killed 616 people was due to contaminated water, linking the water for the very first time, not the air, as the primary source of contamination. Robert Koch, unaware of Pacini's work, independently isolated *Vibrio cholera* in 1884 and became the acknowledged discoverer of *Vibrio cholerae*, until the international committee on nomenclature in 1965 adopted *Vibrio cholerae* Pacini 1854 as the correct name of the cholera-causing organism. Vibrios are inhabitants of estuarine and freshwaters, and some species are pathogenic to humans and marine vertebrates and invertebrates. In humans, some species of vibrios can cause gastroenteritis following ingestion of contaminated food or water and septicemia when preexisting cuts or abrasions on the skin are exposed to contaminated water or seafood. Vibrios are of significant concern in both developed and developing countries because of their continued burden of disease resulting from contaminated water and fish products. In the USA, vibriosis causes an estimated 80,000 illnesses and 100 deaths every year. Three major species, *Vibrio cholerae*, *V. parahaemolyticus*, and *V. vulnificus*, are responsible for the majority of human infections; however, several other species are responsible for sporadic infections.

Classification

The genus *Vibrio* is a member of the *Vibrionaceae* family and contains 63 species, and at least 11 of them are pathogenic to humans including *V. cholerae* (O1 and O139), *V. parahaemolyticus*, *V. vulnificus*, *V. mimicus*, *V. hollisae*, *V. fluvialis*, *V. alginolyticus*, *V. damsela*, *V. furnissii*, *V. metschnikovii*, and *V. cincinnatiensis*. Among these, the first three species cause most human infections.

Biology

Vibrio species are Gram-negative curved rods with size ranging from 1.4 to 2.6 μm in length and 0.5–0.8 μm in width. They are motile and generally possess a single polar flagellum. They are facultative anaerobes, most are oxidase-positive, and utilize D-glucose as the main carbon source. *Vibrio* species produce many extracellular enzymes: amylase, gelatinase, chitinase, and DNase. Some *Vibrio* species are halophilic (tolerance up to 18% NaCl), and sodium ions stimulate their growth. Vibrios grow well in neutral to alkaline pH (~9.0) and are acid-sensitive. The optimum pH range is 8.0–8.8, and the optimum growth temperature range is 20–37 °C. Water temperatures on either side of the range severely affect bacterial growth. Nutrient deficiency, salinity, and changes in temperature promote stress

© Springer Science+Business Media, LLC, part of Springer Nature 2018
A. K. Bhunia, *Foodborne Microbial Pathogens*, Food Science Text Series,
https://doi.org/10.1007/978-1-4939-7349-1_18

resulting in the viable but nonculturable state (VBNC) especially for *V. cholerae* and *V. vulnificus*. Two circular chromosomes, usually one large and the other small, are present in vibrios, which provide diversity in gene structure and gene content. Chromosomes 1 and 2 in *V. cholerae* are 3.0 and 1.1 Mb, respectively, in *V. parahaemolyticus* 3.3 and 1.9 Mb, and in *V. vulnificus* 3.3 and 1.85 Mb.

Source and Transmission

Vibrio species are isolated from fresh, brackish, and marine waters. Vibrios are found as free-living in water or are associated with inanimate surfaces or aquatic organisms (zooplankton, phytoplankton), aquatic animals (seabirds), sewage water, sediments, seafood, fish, and shellfish. Vibrios are associated with chitinous zooplankton and shellfish, forming biofilms to help bacterial prolonged survival in the aquatic environment. Bivalve shellfish such as clams, oysters, and mussels accumulate bacteria because of their filter-feeding habit. Natural disasters, like floods cyclones and hurricanes, cause the failure of sewer systems and result in contamination of the aquatic environment. Foods washed in contaminated water can transmit vibrios. Water temperature, nutrient availability, salinity, and an association with marine organisms influence the *Vibrio* loads in water. *Vibrio* counts are very high during summer and autumn months, and counts appreciably diminish (or are generally absent) at temperatures below 10 °C. Vibrios are obligate halophiles (except *V. cholerae*), and depending on the species preference for water salinity varies.

Vibrio cholerae

Introduction

One of the most important members of the genus *Vibrio* is *V. cholerae*. The disease caused by toxigenic strains of the two serotypes (O1 and O139) of *V. cholerae* is known as cholera. Though

Filippo Pacini and Robert Koch independently isolated *V. cholerae* in 1854 and 1884, respectively, the outbreaks of cholera dates back to 460–377 BC during the times of Hippocrates. In modern history, epidemic and pandemic cholera occur with global implications, and frequent outbreaks are reported in Asia, Africa, and South and Central America. Cholera is endemic in many parts of Asia and Africa. According to the World Health Organization (WHO), *V. cholerae* infect about 3–5 million people with 100,000–120,000 deaths each year worldwide.

Biology

Vibrio cholerae is a Gram-negative rod or curved-shaped bacterium (0.7–1.0 × 1.5–3.0 μm). It is a facultative anaerobe and produces pale-yellow, translucent colonies that are about 2–3 mm in diameter on a special medium known as thiosulfate citrate bile salts sucrose (TCBS) agar. *Vibrio cholerae* is able to grow within a temperature range of 15–45 °C, a pH range of 6–10, and a salt (NaCl) concentration of up to 6%. However, it does not require salt for growth. *V. cholerae* is often associated with zooplankton and crustaceans (Fig. 18.1). *V. cholerae* forms, biofilms on zooplankton and phytoplankton, which both contain chitin, and the bacterium can use chitin as a carbon and nitrogen source. There are 206 known serotypes based on O antigen of the lipopolysaccharide (LPS), of which two major serotypes, O1 and O139, are responsible for epidemic cholera. A major difference between O1 and the O139 is the presence of a thin capsule in O139 and its absence in O1. This difference can be observed during bacterial growth on solid agar media, where O1 produces translucent while O139 produces opaque colonies. Furthermore, the LPS of the O1 serotype is smooth while it is semi-rough in O139. The O139 LPS has a highly substituted core oligosaccharide and shorter side chains of O antigen, which are responsible for the rough phenotype. The LPS and capsule of O139 also share a unique sugar, 3,6-dideoxyhexose (colitose). The serotype O1 is subclassified into the classical and El Tor biotypes based on a set of phenotypic

Fig. 18.1 The lifecycle of *Vibrio cholerae* in the aquatic environment and host intestine. Biofilm formation is critical for their persistence in both the aquatic environment and intestine. *TCP* toxin-coregulated pili, *CT* cholera toxin

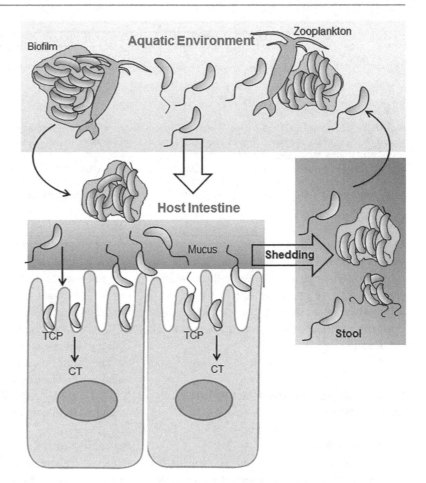

traits. The two biotypes are further classified as Inaba, Ogawa, and Hikojima subserotypes. The El Tor biotype was originally isolated from an outbreak in El Tor, Egypt, in 1905.

All cholera-causing strains carry virulence genes for cholera toxin (CT) and toxin-coregulated pili (TCP). Virulence genes are located in *Vibrio* pathogenicity islands (VPI-1 and VPI-2). The gene (*ctxAB*) encoding CT is located in a lysogenic filamentous bacteriophage, CTXΦ (Fig. 18.2), which also contains several accessory virulence genes clustered in two regions, RS and core. The RS region constitutes *rstA*, *rstB*, and *rstR* genes that are responsible for the site-specific integration, replication, and regulation of the phage into the chromosome. The core region carries *ctxAB* which encodes CT and genes for Psh, Cep (core-encoded pilin), Ace (accessory cholera enterotoxin), and Zot (zonula

occludens) which are required for phage coat synthesis and morphogenesis.

Other factors including outer membrane porins, biotin and purine biosynthetic enzymes, iron-regulated outer membrane proteins (IrgA), and O antigen of LPS are also known virulence factors. The non-O1/O139 also cause diarrhea but generally milder than O1, and these serotypes are common in the USA. *Vibrio cholerae* has a single polar flagellum, which helps the bacterium to reach the intestinal mucosa and aids in colonization. It maintains two lifestyles: one in the aquatic environment where it is free-living or attached to zooplanktons and the other is inside the host gastrointestinal tract (Fig. 18.1). The common survival strategies for *Vibrio* either in the aquatic environment or in human host include (1) the activation of stress response, (2) expression of flagella for motility and chemotaxis, (3)

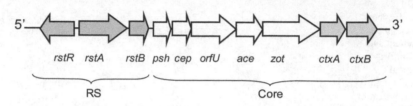

Fig. 18.2 Schematic of CTXΦ bacteriophage genome in *Vibrio cholerae* classical strain. Genes encoding cholera toxin, *ctxAB* appears at the 3′ end of the genome. The *rstR*, *rstA*, and *rstB* genes constitute the RS region (*shaded arrows*) that is responsible for the site-specific integration (*rstA*), replication (*rstB*), and regulation (*rstR*) of the phage. The core region carries *ctxAB* which encodes CT and genes for Psh, Cep (core-encoded pilin), OrfU, Ace (accessory cholera enterotoxin), and Zot (zonula occludens) which are required for phage coat synthesis and morphogenesis (Adapted and redrawn from Safa et al. 2010. Trends Microbiol. 18, 46–54)

attachment to abiotic and biotic surfaces, (4) biofilm formation, and (5) detachment from the surface. Proficient switching between the planktonic (motile) and biofilm (sessile) lifestyles in response to chemical and physical changes in the extracellular milieu is key to the survival and colonization.

Virulence Factors and Pathogenic Mechanism

Gastroenteritis

Vibrio cholerae is the most studied organism that is responsible for acute secretory diarrhea known as cholera. The infectious dose is 10^4–10^{10} cfu g^{-1} and the disease is spread through contaminated water and food, and transmission is through the fecal–oral route. Individuals with the blood group O are more susceptible to cholera than the other blood groups. Serotype O1 causes a fatal form of cholera, while the non-O1 (the O139 form) is generally less virulent. There are two biotypes, classical and El Tor, and they share a common LPS O antigen. A strain of *V. cholerae* O139 Bengal, originating around 1992, was responsible for several outbreaks in India and Bangladesh. It is capable of causing serious diarrhea. Following ingestion, *V. cholerae* overcomes low pH, bile acids, elevated osmolarity, iron limitation, antimicrobial peptides, and natural microflora and can grow to high titers in the human gut. Stress response regulators RpoS (σS) and RpoE (σE) help cope with the intrinsic stressors in the gut. Cholera patients can shed 10^7–10^9 vibrios mL^{-1} in so-called rice-water stools. *Vibrio cholerae* adherence and colonization of small intestinal mucosa are facilitated by toxin-coregulated pili (TCP), flagella, and neuraminidase. Bacteria then produce several toxins such as cholera toxin, ZOT (zona occludin toxin), ACE (accessory cholera enterotoxin), and HlyA (hemolysin), which act on mucosal cells. Toxins alter the ion balance by affecting the ion transport pumps for Na$^+$, Cl$^-$, HCO$_3^-$, and K$^+$ in the cell, resulting in extensive fluid and ion losses.

Adhesion and Colonization

Bacterial adherence and colonization are positively influenced by motility and chemotaxis. A single polar flagellum helps each bacterium to penetrate the mucus layer. Flagellin mutants are nonmotile and are less virulent. The long filamentous pili called TCP (a type IV bundle-forming pilus) help form microcolonies and are involved in colonization. The genes encoding TCP pili are regulated similarly to genes encoding for the cholera toxin. TCP mutants are avirulent in humans. TCP pilin is encoded in *tcpA* and is located in the TCP pathogenicity island. *N*-acetylglucosamine (GlcNAc)-binding protein A (GbpA: 53 kDa) encoded by *gbpA* has been shown to be involved in bacterial colonization by interacting with mucin. Other colonization factors include mannose–fucose-resistant cell-associated hemagglutinin (MFRHA: 26.9 kDa) and some outer membrane proteins (OMPs). TCP and other colonization factors are regulated by regulatory proteins (ToxR/ToxS and ToxT). Immediately adjacent to the *tcp* cluster is the *acf* gene for accessory colonization factor (ACF), a lipoprotein. The exact role of ACF is not known, but it is believed to be involved in bacterial colonization.

Vibrio cholerae Biofilm

The ability to form biofilms on the biotic and abiotic surface is an important survival and colonization strategy for *V. cholerae* (Fig. 18.1). In the aquatic environment, biofilms enhance *V. cholerae* persistence and provide protection against stress, nutrient limitation and predation by protozoa, and attack by bacteriophages. *Vibrio cholerae* forms a biofilm on phytoplankton and zooplankton. Type IV mannose-sensitive hemagglutinin (MSHA) pili help in biofilm formation on plankton, and the extracellular matrix helps maturation of biofilms. After ingestion of biofilms or planktonic cells by humans, bacteria in the intestine express TCP and form aggregates or biofilms aiding bacterial colonization on intestinal mucosa. Biofilms also help bacteria to avoid the host's innate immune response.

Quorum-sensing molecule, cholera autoinducer 1 (CA-1), accumulates in biofilms and promotes expression of the quorum-sensing regulator, HapR, which in turn enhances expression of sigma factor, RpoS. RpoS helps bacteria to cope with the environmental stressors. During biofilm formation, cells also accumulate intracellular cyclic diguanylic acid (c-di-GMP), a second messenger that controls the transition between planktonic and biofilm lifestyles.

Cholera Toxin

Cholera toxin (CT) is the most important virulence factor in *V. cholerae*. The gene encoding CT is located in a lysogenic filamentous bacteriophage, CTXΦ (Fig. 18.2). The bacterial cell surface receptor for CTXΦ interaction is TCP. CT-mutant strains are either avirulent or may cause milder diarrhea because of the presence of other toxins. CT is the best-studied bacterial toxin. It is an A–B type "ADP-ribosylating toxin." The A subunit is a 27 kDa protein encoded by *ctxA*, and the B subunit consists of five identical proteins of 11.7 kDa and is encoded by *ctxB*. A and B subunits are secreted into the periplasm, where they are assembled.

The B subunit of toxin first binds to the host mucosal cell by binding to the ganglioside GM_1 receptor. It is a sialic acid containing oligosaccharide covalently attached to the ceramide lipid. It is found on the surface of many cells. The toxin is internalized and the A subunit is detached. The A subunit has the enzymatic activity; it ADP-ribosylates the Gs proteins (composed of three subunits: α, β, γ) also known as "GTP hydrolyzing proteins." Gs proteins regulate the activity of host cell adenylate cyclase and serve as "off" and "on" switches. Binding of A subunit to Gs subunit α locks it in the "on" position and stimulates the production of the cyclic adenylate cyclase (cAMP). cAMP activates the protein kinase A, which in turn causes phosphorylation of protein especially the CFTR (cystic fibrosis transmembrane conductance regulator) protein in the ion pump and thus alters the function of sodium and chloride ion transport, resulting in the increased Cl^- and HCO_3^- secretion by crypt cells and decreased absorption of Na^+ and Cl^- by absorptive cells (Fig. 18.3).

Regulation of Cholera Toxin Production

Cholera toxin production is regulated by transmembrane proteins ToxR, ToxS, TcpP, and TcpH. ToxR, a 32 kDa transmembrane protein, binds to a 7 bp DNA sequence located in the upstream of *ctxAB* and increases the expression of CT. ToxR, TcpP, and TcpH activate ToxT, which in turn activates CT and TCP expression. ToxR and ToxT also regulate the expression of ACF, outer membrane proteins OmpU and OmpT, and other lipoproteins. The quorum-sensing regulatory proteins LuxO and HapR also control CT and TCP expression.

Other Toxins

Vibrio cholerae also produces two other toxins, ZOT (zonula occludin toxin: 44.8 kDa) and ACE (accessory cholera enterotoxin). The ZOT disrupts the "tight junction" that binds mucosal cells together and preserves the integrity of the mucosal membrane. Normally, the tight junction maintains cellular integrity and prevents the loss of ions or water molecules. ZOT induces a reorganization of F-actin and decreases the G-actin and

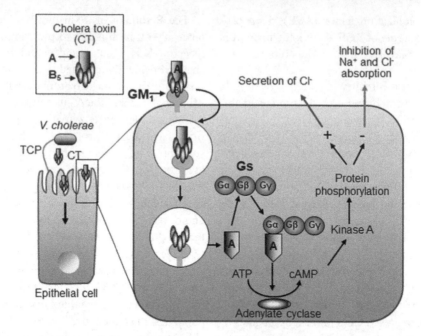

Fig. 18.3 Schematic showing the action of cholera toxin (CT) on enterocyte. *Vibrio cholerae* first bind to the cells using pili (toxin-coregulated pili: TCP) or other colonization factors and produce CT. The CT is an A–B type toxin composed of one A subunit and five B subunits. The B subunits bind to the GM1 receptor, and the CT is transported inside the cell. The A subunit ADP-ribosylate Gs protein (GTP hydrolyzing protein) increases the catalysis of ATP (adenosine triphosphate) to form cyclic adenosine monophosphate (cAMP). cAMP mediates phosphorylation of the CFTR protein, which is involved in ion transport (pump), thus affecting ion losses (Cl^-, HCO_3^-, Na^+) and fluid flow

affects the cytoskeletal rearrangements possibly mediated by protein kinase C. Consequently, the tight junction loses its barrier function and enhances pericellular permeability. ZOT also disrupts the ion balance and promotes diarrhea.

The ACE toxin is responsible for diarrhea in animals but probably has no role in human diarrhea. A hemolysin (HlyA: 65 kDa), also known as El Tor hemolysin, is responsible for enterotoxicity. It binds to cholesterol and oligomerizes in the membrane forming a pore of 1.2–1.6 nm. Other virulence factors like siderophores also help bacterial iron acquisition from host cells.

Immune Response to Cholera Toxin

Vibrio cholerae induces a strong immune response by the production of the secretory immunoglobulin A (sIgA). sIgA acts as an opsonin, and opsonization may aid in the transcytosis of *V. cholerae* through M cells by interacting with the CT receptor ganglioside GM_1 on M cells. CT aids in intestinal dendritic cell (DC) maturation through the production of prostaglandin E_2 (PGE_2) and nitric oxide (NO); thus, CT has been used as an adjuvant in various vaccine formulations. The CT also influences the immune response of lymphocytes and monocytes by altering the expression of several genes. CT induces cAMP, which regulates the expression of genes that regulate various functions including immune response.

Symptoms of *V. cholerae* Infection

The symptoms of diarrhea appear within 6 h–5 days and can last for 2–12 days. Diarrhea looks like "rice-water" with a fishy odor, and a patient may defecate up to 1 L per hour leading to hypotensive shock and death within hours. The patients show high pulse rate, dry skin, sunken eyes, lethargy, low urine volume, nausea, and vomiting in the early phase of infection and may exhibit abdominal cramps. Fever is uncommon, but high fever is indicative of secondary infection. Watery diarrhea causes severe dehydration, loss of electrolytes, and ions causing hypertension that can be life-threatening. The death rate among untreated patients is about 70%. Infants and children are highly susceptible. Recovering patients develop immunity against cholera.

Control, Prevention, and Treatment of *Vibrio cholerae*

Contaminated water and food are the primary sources of *V. cholerae*; therefore, proper processing of water and food and sanitary hygienic practices would lessen the spread of the disease. Moreover, *V. cholerae* remain attached to zooplankton and phytoplankton as biofilms; therefore, water filtration practices that remove particles larger than 20 μm can also significantly reduce cholera cases.

Hallmark of cholera is profuse diarrhea. Loss of water and electrolytes leads to dehydration; thus, fluid therapy is the most effective treatment to prevent dehydration. Oral rehydration with boiled potable water containing table salt and glucose (glucose stimulates water uptake and is a source of energy) is effective in resource-constrained situation. Besides oral rehydration, intravenous fluid therapy is recommended in patients showing advanced signs of dehydration including sunken eyes, high pulse rate, lethargy, low urine volume, and comatose condition. Antibiotics including tetracycline, cotrimoxazole, erythromycin, doxycycline, chloramphenicol, and furazolidone can be used to treat cholera, but concern for antibiotic resistance is very high.

Cholera vaccine is used to prevent infection in the population in the endemic zone. sIgA generated from vaccination prevents bacterial colonization on the mucosal surface. Since there is no cross-protection between serotype O1 and O139, a bivalent vaccine is needed to provide protection against both serotypes. Injectable heat-killed bacterial cells were used for many years, but this vaccine exhibited toxic side effects due to the presence of endotoxin (LPS); hence, such immunization practices have been discontinued. Several strategies have been undertaken to develop a safe yet protective vaccine against cholera.

Oral vaccination with killed bacteria together with a purified B subunit of cholera toxin is widely used and is recommended by the WHO. Two killed vaccines are now used for oral administration, Dukoral (Sweden) that contains several biotypes and serotypes of *V. cholerae* O1, supplemented with 1 mg per dose of recombinant cholera toxin B subunit, and Shanchol (India), a bivalent toxin containing several O1 and O139 biotypes and serotypes without the supplemental cholera toxin B subunit. These vaccines are administered two or three times depending on the age of the patient. The overall protection is 60–85% lasting for 2–3 years. Live attenuated oral vaccine underdevelopment includes CVD 103-HgR (USA) and Peru-15 (China) use genetically modified CT-negative strains; however, their efficacy in a clinical trial has yet to be fully evaluated.

Vibrio parahaemolyticus

Introduction

Vibrio parahaemolyticus was first discovered by a Japanese scientist, Tsunesaburo Fujino, in 1950 when the organism caused an outbreak affecting 272 people, resulting in 20 deaths due to consumption of shirasu (a Japanese fish dish generally prepared with small fish, such as sardine, anchovy, etc.). *V. parahaemolyticus* is distributed in the marine environment (estuarine), and it is one of the major foodborne pathogens that are associated with seafood worldwide. Not all strains are considered pathogenic. The spread and dissemination of *V. parahaemolyticus* depend on the water temperature, zooplankton blooms, and dissolved oxygen. Countries located in a temperate climate experience higher numbers of outbreaks during summer months than the tropical countries. Countries located in the tropical zone maintain a warmer overall temperature, which is conducive for year-round outbreaks. The outbreaks of gastroenteritis are associated with consumption of contaminated seafood including raw oysters and other shellfish, and it is responsible for 20–30% of food poisoning cases related to seafood.

Biology

Vibrio parahaemolyticus is a moderate halophilic enteropathogen and requires salt (1–9%) for survival and growth. The bacterium primarily causes gastroenteritis; however, it can cause extraintestinal infections like eye and ear infections and wound infections affecting extremities. *V. parahaemolyticus* grows at a minimum temperature of 15 °C and a maximum temperature of 44 °C. On liquid medium, the bacterium expresses single polar flagellum (*fla*) exhibiting "swimming" motility, while on solid media, it expresses peritrichous flagella or lateral flagella (encoded by the *laf* gene) exhibiting "swarming" phenotype. The bacterium also expresses capsule (K antigen). All strains of *V. parahaemolyticus* produce H_2S in triple sugar iron (TSI) medium. Urease production is an unusual phenotype for *V. parahaemolyticus*, and a majority of clinical and environmental isolates is urease-negative; however, some clinical isolates are urease-positive. There is a strong correlation between urease production and the presence of the *trh* (heat labile hemolysin) gene among clinical isolates (see below).

Vibrio parahaemolyticus strains are classified based on their somatic (O) and capsular (K) antigen patterns, and the predominant serotype is O3:K6, which is distributed globally. The O3:K6 serotype is thought to have originated from Japan, and it was responsible for a major outbreak in Calcutta (now Kolkata, India) in 1996. The serovariants of O3:K6 exist including O4:K12, O4:K68, O1:K41, O1:K25, O1:KUT, and several others which are responsible for regional outbreaks such as those that occurred in South and South East Asia. Serotype O4:K12, O6:K18, O1:K56, O4:K63, O3:K36 and

O12:K12 have been associated with outbreaks in the US Pacific Northwest of which raw shellfish were associated with many of the outbreaks. Three coastal states (New York, Oregon, and Washington) were involved with 177 cases from the O4:K12 serotype without any fatalities.

Virulence Factors and Pathogenesis

Vibrio parahaemolyticus is distributed widely in the estuarine environment and is a major seafood-associated pathogen, but not all strains are pathogenic. *V. parahaemolyticus* is infectious, and a dose of about 2×10^5–3×10^7 cfu is required to cause disease. The incubation period is about 15 h (range, 4–96 h), and the disease may last for 2–3 days. Bacteria colonize the gut and produce toxins, and the pathogenesis depends on the production of a set of toxins (see below) that cause cell damage resulting in membrane pore formation and loss of fluids and electrolytes. Bacteria also induce a strong inflammatory response in the intestine, which is more severe than the infection caused by the *V. cholerae*. Though the *V. parahaemolyticus* infection induces a strong immune response, the detailed mechanism of pathogenesis is still unclear. The symptoms of *V. parahaemolyticus* infection include acute abdominal pain, nausea, vomiting, headache, low-grade fever, and diarrhea (watery or bloody). The stool is described as "meat washed" due to the presence of blood. The disease is usually self-limiting.

Adhesion and Colonization

Vibrio parahaemolyticus expresses multivalent adhesion molecule 7 (MAM7) for adhesion to host cells for colonization and delivery of effector molecules including toxins. The host cell receptor for MAM7 is fibronectin and phosphatidic acid.

Toxins

Vibrio parahaemolyticus produces four hemolysins: a thermostable direct hemolysin (TDH), a heat-labile TDH-related hemolysin (TRH), a thermolabile hemolysin (TLH), and δ-VPH toxin and an enzyme, hemagglutinin protease (HAP). Properties of toxins and enzymes are summarized below.

1. TDH toxin production in *V. parahaemolyticus* was originally detected by the formation of β-hemolysis on Wagatsuma agar (special kind of blood agar medium), and this phenomenon was called the Kanagawa phenomenon (KP), and later, the toxin was termed the thermostable direct hemolysin (TDH). *Vibrio parahaemolyticus* acquired the *tdh* gene through horizontal gene transfer. TDH is a 21 kDa protein that causes zones of β-hemolysis on Wagatsuma blood agar. TDH is heat-stable (100°C for 10 min) and is produced by KP⁺ strains. The KP⁻ strains generally carry the heat-labile TRH toxin. TDH acts as a porin and allows the influx of ionic species: Ca^{2+}, Na^+, and Mn^{2+} from enterocytes. The porin channels increase with increased concentrations of TDH and increase ionic influx, cell swelling, and death of cells due to ionic imbalance. TDH also disrupts the epithelial barrier function by affecting the tight junction proteins, such as claudin and occludin as well as cytosolic zonula occludin proteins (ZO-1, ZO-2, ZO-3), cingulin, and 7H6. Toxin action can also be detected by a ligated mouse ileal loop model. During gastroenteritis, a strong humoral immune response to TDH and LPS occurs, and the predominant immunoglobulin is IgM.

2. TRH is a heat-labile TDH-related hemolysin (inactivated at 60 °C for 10 min) and induces fluid accumulation when tested in a rabbit ileal loop model. TRH also induces chloride secretion, and its involvement in diarrhea has been strongly implicated.

3. TLH is a thermolabile hemolysin and possesses phospholipase A2/lysophospholipase activity. TLH consists of 2 molecular weight species of 47.5 and 45.3 kDa and is present in all *V. parahaemolyticus* strains. The role of this hemolysin in pathogenesis is unknown.

4. δ-VPH is a heat-stable hemolysin of 22.8 kDa. It is present in all strains including KP-negative *V. parahaemolyticus* strains. The role of this hemolysin in pathogenesis is also unknown.

5. Hemagglutinin protease (HAP): *V. parahaemolyticus* has the ability to detach from damaged/sloughed mucosal cells and reattach to new mucus surfaces. The detachment factor is a zinc- and calcium-dependent protease, and it is called hemagglutinin protease because it has hemagglutination activity.

In a study, a majority of clinical isolates obtained from the Pacific Northwest (USA) belonged to serotype O4:K12, which expressed both TDH and TRH. The serotype O3:K6 that caused outbreaks in the Gulf of Mexico and the East Coast of the USA had only TDH. Environmental isolates may lack both TDH and TRH, yet those isolates may express other virulence factors, such as extracellular proteases, biofilm, and siderophore, and may exhibit cytotoxicity toward intestinal cells.

Type III Secretion System

Vibrio parahaemolyticus has two sets of genes for the synthesis of a type III secretion system (T3SS), which is necessary for injecting virulence proteins directly into the host cell to target the actin cytoskeleton, innate immune signaling, and autophagy leading to apoptosis. The first cluster (T3SS-1) is located on the large chromosome (3.3 Mb), and the second one (T3SS-2) is located on the small chromosome (1.9 Mb). Virulence factors that are secreted by T3SS-1 are responsible for cytotoxicity, mouse lethality, and induction of autophagy, while T3SS-2 is responsible for enterotoxicity and environmental persistence.

Type VI Secretion System

The genes encoding type VI secretion system (T6SS) are located in both chromosomes 1 and 2. T6SS secretes effector proteins that are involved in adhesion to host cells and may work coordinately with proteins secreted by T3SS.

Vibrio vulnificus

Introduction

Vibrio vulnificus is considered the most infectious and invasive of all the human pathogenic vibrios primarily in the immunocompromised host. It is the leading cause of seafood-related mortality causing 95% (~40 cases) of seafood-related deaths in the USA. *Vibrio vulnificus*-related infections are thought to be very high in Japan because of the warmer coastal water and increased consumption of raw seafood. In the USA, most cases are reported during the summer months (May–October). Seafood including filter-feeding mollusks (mussels, oysters, and clams), eels, and fish is the major source of this pathogen. Salinity and temperature of water have a significant impact on *V. vulnificus* persistence in water.

Biology

Vibrio vulnificus is a Gram-negative obligate halophilic bacterium (optimum salt requirement is 10–18%) and widespread in coastal warm waters (≥ 20 °C). The bacterium enters into viable but nonculturable (VBNC) state at or below 13 °C, where the bacterium remains dormant, metabolically inactive, and is unable to readily grow on media to form colonies. Climate change due to global warming has significantly influenced *Vibrio* growth and persistence and has led to an increased incidence of *Vibrio* infections in countries around the globe. A majority of *V. vulnificus* strains are considered nonpathogenic, while some are pathogenic. *V. vulnificus* has been classified into biotype 1 (BT1), biotype 2 (BT2), and biotype 3 (BT3). BT1 contains human clinical and environmental strains, BT2 is mostly eel pathogen, and BT3 is a hybrid strain that is limited to a geographic location, primarily in Israel. Classification is often misguided due to horizontal gene transfer and continuous evolution in *V. vulnificus* strains. Many strains produce several virulence factors: capsules, siderophores, and toxins such as hemolysins, collagenase, protease, elastase, DNase, mucinase, hyaluronidase, fibrinolysin, lipase, and phospholipase.

Virulence Factors and Pathogenic Mechanism

Vibrio vulnificus concentration in water could be less than 10 cells ml^{-1}, but bacteria accumulate inside the filter-feeding mollusks, such as oyster, mussels, and clams, reaching a concentration of 10^5 cfu g^{-1} of tissue. Consumption of raw and/undercooked oysters and other seafood is implicated in outbreaks. Bacteria also can penetrate through cuts or abrasions on skin during recreational or occupational activities associated with the marine environment or the seafood industry (Fig. 18.4). The infectious dose of *V. vulnificus* is unknown but is estimated to be as low as 100 cfu. The average incubation period of the disease is 26 h, but in the case of wound infection, the incubation period is much shorter, about 16 h. *V. vulnificus* is responsible for septicemia, wound infection, and gastroenteritis in humans. Immunocompromised individuals or persons with underlying conditions, such as diabetes, chronic renal disease, cirrhosis of the liver, and hemochromatosis (iron overload in the body), are predisposed to severe infection. Furthermore, males over the age of 40 are at the greatest risk (86%), possibly due to preexisting risk factors.

Adhesion Factors

Vibrio vulnificus expresses an N-acetylglucosamine (GlcNAc)-binding protein A (GbpA; 53 kDa) to attach to chitin on plankton. Filter-feeding mollusks feeding on plankton accumulate *V. vulnificus* in their tissues and serve as a vehicle for transmission to humans. GbpA may also aid in *V. vulnificus* adhesion to host mucin similar to *V. cholerae*, which also expresses GbpA.

Capsular Polysaccharide

Vibrio vulnificus produces a capsular polysaccharide (CPS) which is the primary virulence factor, helping bacteria to avoid phagocytosis by macrophages. The presence of CPS correlates with the opaque colony phenotype (Op) and

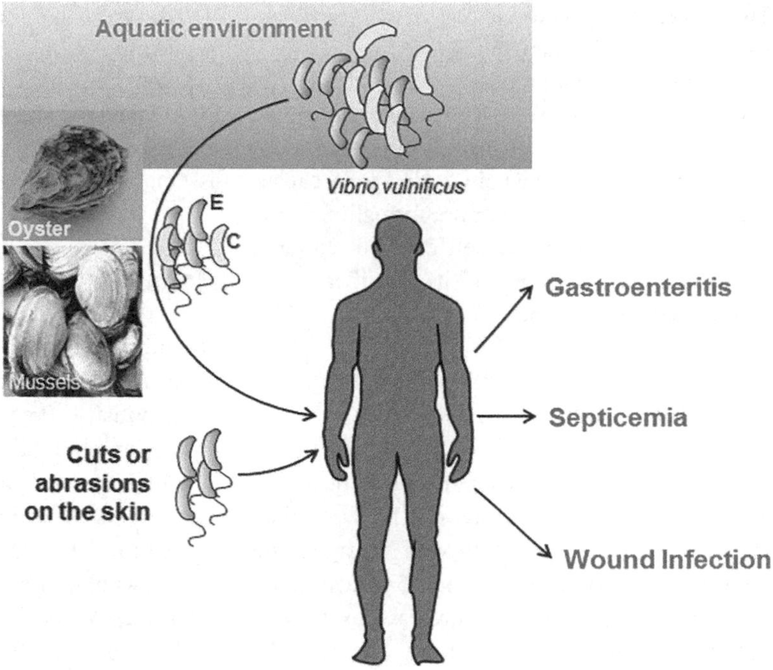

Fig. 18.4 *Vibrio vulnificus* transmission of clinical (C) and environmental (E) genotypes (strains) to humans

pathogenicity. Strains with translucent colony phenotype (Tr) are less virulent. CPS synthesis is encoded by four genes: *wcvA*, *wcvF*, *wcvI*, and *orf4*. Mutation in any of these genes results in the loss of capsule biosynthesis, translucent colony phenotype, and loss of virulence. Furthermore, spontaneous phenotypic switching from the Op phenotype to the Tr phenotype can lead to the development of a less virulent strain. All strains can potentially switch their phenotype from Op to Tr, but switching is much less frequent in clinical strains than in environmental strains.

Iron Acquisition

Vibrio vulnificus expresses siderophores, phenolate, hydroxamate, and vulnibactin to scavenge iron from the host transferrin, hemin, and lactoferrin. Host with high levels of iron such as those who suffer from chronic hemochromatosis or cirrhosis of the liver is highly susceptible to *V. vulnificus* infection. In *V. vulnificus*, *hupA* and *fur* genes regulate iron acquisition. Avirulent strains are unable to acquire iron from their host.

Flagella and Motility

The flagellum is an important adhesion and colonization factor encoded by the *flgC* gene, and it is responsible for cytotoxicity in host cells. The mutant strains exhibit defective motility and are attenuated for infection in suckling mice. Flagella-negative strains lose the ability to adhere to host cells, thus preventing bacteria from delivering toxic effectors to host cells.

Hemolysin

Vibrio vulnificus produces three types of hemolysin/cytolysins. The most widely studied hemolysin, VVH (*Vibrio vulnificus* hemolysin), is a water-soluble polypeptide (51 kDa) that binds to cholesterol on the membrane and forms small pores in the erythrocyte membrane. It can induce apoptosis by elevating cytosolic free Ca^{2+}, releasing cytochrome C from mitochondria, activating caspase-3, and degrading poly-ADP ribose-polymerase (PARP) to cause DNA fragmentation (all hallmarks of apoptosis). Toxin increases vascular permeability and causes skin damage.

Two other hemolysins, encoded by *hlyIII* and *trkA*, are not well characterized.

V. vulnificus also produces an RTX (repeat in toxin) family of the toxin. They share a repeated nine amino acid sequence motif among several Gram-negative bacterial hemolysins. RTX causes depolymerization of actin, pore formation in red blood cells, and necrotic cell death in cultured mammalian cells; however, its significance in pathogenesis remains unresolved.

Metalloprotease

Vibrio vulnificus secretes a metalloprotease, a 45 kDa zinc containing protease. The C-terminal 10 kDa segment binds to protein substrates on erythrocytes, and the N-terminal 35 kDa segment facilitates proteolysis. This protease exerts two functions: membrane permeability enhancement and tissue hemorrhage.

Septicemia

Vibrio vulnificus causes a systemic disease in hosts with underlying preexisting conditions, such as liver or kidney disease, malignancies, and diabetes. Infection is also severe in people with high levels of iron in serum caused by liver cirrhosis or genetic disorders, like hemochromatosis. From the intestinal tract, the bacterium invades epithelial cells. It first binds to epithelial cells with the aid of pili; produces hemolysin, which induces apoptosis; and facilitates bacterial invasion and translocation into the bloodstream. Bacteria acquire iron from host cells using siderophores and rapidly proliferate causing septicemia. Bacterial LPS and capsules activate the complement cascade; however, the capsule helps the bacterium escape neutrophil- or macrophage-mediated phagocytosis. From blood circulation, the bacterium can invade cutaneous tissue with the help of toxins, such as hemolysins, collagenase, protease, lipase, and phospholipase. These toxins also aid in the development of edematous hemorrhagic skin lesions, also known as buboes. Induction of proinflammatory cytokines, TNF-α,

IL-1β, and IL-6, can lead to septic shock. Septicemia is manifested by fever, prostration, hypotension, chills, nausea, and occasional vomiting, diarrhea, and abdominal pain. In the USA, the mortality rate is 40–60% and is generally associated with underlying conditions.

Wound Infection

Wound infection is associated with recreational or occupational activities in seawater and the seafood industry. Preexisting cuts, skin lesions, or injuries and trauma resulting from activities from handling seafood and marine animals or from recreational activities are risk factors. *V. vulnificus* is thought to be associated with most *Vibrio*-related wound infections. Other species involved are *V. parahaemolyticus*, *V. alginolyticus*, and *V. cholerae* non-O1. The average incubation period for this type of infection is about 16 h. Vibrios produce collagenase and metalloprotease that allow the bacteria to colonize the wound. Protease evokes two types of reactions, increased vascular permeability, and hemorrhagic actions. The protease is bifunctional: the N-terminal portion mediates enzymatic activity to degrade type IV collagen located in the vascular basement membrane and causes tissue damage leading to hemorrhage. The C-terminal end binds to mast cell receptors, activating and aiding the release of histamine and bradykinin. These biologically active amines increase membrane permeability and cause wound edema. Infection results in high fever, chills, wound edema, vesicle formation, cellulitis, erythema, and tissue necrosis and may require hospitalization and amputation of extremities under severe conditions. Antibiotic therapy is needed to clear the infection. The fatality rate for wound infection is 20–30% in patients with underlying conditions.

Symptoms of *V. vulnificus* Infection

Three major conditions are associated *V. vulnificus* infection: gastroenteritis, septicemia, and wound infections. Gastroenteritis symptoms are associated with abdominal pain, vomiting, and diarrhea. Septicemia and wound infections progress very rapidly, and a patient may die within 24 h of exposure due to toxic shock. Clinical symptoms include fever, chills, nausea, abdominal pain, hypotension, and the development of secondary lesions, which typically develop in arms and legs. The mortality rate is very low with gastrointestinal illness but about 50% for septicemia and 20–30% for wound infection.

Prevention and Control of *V. parahaemolyticus* and *V. vulnificus* Infection

Vibrios are becoming one of the most dangerous emerging foodborne pathogens, due in part of increased high-risk human population, the popularity of seafood, and global warming with the consequent bloom of bacteria and their habitats within zooplankton and phytoplankton. Aquaculture is one of the fastest growing industries worldwide. The risk of contamination of seafood is much greater in coastal waters and freshwaters than in the open ocean. Seafood safety may be enhanced by harvesting from water when the water temperature is low especially during the winter months and from unpolluted water. In addition, products need to be placed in ice or chilled, especially molluscan shellfish at harvest through shipping and processing to prevent bacterial growth. Workers' safety should be addressed especially for those who have cuts or wounds or abrasions (lacerations) of the skin. They should take precautions to avoid contact with water or seafood. Consumer education should be a part of the seafood safety program. That would include alerting consumers to the dangers of eating raw or undercooked seafood especially for persons with underlying conditions including liver disease, diabetes, kidney disease, and immunocompromised immune systems.

In diarrheal patients, loss of water and electrolytes lead to dehydration, and fluid therapy is the most effective treatment to prevent dehydration. Rehydration with water containing table salts and glucose is effective in resource-constrained situations, but intravenous fluid replacement is essential in patients showing severe signs of dehydration. Antibiotics including tetracycline and fluoroquinolones (ciprofloxacin, levofloxacin, moxifloxacin) are effective.

Detection of *Vibrio* Species

Culture and Serological Methods
Selective enrichment is performed in alkaline peptone water (APW) containing 1–3% NaCl, and colonies are isolated by streaking enriched samples onto the selective thiosulfate citrate bile salts sucrose (TCBS) agar. Bacteria have been also isolated on cellobiose polymyxin B colistin (CPC) and mannitol–maltose agar. In addition, *V. vulnificus* is also isolated using sodium dodecyl sulfate-polymyxin B-sucrose agar (SPS), *Vibrio vulnificus* agar (VVA), and modified CPC (mCPC) agar. Biochemical characterizations can be performed to determine the species of *Vibrio*. *V. parahaemolyticus* has been tested for Kanagawa phenomenon (hemolytic activity) by growing them on Wagatsuma agar containing high salt (7%) and blood for TDH activity. Identification is further accomplished by serotyping for somatic O antigen and capsular K antigens. Immunological assays including ELISA have been used that target intracellular and TDH antigens.

Molecular Techniques
Single gene or multigene-specific PCR assays have been developed targeting the 16S rRNA *tdh*, *trh*, *gyrB*, *toxR*, *ctxB*, *ctxAB*, and *tcpA* genes for detection of *Vibrio* spp. A detection limit of 10^1–10^2 cfu has been reported when used in a multiplex format targeting two to three genes. A PCR assay targeting the *vvhA* gene (specific for *V. vulnificus*) has been used to differentiate clinical (C) and environmental (E) strains. For genomic typing and identification, ribotyping, restriction fragment length polymorphism (RFLP), amplified restriction fragment length polymorphism (AFLP), randomly amplified polymorphic DNA, and enterobacterial intergenic consensus sequence-PCR (ERIC-PCR) have been used. Multilocus

enzyme electrophoresis and multilocus sequence typing of housekeeping genes have been performed for identification and typing purposes.

Summary

Although cholera is considered an old-world disease, it continues to be a serious problem in developing and economically impoverished countries. The infections caused by other vibrios are also increasing worldwide especially in developed countries and are increasingly being recognized as emerging diseases. *Vibrio cholerae* is known for its epidemic and pandemic outbreaks, especially in countries throughout Asia, Africa, and South and Central America, where the fecal–oral transmission mode spreads the disease, often through the consumption of contaminated drinking water. Upon entry into the intestine, the bacterium produces several adhesion factors including toxin-coregulated pili (TCP), flagella, neuraminidase, and accessory colonization factor (ACF) for colonization. The bacterium produces cholera toxin (CT) and zona occludin toxin (ZOT), which affect the ion transport pumps for Na^+, Cl^-, HCO_3^-, and K^+ and junctional integrity and results in extensive fluid and ion losses. Diarrhea appears within 6 h–5 days and lasts for 2–12 days. Oral vaccination with killed bacteria together with a purified B subunit of cholera toxin is widely used and is recommended by the WHO. *Vibrio parahaemolyticus* and *V. vulnificus* infections are associated with seafood harvested from estuarine or freshwaters. They produce several heat-stable (TDH) and heat-labile TDH-related hemolysins (TRH) and phospholipases, which are responsible for membrane pore formation, apoptosis, and fluid loss resulting in diarrhea. Additionally, *V. vulnificus* causes septicemia and wound infections, which could be fatal. *Vibrio vulnificus* is the most invasive of all vibrios in immunocompromised high-risk population. In addition to hemolysin, it produces collagenase, metalloprotease, lipase, and phospholipases, which promote rapid tissue destruction resulting in death within 24 h. The mortality rate of septicemic infection is about 50%, and wound infection is 22%.

Further Readings

1. Al-Assafi, M.M.K., Mutalib, S.A., Ghani, M.A. and Aldulaimi, M. (2014) A review of important virulence factors of *Vibrio vulnificus*. *Curr Res J Biol Sci* **6**, 76–88.
2. Albert, M.J. and Nair, G.B. (2005) *Vibrio cholerae* O139 Bengal - 10 years on. *Rev Med Microbiol* **16**, 135–143.
3. Ali, M., Lopez, A.L., You, Y., Kim, Y.E., Sah, B., Maskery, B. and Clemens, J. (2012) The global burden of cholera. *Bull World Health Organ* **90**, 209–218.
4. Elmahdi, S., DaSilva, L.V. and Parveen, S. (2016) Antibiotic resistance of *Vibrio parahaemolyticus* and *Vibrio vulnificus* in various countries: A review. *Food Microbiol* **57**, 128–134.
5. Faruque, S.M., Albert, M.J. and Mekalanos, J.J. (1998) Epidemiology, genetics, and ecology of toxigenic *Vibrio cholerae*. *Microbiol Mol Biol Rev* **62**, 1301–1314.
6. Gulig, P.A., Bourdage, K.L. and Starks, A.M. (2005) Molecular pathogenesis of *Vibrio vulnificus*. *J Microbiol* **43**, 118–131.
7. Harwood, V.J., Gandhi, J.P. and Wright, A.C. (2004) Methods for isolation and confirmation of *Vibrio vulnificus* from oysters and environmental sources: a review. *J Microbiol Methods* **59**, 301–316.
8. Kanungo, S., Sur, D., Ali, M., You, Y.A., Pal, D., Manna, B., Niyogi, S.K., Sarkar, B., Bhattacharya, S.K. and Clemens, J.D. (2012) Clinical, epidemiological, and spatial characteristics of *Vibrio parahaemolyticus* diarrhea and cholera in the urban slums of Kolkata, India. *BMC Pub Health* **12**, 830.
9. Letchumanan, V., Chan, K.-G. and Lee, L.-H. (2014) *Vibrio parahaemolyticus*: a review on the pathogenesis, prevalence, and advance molecular identification techniques. *Front Microbiol* **5**, 705.
10. Levin, R.E. (2005) *Vibrio vulnificus*, a notably lethal human pathogen derived from seafood: A review of its pathogenicity, subspecies characterization, and molecular methods of detection. *Food Biotechnol* **19**, 69–94.
11. Levin, R.E. (2006) *Vibrio parahaemolyticus*, a notably lethal human pathogen derived from seafood: A review of its pathogenicity, characteristics, subspecies characterization, and molecular methods of detection. *Food Biotechnol* **20**, 93–128.
12. Lippi, D. and Gotuzzo, E. (2014) The greatest steps towards the discovery of *Vibrio cholerae*. *Clin Microbiol Infect* **20**, 191–195.
13. Nair, G.B., Ramamurthy, T., Bhattacharya, S.K., Dutta, B., Takeda, Y. and Sack, D.A. (2007) Global dissemination of *Vibrio parahaemolyticus* serotype O3:K6 and its serovariants. *Clin Microbiol Rev* **20**, 39–48.
14. Oliver, J.D. (2013) *Vibrio vulnificus*: death on the half shell. A personal journey with the pathogen and its ecology. *Microb Ecol* **65**, 793–799.
15. Oliver, J.D. (2015) The biology of *Vibrio vulnificus*. *Microbiol Spectrum* **3**.

16. Raghunath, P. (2015) Roles of thermostable direct hemolysin (TDH) and TDH-related hemolysin (TRH) in *Vibrio parahaemolyticus*. *Front Microbiol* **5**, 805.

17. Safa, A., Nair, G.B. and Kong, R.Y.C. (2010) Evolution of new variants of *Vibrio cholerae* O1. *Trends Microbiol* **18**, 46–54.

18. Silva, A.J. and Benitez, J.A. (2016) *Vibrio cholerae* biofilms and cholera pathogenesis. *PLoS Negl Trop Dis* **10**, e0004330.

19. Teschler, J.K., Zamorano-Sánchez, D., Utada, A.S., Warner, C.J., Wong, G.C., Linington, R.G. and Yildiz, F.H. (2015) Living in the matrix: assembly and control of *Vibrio cholerae* biofilms. *Nat Rev Microbiol* **13**, 255–268.

20. Turner, J.W., Paranjpye, R.N., Landis, E.D., Biryukov, S.V., González-Escalona, N., Nilsson, W.B. and Strom, M.S. (2013) Population structure of clinical and environmental *Vibrio parahaemolyticus* from the Pacific Northwest coast of the United States. *PLoS One* **8**, e55726.

21. Whitaker, W.B., Parent, M.A., Boyd, A., Richards, G.P. and Boyd, E.F. (2012) The *Vibrio parahaemolyticus* ToxRS regulator is required for stress tolerance and colonization in a novel orogastric streptomycin-induced adult murine model. *Infect Immun* **80**, 1834–1845.

22. Zhang, X.H. and Austin, B. (2005) Haemolysins in *Vibrio* species. *J Appl Microbiol* **98**, 1011–1019.

Shigella Species

19

Introduction

Shigella bacterium was first discovered in 1896 by a Japanese physician, Kiyoshi Shiga, who was investigating an outbreak of *sekiri* (means "dysentery" in Japanese). He isolated a bacillus from the stool sample and called it *Bacillus dysenteriae*; now it is known as *Shigella dysenteriae* type 1. The toxin produced by this organism is called Shiga toxin and the disease is called shigellosis. *Shigella* species are members of the *Enterobacteriaceae* family and cause shigellosis characterized by bacillary dysentery (mucoid bloody stool). *Shigella* species are commonly found in water contaminated with human feces, and fecal–oral route is the primary mode of transmission. *Shigella* is responsible for health problem worldwide; however, it is a serious concern in the developing countries. *Shigella* accounts for 125 million cases of dysentery annually, and 13.2 per 1000 children under the age of five suffer from shigellosis worldwide. Malnourished children are highly susceptible, and *Shigella* infection further promotes impaired nutrition, recurring infection and stunted growth. Antibiotic-resistant strains are continuously emerging; thus treatment regimens become very difficult against shigellosis. The Centers for Disease Control and Prevention (CDC) estimates that annually there are 131,000 reported cases of shigellosis, 1456 hospitalizations, and 10 deaths in the USA. Shigellosis sometimes appears as mild; therefore, many cases are not reported.

Biology

Shigella spp. are Gram-negative intracellular pathogens and transmitted by fecal–oral route. *Shigella* spp. are nonsporulating nonmotile rod and are facultative anaerobe. They are generally catalase-positive, and oxidase- and lactose-negative, and ferment sugars, usually without forming gas. The bacterium grows between 7 ° and 46 °C, with an optimum temperature of 37 °C. The bacterial cells are not as fragile as once thought and can survive for days under harsh physical and chemical exposures, such as refrigeration, freezing, 5% NaCl, and media with a pH of 4.5. *Shigella* is sensitive to heat treatments and is killed by pasteurization. *Shigella* can multiply in many types of food when stored at their growth temperature ranges. *Shigella* is biochemically and serologically related to enteroinvasive *E. coli* (EIEC), which also induces diarrhea and dysentery (see Chap. 14).

Classification

The genus *Shigella* has four species, which sometimes referred to as subgroups: *Shigella dysenteriae* is designated subgroup A and has 15 serotypes; *S. flexneri* is subgroup B with 19 serotypes; *S. boydii* is subgroup C with 20 serotypes; and *S. sonnei* is subgroup D with one serotype. Of these, *S. dysenteriae* type 1 is known to cause

© Springer Science+Business Media, LLC, part of Springer Nature 2018
A. K. Bhunia, *Foodborne Microbial Pathogens*, Food Science Text Series,
https://doi.org/10.1007/978-1-4939-7349-1_19

deadly epidemics; *S. flexneri* and *S. sonnei* are responsible for endemic outbreaks; and *S. boydii* causes rare disease. *S. boydii*-related outbreak is common in the Indian subcontinent. *Shigella dysenteriae* and *S. flexneri* are responsible for shigellosis in developing countries, in parts of Asia and sub-Saharan Africa, and *S. sonnei* causes sporadic outbreaks in industrialized countries transmitted by contaminated water or undercooked food. In the USA, *S. sonnei* accounts for two-thirds of shigellosis, while *S. flexneri* the rest. However, *S. flexneri* is the most extensively studied species among the shigellae, and the molecular study of this pathogen is beginning to shed light on its mechanism of infection and the disease process.

Source and Transmission

Drinking water and food contaminated with human feces serve as the primary source of shigellae (Fig. 19.1). Vegetables can be contaminated in the field by farm workers defecating in the fields, exposure to sewage, or the polluted or contaminated irrigation water. *Shigella* infection is a major problem in the developing countries due to inadequate sanitary facilities, and flies can transmit bacteria from human feces in their legs and wings to contaminate food. Children and infants are highly susceptible to shigellosis, and moreover, the chance of contracting the infection is very high because of lack of sanitary toilets and poor hygienic practices. Adults are less susceptible, because of improved hygienic practices and immunity to the bacterium. In the day-care center, person-to-person contact results in the spread of the disease. Homosexuals, migrant workers, and travelers to the endemic zone are often infected with *Shigella* and become the source of infection for others.

Pathogenesis

Shigella spp. are highly infectious and a dose of 10–100 cells can cause infection (Fig. 19.1). The incubation period is 1–4 days or as many as 8 days. *Shigella* spp. are resistant to low pH

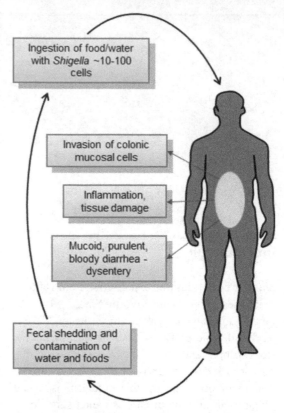

Fig. 19.1 Flow diagram showing the general outline of *Shigella* pathogenesis

(pH 2.5 for 2 h), thus facilitating survival during transit through the stomach. The primary site of infection is the large intestine, colon, and rectum. The bacterium is nonmotile and thus cannot persist in the small intestine because of increased ciliary movement of the epithelial cells, fast flow of liquids, and increased intestinal peristaltic movement. The invasive traits are expressed at 37 °C but not at 30 °C. Thus shigellae growing at 30 °C need a few hours of conditioning at 37 °C before bacteria can invade intestinal epithelial cells. Shigellae colonize and invade mucosal epithelial cells, and the stages of pathogenesis include (1) invasion, (2) intracellular multiplication, (3) inter- and intracellular (cell-to-cell) movement, and (4) the host cell killing (Fig. 19.2). These events ultimately provoke a pronounced epithelial inflammation and ulcerative lesions with mucopurulent bloody stool that represents characteristics clinicopathological symptoms of dysentery. The pathogenic mechanism is complex, involving an array of gene

Fig. 19.2 Diagram
showing *Shigella* entry
through mucosal
membrane in the
intestine and the
induction of
inflammation. *M*,
Microfold cell, *PMNL*,
polymorphonuclear
leukocytes; *NK cell*,
natural killer cell

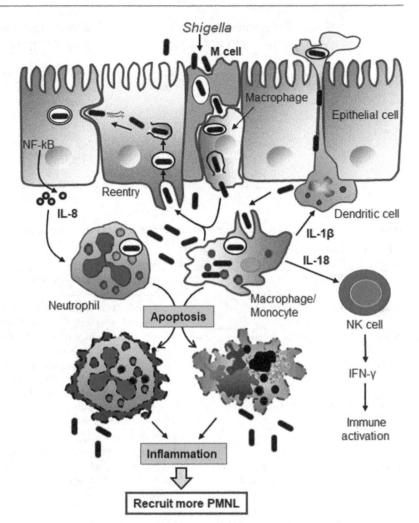

products that facilitate bacterial adhesion, invasion, cell-to-cell movement, cell death, and inflammatory immune response (Fig. 19.2).

All pathogenic isolates carry the 220 kb virulence plasmid, and the major virulence genes required for bacterial invasion and the protein export system are located in a 30 kb pathogenicity island (PAI) (Fig. 19.3). The invasion-associated virulence genes are referred to as invasion plasmid antigens (*ipa*), and the gene products are designated as IpaB, IpaC, IpaD, and IpaA. Expression of *ipa* genes is regulated by *virB* located in the PAI. The other genes present in the 30 kb PAI include *icsA* required for actin-based intracellular bacterial motility, *icsB* for growth inside cytoplasm, *ipg* for bacterial invasion, and *spa* (surface presentation antigens) and

mxi (membrane excretion proteins), consisting of genes coding for the components of type III secretion system (T3SS) (Table 19.1).

Several virulence genes are located on the chromosome, and they are involved in the regulation of plasmid-encoded virulence genes. For example, the *virR* gene controls the temperature-dependent expression of Spa/Mxi proteins.

Shigellae produce three types of enterotoxins: *Shigella* enterotoxin 1, found in *S. flexneri*; *Shigella* enterotoxin 2, found in many shigellae but not all; and Shiga toxin (Stx), which is produced by *S. dysenteriae* type 1. The gene for *Shigella* enterotoxin 1 is located on the chromosome, *Shigella* enterotoxin 2 on the virulence plasmid, and the Stx on the chromosome (a bacteriophage-borne gene).

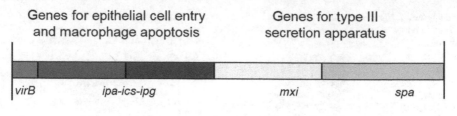

30-kb Pathogenicity Island (PAI)

Fig. 19.3 Schematic representation of virulence gene cluster in the 30 kb pathogenicity island of 220 kb *Shigella* plasmid

Table 19.1 Virulence factors acting as effector proteins secreted via type III secretion system (T3SS) in *Shigella*

Virulence factors	Size (kDa)	Target protein	Functions
IpaA	70	Vinculin, β1-integrins, Rho signaling	Actin cytoskeleton rearrangement, loss of actin stress fibers, membrane ruffling, invasion
IpaB	62	Cholesterol, CD44, caspase-1	Hemolysin, release from phagosome, lysis of protrusion in cell–cell spreading, CD44 stimulation, translocator for T3SS
IpaC	42	Actin, β-catenin	Actin polymerization, translocon formation, phagosome escape, β1-integrin stimulation, disruption of tight junction
IpaD	37	–	Control of type III secretion, regulation of T3SS translocator, membrane insertion of translocon; targets TLR2 on B cell and induces apoptosis, independent of invasion
IcsA (VirG)	116	N-WASP, vinculin	Recruitment of actin-nucleating complex for actin-based motility and intercellular spread
IcsB	52	–	Escape from autophagy
IpgB1	23	–	Membrane ruffling, actin polymerization
IpgD	66	Phosphatidylinositol 4,5-bisphosphate	Phosphatidylinositol (4,5) bisphosphate phosphatase activity; impedes migration of activated T cells
IpaH9.8	62	Splicing factor U2AF, MAPK kinase	E3 ubiquitin ligase activity, suppression of inflammatory response
OspF	28	MAPKs Erk and p38	MAPK phosphatase activity, suppression of NF-κB-dependent inflammatory response
OspG	23	Ubiquitin-conjugating enzymes	Kinase activity for ubiquitin E2 molecule, suppression of NF-κB-dependent inflammatory response
VirA	45	α-Tubulin	Degradation of microtubule, membrane ruffling, intracellular bacterial motility

Invasion

In the intestine, *Shigella* is unable to adhere to the epithelial cells, since the bacterium does not possess any adhesion factor or flagella. Instead, *Shigella* delivers the effectors via T3SS such as IpaA, IpaB, IpaC, IpgB1, IpgD, and VirA, and these proteins are involved in the induction of membrane ruffles to form macropinocytosis to allow bacterial entry into the host cells. *Shigella*

invasion of epithelial barrier is in the intestine is exclusively limited to the M-cell-mediated pathway (Fig. 19.4). M cells are distributed in the Peyer's patches and solitary lymphoid nodules in the small and large intestine. M cells have the endocytic activity for foreign antigens including microbes and transport them to antigen-presenting cells such as macrophages and dendritic cells present underneath the M cells. Polymorphonuclear leukocytes (PMNL) attracted

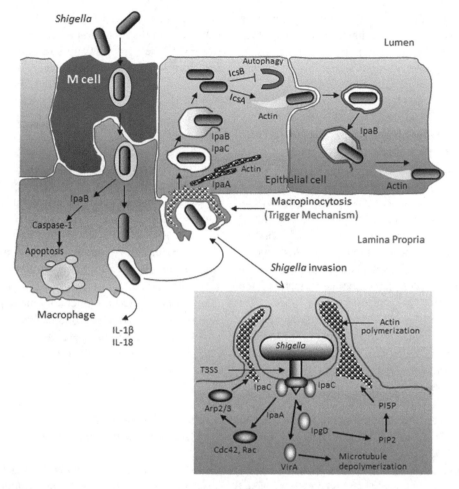

Fig. 19.4 Diagram showing various stages of cellular infection mechanism of *Shigella*. The steps include entry, lysis of vacuole, replication, blockage of autophagy, actin polymerization, and cell-to-cell movement. *Shigella* entry into epithelial cells is mediated by trigger mechanism (macropinocytosis), and the details of which is presented inside the box (this is also presented in Fig. 4.3 in Chap. 4). *PIP2*, phosphatidylinositol 4,5-bisphosphate, *PI5P*, phosphatidylinositol 5-phosphate

to the lumen of intestine upon bacterial infection can also transport bacteria across the mucosal barrier. Once *Shigella* passes through the epithelial barrier via the M cells, they infect resident macrophages and subsequently escape from the phagosome into the cytoplasm, where they multiply and induce rapid cell death (apoptosis) via the caspase-1-dependent pathway. *Shigella* is released from dead macrophages and actively enters into the enterocytes through the basolateral surface by inducing macropinocytosis by a process called trigger mechanism (Fig. 19.4). Once inside the epithelial cell, the bacterium is trapped inside the vacuole. *Shigella* disrupts the vacuolar membrane using IpaB and escapes into the cytoplasm, in which the bacterium multiplies and induces actin polymerization at one pole of the bacterium with the help of IcsA (intracellular spread) to move within the cytoplasm as well as into the adjacent cells. The bacterium uses IcsB to avoid host cell autophagy system from being degraded (see below). Proinflammatory cytokines (IL-1β, IL-8, IL-18) produced by macrophages and epithelial cells recruit neutrophils, and their degranulation invokes tissue damage, which in turn destabilizes intestinal barrier integrity, thus further promoting additional shigellae entry into the subepithelial region.

Cytokine (IL-8) released from infected epithelial cells also recruits more inflammatory cells, which exacerbate the infection by inducing tissue destruction, mucus secretion, and blood and pus formation.

The large 220 kb plasmid of *Shigella* encoding the invasion-associated proteins is essential for bacterial invasion of epithelial cells (Table 19.1). IpaB–C complex interacts with β1-integrin, while IpaB interacts with CD44, those located at the basolateral surface of polarized epithelial cells (Fig. 19.4). The IpaB binding to CD44 plays an important role in *Shigella* uptake. Interaction of IpaB with CD44 initiates a cascade of signaling event that allows recruitment of ezrin, which forms crosslink with actin and possibly aids in the membrane ruffles, lamellipodia formation, and the formation of macropinocytic pocket for engulfment of bacteria. *Shigella* also injects the invasion-associated proteins (IpaA, IpaB, IpaC, IpgB1, IpgD, and VirA) into the host cell cytoplasm through the T3SS needle (60 nm long). IpaC and IpaA initiate signaling cascade to induce cytoskeletal rearrangement and to form a focal adhesion-like structure that is required for bacterial entry by membrane ruffling. Rho GTPases (Cdc42, Rac, and RhoA) allow recruitment of actin, vinculin, and ezrin through activation of nucleator proteins Arp2/3 and N-WASP (neuronal Wiskott–Aldrich syndrome protein) to form the membrane ruffling structure. IpgD, a phosphoinositide phosphatase that converts phosphatidylinositol 4,5-bisphosphate (PIP2) into phosphatidylinositol 5-phosphate (PI5P), participates in actin polymerization through PI3K (phosphatidylinositol 3-kinase) activation and promotes bacterial invasion. IpgB1 also plays important role in inducing large-scale membrane ruffling, and IpaA binds to vinculin and stimulates RhoA to promote local actin cytoskeleton rearrangements necessary for bacterial entry.

Intracellular Multiplication

Shigella encased inside the phagosome is released with the help of IpaB, which forms membrane pore (Fig. 19.4). IpaB together with IpaC acts like a hemolysin, which oligomerizes in the vacuolar membrane to form a pore. *Shigella* escapes from the phagosome and multiplies rapidly inside the cytoplasm utilizing the cytoplasmic nutrients rich in amino acids. The bacterial generation time inside the cytoplasm is about 40 min. Bacteria growing inside, however, can occasionally be the target for degradation by autophagy. Normally, autophagy mediates the bulk degradation of undesirable cytoplasmic proteins and organelle in the cytoplasm. In addition, autophagy also plays a critical role in the host innate defense by degrading intracellular bacterium, which multiplies in the cytoplasm. *Shigella* upon release from the phagosome secretes IcsB via T3SS, and IcsB blocks the recognition of intracellular *Shigella* by autophagy, thus allowing the bacteria to escape autophagic degradation and promote bacterial replication inside the cell. Strains that are lacking *icsB* gene are trapped by autophagosome, which is eventually fused with the lysosome for degradation in the autolysosomal compartment.

Bacterial Movement: Inter- and Intracellular Spreading

Shigella produces IcsA/VirG protein, a 116 kDa surface-exposed outer membrane protein, which accumulates at one pole of the bacterium and triggers actin polymerization to aid in bacterial movement inside the host cell, thus referring to actin-based bacterial motility (Fig. 19.4). The bacterium multiplies inside the host cell, and induces actin polymerization that propels the bacterium forward through the cytoplasm at a rate of 10–15 μm min^{-1}. Indeed, IcsA directly binds to nucleation promoting factor N-WASP, one of the members of WASP family, to polymerize actin with the aid of Arp2/3 complex, which initiates actin polymerization. As the bacterium propels inside the cell, it eventually makes contact with the cytoplasmic membrane and pushes the cell membrane forward to form a protrusion. The protrusion containing *Shigella* is engulfed by the neighboring cell. IpaB aids in the lysis of the double membrane to allow bacterial release into the neighboring cell. *Shigella* again initiates actin polymerization and continues cell-to-cell movement without ever leaving the intracellular location, by which it can continuously expand its own replicative niche.

Cell Death and Inflammation

After the arrival of shigellae in the subepithelial region, they are rapidly phagocytosed by resident macrophages. *Shigella* escapes phagocytic vacuole with the help of IpaB, IpaC, IpaD, and IpaH and avoids degradation by the lysosomal antimicrobial system. *Shigella* replicates and induces macrophage apoptosis by two distinct pathways: (1) activation of caspase-1 by IpaB to release IL-1β and (2) translocation of cytosolic lipid A component of LPS. Apoptosis is a noninflammatory process; however, during *Shigella* infection, the bacterium appears to promote inflammation. Caspase-1 also activates proinflammatory cytokines IL-1β and IL-18, which recruit inflammatory cells at the site of infection (Fig. 19.2).

Intracellular shigellae growth inside the epithelial cells also triggers an inflammatory response. During intracellular growth, bacterial peptidoglycan (PGN) composed of diaminopimelate (DAP) containing *N*-acetylmuramic acid and *N*-acetylglucosamine is recognized by Nod1 (nuclear oligomerizing domain protein) and induces inflammatory cytokine IL-8 production through activation of nuclear transcription factor, NF-κB. In addition, LPS released by intracellular bacteria also provokes sustained activation of NF-κB. IL-8 helps recruit neutrophils to the site of infection for destruction of shigellae. However, bacteria-induced neutrophil death and the resulting enzyme release destabilize the mucosal membrane integrity, thus permitting entry of more shigellae during this early phase of infection. In the later stages of infection, a large number of activated infiltrating neutrophils aid in the resolution of disease by actively phagocytosing the shigellae. Still, it is rather mysterious why bacteria would induce such a strong inflammatory response to promote its own clearance by immune cells. IL-18 activates natural killer (NK) cells and stimulates the production of IFN-γ to aid in innate immune response (Fig. 19.2).

Killing of mucosal epithelial cells is mediated by bacterial growth inside and is not due to the production of Shiga toxin (Stx) since a toxin-negative mutant strain also kills epithelial cells. Intracellular bacterial growth results in the blockage of protein synthesis, depletion of essential small molecules, interference of respiration, drop in ATP levels, increase in pyruvate levels, and disruption of energy metabolism and apoptosis.

Shiga Toxin and Hemolytic Uremic Syndrome

Among *Shigella* spp., only *Shigella dysenteriae* type 1 produces Stx and the gene (*stxAB*) is encoded on the chromosome (bacteriophage-borne). Stx is an A–B type toxin with an approximate molecular mass of 70 kDa. The A subunit is 32 kDa, and the B subunit is a pentamer, each consisting of 7.7 kDa. Stx is released upon lysis of the cell following activation of the lytic phage. The B subunit of Stx binds to the receptor, globotriaosylceramide (Gb$_3$), a glycolipid and allows the entry of A subunit into the cell. The A subunit inhibits protein synthesis in the 28S RNA of the 60S host cell ribosome (see the mechanism of Stx action in Chap. 14). Shiga toxin exhibits a multitude of activity: (1) it acts as an enterotoxin and induces fluid accumulation; (2) it acts as a neurotoxin, blocks nerve impulses, and elicits paralysis; and (3) it acts as a cytotoxin and kills cells by inhibiting protein synthesis and by triggering apoptosis.

The major site of action of Shiga toxin is on the kidney tubule, which is rich in Gb$_3$ receptor. The toxin damages the tubule resulting in acute kidney failure, and blood is excreted into urine causing hemorrhagic uremic syndrome (HUS). The Stx also causes damage to the colonic endothelium of blood vessels causing bloody stool. Another sequela of shigellosis is the development of Reiter's syndrome or reactive arthritis due to an autoimmune disease. In addition, the release of LPS induces increased production of IL-1 and TNF-α, which provoke toxic shock syndrome. Increased cytokine production also induces vascular damage and kidney failure.

Regulation of Virulence Genes

The plasmid-encoded virulence gene expression in *Shigella* is temperature dependent with a maximum expression occurring at 37 °C and virtually no expression is seen at 30 °C. These virulence genes are regulated by *virF*, which is not transcribed at 30 °C, because a repressor gene, *virR*, is active at 30 °C, and, furthermore, it inhibits the expression of *virF*. Another transactivator of virulence genes is *virB*, which is controlled by *virF* and the temperature. The *virB* is located directly downstream from *ipaA* and is responsible for expression of Ipa proteins necessary for bacterial invasion and cell-to-cell movement. Another important regulatory gene, *mxiE*, located in the 30 kb PAI on the plasmid regulates the expression of genes that are necessary for intracellular survival. When *Shigella* enters into the epithelial cell, the MxiE protein acts as a transcriptional activator for expression of a subset of genes such as *ipaH9.8*, *ospF*, and *ospG*. Indeed, under intracellular environment, MxiE induces the expression of several virulence-associated genes, carrying so-called MxiE boxes (GTATCGTTTTTTANAG) located between position −49 and −33 with respective transcriptional start site in the upstream region of the MxiE target genes. This strategy allows the intracellular *Shigella* to induce expression of specific genes that are necessary for survival inside the cell. Furthermore, MxiE-regulated genes (*ipaH9.8*, *ospF*, and *ospG*) also are involved in modulating the host innate immune response.

Immunity Against Infection

Host defense against mucosal pathogens like *Shigella* is accompanied by the innate immune response, which provides an early defense against infection (Fig. 19.2). Phagocytic cells including macrophages, monocytes, dendritic cells, and neutrophils provide the bulk of resistance. Upregulation of proinflammatory cytokines including IL-1β, IL-6, IL-8, TNF-α, and TNF-β allows recruitment of neutrophils. Invading intracellular bacteria can be recognized by various innate immune systems. For instance, invasion of epithelial cells by *Shigella* is recognized by pattern recognition molecules such as Nod1, which ultimately activates NF-κB and produces IL-8, thus leading to recruitment of neutrophils. In addition, IFN-γ production and complement activation as byproducts are also involved in the innate immunity. Defense against *Shigella* is also mediated by antimicrobial peptides, designated LL-37 belonging to the cathelicidin family, and β-defensin-1 secreted by neutrophils and macrophages. However, *Shigella* delivers the effector proteins such as IpaH9.8, OspG, and OspF via T3SS to host cells and downregulates the host innate immune response including the expression of antimicrobial peptides in human colonic tissues to overcome the host innate defense. As mentioned above, survival of shigellae inside the intestinal epithelial cells is also promoted by several strategies such as escape from phagocytic vacuoles by IpaB, escape from autophagic compartments by IcsB, and intracellular movement by IcsA. In addition, bacterial LPS may play a pivotal role in bacterial survival. It protects against complement-mediated lysis and induces apoptosis in phagocytes. Glucosylation of LPS also promotes bacterial invasion and evasion of innate immunity.

Shigella spp. are present in both extracellular and intracellular locations during infection, and thus humoral and cellular immune systems are thought to be important for eradicating the pathogen. The adaptive humoral immune response has been observed in experimental animal models to produce immunoglobulins (IgG, IgM, and sIgA) directed against the LPS. In addition, activated T cells are also induced upon *Shigella* infection of animals.

Animal and Cell Culture Models

Shigella does not cause infection in adult laboratory rodents when administered orally. In guinea pigs, there is no infection, when administered orally – but it can cause infection when applied to the eye causing the severe inflammatory reaction called "sereny keratoconjunctivitis test." Rabbit ileal loop assay (RIL) has been used to test the diarrheagenic action of toxins. A newborn mouse model has been developed to study shigellosis showing inflammation and tissue damage in the large intestine. Monkeys have been used primarily to study bacterial colonization; however, their use is restricted because of ethical consideration and the cost.

Various cultured cell lines are used to study bacterial invasion and bacterial cell-to-cell spread. Plaque assay has been used to study bacterial ability to move from cell to cell by forming a focal lesion on a cultured cell monolayer (see Chap. 5).

Symptoms

Shigellosis is characterized by mucoid bloody diarrhea, which is generally self-limiting. The symptoms appear within 12 h to 7 days, but generally in 1–3 days. In the case of mild infection, symptoms last for 5–6 days, but in severe cases, symptoms can linger for 2–3 weeks. Typical symptoms of dysentery include anorexia, fever, colitis, mucopurulent bloody stool, abdominal cramp, and tenesmus, a sense of incomplete evacuation of bowel with rectal pain. Inflammation in lamina propria and mucosal layer results in edema, erythema, and mucosal hemorrhage. In adults, the disease is self-limiting and resolves within 5–7 days. Some individuals may not even develop symptoms. An infected person sheds the pathogen long after the symptoms have stopped. Generally, children are more susceptible to shigellosis than the adults are, and it is fatal in children especially when suffering from malnourishment. Children with malnourished conditions show dehydration, megacolon, rectal prolapse, intestinal perforations, and infection which are life-threatening. Neurological disorders including lethargy, headache, and convulsions are seen. In some patients, HUS may develop resulting in kidney failure.

Prevention and Control

In the economically rich countries, foodborne shigellosis is caused by contamination of foods including fresh produce (fruits and vegetables) by food handlers shedding the pathogen in the feces and having poor personal hygiene. Therefore, those food handlers should be forbidden to prepare or serve ready-to-eat foods. In addition, proper education of the food handlers about the importance of good personal hygiene, and the digestive disease suspect from serving to others is important. Use of rigid sanitary standards to prevent cross-contamination of ready-to-eat food, use of properly chlorinated water to wash vegetables to be used for salads, and refrigeration of foods are necessary to reduce foodborne shigellosis.

In the economically poor countries, water and contaminated food are the major sources, and the disease is primarily associated with unsanitary living conditions. Many communities do not have access to sanitary toilet facilities; thus outdoor open-air toilet practices allow the direct bacterial transmission to food and water supplies. Children under the age of 5 are the most susceptible, and this group has the least awareness of sanitary and hygienic practices. Thus, they need most attention if the *Shigella*-induced disease needs to be controlled in these countries. Handwashing should be practiced to prevent the spread of infection.

The most important therapy is rehydration, and the electrolyte-containing fluid is administered orally or intravenously. Antibiotic treatment is controversial since *Shigella* develops resistance against antibiotics. Treatment of shigellosis varies depending on the severity of the infection. However, for severe cases antibiotics are normally administered along with the fluid therapy to prevent dehydration. The most common antibiotics used are ampicillin, trimethoprim/sulfamethoxazole, nalidixic acid, or ciprofloxacin.

Vaccination strategy seems to be a probable solution for the developing countries. The vaccine must be effective against 17 epidemiologically important serotypes of *Shigella* including

S. dysenteriae type 1, 15 *S. flexneri* serotypes, and *S. sonnei*. Several vaccines are currently under development that include the use of attenuated wild-type strain without virulence genes and a conjugate vaccine that includes *Shigella*-O polysaccharide conjugated to a carrier protein. Many of these vaccines are now gone through phase I and II trials, and some are even in phase III clinical trials and should be available for use in the near future.

Diagnosis and Detection

Bacterial Culture Methods

Shigellae can be isolated from the stool sample, but bacteria remain viable for a short period outside the human body; therefore, stool should be tested immediately or stored in appropriate media. Furthermore, the stool should be collected in the early phase of infection, before the antibiotic treatment has begun, to ensure isolation of bacteria. Shigellae have been isolated by streaking the diluted stool sample or rectal swab on MacConkey, *Salmonella-Shigella*, xylose lysine deoxycholate (XLD), and Hektoen enteric agars. Isolated colonies can be tested by slide agglutination test for serotyping and for presumptive identification. Shigellae have been isolated from food samples in a similar way.

Immunological Methods

Immunoassays including enzyme immunoassay (EIA), latex agglutination (LA) test, and dipstick immunoassays are used for detection of *Shigella* species. Commercially available LA test designated Wellcolex Color *Shigella* test (WCT-Shigella) has greater than 90% accuracy in detection of *Shigella*. Similarly, commercial EIA kits are available for *S. dysenteriae* (Shigel-Dot A), *S. flexneri* (Shigel-Dot B), *S. boydii* (Shigel-Dot C), and *S. sonnei* (Shigel-Dot D). These EIA assays are performed on the membrane and are also known as dot blot assay and have greater than 94% success rate.

Molecular Techniques

Conventional or nested PCR methods have been developed to detect various species of *Shigella*, and the gene targets include *ipaH*, *virA*, *ial* (invasion-associated locus), LPS, and plasmid DNA. The detection limit for the majority of these assays is from 1 to 1×10^4 cfu g^{-1} of food samples. Whole genome sequencing is now used for identification, strain typing, and source tracking.

Summary

Shigella is highly infective and causes disease worldwide. In particular, it causes frequent outbreaks in economically poor countries, parts of Asia, sub-Saharan Africa, and South America and is mostly associated with unhygienic and poor sanitary living conditions. *S. dysenteriae* type 1 causes epidemic outbreak and *S. flexneri* and *S. sonnei* cause endemic outbreaks. *Shigella* expresses multiple virulence factors encoded in a large plasmid. The major virulence genes are encoded in a 30 kb pathogenicity island as well as are scattered on a large plasmid. The virulence proteins are required for bacterial invasion, intracellular growth, cell-to-cell spread, killing of host cells, and evasion of the host innate defense system. The virulence proteins are delivered to the host cell by type III secretion system (T3SS). After being ingested with contaminated drink or food, bacteria colonize the colon and rectal mucus membrane. The infection dose is very low, 10–100 cells, because, in part, of their ability to survive gastric acid (pH 2.5). Bacteria cross the epithelial barrier by passing through naturally phagocytic M cells. After arrival in the subcellular location, they are engulfed by macrophages, neutrophils, and dendritic cells. Bacteria kill these cells by apoptosis and invade epithelial cells in the basolateral side by inducing membrane ruffling and macropinocytosis, a mechanism termed as "trigger mechanism." This process is aided by bacterial ability to deliver many virulence proteins called effectors by T3SS. Bacteria are then released from phagosome with the aid of IpaB, avoid autophagy using IcsB, move inside the cell by inducing actin polymerization with the help of IcsA, and infect the neighboring cell. Infection results in the release of high levels of proinflammatory cytokines (IL-1β, IL-8, IL-18) that recruit neutrophils and NK cells. Bacteria-induced epithelial cell damage and the activation of neutrophils result in massive inflammation characterized by ulceration and hemorrhage, and patients show signs of mucopurulent bloody stool, abdominal cramp, and tenesmus. Children under the age of 5 are most susceptible, and the infection could be fatal in malnourished children.

Further Readings

1. Ajene, A.N., Fischer Walker, C.L. and Black, R.E. (2013) Enteric pathogens and reactive arthritis: a systematic review of *Campylobacter, Salmonella* and *Shigella*-associated reactive arthritis. *J Health Pop Nutr* **31**, 299–307.

2. Anderson, M., Sansonetti, P.J. and Marteyn, B.S. (2016) *Shigella* diversity and changing landscape: insights for the twenty-first century. *Front Cell Infect Microbiol* **6**.

3. Ashida, H., Mimuro, H. and Sasakawa, C. (2015) *Shigella* manipulates host immune responses by delivering effector proteins with specific roles. *Front Immunol* **6**, 219.

4. Levin, R.E. (2009) Molecular methods for detecting and discriminating *Shigella* associated with foods and human clinical infections — A review. *Food Biotechnol* **23**, 214–228.

5. Levine, M.M. (2006) Enteric infections and the vaccines to counter them: Future directions. *Vaccine* **24**, 3865–3873.

6. Niyogi, S.K. (2005) Shigellosis. *J Microbiol* **43**, 133–143.

7. Ogawa, M. and Sasakawa, C. (2006) Intracellular survival of *Shigella*. *Cell Microbiol* **8**, 177–184.

8. Phalipon, A. and Sansonetti, P.J. (2007) Shigella's ways of manipulating the host intestinal innate and adaptive immune system: a tool box for survival? *Immunol Cell Biol* **85**, 119–129.

9. Philpott, D.J., Edgeworth, J.D. and Sansonetti, P.J. (2000) The pathogenesis of *Shigella flexneri* infection: lessons from *in vitro* and *in vivo* studies. *Phil Transact Royal Soc London Series B: Biol Sci* **355**, 575–586.

10. Picking, W.L. and Picking, W.D. (2016) The many faces of IpaB. *Front Cell Infect Microbiol* **6**.

11. Sansonetti, P.J. (2001) Rupture, invasion and inflammatory destruction of the intestinal barrier by *Shigella*, making sense of prokaryote-eukaryote cross-talks. *FEMS Microbiol Rev* **25**, 3–14.

12. Schroeder, G.N. and Hilbi, H. (2008) Molecular pathogenesis of *Shigella* spp.: Controlling host cell signaling, invasion, and death by type III secretion. *Clin Microbiol Rev* **21**, 134–156.

13. Sur, D., Ramamurthy, T., Deen, J. and Bhattacharya, S. (2004) Shigellosis: challenges & management issues. *Ind J Med Res* **120**, 454.

14. The, H.C., Thanh, D.P., Holt, K.E., Thomson, N.R. and Baker, S. (2016) The genomic signatures of *Shigella* evolution, adaptation and geographical spread. *Nat Rev Microbiol* **14**, 235–250.

15. Warren, B.R., Parish, M.E. and Schneider, K.R. (2006) *Shigella* as a foodborne pathogen and current methods for detection in food. *Crit Rev Food Sci Nutr* **46**, 551–567.

16. Zaidi, M.B. and Estrada-García, T. (2014) *Shigella*: a highly virulent and elusive pathogen. *Curr Trop Med Reports* **1**, 81–87.

20

Opportunistic and Emerging Foodborne Pathogens: *Aeromonas hydrophila, Plesiomonas shigelloides, Cronobacter sakazakii,* and *Brucella abortus*

Introduction

The Centers for Disease Control and Prevention (CDC) estimates that 38.6 million foodborne illnesses in the USA each year are caused by unknown foodborne pathogens, many of which are caused by infectious agents, opportunistic pathogens, emerging pathogens, or noninfectious agents. This chapter summarizes microbial pathogens that are considered opportunistic or emerging. The biology and the mechanism of pathogenesis of these microbes are not well understood, or their implication as foodborne pathogens have not been fully established. These pathogens have potential to cause a widespread problem in the future; thus, scientific awareness is essential. These microbes include but not limited to *Aeromonas hydrophila, Brucella abortus, Cronobacter sakazakii, Plesiomonas shigelloides, Klebsiella pneumoniae, Enterobacter cloacae,* and *Citrobacter freundii.* Properties of the first five pathogens are discussed in this chapter. The remaining three that include *Klebsiella pneumoniae, Enterobacter cloacae,* and *Citrobacter freundii* belong to the family of *Enterobacteriaceae* and share some common characteristics. These three pathogens can colonize the human gut and produce potent enterotoxins and cause acute and chronic diarrhea. The enterotoxins produced by these pathogens are similar to heat-labile (LT) or heat-stable (ST) toxins of enterotoxigenic *E. coli* (ETEC). Discussion on these three pathogens are not included in this chapter.

Aeromonas hydrophila

Biology

The genus *Aeromonas* belongs to *Aeromonadaceae* family, and it has 24 species including *Aeromonas hydrophila, A. caviae, A. veronii* biovar sobria, *A. piscicola,* and others. *Aeromonas* spp. are Gram-negative motile rods, facultative anaerobe, and are pathogenic to amphibians, fish, reptiles, and humans. Being aquatic bacteria, the primary habitat is saltwater and freshwater, but bacteria are also found in the intestinal tracts of humans and animals.

Aeromonas hydrophila is the most studied species, and the chromosome size (strain ATCC 7966^T) is 4.7 Mb and it contains 4128 protein coding genes; however, a significant strain-to-strain variation exists. *A. hydrophila* is motile and expresses a single polar flagellum (Pof) responsible for swimming motility in liquid, but it can also express lateral flagella (Laf) for swarming motility on the solid agar surface. Aeromonads express type IV pili also known as bundle-forming pili (BFP) and produce capsules and surface layer (S-layer) proteins and form biofilms. The growth temperature is broad, 3–42 °C, with an optimum between 15 °C and 20 °C; however, a few strains can grow at 1 °C. Aeromonads are facultative anaerobes, but grow better in the aerobic environment. Factors such as pH (below 4.5), NaCl (above 4%), and low temperature (below 3 °C) can significantly slow their growth.

Source

Aeromonads are primarily associated with water and aquatic environments and are isolated from rivers, lakes, ponds, seawater (estuaries), drinking water, groundwater, wastewater, and sewage. Aeromonads are frequently isolated from animals; thus, the bacteria are found in many foods, especially in foods of animal origin. It is also isolated from milk, seafood, red meat, and poultry. In some foods, the contamination level could be 10^5 cells g^{-1} or ml^{-1}. No confirmed cases of foodborne illness caused by *A. hydrophila* have been reported. Many strains, especially those isolated from foods, produce cytotoxins and hemolysins; however, their role in gastroenteritis is not clearly established. Aeromonads also cause food spoilage and psychrotrophic property aid in their growth in foods at refrigerated temperature during storage.

Disease

Aeromonads are pathogenic to fish, cold-blooded animals (amphibians and reptiles), cows, dogs, and humans. Bacteria cause ulcerative stomatitis in snakes and lizards, "red leg" disease in frogs, septic arthritis in calves, and septicemia in dogs. *Aeromonas* also infects immunocompromised and immunocompetent humans. The clinical symptoms associated with *Aeromonas* infection include gastroenteritis, septicemia, and wound infection. Gastroenteritis (diarrhea) is associated with consumption of large numbers of bacteria by individuals with impaired health (immunosuppressed condition) (Fig. 20.1). However, gastroenteritis symptom in humans still remains

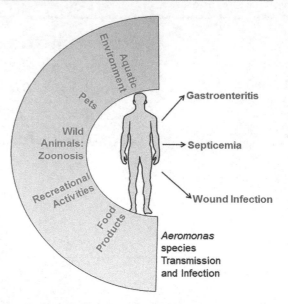

Fig. 20.1 Schematic showing *Aeromonas* species transmission and infection in humans

controversial. In a study, human volunteers challenged with *A. hydrophila* strains up to 10^{10} cfu per person, only 2 of 57 individuals developed diarrhea, and bacterial colonization in the gut was low. Nevertheless, epidemiological data and clinical symptoms (diarrhea) and stool analysis have established *Aeromonas* as enteropathogen, and the disease episode depends on the strains ingested. Gastroenteritis could be characterized as watery diarrhea, cholera-like (loss of >10 liters of rice-water feces per day), dysentery-like bloody mucoid stool, chronic colitis, or ulcerative colitis. The complication such as hemolytic uremic syndrome (HUS) is also reported in some *Aeromonas* infections.

Aeromonas can cause blood-borne disease such as septicemia, and immunocompromised individuals are at the greatest risk. In addition, *Aeromonas* is also involved in the skin and soft tissue infections. Necrotizing fasciitis or myonecrosis can happen in persons with liver disease or malignancy, and the mortality rate in those patients is 60–77%. *Aeromonas* can also cause peritonitis, respiratory tract infection (pneumonia), urinary tract infection (UTI), and eye infection.

Pathogenesis

Aeromonas hydrophila pathogenesis is mediated by the production of several virulence factors, such as adhesins, cytotoxins, hemolysins, enolase, collagenase, metalloprotease, lipases, protease, DNase, RNase, and other proteases, as well as the bacterial capacity to form biofilms.

Adhesion

Aeromonas hydrophila adheres to the host cells by using flagella and pili and by forming biofilms. The bacterium produces two types of flagella, a polar flagellum (Pof) and lateral flagella (Laf). Laf possibly contributes toward adhesion, biofilm formation, and colonization. Adherence to enterocytes may be mediated by two types of pili consisting of short and rigid pili similar to type I and Pap pili of *E. coli* and long, wavy, type IV pili similar to BFP. Another type IV *Aeromonas* pilus (Tap) may also participate in adhesion. Biofilm formation also helps the bacterium to adhere and to colonize. *Aeromonas* secrets acyl-homoserine lactone (acyl-HSL)-dependent transcriptional activator during biofilm formation. Bacterial capsules and S-layer proteins promote adhesion to host cells and these employ resistance to complement activation and antiphagocytic activity.

Cytotoxins

Aeromonas hydrophila secretes several cytotoxins and virulence factors using type II secretion system (T2SS) that include hemolysins (aerolysin), lecithinase, chitinase, gelatinase, amylases, DNases, and proteases. The bacterium produces three types of hemolysins: (1) Aerolysin is a broad category of the cytolytic pore-forming toxin with hemolytic activity and cytotoxic enterotoxins, often referred to as Bernheimer's aerolysin. More than 75% of *A. hydrophila* strains produce aerolysin. (2) HlyA is another hemolysin and is produced by most *A. hydrophila* strains. (3) *Aeromonas* cytotoxic enterotoxin (Act) is a type II secreted pore-forming hemolytic toxin that induces fluid accumulation and stimulates proinflammatory cytokine production (TNF-α, IL-1, and IL-6).

Aeromonas hydrophila also expresses type III secretion system (T3SS), which is predominantly found in clinical isolates, but not in aquatic isolates. T3SS is thought to inject *Aeromonas* enterotoxin into host cells. Some *A. hydrophila* strains also express T6SS, but may not be an obligatory virulence factor.

Symptoms

Aeromonas-induced diarrhea may last up to 14 days. The diarrhea is characterized as watery, cholera-like rice water, and dysentery-like bloody mucoid. Other symptoms associated with *Aeromonas* infection include chronic colitis, ulcerative colitis hemolytic anemia, renal dysfunction, hemolytic uremic syndrome, septicemia, and wound infection.

Plesiomonas shigelloides

Biology

Plesiomonas shigelloides is a Gram-negative facultative anaerobic motile, oxidase-positive, and nonspore forming rod-shaped bacterium. *P. shigelloides* shares many characteristics with *Aeromonas* spp.; thus, it was previously classified as *Aeromonas shigelloides*. Since *Plesiomonas shigelloides* shares antigens with *Shigella*, it was also called *Shigella*-like hence shigelloides. *Plesiomonas shigelloides* belongs to *Enterobacteriaceae* family, and it is the single species in the genus *Plesiomonas*. The bacterium carries a single chromosome and the genome size is 3.4 Mbp. Most strains grow between 8 °C and 45 °C, with optimum growth at 25–35 °C, pH 4–9, and a salt concentration of 2–3%.

Source

Plesiomonas shigelloides strains are found in fresh and brackish water and in fish and oysters harvested from the contaminated water. The bacterium is also isolated from the intestinal contents of humans and warm- and cold-blooded animals including fish, shellfish, frogs, reptiles, marine mammals (dolphins, sea lions, sea otters), water-fowls (gulls, heron, penguins), alligators, cats, dogs, foxes, hares, lizards, and wolves. The bacterial level in oysters is very high when collected during the warmer months and from muddy beds. Fecal contamination of foods of animal, bird, and plant origin can be a source. Bacteria can grow rapidly in most foods under optimum growth conditions. Disease outbreaks are associated with fish salad, potato salad, salt mackerel, tap water, oysters, shellfish, chicken, and so forth.

Pathogenesis and Disease

Plesiomonas shigelloides strains were implicated in many human gastroenteritis outbreaks associated with contaminated drinking water and raw or under-cooked seafood (oysters, crabs, and fish). The incubation period of the disease varies from 24 h to 50 h, and the disease lasts for 1–9 days or longer. The pathogenesis of the disease is not well understood because of lack of a suitable animal model. However, many virulence factors have been associated with the infection, including β-hemolysin, enterotoxins, and LPS. *P. shigelloides* produces three types of enterotoxins: a cholera-like, a thermostable (TS), and a thermolabile (TL) enterotoxin that may be responsible for diarrhea. *P. shigelloides* is also shown to adhere and invade cultured enterocyte cell lines.

Symptoms

The typical symptoms of *P. shigelloides* infection are diarrhea which could be an acute secretory form of gastroenteritis (most common), bloody or dysenteric colitis, or chronic or persistent diarrhea lasting for more than 14 days. In addition, patients show symptoms of nausea and abdominal pain and may be associated with vomiting, fever, and chills. Beside gastroenteritis, several strains are associated with bacteremia and septicemia. Immunocompromised individuals, young, old, and persons with underlying conditions such

as cancer, liver cirrhosis, HIV, and sickle cell anemia and thalassemia show symptoms of septicemia and gastroenteritis; thus, this pathogen is considered an opportunistic pathogen.

Cronobacter sakazakii

Biology

Cronobacter sakazakii, formerly *Enterobacter sakazakii*, is considered an emerging opportunistic pathogen that primarily infects infants; now it is implicated to infect adults. *C. sakazakii* is a Gram-negative, facultatively anaerobic, nonsporulating, motile rod-shaped bacterium and is the member of the *Enterobacteriaceae* family. Seven species of *Cronobacter* have been identified: *C. sakazakii*, *C. malonaticus*, *C. turicensis*, *C. muytjensii*, *C. universalis*, *C. condimenti*, and *C. dublinensis*, which has three subspecies, *dublinensis*, *lactaridi*, and *lausannensis*.

C. sakazakii produces yellow-pigmented colony on tryptic soy agar (TSA) and can grow at a temperature range of 6–45 °C with an optimum temperature of 37–43 °C. In general, these organisms are thermotolerant and can adapt well to osmotic desiccation stress. Bacteria can survive in infant cereal with A_w range of 0.30–0.83 for up to 12 months. *Cronobacter* also forms biofilm possibly aided by the production of capsular polysaccharides. Biofilms protect cells from disinfection and help bacterial persistence in dairy food processing environment, bottles, and enteral feeding tubes used in neonates in the neonatal intensive care.

For isolation of this pathogen, the US-FDA and ISO methods are used that involve pre-enrichment and/or selective enrichment before plating on TSA or chromogenic media. Yellow-pigmented colonies are further confirmed by biochemical testing and by PCR methods.

Source

Cronobacter has been isolated from varieties of food including milk, cheese, dried foods, meats, water, vegetables, rice, bread, tea, herbs, spices, and powdered infant formula (PIF). *Cronobacter* has been associated with a number of outbreaks involving PIF. These organisms are also found in food production factories including powdered infant formula production and livestock production facilities.

Pathogenesis and Disease

Pathogenesis of *C. sakazakii* is not fully understood; however, the bacterium is responsible for the opportunistic infection. Infants under the age of 4 weeks are most susceptible. Adults or the elderly with immunosuppressed conditions are highly susceptible. The pathogen translocates from the gastrointestinal tract to the blood where it multiplies in large numbers and then crosses the blood–brain barrier to reach to central nervous system (brain). Multiple virulence factors are involved in pathogenesis.

Adhesion and Invasion

No specific information is available as to the involvement of gene products that contribute to adhesion. However, the genome sequence information revealed the presence of genes for the type IV pili and P pili, suggesting their involvement in adhesion to the gut mucosa. Bacteria also express outer membrane protein A (OmpA), which interacts with the host cell surface fibronectin. OmpA also promotes host cell invasion through activation of PI3K (phosphatidylinositide 3-kinase) and Akt signaling pathways to alter cytoskeletal architecture through actin microfilament rearrangement. Bacteria-induced disruption of epithelial tight junction increases epithelial permeability and facilitates bacterial translocation to blood circulation.

Toxins

Cronobacter produces a 66 kDa enterotoxin, essential for causing cytotoxicity and diarrhea. The toxin is moderately heat-stable at 90 °C and thus can survive commercial milk pasteurization and remains active in powdered infant formula. Bacteria also produce a zinc metalloprotease

enzyme and endotoxin (LPS). Endotoxin impairs enterocyte migration and epithelial restitution, increases proinflammatory cytokine secretion, and promotes disruption of tight junction and increased bacterial translocation to blood circulation. LPS is heat stable at 100 °C, thus can remain active in PIF.

Symptoms

The symptoms of *Cronobacter sakazakii* infection include necrotizing enterocolitis, neonatal meningitis, and meningoencephalitis and bacteremia. After crossing the blood–brain barrier, the bacterium can cause ventriculitis and brain abscess, which later may develop into hydrocephalus brain. The fatality rate varies from 40% to 80%. Recovering patients from the central nervous system infections develop chronic neurological and developmental disorders. Adults, especially the immunocompromised ones and the elderly, are susceptible to this infection showing clinical symptoms of pneumonia, sepsis, foot ulcers, wound infections, and osteomyelitis.

Prevention and Control

Control strategies for *Cronobacter* include strict hygienic practices during PIF production, monitoring of the processing environments, and raw materials and final product testing. Irradiation, bacteriophages, bacteriocins, and plant phenolics have been proposed for inactivation of *Cronobacter* spp. in foods. To prevent *Cronobacter* infection in infants, (1) PIF should be reconstituted with hot water (>70 °C), which can reduce bacterial counts by 4 logs; (2) utensils for formula preparation must be disinfected; (3) reconstituted PIF should be used within 4 h; (4) and unused reconstituted PIF should be refrigerated at 4 °C immediately. For treatment of *Cronobacter* infection, antibiotics such as trimethoprim/sulfamethoxazole, streptomycin, gentamicin, kanamycin, and ciprofloxacin are effective.

Brucella abortus

Biology

The member of the genus *Brucella* belongs to *Alphaproteobacteria*. *Brucella* spp. are Gram-negative small coccobacilli (0.6–0.8 μm), non-sporulating, nonmotile, aerobic, and harbor a 3.2 Mbp genome divided into two chromosomes. *Brucella* spp. are facultative intracellular bacteria and infect humans and animals such as swine, cattle, goat, sheep, dogs, dolphins, whales, seals, and desert wood rats. *Brucella* genus has ten species, but traditionally, there were six species based on their primary host preferences: *B. abortus* (cattle), *B. melitensis* (sheep and goats), *B. suis* (pigs), *B. ovis* (sheep), *B. canis* (dogs), and *B. neotomae* (wood desert rats). Only three of these species are considered zoonotic, *Brucella abortus*, *B. suis*, and *B. melitensis*, and cause brucellosis, which is endemic in the Middle East, Asia, Africa, and South America and parts of the USA where the disease persists in wildlife.

Source

In infected animals, *Brucella* spp. are located in the uterus of pregnant animals and in the mammary glands of lactating females. Thus, the pathogens can be excreted in milk. People working with animals, animal products, and meat can be infected with *Brucella* spp. Worldwide, more than 500,000 human brucellosis cases are reported annually. Consumption of raw milk or products made from raw milk (such as cheese) has been implicated in foodborne brucellosis. The bacteria survive for a long time in milk and milk products. Pasteurization of milk generally kills *Brucella* cells. Between 1983 and 1987, there were two outbreaks in the USA from the consumption of imported cheese affecting 38 people, with 1 death. The Centers for Disease Control and Prevention (CDC) estimates about 839 cases of brucellosis in the USA annually and 50% of which are foodborne. Consumption of illegally imported cheese products made with unpasteurized milk is responsible for most brucellosis cases in the USA, primarily in the state of Texas and California.

Pathogenesis

Brucella is an intracellular highly infectious pathogen, and the infectious dose for human infection is 10–100 cells. Ingestion of contaminated and unpasteurized milk, cheeses, or contact with infected animals can transmit the disease to humans. Broadly, the disease has three distinct phases: (1) the incubation phase, (2) the acute phase, and (3) the chronic phase. During the incubation phase, there are no clinical signs. In the acute phase, the pathogen invades and propagates to diverse host tissues, and in the chronic phase, the bacterium causes severe damage to host organs and leads to death. In humans, the disease manifests as an undulant fever called Malta fever, and the disease can be chronic and debilitating if left untreated. In animals, it causes abortion in female and sterility in males.

Adhesion and Invasion

Brucella can enter the host through the mucosal surface of respiratory, intestinal, and genital tracts. On the mucosal surface, the bacterium uses several adhesion factors: (1) SP41, a surface protein that interacts with the host eukaryotic receptors containing sialic acid; (2) BmaC (340 kDa *Brucella* monomeric autotransporter) that interacts with host fibronectin; (3) BtaE (*Brucella* trimeric autotransporter) from the type II autotransporter family; and (4) flagella. Though nonmotile, the bacterium can use flagella to attach to the host cells. After attachment to the host cells, activation of small GTPases triggers a signaling cascade that reorganizes the actin cytoskeleton to induce host cell membrane rearrangement that promotes bacterial invasion through the zipper-like mechanism (*see* Chap. 4).

Brucella has tropism for macrophages, dendritic cells (DCs), and placental trophoblasts. Upon engulfment by the macrophage/DC, the bacterium survives inside the vacuole known as *Brucella*-containing vacuole (BCV). In BCV, the bacterium activates T4SS to inject effector

Fig. 20.2 Mechanism of *Brucella abortus* cell invasion and multiplication. *BCV*, *Brucella-* containing vacuole; *T4SS*, type IV secretion system; *ER*, endoplasmic reticulum (Schematics based on de Figueiredo et al. 2015. Am. J. Pathol. 185, 1505–1517)

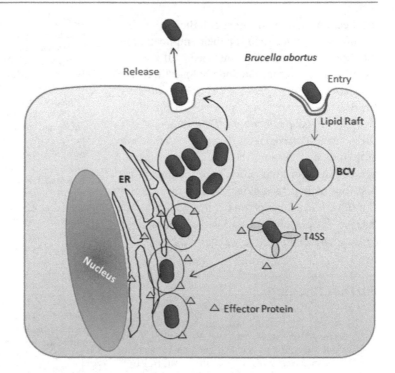

proteins and promotes BCV interaction with the endoplasmic reticulum (ER). At this stage, the bacterium migrates to ER compartment, multiplies, and moves from cell to cell (Fig. 20.2).

Symptoms

Symptoms of brucellosis in humans include undulant fever with irregular rise and fall of temperature, influenza-like symptoms, diaphoresis (profuse sweats), body aches, aching joints, chills, pyrexia (high fever), fatigue, anorexia, myalgia, and weakness. Symptoms appear in 3–21 days following consumption of a contaminated food or animal contact. In chronic brucellosis cases, organs may be affected leading to arthritis, orchitis, hepatitis, encephalomyelitis, and endocarditis. Splenomegaly is associated with increased lymphohistiocytic cells in the spleen.

Prevention and Control

Control measures for foodborne brucellosis include pasteurization of milk, manufacturing of dairy products from pasteurized milk, and proper sanitation to prevent recontamination of pasteurized products. No approved vaccine is available for humans, but the vaccine is available for animals. Since brucellosis is zoonotic, controlling the disease in animals will help reduce human cases. Three vaccines are used extensively in animals: *B. abortus* S19 and RB51 and *B. melitensis* Rev.1. The S19 and Rev.1 are attenuated strains of *Brucella* which still maintain intact LPS and are called the smooth strains. RB51 is a *Brucella abortus* strain that has defective LPS lacking the O-polysaccharide and is called a rough strain. This strain is safe for use in pregnant animals and does not induce antibodies to O-polysaccharide that can be used to distinguish vaccinated animals from the infected animals. In endemic zone (the Middle East, Asia, Africa, and South America and parts of the USA), animals are vaccinated to prevent abortion and brucellosis.

Summary

Opportunistic and emerging pathogens are thought to be involved in large of numbers of foodborne illnesses each year. The biology and the mechanism of pathogenesis of these microbes are not well understood, or their implication as foodborne pathogens have not been fully established. These microbes include but not limited to *Aeromonas hydrophila, Brucella abortus, Cronobacter sakazakii, Plesiomonas shigelloides, Klebsiella pneumoniae, Enterobacter cloacae,* and *Citrobacter freundii.* Pathogenic properties and glimpse on mechanism of pathogenesis and their interaction with the host of the first four pathogens are discussed in this chapter. Continued investigation on the mechanism of the disease process, and prevention and control strategies are needed to reduce human illnesses.

Further Readings

1. Alexander, S., Fazal, M.-A., Burnett, E., Deheer-Graham, A., Oliver, K., Holroyd, N., Parkhill, J. and Russell, J.E. (2016) Complete genome sequence of *Plesiomonas shigelloides* type strain NCTC10360. *Genome Announcements* **4**(5), e01031–16.
2. Chenu, J.W. and Cox, J.M. (2009) *Cronobacter* ('*Enterobacter sakazakii*'): current status and future prospects. *Lett Appl Microbiol* **49**(2), 53–159.
3. de Figueiredo, P., Ficht, T.A., Rice-Ficht, A., Rossetti, C.A. and Adams, L.G. (2015) Pathogenesis and immunobiology of Brucellosis: Review of *Brucella–* host interactions. *Am J Pathol* **185**(6), 1505–1517.
4. Forsythe, S. (2018) Updates on the *Cronobacter* Genus. *Annu Rev Food Sci Technol* **9** (doi.org/10.1146/annurev-food-030117-012246).
5. Galindo, C.L., Sha, J., Fadl, A.A., Pillai, L.L. and Chopra, A.K. (2006) Host immune responses to *Aeromonas* virulence factors. *Curr Immunol Rev* **2**(1):13–26.
6. Gauthier, D.T. (2015) Bacterial zoonoses of fishes: A review and appraisal of evidence for linkages between fish and human infections. *Vet J* **203**(1), 27–35.
7. Godfroid, J., Scholz, H.C., Barbier, T., Nicolas, C., Wattiau, P., Fretin, D., Whatmore, A.M., Cloeckaert, A., Blasco, J.M., Moriyon, I., Saegerman, C., Muma, J.B., Al Dahouk, S., Neubauer, H. and Letesson, J.J. (2011) Brucellosis at the animal/ecosystem/human interface at the beginning of the 21st century. *Prev Vet Med* **102**(2), 118–131.
8. Healy, B., Cooney, S., O'Brien, S., Iversen, C., Whyte, P., Nally, J., Callanan, J.J. and Fanning, S. (2010) *Cronobacter* (*Enterobacter sakazakii*): An opportunistic foodborne pathogen. *Foodborne Pathog Dis* **7**(4), 339–350.
9. Hunter, C. and Bean, J. (2013) *Cronobacter*: an emerging opportunistic pathogen associated with neonatal meningitis, sepsis and necrotizing enterocolitis. *J Perinatol* **33**(8):581–585.
10. Iversen, C., Mullane, N., McCardell, B., Tall, B.D., Lehner, A., Fanning, S., Stephan, R., Joosten, H. (2008) *Cronobacter gen. nov.*, a new genus to accommodate the biogroups of *Enterobacter sakazakii*, and proposal of *Cronobacter sakazakii gen. nov.*, comb. nov., *Cronobacter malonaticus* sp. nov., *Cronobacter turicensis sp. nov., Cronobacter muytjensii sp. nov., Cronobacter dublinensis sp. nov., Cronobacter* genomospecies 1, and of three subspecies, *Cronobacter dublinensis* subsp. *dublinensis* subsp. nov., *Cronobacter dublinensis* subsp. *lausannensis* subsp. nov. and *Cronobacter dublinensis* subsp. *lactaridi* subsp. nov. *Int J Syst Evol Microbiol* **58**(6), 1442–1447.
11. Janda, J.M., Abbott, S.L. and McIver, C.J. (2016) *Plesiomonas shigelloides* revisited. *Clin Microbiol Rev* **29**(2):349–374.
12. Jaradat, Z.W., Al Mousa, W., Elbetieha, A., Al Nabulsi, A. and Tall, B.D. (2014) *Cronobacter* spp.–opportunistic food-borne pathogens. A review of their virulence and environmental-adaptive traits. *J Med Microbiol* **63**(8), 1023–1037.
13. Kalyantanda, G., Shumyak, L. and Archibald, L.K. (2015) *Cronobacter* species contamination of powdered infant formula and the implications for neonatal health. *Front Pediatrics* **3**, 56.
14. Levin, R.E. (2008) Plesiomonas shigelloides - An aquatic food borne pathogen: A review of its characteristics, pathogenicity, ecology, and molecular detection. *Food Biotechnol* **22**(1–2), 189–202.
15. Oyarzabal, O.A. (2012) Emerging and reemerging foodborne pathogens. In *Microbial food safety: An introduction.* eds. Oyarzabal, O.A. and Backert, S. pp. 3–12. Springer Science+Business Media, LLC.
16. Pang, M., Jiang, J., Xie, X., Wu, Y., Dong, Y., Kwok, A.H., Zhang, W., Yao, H., Lu, C. and Leung, F.C. (2015) Novel insights into the pathogenicity of epidemic *Aeromonas hydrophila* ST251 clones from comparative genomics. *Sci Report* **5**, 9833.
17. Rasmussen-Ivey, C.R., Figueras, M.J., McGarey, D. and Liles, M.R. (2016) Virulence factors of *Aeromonas hydrophila*: in the wake of reclassification. *Front Microbiol* **7**.
18. Seshadri, R., Joseph, S.W., Chopra, A.K., Sha, J., Shaw, J., Graf, J., Haft, D., Wu, M., Ren, Q. and Rosovitz, M. (2006) Genome sequence of *Aeromonas hydrophila* ATCC 7966T: jack of all trades. *J Bacteriol* **188**(23), 8272–8282.
19. Smith, J.L. and Fratamico, P. (2005) Look what's coming down the road: Potential foodborne pathogens. In *Foodborne Pathogens: Microbiology and Molecular Biology.* eds. Fratamico, P., Bhunia, A.K. and Smith, J.L. pp. 427–445. Norfolk: Caister Academic Press.

Index

© Springer Science+Business Media, LLC, part of Springer Nature 2018
A. K. Bhunia, *Foodborne Microbial Pathogens*, Food Science Text Series,
https://doi.org/10.1007/978-1-4939-7349-1

CPSIA information can be obtained
at www.ICGtesting.com
Printed in the USA
LVHW061034280719
625625LV00007B/885/P

9 781493 992461